Fundamentals Of Physics, Alternate Edition

Part 3

David Halliday
University of Pittsburgh

Robert Resnick
Rensselaer Polytechnic Institute

Jearl Walker
Cleveland State University

Karen Cummings
Rensselaer Polytechnic Institute

with members of the
Activity Based Physics Group

*Activity*B*ased*

PHYSICS

Improving Learning Through Educational Research and Technology

Patrick J. Cooney (Millersville University)
Priscilla W. Laws (Dickinson College)
Edward F. Redish (University of Maryland)
David R. Sokoloff (University of Oregon)
Ronald K. Thornton (Tufts University)

John Wiley & Sons, Inc.
New York/Chichester/Weinheim/Brisbane
Singapore/Toronto

ACQUISITIONS EDITOR	Stuart Johnson
SENIOR PRODUCTION EDITOR	Elizabeth Swain
DEVELOPMENT EDITOR	Ellen Ford
SENIOR MARKETING MANAGER	Bob Smith
SENIOR DESIGNER	Madelyn Lesure
ILLUSTRATION EDITOR	Anna Melhorn
PHOTO EDITOR	Hilary Newman

This book was typeset by the authors and Progressive and was printed and bound by Bradford & Bigelow, Inc. The cover was printed by Phoenix.

This book is printed on acid-free paper. ∞

To order books or for customer service, call 1(800)-CALL-WILEY (225-5945).

ISBN 0471-39383-5

Printed in the United States of America

10 9 8 7 6 5 4 3 2 1

TABLE OF CONTENTS

Preface

To the Student

The Nature of Physics

Welcome to the study of physics. The study of physics has been a part of the traditional education of scientists, engineers, and technologists for over a century. Why is this so? Why should you study physics and what's it all about? First of all, physics is the most basic of the natural sciences, but what distinguishes physics from the other sciences? It's easy to say what the other sciences study. Biology is about living things, and chemistry is about atoms and molecules. But physics is somehow different. Trying to stay in the same pattern, we could say that physics is about matter and energy. Because living things are matter and energy, and molecules are matter and energy, physics applies to both biological and chemical systems—just as it does to building bridges or electronic chips. Physics is important because it sets the underlying rules both for the other sciences and for engineering.

Physics is also the process of learning about the physical world by finding ways to make sense of what we know through observation. As Nobel Laureate Richard Feynman[1] wrote about science, *"The test of all knowledge is experiment.* But what is the source of knowledge? Where do the laws that are to be tested come from?...Experiment, itself, helps to produce these laws, in the sense that it gives us hints. But also needed is imagination to create from these hints the great generalizations—to guess at the wonderful, simple, but very strange patterns beneath them all, and then to experiment to check again whether we have made the right guess." Progress in all of the natural sciences depends on this interaction between experiment and theory.

In learning physics you may be surprised to find that in some ways it seems much simpler than other courses like biology or chemistry. There are fewer things to consider and the systems we treat are simpler. If you were to write down all the equations that you have to know in a physics course and compare it to the number of organisms you are expected to be familiar with in a general biology class or to the number of chemical reactions you have to know in general chemistry, the physics number is relatively small. Also the situations seem simpler. A ball rolling down an inclined plane or a battery connected to a bulb are a lot simpler than a sea squirt or cyclohexane. But on the other hand, for many students, physics somehow seems harder than those other courses. What's going on?

In part, the reason physics seems hard is that we're not just trying to learn about the properties of specific systems. Very few people really care about balls rolling down inclined planes or simple battery and bulb circuits. What we are trying to learn about is the whole nature of the scientific process. How do we figure out scientific laws? What is the nature of measurement? When we propose a theory, what does it mean and when does it hold? How do we know what we know? These questions are critical to solving any real-world scientific problem such as creating a smaller computer chip that dissipates heat effectively or diagnosing what's really wrong with a patient.

The Art of Idealization

In physics, we want to understand how things work. To do that, we start with the simplest possible system that shows the behavior. If we're studying motion, we start with

---

[1]R. P. Feynman, *The Feynman Lectures on Physics*, Ch. 1, (Addison-Wesley, Reading, MA, 1964).

a small, massive object whose structure we don't need to worry about at first. We pretend a baseball is just a tiny blob and figure out how it moves under the influence of a hit and of gravity—pretending that it is in a vacuum and that it never rotates or deforms. These are clearly not good assumptions for a real baseball! But they provide an excellent starting point for making sense of the motion. Over small distances (a few feet), and for reasonably low speeds (below about 20 miles/hour) the idealized description works very well. As you get up to higher speeds and distances, the effects of the air grow in importance—but now that you know the basic principles, you can put in additional physics that make the examples more realistic and extend the number of cases you can treat.

This approach of starting with the simplest cases, understanding them as completely as possible (inventing physical laws to describe the situations), and then extending our considerations a step at a time, has been greatly successful in building the huge, consistent, and powerful body of knowledge that is physics.

Following this procedure has led to striking and surprising results in contemporary physics theories that have led to improving our understanding of how the world works and to applications that have changed our world. The prediction that the mass of an object would appear to grow and its internal clock would slow down as it approached the speed of light has been strikingly confirmed over and over again—and our understanding of these effects makes the global positioning system possible. Our understanding of the strange quantum behavior of electrons inside atoms and molecules allows us to create new and ever smaller computer chips.

Yet to find the surprises in physics you don't need to wait till you study relativity or quantum mechanics (though they both are really interesting and lots of fun). Even the physics that you'll be studying here produces some really surprising insights into our everyday world. If you take a 1-inch ball made of wood and a similar ball made of lead, the lead ball may weigh 20 times as much as the wooden ball. Yet if you stand on a chair and drop the two of them together, they fall the 10 feet to the ground in almost exactly the same time. Even a Styrofoam ball, 5 times lighter than the wood, will hit only a tiny fraction of a second later than the lead ball. Why does that happen? Clearly, if you hold the lead ball in one hand and the wood in the other, gravity is pulling much harder on the lead than on the wood. Why doesn't the lead ball go faster?

Actually, you know the answer. You just have to put two insights together. Think about kicking a soccer ball and a bowling ball. If you give them the same kick (ouch!) the soccer ball will speed up a lot more than the bowling ball. So to get a heavy object moving (a bowling ball), you need more force than to get a light object (a soccer ball) moving. So why doesn't the lead ball fall more slowly than the wooden one? To make a heavy ball (the lead) fall as fast as a light ball (the wood) you need to pull on it with a bigger force. Well this is just what gravity does! Your hand tells you that gravity pulls harder on the lead ball. It turns out that the extra pull that gravity exerts on the heavy object is just what is needed to balance the tendency of a heavy object not to be moved as easily as a light object.

This book is full of other such examples. When an object is imbedded in water, it seems to weigh less—and the amount less is the weight of the water that it pushed out of the way. What could that water have to do with anything? That water is gone! When I connect two identical bulbs up to a battery, if I connect them in one way they'll both have the same brightness as a single bulb connected up. If I connect them in another way, they'll be much dimmer but both have the same brightness. Huh? Why does that happen?

Pay attention to your observations and intuitions when reading this book. Sometimes they'll be right, sometimes they'll be partially right, and sometimes they'll be dead wrong. Learning not only the results, but learning to see how they make sense and are a useful way to look at things is one of the most important things you can learn from a physics course.

And remember: Physics is supposed to help you make sense of the physical world. If it doesn't appear to make sense at first, you need to keep thinking about your ideas and what they imply. Keep asking questions and looking at experiments. Einstein said, "Physics is the refinement of common sense." The key here is on the word "refinement."

Physics is more than common sense. It's common sense made careful and consistent by the continuous interaction of theory and experiment.

Using this Book as a Learning Tool

This textbook is only one of many resources that you will need to learn physics. It is critical that you make your own observations, do experiments, and understand the logical reasoning that is used to develop physics concepts and theories based on them. It is equally critical that you test and enrich your knowledge by applying it to problem solving. Problem solving involves making predictions about the behavior of physical systems using an appropriate combination of reasoning and your understanding of the equations that describe physical systems. Although we have tried in this alternate version of the textbook to present both the experimental results that support theories and some of the reasoning that has gone into the development of theories, you will understand the physics only when you also engage in making your own observations and are actively engaged in the reasoning process. We hope this book helps you enjoy the study of physics as much as we do.

To the Instructor

Welcome to the alternative edition of *Fundamentals of Physics*. You might well be asking yourself, "Why an alternative edition?" The textbook and its descendants, first written by Halliday and Resnick over 40 years ago, has been among the best-selling introductory physics texts since it appeared. It sets the standard against which all other texts are judged. That's why, when the Activity-Based Physics Group decided a new way of publishing physics materials was needed, we approached John Wiley & Sons.

Why HRW Alternate?

We have chosen to create this alternative to the Sixth Edition of the popular HRW text for several reasons.

First, we responded to student complaints that textbooks are too fragmented. The tendency to insert sidebars, extra boxes, and examples breaks up the flow of expository material. We decided to adopt a more narrative style. For this reason the critical examples have been placed at the end of chapters as Touchstone Examples. The rest have been moved to the problem book.

Second, we wanted to take advantage of what has been learned from scholarship in physics education about student learning difficulties. Thus, presentations that are known to be associated with common student confusions have been rewritten and clarified. Places where common student difficulties had previously been ignored or glossed over have been expanded or elaborated to help students over these traditional barriers.

Third, the topics have been rearranged somewhat (especially in the adoption of the *New Mechanics Sequence*[2]) to provide a more coherent learning path and story line. The story line is reinforced by the use of Reading Exercises that help students focus on thoughtful reading of the text sections in each chapter.

Fourth, the experimental evidence for many of the physical laws and relationships discussed in the narrative has been presented in graphical form. In almost all cases, the experiments described can be easily replicated by introductory physics students with computer-based data gathering tools commonly found in up-to-date physics labs.

Fifth, sections of the text have been reorganized and enhanced so that they reinforce the use of additional elements in a *Suite of Activity Based Physics Materials*.[3] Different Suite elements have been designed for use in lecture, laboratory, and recitation sessions. An electronic version of the text is under development that will allow students and instructors to link to related parts of the text as well as to appropriate Suite materials.

The Activity Based Physics Suite

In addition to this textbook, the *Activity Based Physics Suite* includes materials that can be used in lecture settings—the *Interactive Lecture Demonstration Series*[4] (currently under development). Suite materials that can be used in laboratory settings include the *Workshop Physics Activity Guide*[5] modules as well as the *Real Time Physics Laboratory* modules.[6] Other materials in the collection are suitable for use in recitation sessions such as the well-known University of Washington *Tutorials in Introductory Physics*[7] and a set of *Quantitative Tutorials*[8] developed at the University of Maryland. The student component of the *Activity Based Physics Suite* is rounded out with a set of collaborative problem solving materials, also developed at the University of Maryland.

The teacher's manual, *Teaching Physics with the Physics Suite*,[9] provides an overview of what has been learned about student difficulties with physics and a guide designed to aid instructors in the selection and integration of various *Suite* materials. Included as an Appendix to that volume is an *Action Research Kit* with conceptual and

---

[2]Priscilla W. Laws, "A New Order for Mechanics," *Proceedings of the Rensselaer Polytechnic Institute Conference on Introductory Physics Course*, 125-136, May 1993.

[3]http://physics.dickinson.edu/suite_prototype

[4]David R. Sokoloff and Ronald K. Thornton, "Using Interactive Lecture Demonstrations to Create an Active Learning Environment." *The Physics Teacher*, **35**, 340-347, September 1997.

[5]Priscilla W. Laws, *Workshop Physics Activity Guide*, Modules 1-4 w/ Appendices (John Wiley & Sons, New York, 1997).

[6]David R. Sokoloff, *RealTime Physics*, Modules 1-2, (John Wiley & Sons, New York, 1999).

[7]Lillian C. McDermott and Peter S. Shaffer, *Tutorials in Introductory Physics*, (Prentice-Hall, Upper Saddle River, NJ, 1998).

[8]E.F. Redish, "Implications of cognitive studies for teaching physics," *Am. J. Phys.*, **62**, 796-803, 1994.

[9]E.F. Redish, *Teaching Physics with the Physics Suite*, (under preparation for publication by John Wiley & Sons).

attitudinal surveys to help instructors gauge the effectiveness of their introductory physics teaching.

The Sixth Edition of HRW

This Alternative edition is based on the Sixth edition of Halliday, Resnick, and Walker.[10] Since this edition was just released in 2000, some instructors might be unaware of the many exciting changes that Jearl Walker has made in the Sixth Edition. Highlights of these changes which we have retained and embellished upon are worth mentioning:

Design Changes: more open format, reduction in the number of topics covered, and reduction in the number of sample problems included in the book.

Pedagogical Changes: more emphasis on reasoning, the statement of key ideas in sample problems, and the incorporation of more sample problems that stress applications.

A Final Word

Over the past decade we have learned how valuable it is for us as teachers to focus on what it is the students actually do to learn physics, and how valuable it can be for students to work with research-based materials that promote active learning. We hope you and your students find this book and some of the other Suite materials helpful in your quest to make the study of physics both exciting and understandable to your students.

Beta Version 1, January 2001

Karen Cummings (Rensselaer Polytechnical Institute)

with   Patrick J. Cooney (Millersville University)
Priscilla W. Laws (Dickinson College)
Edward F. Redish (University of Maryland)
David R. Sokoloff (University of Oregon)
Ronald K. Thornton (Tufts University)

---

[10]David Halliday, Robert Resnick, and Jearl Walker, *Fundamentals of Physics*, 6th Ed., (John Wiley and Sons, New York, 2001).

22 Electric Charge

Nothing happens if you place a plastic comb near tiny scraps of paper, but immediately after you comb your hair or stroke the comb with fur, it will attract the paper scraps. In fact, the attractive force exerted on the paper by the small comb is so strong that it overcomes the opposing gravitational pull of the entire Earth. This phenomenon, commonly called "static cling" occurs between many different objects,and is especially easy to observe during dry weather.

What causes these pieces of paper to stick to the comb and to one another?

The answer is in this chapter.

22-1 The Importance of Electricity

If you walk across a carpet in dry weather, you can produce a spark by bringing your finger close to a metal doorknob. Television advertisements alert us to the problem of "static cling" in clothing. On a grander scale, lightning is familiar to everyone. These phenomena represent a tiny glimpse into the vast number of electric interactions that occur everyday.

The phenomenon of electricity plays a major role in modern life. Less than two hundred years ago, fire was almost the only source of heat, the only source of light when the sun or moon were not up, and the only way to cook food. Without electric water pumps, most people did not even have indoor plumbing. It's hard to imagine life without electric lights (not even flashlights), stoves, refrigerators, air conditioners, computers, telephones, radios, televisions, CD players, and a host of other electrical devices. We make extensive use of electricity, but *what is it*? This chapter is a start in the discussion of this very important question.

So far in our study of the physical world we have learned how the forces acting on objects affect motion. We have discussed several different kinds of forces, including the pushes and pulls (called contact forces) one object exerts on another by directly touching it or through contact with it via another object such as a string. We have also learned about an action-at-a-distance force, the gravitational force which an object exerts on another object without any direct or indirect contact. In this chapter, we will undertake an investigation of the *electrostatic* interaction force (another action-at-a-distance force) which will provide a foundation for our understanding of the phenomenon of electricity.

We will begin our study by looking at the nature of electrical interaction forces between some every day objects. We will then develop the concepts of charging and electric charge as tools for explaining our observations of electrostatic forces on a macroscopic level. However, to obtain a more coherent understanding of electrostatic phenomena, we must turn to a microscopic look at the nature of matter found in the theories of atomic physics. According to these theories all matter we sense in our everyday world is made up of atoms, and electrical charge is an intrinsic property of the protons and electrons that are always present in atoms. By understanding the electrostatic forces between charges, we can easily explain why the contact forces objects exert on each other are basically electrostatic forces in disguise.

In addition, an understanding of electrostatic interactions will give you insight into the fundamental relationship between electricity and magnetism. In Chapter 30 on magnetic fields due to currents, you will discover although magnetic forces are generated by the interaction between moving charges, they have distinctly different properties than electrostatic forces. Later, in Chapters 32 and 38, you will see how electricity and magnetism are fundamentally related to each other and obey the principles of relativity postulated by Albert Einstein.

READING EXERCISE 22-1: List all the electrical devices that you use in a typical week.

22-2 The Discovery of Electric Interactions

Amber, which is resin that oozed from trees long ago and hardened, has been admired both for its beauty and its ability to preserve insects and other early life forms mired in it (Fig. 22-1). The semi-transparent, yellowish amber has electrical properties of interest to scientists. The early Greek philosophers knew that if one rubbed a yellow-brown piece of amber with fur, it would attract bits of straw. The strength of the attraction was known to fade over time (and was affected by the weather), but could be renewed with additional rubbing. The strength of the attraction decreased as the distance between the amber and the straw was increased.

By the 1600s, this ancient observation of a strange force that was sometimes present in amber and sometimes not, prompted more careful studies. It was subsequently discovered that other materials such as glass can also attract small bits of matter by being rubbed with silk. Like amber, the attractive force diminished with time, especially on humid days, and was not present if the glass had not been rubbed or stroked. Also like amber, the strength of the attraction decreased as the distance between the glass and small bits of matter increased.

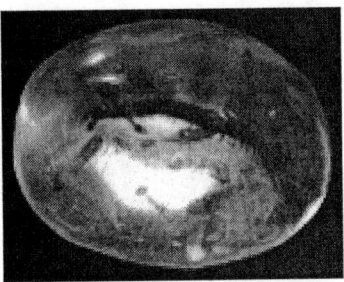

Fig 22-1: Fossilized resin, known as amber, is popular both for its beauty and its ability to preserve ancient vegetation and insects, such as this millipede. Amber also has electrical properties. Courtesy Swedish Amber Museum.

Because glass, amber, and other materials have been found to attract certain small objects after being stroked with certain kinds of cloth, the interaction phenomenon created by rubbing or stroking an object with cloth was called *electrification*. In fact, the term electrification is derived from the Greek word for amber which is *electron*. Any object (not just glass or amber) is defined as becoming *electrified* if it interacts with other objects so that:

1. An interaction force between two objects is present only if at least one of the objects has been stroked or rubbed by a third object;

2. The magnitude of the interaction force diminishes with time and is affected by the weather; and

3. The magnitude of the force decreases with increasing distance between objects.

Although the similarities between rubbed glass and rubbed amber were interesting, it was not until the 1700s, one hundred years later, that anyone thought to investigate how rubbed amber and rubbed glass interact with one another. In 1733, it was observed that:

Two glass rods stroked with silk always repel one another;

Two amber rods stroked with fur always repel one another;

A stroked amber rod *attracts* a stroked glass rod.

Fig. 22-2: (*a*) Two glass rods electrified in the same way repel each other. (*b*) Two amber rods electrified in the same way also repel each other. (*c*) An electrified glass rod and an electrified amber rod attract each other.

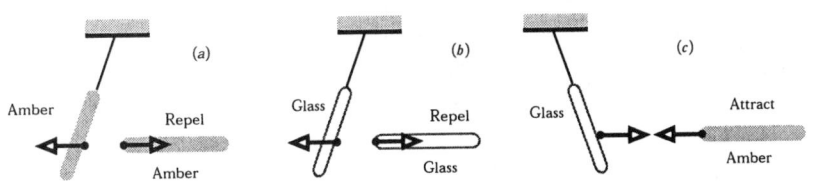

Provided the weather is not hot and humid where you are, you may be able to repeat this experiment yourself by replacing the amber and glass rods (which are difficult to find outside of a physics laboratory) with Styrofoam cups and plastic sandwich bags. Place your hand inside a plastic bag and use a rubbing motion to assure that the entire surface of the Styrofoam cup comes in contact with the entire surface of the plastic bag. Then rub another Styrofoam cup with a second sandwich bag in the same manner. If you put one of the cups on its side on a smooth, level, non-metallic surface and bring the other cup near it, the first cup should roll away. Note that after the the two cups have been electrified in a *like manner* they *repel* one another. Now hold the two plastic bags together at the top end. Both plastic bags have also been electrified in a like manner and they repel one another as well. However, an electrified sandwich bag and an electrified Styrofoam cup will be *attracted* to each other. Note that the bag was electrified by its contact with a cup and the cup was electrified by its contact with a bag. Thus, these two different types of materials have not been electrified in the same manner.

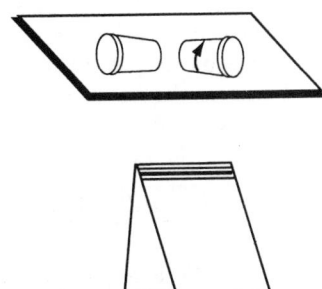

Fig. 22-3: Two Styrofoam cups electrified in the same way repel each other (upper). The two sandwich bags used to electrify the cups also repel each other (lower).

Not all types of materials can be electrified. However, whenever objects can be electrified, additional observations with electrified materials lead to the following general statements:

➤Two identical objects electrified by the same process (for example, by being stroked with two different types of objects that are identical to each other) always repel one another.

➤Two electrified objects made of different types of materials will always interact, but they may either repel or attract one another depending on the material each is made of. (For example, an electrified glass rod will *attract* an electrified sandwich bag. On the other hand, an electrified glass rod will *repel* an electrified Styrofoam cup.)

➤Any two objects that have not been electrified will neither repel nor attract one another. In other words, they will not interact except by means of relatively weak gravitational forces.

You can also do an additional experiment with Styrofoam cups and plastic sandwich bags to discover another important characteristic of electric interactions. If we take one of the electrified Styrofoam cups and stroke it with a plastic sandwich bag for a longer time, the magnitude of the interaction forces between the cups increases. If we think of the first cup that has not received more stroking than our standard cup, we can determine the degree of electrification of an object by measuring the magnitude of the electric force exerted on it by a standard object.

Today, we say that electrification changes the *state* of an object. We give this new state a name: we say the object has become **electrically charged** or that it carries **electrical charge**. Furthermore, any process of electrification (not just rubbing) is called **charging**. Thus, in the example above, an object with a greater quantity of charge will experience more force in the presence of a standard electrified object than one with a smaller quantity of charge. Note that at this point we have no idea what charge is based where from direct observation.

READING EXERCISE 22-2: The creation of electrified objects can also be done with strips of Scotch™ Magic Tape using a peeling action rather than stroking. In order to charge the Magic Tape, cut 2 strips of tape to a length of about 10 cm each. (a) If you were to stick the tapes side-by-side on a table and peel them both off, what do you predict would happen if you then bring the tapes together? Would the tapes attract each other? Repel each other? Or not interact? Explain the reasoning for your prediction. (b) Perform the experiment and describe what happens. Is this consistent with your prediction? If not, explain what you think is going on.

22-3 The Concept of Charge

Various observations about the interaction forces between charged objects can be explained if we assume that there are two (and only two) different types of charge states. The type of charge state that occurs when glass is rubbed with silk is one and the type of charge state that occurs when amber or plastic is rubbed with fur is the other. We cannot prove directly there are no more than two charge states. However, the fact that so far no one has found a charged object that attracts both charged glass and charged amber (or plastic) leads us to believe that there is no third type of charge.

Today, the terms we associate with these two charge states are *positive* and *negative*. Benjamin Franklin is responsible for assigning these names. Benjamin Franklin introduced the following definitions:

➤An object that is repelled by glass stroked with silk is *positively charged*.

➤An object that is repelled by amber (or rubber or plastic) stroked with fur is *negatively charged*.

➤Two objects that do not interact with each other except by gravitational forces are *electrically neutral*.

Note that the names given to the two varieties of charge are arbitrary. Benjamin Franklin could just as easily have called the charge state created by stroking glass with silk negative as positive. Further, he could have used other words to distinguish between the two charges such as light and dark, male and female, or ugh and glug.

If we find two objects are each repelled by a piece of glass that has been rubbed with silk, then we say both objects are in the same charge state as the rubbed glass is, which we call a positive charge state. Thus, we hypothesize that objects repel because they contain **like charges**. On the other hand, if we find two objects made of different materials attract after being stroked, we hypothesize one object has a positive charge while the other has a negative charge. We conclude objects with **unlike** or **opposite charges** attract.

READING EXERCISE 22-3: How do we know there are only two kinds of charge? Suppose you stroked a hypothetico rod with a kang and announced that you had created a type of charge you decide to call *kang charge*. If a skeptic asked you to prove that kang was really a new type of charge, how would you do it? Specifically what would have to happen if you were to bring two charged hypothetico rods together? If you brought a charged hypothetico rod near a charged glass rod? Near a charged amber (or plastic) rod?

22-4 Using Atomic Theory to Explain Charging

We still don't know what charge is. How can we account for the fact that when certain objects are brought into close contact they acquire opposite charge states? Our preliminary model of electric interactions does not account for objects becoming oppositely charged. One way to make sense of this observed fact is to use a contemporary understanding of the atomic structure of matter as a plausible model for how electrification can occur when different types of materials are brought into contract. The atomic model which we present has been developed over the past century. We use it here as an explanatory tool without presenting evidence for it.

The Atomic Model

The modern theory of the atomic structure states atoms consist of positively charged *protons*, negatively charged *electrons*, and electrically neutral *neutrons*. The protons and neutrons are massive and packed tightly together in a central *nucleus*. Most of the atoms that are contained in matter have equal numbers of electrons and protons so whenever a charged object is at some distance away from the atom, the atom appears to be electrically neutral.

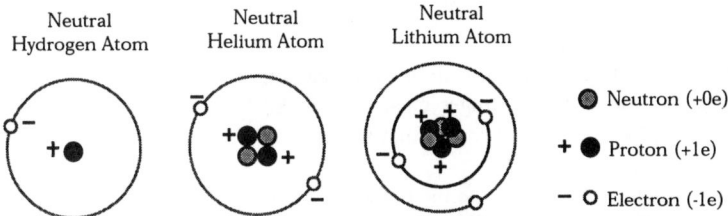

Fig. 22-4: The structure of the atoms representing the three lightest chemical elements. The number of protons that define the element along with the typical number of neutrons in each element's nucleus are shown. The darker circles represent protons, the lighter circles neutrons, and the white circles electrons. The diagram is simplified as physicists do not believe that electrons orbit nucleii in nice neat circles.

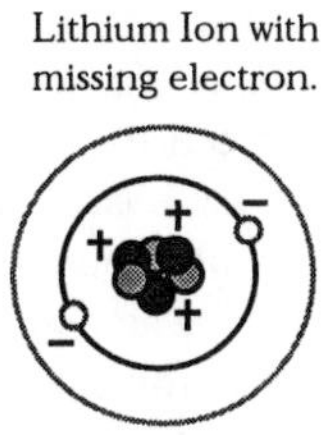

Fig. 22-5: A lithium atom has lost an outer electron and is now an ion with a net charge of +1e because it has more protons than electrons.

According to contemporary atomic theory, **electric charge** is an intrinsic characteristic of the electrons and protons. They would not be the same entities without their charge. You often encounter casual phrases—such as "the charge on a sphere," "the amount of charge transferred," and "the charge carried by the electron"—that suggest charge is a substance. You should, however, keep in mind we believe electrons and protons are basic substances and charge and mass happen to be two of their fundamental properties.

The mass of an electron is about 2000 times smaller than that of a neutron or proton. Electrons are attracted to the nucleus because electrons and the protons within the nucleus are unlike charges. However, the electrons that are farthest away from the nucleus are only weakly attracted to the nuclear protons and so they don't always remain associated with individual atoms. In many types of materials the electrons are free to wander about within the material if they experience other forces. If the atom loses an electron it is no longer neutral, but positively charged because there are now more protons than electrons. Such positively charged atoms are called *positive ions*. We call the mobile electrons *conduction electrons*. If an electric or other force is applied to the atom, only the conduction electrons, with their negative charges, move appreciably. The much more massive positive ions stay fixed in place by their attraction to nearby atoms.

Charge is Quantized

In Benjamin Franklin's day, electric charge was thought to be a continuous fluid that could have arbitrary magnitude. Today we know fluids, such as air and water, are not continuous but are made up of atoms and molecules; matter is discrete. In 1909 an American physicist, Robert Millikan, used opposing electric and gravitational forces to balance drops of oil between two electrified metal plates. His famous oil drop experiment and others that followed, showed that the "electrical fluid" is not continuous but is made up of multiples of an elementary charge.

TABLE 22-1: Charges of the 3 Fundamental Atomic Particles

Particle	Symbol	Charge
Electron	e or e⁻	$-e$
Proton	p	$+e$
Neutron	n	0

Any positive or negative charge q that has ever been detected can be written as

$$q = ne, \quad n = \pm 1, \ \pm 2, \ \pm 3, ..., \qquad (22\text{-}1)$$

in which e, the **elementary charge,** has the value

$$e = 1.60 \times 10^{-19} \text{ C.} \qquad (22\text{-}2)$$

The SI unit of charge is the **coulomb** (C), named for Charles Augustin Coulomb, who studied electric forces in the late 1700s. The electron and proton both have a charge of magnitude e (Table 22-1). When a physical quantity such as charge can have only discrete values rather than any arbitrary value, we say that the quantity is **quantized.** It is possible, for example, to find a particle that has no charge at all or a charge of $+10e$ or $-6e$, but not a particle with a charge of, say, $3.57e$.

▶Charge is quantized. It has never been measured to have a magnitude other than an integral multiple of 1.60×10^{-19} C.

As we noted in the introduction, if you drag your feet as you walk across a carpet, you can produce a spark of electric charge by bringing your finger close to a metal doorknob. This is a demonstration of the vast amount of electric charge that is stored in the familiar objects that surround us.

However, the charge within an object is usually hidden because the object contains equal amounts of material that is positively charged and negatively charged. This was hinted at above when we stated electrically neutral atoms contain equal numbers of protons and electrons. With such an equality—or *balance*—of charge there is no *net* charge. If the two types of charge are not in balance, then there *is* a net charge. This is the case in what we have been calling charged objects. We say an object is *charged* to indicate that it has a charge imbalance, or **net charge.**

▶Electrically neutral objects are *not* devoid of charge. Instead, they contain equal amounts of positively and negatively charged material. Combining equal amounts of oppositely charged materials results in a cancellation of their effects.

Charging as a Transfer of Electrons

We can use our modern understanding of the atom to explain why it is plausible objects like glass and silk become oppositely charged when they are brought into close contact. Since we observe glass becomes positively charged while the silk becomes negatively charged, it is logical to assume that outer electrons associated with atoms of silicon and oxygen in the glass are attracted to the silk atoms and hop over to the silk. The glass is now missing electrons so there is a net positive charge on the glass. The silk now has excess electrons and has a net negative charge.

In general we believe an object becomes charged when a tiny percentage of the mobile electrons that always carry negative charge are transferred from one object to another. This is why we must rub, stroke or otherwise make significant contact between two objects in order for the objects to become charged. Thus, we believe when a glass rod is stroked with silk cloth, a very tiny fraction of the electrons near the surface of the glass rod are transferred to the silk cloth.

Why doesn't silk cloth get heavier when electrons are transferred to it? Using modern atomic theory, we understand that even if we transfer a typical number of electrons to the silk cloth (about a trillion or 10^{12}) the increase of its mass would not be measureable. This is because only a small fraction of the electrons are transferred and because electrons are so much less massive than protons. Because positive charge is not mobile, an object becomes positively charged only through the *removal of negatively charged electrons.*

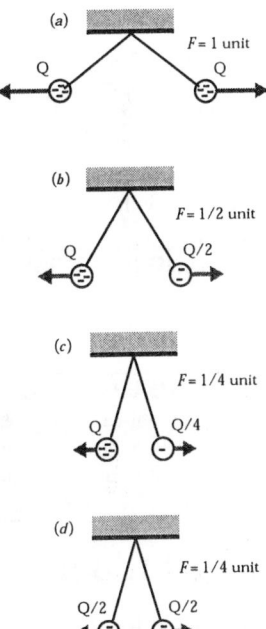

Charge is Conserved

We believe sometimes when two electrically neutral objects are in close contact, charging or electrification occurs because electrons transfer from one body to another, upsetting the electrical neutrality of each body. Careful measurements reveal whenever two electrically neutral objects are brought into contact, the magnitude of excess charge that is found in each object is the same. This indicates when electrons are transferred from one object to another, no additional electrons are destroyed or created in the process. The amount of charge contained in the two objects is conserved.

This hypothesis of **conservation of charge** was first proposed by Benjamin Franklin based on experiments he performed. It has stood the test of time. It is observed to hold both for large-scale charged bodies and for atoms, nuclei, and elementary particles. No exceptions have ever been found. We add electric charge to our list of quantities—including energy, linear momentum, and angular momentum—that are conserved quantities. In summary, we believe:

▶The total amount of electric charge in the universe is conserved. Although charge can be transferred from one object to another, it cannot be created or destroyed.

Fig. 22-6: Depiction of an idealized experiment to measure the forces between small metal-coated spheres that hold different fractions of charge. Note: In order to make valid measurements, the distance between the centers of the two balls must be more than twice the diameter of a ball.

Force and Quantity of Charge on Particle-like Objects

The ability to transfer charge from one object to another without the loss or addition of charge allows us to perform experiments that indicate how the interaction force between charged objects depends on the magnitudes of the charge on each object. These experiments lead to surprisingly simple results when the charged objects are symmetric, made of metal, and are particle-like so their dimensions are small compared to the distances between their centers. For example, consider the experiment shown in Fig. 22-6. In this experiment there are two identical uncharged metal spheres. We electrify one by stroking it with a charged plastic rod. We then touch the two spheres together. As we will see in Section 22-6, metal is a good electrical conductor and electrons can transfer easily from the charged sphere to the uncharged one. Since the spheres are identical and the excess electrons repel each other, we expect electrons to travel between the spheres until they both have the same number of excess electrons. Next we measure the force exerted by one sphere on the other and record it (Fig. 22-6a).

Next we leave sphere 1 alone and move sphere 2 a long distance from sphere 1 to place it in contact with a third sphere that is uncharged. The excess electrons on sphere 2 will now be shared equally between spheres 2 and 3 so the number of excess electrons on sphere 2 will now be half of what it was before. If we measure the force between spheres 1 and 2, we find that it is one-half of the force we first measured (Fig. 22-6b). If we repeat this process so we reduce the amount of charge on sphere 2 to one-fourth of what it was originally, then the interaction force between the spheres is also reduced to one-fourth of what it was originally (Fig. 22-6c). In a similar experiment we can reduce the charge on both spheres 1 and 2 to half their original values and then the force measures one-fourth the original force between them (Fig. 22-6d). These observations indicate the force is dependent on the *product* of the charges on the two spheres and give us a way to measure the amount of charge on an object.

▶The magnitude of the charge on a particle-like object can be quantified through measurement of the interaction force between it and a standard charged object that is also particle-like.

The Electroscope

The fact that like charges repel has been used in the development of the electroscope, a sensitive charge-measuring device. A net charge can be transferred to an electroscope by touching a conducting ball with a positively or negatively charged rod. Some of its excess charge will spread throughout a metal bar and the foil attached to the ball. If a flexible metal leaf is attached to the central conducting bar, the flexible conductor will be repelled from the central charges and rise. As more charges are transferred to the electroscope, the metal leaf will rise higher.

READING EXERCISE 22-4: Assuming that solid objects are made up of atoms rather than being continuous, can you think of a plausible way to explain why it is so difficult to pull solids apart or push them together?

READING EXERCISE 22-5: Suppose you have measured the repulsion force between two identical metal-coated spheres that each have an excess negative charge of Q. Next, you would like to measure the force on the metal-coated spheres that each have one-fourth of the excess charge they originally had (as depicted in Fig. 22-6d). Describe how you could use similar uncharged spheres to reduce the excess charge on each of the original spheres to $Q/4$.

22-5 Induction

Let's consider some additional observations that can be made that involve electrical interactions. Typically, bits of straw, bits of paper, and other materials that have not been rubbed do not either attract or repel one another. We say they are **electrically neutral**. Thus, it is surprising to find that a plastic comb rubbed with fur to become *negatively charged* (like the comb shown in the photograph at the beginning of this chapter) can attract bits of electrically neutral paper. It is even more surprising to find that a *positively charged glass rod will also attract bits of paper*.

▶The attraction that we always observe between a charged and uncharged object regardless of the state of the charged object is called **induction**.

One of the most familiar experiences with induction is the static cling phenomenon described at the beginning of the chapter. It is literally impossible to prevent clothes that have been tumbling in a dryer, or plastic wrap that is peeled off its roll from sticking to neutral objects they come into contact with. How can we explain induction—the attraction of an electrically neutral object to a charged object? We can turn to atomic theory to develop an explanation of induction. Although, you cannot use direct observations to verify that the atomic model is "correct," it is nonetheless a plausible model.

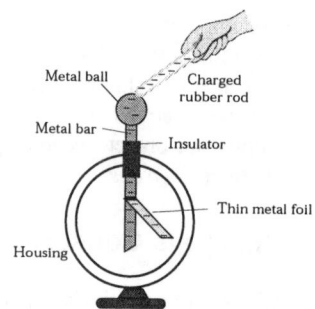

Fig. 22-7: The electroscope can be used to measure charge. The rise of a metal foil is caused by the repulsion of like charges distributed in a metal conductor. The foil rises in proportion to the charge contained in the conductor.

Since the electrical phenomena we have observed so far appear to be due to interactions between charges, induction seems mysterious. The idea that electrically neutral materials are not devoid of charge, but rather composed of atoms that have the same number of positive protons and negative electrons, is the first step in developing a viable explanation for induction. The second important idea, mentioned in the last section, is that the negative electrons present in objects are more mobile than the protons that carry positive charge. Let's consider what could happen to a neutral *metal* object where electrons are quite mobile. We will consider how to explain induction in non-metals in the next section.

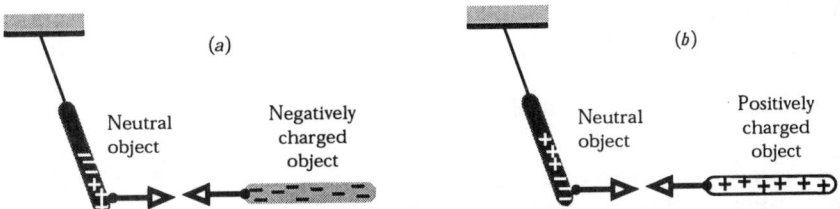

Fig. 22-8: A tiny neutral metal rod is suspended on a nonconducting thread. The part of the neutral rod that is closest to a charged object will be attracted by either: (*a*) a negatively charged object (such as amber) or (*b*) a positively charged rod (such as glass).

Suppose we dangled a very small rod from a strip as shown in Fig. 22-8. According to our atomic model, the mobile negative electrons would be repelled from a negatively charged object which has extra electrons on it. When the electrons in the neutral object are repelled from the negatively charged object, protons that are not completely neutralized by surrounding electrons are left behind. The unlike charges left near the surfaces of the charged object and the neutral one will attract as shown in Fig. 22-8a. An attraction between a neutral object and a positively charged object as shown in Fig. 22-8b also occurs. Recall a positively charged object has missing electrons. In this case the electrons in the neutral object are attracted to the un-neutralized protons in the positively charged object.

Although the rod dangling from the string in both Fig. 22-8a and Fig. 22-8b remains neutral, our hypothesis is its positive and negative charges have been separated by the proximity of a charged object. This charge separation process is called **polarization**.

The ideas of induction and polarization can be used to explain how the charged comb pictured in the puzzler at the beginning of the chapter can be used to pick up one piece of paper, which then picks up another piece of paper and so on. We assume the first piece of paper is attracted to the charged comb by induction and becomes polarized in the process. Then the excess negative charge at the bottom end of the first paper bit attracts the second paper bit by induction. Since the second paper bit is now polarized, the process continues. Although we believe magnetic forces behave differently than electrostatic forces, this process of electric induction is not unlike a similar process in which a magnet can induce magnetic polarization in a steel paper clip which then attracts and polarizes a second clip and so on.

READING EXERCISE 22-6: (a) If we state two bits of paper are electrically neutral, what observation can you make to verify this is the case? Explain. (b) Can induction (the observation of an attraction between a charged and uncharged object) be used to determine whether the charged object is positive or negative? Why or why not?

READING EXERCISE 22-7: (a) In the figure that follows, objects made of different materials are arranged in pairs. *A*, *B*, and *D* are charged plastic and *C* is an electrically neutral metal plate. The arrows showing the electrostatic interaction forces between pairs (1), (2) and (4) are shown. Do the objects in pairs (3) and (5) repel or attract each other? Explain your reasoning.

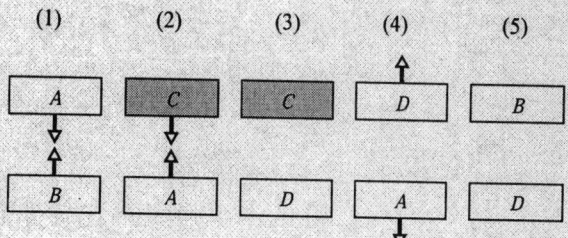

(b) According to Benjamin Franklin's convention, which types of objects are positive? Which are negative? Explain.

(c) Suppose you were told that *A*, *B*, and *D* were made of the same electrified material but you were not told what that material was. Would it be possible to determine whether they were negatively charged or positively charged by observing how pairs (1), (2) and (4) interact as shown in the diagram above? Why or why not?

22-6 Conductors and Insulators

Whenever a charged object is near an electrically neutral object, induction occurs. However, the attractive forces are stronger for neutral metal objects than for non-metal objects. Why? Let's summarize the outcomes of observations when we touch or hold charged objects. These new observations when coupled with the application of the atomic model will serve to give us a more complete understanding of the nature of electric interaction forces in metals and non-metals.

Observation 1. (a) The electrification created on non-metal objects such as plastic or glass does not spread out. Instead charge seems to remain in regions where the object is rubbed, and (b) touching charged non-metal objects removes the electrification only at locations where an object is touched.

Observation 2. (a) Metal objects can be charged when mounted on non-metal objects such as glass or plastic but they cannot be charged while being held in someone's hand, and (b) metal objects that are touched anywhere will immediately lose *all* of their charge.

Atomic Theory and the Behavior of Conductors and Insulators

Once again we can turn to atomic theory to develop a plausible explanation for these new observations. We believe in materials, such as metals, tap water, water droplets in air and the human body, some of the negative electrons move easily. We call such materials **conductors**. We observe charge can flow onto or off conductors quite quickly. In other materials, such as glass, chemically pure water, and plastic, the electrons can reposition themselves within an atom but cannot migrate between atoms. We call these materials **nonconductors** or **insulators**. Electrons do not travel from atom to atom very easily in insulators. An exception is some of the electrons at the surface of an insulator can have a greater affinity for another type of surface. For example, electrons can travel from the surface of a glass rod to the surface of a piece of silk cloth brought into contact with it, leaving the glass rod positively charged.

One ramification of this difference in the mobility of charge in insulators and conductors is the polarization process discussed in Section 22-5 is not as strong in insulators. Neutral conductors and neutral insulators both undergo charge separation (become polarized) when they are brought close to a charged object. But in insulators, the charge separation occurs within atoms, while in conductors the charges move through the material; they become separated from the atoms to which they were originally bound. This difference is shown in the comparison of Figs. 22-4 and 22-5. This atomic model

Fig. 22-9: This is not a parlor stunt but a serious experiment carried out in 1774 to prove that the human body is a conductor of electricity. The etching shows a person suspended by nonconducting ropes while being charged by a charged rod (which probably touched flesh instead of the trousers). When the person brought his face, left hand, or the conducting ball and rod in his right hand near bits of paper on the plates, charge was induced on the paper, which flew through the intermediate air to him.

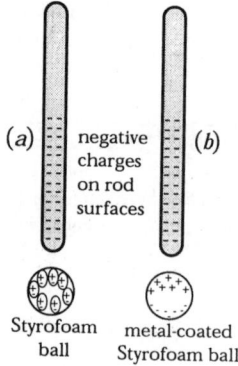

Fig. 22-10: According to atomic theory: (*a*) polarization induced in an insulator involves atomic scale charge separation as shown for a Styrofoam ball, while (*b*) polarization in a conductor involves larger scale migration of electrons as shown in a metal coated ball.

provides a plausible explanation for the observed fact induction is much stronger between a charged object and a conductor than between the same charged object and an insulator.

Another outcome of the difference in mobility of charge carriers in conductors and insulators explains the observation you cannot charge a metal rod by rubbing if you are holding it, because both you and the rod are conductors. Although the rubbing will cause a charge imbalance on the rod, the excess charge will immediately move from the rod through you to the floor (which is connected to Earth's surface), and the rod will quickly be neutralized. Setting up a pathway for electrons between an object and Earth's surface is called *grounding* the object, and always results in electrically neutralizing the object (by eliminating an unbalanced positive or negative charge). If instead of holding the metal rod in your hand, you hold it by an insulating handle, you eliminate the conducting path to Earth, and rubbing can then charge the rod, as long as you do not touch it directly with your hand.

These ideas give us a very functional way to determine if a material is a conductor or an insulator. If you have two interacting charged objects and you touch of one of them, do the objects stop interacting? If so, the object you touched underwent a change in charge as a result of contact with the Earth. Charge must have flowed onto or off the object. Hence, the object must be a conductor. If the change in charge does not occur, the object must be an insulator.

Charging by Induction

The polarization model which was devised to explain induction can also be used to explain a related process known as **charging by induction**. Suppose an electrically neutral metal rod is polarized while it is in the presence of the positively charged object as shown in Fig. 22-8b. What happens if you then touch the top of the metal rod when there is a shortage of electrons? Since you and the metal rod are both conductors, some of the electrons in your body are attracted to the top of the rod. Since they are free to move they will transfer from your hand to the rod to balance the charges near the top of the rod. This leaves the metal rod negatively charged, since it started with an overall balance of electrons and protons before your electrons were added. What happens if you now stop touching the rod? The return pathway for your electrons is removed. So even if you remove the positively charged object from the region around the metal rod, the additional electrons are trapped on the metal rod and so it will remain charged. If you now bring a negatively charged object close to the metal rod, these two objects will repel one another. This provides confirmation that the metal rod has become negatively charged by induction.

READING EXERCISE 22-8: (a) Make the observation described in Reading Exercise 22-2. Is Scotch™ Magic Tape best described as an insulator or a conductor? Explain your reasoning. (b) Is a balloon an insulator or a conductor? Explain your reasoning.

READING EXERCISE 22-9: (a) Can you charge an insulator by induction? Explain your reasoning. (b) Describe the steps that you would take to give an object a positive charge using the process of charging by induction.

22-7 Coulomb's Law

So far all the observations we have reported on and tried to explain have been qualitative. Can a mathematical law be formulated to describe the interaction forces between electric charges? Initial attempts to formulate such a law involved a consideration of charged objects that are particle-like in the sense that the distance between the centers of such objects are much larger than the sizes of the objects.

In Figure 22-6 we depicted observations on particle-like charged objects that led us to the conclusion that the interaction forces between particle-like objects is proportional to the product of the charges on them, so that $|\vec{F}| \propto |q_1 q_2|$. But, what observations have been made that would lead to a mathematical relationship that also describes how interaction forces are related to the distance between charged particle-like objects?

One rather odd observation made in about 1775 by Benjamin Franklin led Joseph Priestley to speculate the magnitude of the electrical force between particle-like objects falls off as the inverse square of the distance between objects just as it does for gravitational forces between masses. Franklin observed a small cork hanging from a silk thread is attracted by induction to the outside of a charged metal can. However, if the cork is dangled inside the can, there are no apparent forces on it. Recall in Chapter 14 on gravitation we presented a shell theorem Newton derived mathematically from the assumption gravitational forces fall off as the inverse square of the distance between masses. Priestly reasoned if an analogous shell theorem seems to hold for electric interactions, then the inverse square law ought to hold for electric forces.

In 1785, Priestley's hypothesis regarding the dependence of electric forces on distance was verified by the experiments performed by Charles Augustin Coulomb using a torsion balance to measure the forces between charged spheres. Newton applied the shell theorem to deduce two spherical masses such as the Earth and the Moon would exert forces on each other as if they were each particle-like objects with mass concentrated at their centers.

Coulomb also assumed the forces between charged spheres that are some distance apart would be the same as if the charge of each object was concentrated at its center, and he found the forces between spherical objects lie along a line between their centers. Coulomb also used a method like the one we described in Section 22-4 to reduce the charges on the metal spheres in his torsion apparatus by known fractions. Thus, he was able to verify that the interaction forces were proportional to the product of the charges on the interacting objects.

As a result of his careful experiments, Coulomb verified the mathematical relationship between the magnitude of the interaction force between two particle-like objects,

$$\left|\vec{F}\right| = k\frac{|q_1||q_2|}{r^2} \qquad \text{(Coulomb's Law)}, \qquad (22\text{-}3)$$

where k is a positive constant of proportionality, r is the distance between the centers of the two objects, q_1 is the charge on one of the objects and q_2 is the charge on the other object. The full name of this relationship is Coulomb's law of *electrostatic force* (or *electric force*). The term *electrostatic* is used to emphasize that, relative to each other, the charges are either stationary or moving only very slowly.

Using modern tools available in many introductory physics laboratories, we can verify Coulomb's Law. The experiment is pictured in Figs. 22-11 and 22-12. Two ping-pong balls that are covered with conducting paint can be stroked with a fur-charged rubber rod and touched together to equalize their charges. One of the negatively-charged balls is hung as a pendulum from a long, non-conducting string. The other, which serves as a prod, is attached to a non-conducting rod, as shown in Fig. 22-11.

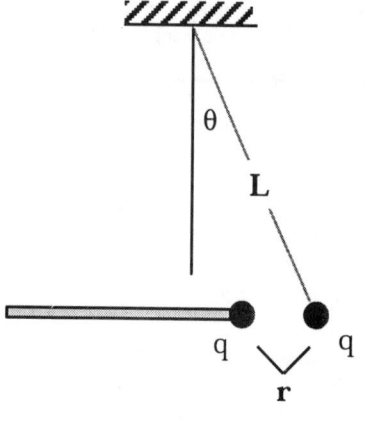

Fig. 22-11: A charged metal-coated ping pong ball is repelled from an equally charged prod. At equilibrium, the vector sum of the gravitational force, the tension in the string, and the Coulomb force on the hanging ball is zero.

The hanging ball moves in two dimensions, pushed to larger angles and rising higher as the prod is brought closer to it. This demonstrates qualitatively the force exerted by the prod on the hanging ball is greater when the distance, r, between the centers of the two charged balls is smaller. It also indicates the electrostatic force acts along the line connecting the two charges. We know this because otherwise the hanging ball would be pushed off to the side.

Video technology allows us to take this experiment a step further. The motion of the prod inching forward can be captured with a video camera and digitized. Then computer software can be used to perform a frame-by-frame analysis of the angular displacement of the hanging ball and of the distance between the balls. This yields the information needed to find the magnitude of the Coulomb force, $\left|\vec{F}_c\right|$, as a function of the distance between the centers of the hanging ball and the ball on the prod. When the ball is stationary, the next force consists of the vector sum of the gravitational force acting vertically downward, the tension force exerted by the string and the electric force acting in the horizontal direction. Thus, the magnitude of the electric force on the ball can be calculated from the mass of the ball and its angle of rise, θ, with respect to the vertical.

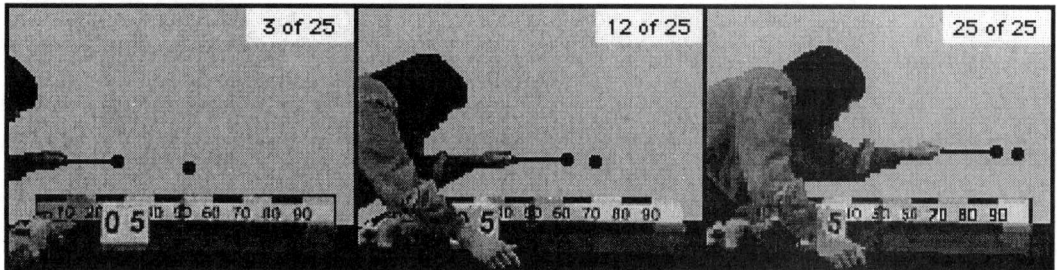

Fig. 22-12: Three of twenty-five digitized video frames depicting the forces between two charged balls. The string holding up the hanging ball is not visible. (Coulomb004.mov)

A plot of the data is shown in Fig. 22-13. If we try to fit the data we find that the Coulomb force fits as $1/r^2$ just as the gravitational force does.

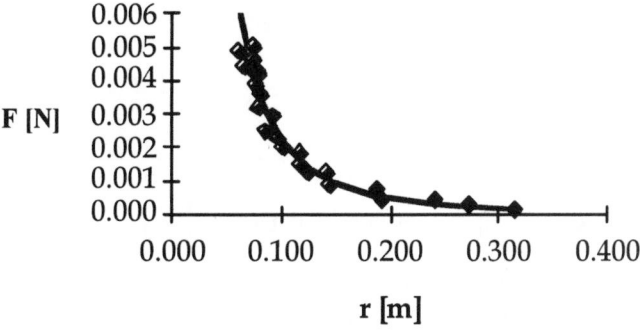

Fig. 22-13: A graph of the Coulomb force vs. the distance between two charged ping-pong balls. The gray line is based on a fit of $\vec{F}_c = \text{constant}/r^2$ that shows the inverse square law holds quite well. Assuming that Coulomb's Law holds, it can also be shown each ball carries about 5×10^{-8} coulombs of excess charge. VideoPoint® software was used to obtain the data from video frames.

Combining all we have learned about the electrostatic force allows us to express Coulomb's law in the following (more formal) terms: Let two charged particles (also called *point charges*) have charge magnitudes q_1 and q_2 and be separated by a distance r. The **electrostatic force** of attraction or repulsion q_1 exerts on q_2 has the magnitude

$$\left|\vec{F}_{12}\right| = k\frac{|q_1||q_2|}{r^2} \qquad \text{(Coulomb's Law)}, \qquad (22\text{-}4)$$

in which k is a constant.

We use the absolute value signs on the charges in this expression to remind ourselves that the sign on the force (a vector) indicates the *direction* of the force and not simply whether we are multiplying like or unlike charges. Hence, we should calculate the magnitude of the force between two charged objects using the magnitudes of the charges. We determine the direction of the force using the attraction and repulsion rules we discussed earlier, remembering the force always acts along the line connecting the two charges. If the particles *repel* each other, the force on each particle is directed *away from* the other particle (as in Figs. 22-14a and b). If the particles *attract* each other, the force on each particle is directed *toward* the other particle (as in Fig. 22-14c).

As an example of the need for absolute values in

$$\left|\vec{F}_{1\to 2}\right| = k\frac{|q_1||q_2|}{r^2},$$

consider the two unlike charges in Fig. 22-14c. Inspection of the expression for force above indicates that each particle exerts a force of the same magnitude on the other particle,

$$\left|\vec{F}_{1\to 2}\right| = \left|\vec{F}_{2\to 1}\right| \qquad (22\text{-}5)$$

However, as we would expect for a Newton's Third Law force pair, the forces point in opposite directions. The force on the positive charge (in Fig. 22-11c) due to the negative charge points to the right. The force on the negative charge due to the positive charge points to the left. If we use explicit positive and negative signs on the charges and don't make use of absolute values, the product q_1q_2 (or q_2q_1) is always negative and so the force is negative, regardless of whether we are calculating the force on the positive charge or the force on the negative charge. This cannot be correct, since these two forces point in opposite directions and the sign associated with the force vector is to denote that direction. The force cannot be negative in both cases. In this and every other situation, we avoid this pitfall if we use the absolute values of the charges in our calculations and then determine the sign associated with the force *by thinking* about our coordinate system and the issues of attraction and repulsion.

Curiously, the form of Eq. 22-4 is the same as Newton's law of gravitation presented in Chapter 14 that relates the gravitational force between two particles with masses m_1 and m_2 to the distance, r, between their centers

$$\left|\vec{F}_{1\to 2}\right| = G\frac{m_1m_2}{r^2} \qquad (14\text{-}1)$$

in which G is the gravitational constant. The constant k in Eq. 22-4, by analogy with the gravitational constant G in Eq. 14-1, may be called the *electrostatic constant.*

Coulomb's law has survived every experimental test; no exceptions to it have ever been found. It holds even within the atom, correctly describing the force between the positively charged nucleus and each of the negatively charged electrons, even though classical Newtonian mechanics fails in that realm and is replaced there by quantum physics. This simple law also correctly accounts for the forces that bind atoms together to form molecules, and for the forces that bind atoms and molecules together to form solids and liquids.

For historical reasons (and because doing so simplifies many other formulas), the electrostatic constant k is often written $1/4\pi\varepsilon_0$. Then Coulomb's law becomes

$$\left|\vec{F}_{1\to 2}\right| = \frac{1}{4\pi\varepsilon_0}\frac{|q_1||q_2|}{r^2}, \qquad \text{(Coulomb's law),} \qquad (22\text{-}6)$$

where the constant k has the value

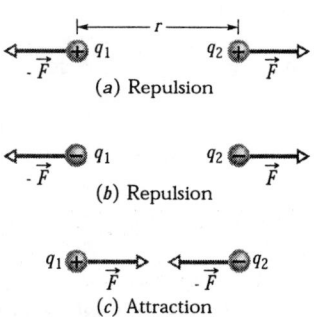

Fig. 22-14: Two charged particles, separated by distance r, repel each other if their charges are (a) both positive and (b) both negative. (c) They attract each other if their charges are of opposite signs. In each of the three situations, the force acting on one particle is equal in magnitude to the force acting on the other particle but has the opposite direction.

$$k = \frac{1}{4\pi\varepsilon_0} = 8.99 \times 10^9 \ \text{N} \cdot \text{m}^2/\text{C}^2. \tag{22-7}$$

The quantity ε_0, called the **permittivity constant,** sometimes appears separately in equations and is

$$\varepsilon_0 = 8.85 \times 10^{-12} \ \text{C}^2/\text{N} \cdot \text{m}^2. \tag{22-8}$$

READING EXERCISE 22-10: Use the information provided at the end of Section 22-4 and in Fig. 22-6 to explain why the following statements can not be true: (a) The force between two charged particle-like objects is independent of the charge on the objects. (b) The force between two charged particle-like objects is proportional to $1/q_1 q_2$. (c) The force between two charged objects is proportional to $q_1 + q_2$.

22-8 Solving Problems Using Coulomb's Law

Coulomb's law can be used to find the forces on particle-like objects having excess charge on them. By particle-like, we mean the distance between charged objects is much larger than the sizes of the object. When solving quantitative (numerical) problems using Coulomb's law to determine forces between charged particle-like objects having excess charge on them, there are several issues to keep in mind. For example, we must be sure to express the charges in coulombs and the distance between the charges in meters. These are the SI units for distance and charge and are required for use if we are to use the standard value for k,

$$k = \frac{1}{4\pi\varepsilon_0} = 8.99 \times 10^9 \ \text{N} \cdot \text{m}^2/\text{C}^2 \ ,$$

which is written in terms of these units.

As we discussed in Section 22-7, we must calculate the magnitude of the force using the (positive) magnitude of the charges in Coulomb's law. We should then make a sketch of the situation, showing the direction of the force and adopting a coordinate system. If the force acts in the negative direction, then we associate a negative sign with the magnitude of the force.

But what happens if there are more than two charges interacting? This situation is depicted in Fig. 22-15.

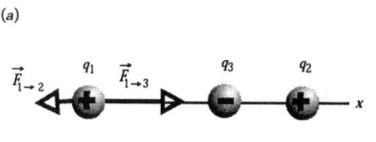

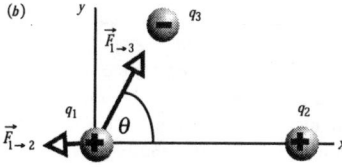

Fig. 22-15: (*a*) A set of particle-like objects with excess charge on them that lie along a line. The force vectors depict the forces of charges 2 and 3 on charge 1. (*b*) A similar diagram showing the force vectors on charge 1 when the other charges do not lie along the same line.

Superposition of Forces

As is the case with all other forces (including the gravitational force), the electrostatic force obeys the principle of superposition. That is, if we have n charged particles, they interact *independently in pairs*, and the force on any one of the charges, let us say particle 1, is given by the vector sum

$$\vec{F}_{1,\text{net}} = \vec{F}_{2\to1} + \vec{F}_{3\to1} + \vec{F}_{4\to1} + \vec{F}_{5\to1} \ldots \vec{F}_{n\to1} \tag{22-9}$$

in which, for example, $\vec{F}_{4\to1}$ is the force acting on particle 1 owing to the presence of particle 4.

Often, as is the case for the charges in Fig. 22-15, the various forces acting on a particle do not all act along the same line. We know how to combine forces such as these, but let's review the process. First we must calculate the magnitudes of the individual forces (in this case, using Coulomb's law), adopt a coordinate system and determine the directions of the forces. We then calculate the orthogonal (perpendicular) components of each force and determine the direction of these components. For example, this might mean determining the *x*- and *y*- components of each of the forces as well as determining if those components are in the positive or negative direction. We then add or subtract all of the components of forces that act along the same line. We add or subtract depending on whether the components are in the same or opposite directions. This gives us the components of the net (resultant) force. We then use trigonometry to get the magnitude and direction of the resultant force. These steps are presented in brief below.

Steps to Solving Quantative Problems using Coulomb's Law

1. Calculate the *magnitudes* of the individual forces using Coulomb's law;

2. Adopt a coordinate system;

3. Determine the directions of the forces;

4. Calculate the orthogonal (perpendicular) components of each force using $F_y = \left|\vec{F}\right|\sin\theta$ and $F_x = \left|\vec{F}\right|\cos\theta$, or equivalent expressions;

5. Determine the direction of these components;

6. Combine all the force components that act along the same line, adding and subtracting as appropriate;

7. Combine the resultant components to get the magnitude of the resultant force using

$$F_{net}^2 = F_{x,net}^2 + F_{y,net}^2;$$
(22-10)

8. Determine the angle at which the force acts (relative to the positive *x* axis) using

$$\tan\theta = \frac{F_{y,net}}{F_{x,net}}$$
(22-11)

or some equivalent expression.

READING EXERCISE 22-11: The figure shows two protons (symbol p) and one electron (symbol e) on an axis. What are the directions of (a) the electrostatic force on the central proton due to the electron, (b) the electrostatic force on the central proton due to the other proton, and (c) the net electrostatic force on the central proton? Are there points off the line connecting the charges where the force is also zero? Explain your reasoning and how your answers relate to superposition for forces.

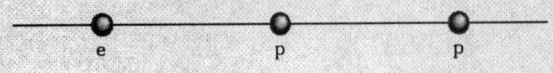

Touchstone Examples 22-8-1, 22-8-2, and 22-8-3, at the end of this chapter, illustrate how to use what you learned in this section.

| TE |

22-9 Comparing Electrical and Gravitational Forces

As we noted earlier, Coloumb's law

$$\left|\vec{F}_{1\rightarrow2}\right| = k\frac{|q_1||q_2|}{r^2},$$

has the same form as that of Newton's equation for the gravitational force between two particles with masses m_1 and m_2 that are separated by a distance r:

$$\left|\vec{F}_{1\rightarrow2}\right| = G\frac{m_1m_2}{r^2}.$$

Both of these equations have the distance between the two interacting objects squared and in the denominator of the fraction. That is, they are both "inverse square laws." Both also involve a property of the interacting particles—the mass in one case and the charge in the other. Both the gravitational force and the electrostatic force are conservative forces—the work done by these forces around a closed path is zero. Both forces act along the line connecting the two objects—such forces are called "central" forces.

However, as similar as these forces are, they are not the same. They are not different aspects of one force. How do we know this? Electrostatic forces are intrinsically much stronger than gravitational forces. For example, the gravitational attraction between a plastic comb and a small piece of paper is not large enough to overcome the opposing gravitational attraction of the Earth on the paper. However, if you rub the comb with fur, the resulting electrostatic force *is* large enough to overcome the gravitational attraction of the Earth. Furthermore, the electrostatic force differs from the gravitational forces because the gravitational force is always attractive but the electrostatic force may be *either* attractive or repulsive, depending on the signs of the two charges. This difference arises because although there is only one kind of mass, there are two kinds of charge. That is why absolute value signs are needed in

$$\left|\vec{F}_{1\rightarrow2}\right| = k\frac{|q_1||q_2|}{r^2}$$

but not in

$$\left|\vec{F}_{1\rightarrow2}\right| = G\frac{m_1m_2}{r^2}.$$

Before concluding our discussion of the electrostatic force, let's compare it to another somewhat similar force—the force associated with magnets.

In addition to amber, the early Greeks knew of another special material that had the ability to attract other objects. They had recorded the observation some naturally occurring "lodestones," known today as the mineral magnetite, would attract iron. Lodestones were the first known magnets. Could the phenomena of amber (electricity) and lodestones (magnetism) be related?

Observation of the interactions between two magnets and two electrified objects shows that the phenomena of electricity and magnetism are *not* the same. Two magnets will either attract or repel one another, depending on their orientation. Two pieces of rubbed amber (or glass) always repel one another, regardless of their orientation. Further, the forces associated with magnets do not fade with time like electric forces and are unaffected by weather.

Hence, the study of electricity and magnetism developed separately for centuries—until 1820, in fact, when Hans Christian Oersted found a connection between them: an electric current in a wire can deflect a magnetic compass needle. The new science of *electromagnetism* (the combination of electrical and magnetic phenomena) was developed further by Michael Faraday, a truly gifted experimenter with a talent for physical intuition and visualization. That talent is attested to by the fact his collected laboratory notebooks do not contain a single equation. In the mid-19th century, James Clerk Maxwell put Faraday's ideas into mathematical form, introduced many new ideas of his own, and put electromagnetism on a sound theoretical basis.

Table 32-1 shows the basic laws of electromagnetism, now called Maxwell's equations. We plan to work our way through them in the chapters between here and there, but you might want to glance at them now, to see our goal.

22-10 Many Everyday Forces are Electrostatic

In Chapter 6 we presented an idealized model of a solid as an array of atoms held together by electrical forces that act like tiny springs that resist either stretching or compression forces. We then used this spring model to help explain the nature of most of the everyday forces encountered in the study of motion including normal forces, friction forces, and tension forces. We made the claim that all of these forces are basically electrical.

Let's look once again at our spring model of solids in light of our new understanding of the nature of the electrostatic forces between the protons and electrons in atoms. Since protons and electrons have opposite charges they attract each other. This is what holds individual atoms together. But why do atoms combine with each other into molecules in gases, liquids, and solids? And why are solids spring-like? The answer to this question cannot be fully understood solely on the basis of our study of classical mechanics and our observations of electrostatic interactions. We must turn to quantum theory to see electrically neutral atoms of different types become more stable when they share some of their electrons and come together in various arrangements. However, let's use our understanding of electrostatic forces to imagine in a simplified view what might be happening to the electrons and protons in a solid when a compressed block exerts a normal force or a stretched string exerts a tension force.

Under compression the outer electrons in the atoms of one object are repelling the outer electrons in the other object. Although we imagine this repulsion starting at the surface, we believe it is happening in other layers of atoms. As the electrons from one layer of atoms are being moved closer to those in the next layer, the repulsion forces increase sharply as the electrons are forced closer together. This electrostatic repulsion gets greater as the objects are pushed together more.

What happens when a string is stretched? When a string is stretched we imagine there is a tug-of-war taking place as we try to move the outer electrons farther away from their nuclei. The electrostatic attraction between the positively charged protons in the nuclei and the negatively charged electrons prevents much stretching.

Thus, we think of a solid as having a delicately balanced equilibrium in which the electron glue holds the atoms together at just the right spacing. A pull on a solid will lead to a pull back by the solid due to the electrostatic attractions between electrons and protons in the nuclei. A push leads to electrostatic repulsions between outer electrons that surround each atom.

So it is we believe all of the everyday forces we encounter are either gravitational, electrical, or magnetic. We will explore the relationship between electrostatic and magnetic interactions further in Chapter 32.

Touchstone Example 22-8-1

(a) Figure 22-1*a* shows two positively charged particles fixed in place on an *x* axis. The charges are $q_1 = 1.60 \times 10^{-19}$ C and $q_2 = 3.20 \times 10^{-19}$ C, and the particle separation is $R = 0.0200$ m. What are the magnitude and direction of the electrostatic force $\vec{F}_{12}$ on particle 1 from particle 2?

SOLUTION: The **Key Idea** here is that, because both particles are positively charged, particle 1 is repelled by particle 2, with a force magnitude given by Eq. 22-6. Thus, the direction of force $\vec{F}_{12}$ on particle 1 is *away from* particle 2, in the negative direction of the *x* axis, as indicated in the free-body diagram of Fig. TE22-1*b*. Using Eq. 22-6 with separation R substituted for r, we can write the magnitude $\vec{F}_{12}$ of this force as

$$|\vec{F}_{12}| = \frac{1}{4\pi\varepsilon_0} \frac{|q_1||q_2|}{R^2}$$

$$= (8.99 \times 10^9 \text{ N} \cdot \text{m}^2/\text{C}^2)$$

$$\times \frac{(1.60 \times 10^{-19} \text{ C})(3.20 \times 10^{-19} \text{ C})}{(0.0200 \text{ m})^2}$$

$$= 1.15 \times 10^{-24} \text{ N}.$$

Thus, force $\vec{F}_{12}$ has the following magnitude and direction (relative to the positive direction of the *x* axis):

$$1.15 \times 10^{-24} \text{ N} \quad \text{and} \quad 180°. \qquad \text{(Answer)}$$

We can also write $\vec{F}_{12}$ in unit-vector notation as

$$\vec{F}_{12} = -(1.15 \times 10^{-24} \text{ N})\hat{\imath}. \qquad \text{(Answer)}$$

(b) Figure TE22-1*c* is identical to Fig. TE22-1*a* except that particle 3 now lies on the *x* axis between particles 1 and 2. Particle 3 has charge $q_3 = -3.20 \times 10^{-19}$ C and is at a distance $\frac{3}{4}R$ from particle 1. What is the net electrostatic force $\vec{F}_{1,\text{net}}$ on particle 1 due to particles 2 and 3?

SOLUTION: One **Key Idea** here is that the presence of particle 3 does not alter the electrostatic force on particle 1 from particle 2. Thus, force $\vec{F}_{12}$ still acts on particle 1. Similarly, the force $\vec{F}_{13}$ that acts on particle 1 due to particle 3 is not affected by the presence of particle 2. Because particles 1 and 3 have charge of opposite sign, particle 1 is attracted to particle 3. Thus, force $\vec{F}_{13}$ is directed *toward* particle 3, as indicated in the free-body diagram of Fig. TE22-1*d*.

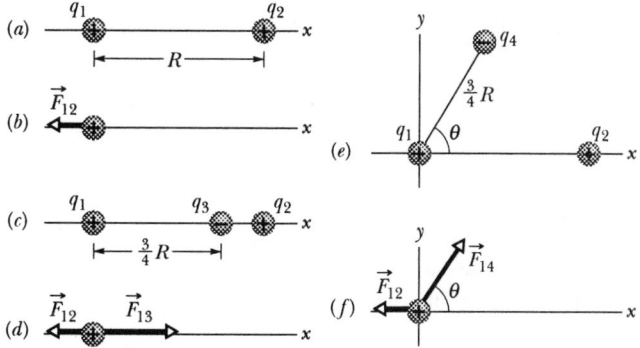

Fig. TE22-1 (*a*) Two charged particles of charges q_1 and q_2 are fixed in place on an *x* axis, with separation R. (*b*) The free-body diagram for particle 1, showing the electrostatic force on it from particle 2. (*c*) Particle 3 is now fixed in place on the *x* axis, along with particles 1 and 2. (*d*) The free-body diagram for particle 1. (*e*) Particle 4 is fixed in place on a line at angle θ to the *x* axis, again with particles 1 and 2. (*f*) The free-body diagram for particle 1.

To find the magnitude of $\vec{F}_{13}$, we can rewrite Eq. 22-6 as

$$|\vec{F}_{13}| = \frac{1}{4\pi\varepsilon_0} \frac{|q_1||q_3|}{(\frac{3}{4}R)^2}$$

$$= (8.99 \times 10^9 \,\text{N}\cdot\text{m}^2/\text{C}^2)$$

$$\times \frac{(1.60 \times 10^{-19}\,\text{C})(3.20 \times 10^{-19}\,\text{C})}{(\frac{3}{4})^2(0.0200\,\text{m})^2}$$

$$= 2.05 \times 10^{-24}\,\text{N}.$$

We can also write $\vec{F}_{13}$ in unit-vector notation:

$$\vec{F}_{13} = (2.05 \times 10^{-24}\,\text{N})\hat{i}.$$

A second **Key Idea** here is that the net force $\vec{F}_{1,\text{net}}$ on particle 1 is the vector sum of $\vec{F}_{12}$ and $\vec{F}_{13}$; that is, from Eq. 22-9, we can write the net force $\vec{F}_{1,\text{net}}$ on particle 1 in unit-vector notation as

$$\vec{F}_{1,\text{net}} = \vec{F}_{12} + \vec{F}_{13}$$

$$= -(1.15 \times 10^{-24}\,\text{N})\hat{i} + (2.05 \times 10^{-24}\,\text{N})\hat{i}$$

$$= (9.00 \times 10^{-25}\,\text{N})\hat{i}. \qquad \text{(Answer)}$$

Thus, $\vec{F}_{1,\text{net}}$ has the following magnitude and direction (relative to the positive direction of the x axis):

$$9.00 \times 10^{-25}\,\text{N} \quad \text{and} \quad 0°. \qquad \text{(Answer)}$$

(c) Figure TE22-1*e* is identical to Fig. TE22-1*a* except that particle 4 is now positioned as shown. Particle 4 has charge $q_4 = -3.20 \times 10^{-19}$ C, is at a distance $\frac{3}{4}R$ from particle 1, and lies on a line that makes an angle $\theta = 60°$ with the x axis. What is the net electrostatic force $\vec{F}_{1,\text{net}}$ on particle 1 due to particles 2 and 4?

SOLUTION: The **Key Idea** here is that the net force $\vec{F}_{1,\text{net}}$ is the vector sum of $\vec{F}_{12}$ and a new force $\vec{F}_{14}$ acting on particle 1 due to particle 4. Because particles 1 and 4 have charges of opposite sign, particle 1 is attracted to particle 4. Thus, force $\vec{F}_{14}$ on particle 1 is directed *toward* particle 4, at angle $\theta = 60°$, as indicated in the free-body diagram of Fig. TE22-1*f*.

To find the magnitude of $\vec{F}_{14}$, we can rewrite Eq. 22-6 as

$$|\vec{F}_{14}| = \frac{1}{4\pi\varepsilon_0} \frac{|q_1||q_4|}{(\frac{3}{4}R)^2}$$

$$= (8.99 \times 10^9 \,\text{N}\cdot\text{m}^2/\text{C}^2)$$

$$\times \frac{(1.60 \times 10^{-19}\,\text{C})(3.20 \times 10^{-19}\,\text{C})}{(\frac{3}{4})^2(0.0200\,\text{m})^2}$$

$$= 2.05 \times 10^{-24}\,\text{N}.$$

Then from Eq. 22-9, we can write the net force $\vec{F}_{1,\text{net}}$ on particle 1 as

$$\vec{F}_{1,\text{net}} = \vec{F}_{12} + \vec{F}_{14}.$$

To evaluate the right side of this equation, we need another **Key Idea**: Because the forces $\vec{F}_{12}$ and $\vec{F}_{14}$ are not directed along the same axis, we *cannot* sum by simply combining their magnitudes. Instead, we must add them as vectors, using one of the following methods.

Method 1. *Summing directly on a vector-capable calculator.* For $\vec{F}_{12}$, we enter the magnitude 1.15×10^{-24} and the angle $180°$. For $\vec{F}_{14}$, we enter the magnitude 2.05×10^{-24} and the angle $60°$. Then we add the vectors.

Method 2. *Summing in unit-vector notation.* First we rewrite $\vec{F}_{14}$ as

$$\vec{F}_{14} = (F_{14}\cos\theta)\hat{i} + (F_{14}\sin\theta)\hat{j}.$$

Substituting 2.05×10^{-24} N for F_{14} and $60°$ for θ, this becomes

$$\vec{F}_{14} = (1.025 \times 10^{-24} \text{ N})\hat{i} + (1.775 \times 10^{-24} \text{ N})\hat{j}.$$

Then we sum:

$$\begin{aligned}
\vec{F}_{1,\text{net}} &= \vec{F}_{12} + \vec{F}_{14} \\
&= -(1.15 \times 10^{-24} \text{ N})\hat{i} \\
&\quad + (1.025 \times 10^{-24} \text{ N})\hat{i} + (1.775 \times 10^{-24} \text{ N})\hat{j} \\
&\approx (-1.25 \times 10^{-25} \text{ N})\hat{i} + (1.78 \times 10^{-24} \text{ N})\hat{j}. \quad \text{(Answer)}
\end{aligned}$$

Method 3. *Summing components axis by axis.* The sum of the x components gives us

$$\begin{aligned}
F_{1,\text{net},x} &= F_{12,x} + F_{14,x} = F_{12} + F_{14} \cos 60° \\
&= -1.15 \times 10^{-24} \text{ N} + (2.05 \times 10^{-24} \text{ N})(\cos 60°) \\
&= -1.25 \times 10^{-25} \text{ N}.
\end{aligned}$$

The sum of the y components gives us

$$\begin{aligned}
F_{1,\text{net},y} &= F_{12,y} + F_{14,y} = 0 + F_{14} \sin 60° \\
&= (2.05 \times 10^{-24} \text{ N})(\sin 60°) \\
&= 1.78 \times 10^{-24} \text{ N}.
\end{aligned}$$

The net force $\vec{F}_{1,\text{net}}$ has the magnitude

$$|\vec{F}_{1,\text{net}}| = \sqrt{F_{1,\text{net},x}^2 + F_{1,\text{net},y}^2} = 1.78 \times 10^{-24} \text{ N}. \quad \text{(Answer)}$$

To find the direction of $\vec{F}_{1,\text{net}}$, we take

$$\theta = \tan^{-1} \frac{F_{1,\text{net},y}}{F_{1,\text{net},x}} = -86.0°.$$

However, this is an unreasonable result because $\vec{F}_{1,\text{net}}$ must have a direction between the directions of $\vec{F}_{12}$ and $\vec{F}_{14}$. To correct θ, we add $180°$, obtaining

$$-86.0° + 180° = 94.0°. \quad \text{(Answer)}$$

Touchstone Example 22-8-2

Figure TE22-2a shows two particles fixed in place: a particle of charge $q_1 = +8q$ at the origin and a particle of charge $q_2 = -2q$ at $x = L$, where q represents a positive charge. At what point (other than infinitely far away) can a proton be placed so that it is in *equilibrium* (meaning that the net force on it is zero)? Is that equilibrium *stable* or *unstable*?

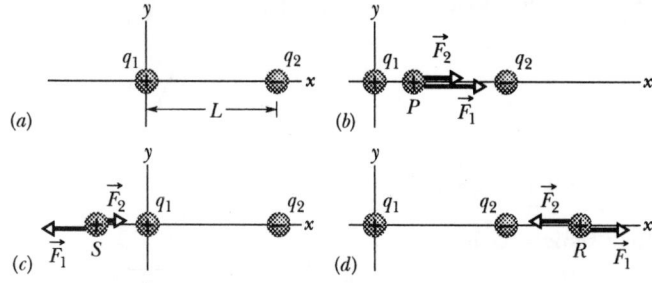

Fig. TE22-2 (a) Two particles of charges q_1 and q_2 are fixed in place on an x axis, with separation L. (b)–(d) Three possible locations P, S, and R for a proton. At each location, $\vec{F}_1$ is the force on the proton from particle 1 and $\vec{F}_2$ is the force on the proton from particle 2.

SOLUTION: The **Key Idea** here is that, if $\vec{F}_1$ is the force on the proton due to charge q_1 and $\vec{F}_2$ is the force on the proton due to charge q_2, then the point we seek is where $\vec{F}_1 + \vec{F}_2 = 0$. This condition requires that

$$\vec{F}_1 = -\vec{F}_2. \tag{TE22-1}$$

This tells us that at the point we seek, the forces acting on the proton due to the other two particles must be of equal magnitudes,

$$|\vec{F}_1| = |\vec{F}_2|, \tag{TE22-2}$$

and that the forces must have opposite directions.

A proton has a positive charge. Thus, the proton and the particle of charge q_1 are of the same sign, and force $\vec{F}_1$ on the proton must point away from q_1. Also, the proton and the particle of charge q_2 are of opposite signs, so force $\vec{F}_2$ on the proton must point toward q_2. "Away from q_1" and "toward q_2" can be in opposite directions only if the proton is located on the x axis.

If the proton is on the x axis at any point between q_1 and q_2, such as P in Fig. TE22-2b, then $\vec{F}_1$ and $\vec{F}_2$ are in the same direction and not in opposite directions as required. If the proton is at any point on the x axis to the left of q_1, such as point S in Fig. TE22-2c, then $\vec{F}_1$ and $\vec{F}_2$ are in opposite directions. However, Eq. 22-6 tells us that $\vec{F}_1$ and $\vec{F}_2$ cannot have equal magnitudes there: $|\vec{F}_1|$ must be greater than $|\vec{F}_2|$, because $|\vec{F}_1|$ is produced by a closer charge (with lesser r) of greater magnitude ($8q$ versus $2q$).

Finally, if the proton is at any point on the x axis to the right of q_2, such as point R in Fig. TE22-2d, then $\vec{F}_1$ and $\vec{F}_2$ are again in opposite directions. However, because now the charge of greater magnitude (q_1) is *farther* away from the proton than the charge of lesser magnitude, there is a point at which $|\vec{F}_1|$ is equal to $|\vec{F}_2|$. Let x be the coordinate of this point, and let q_p be the charge of the proton. Then with the aid of Eq. 22-6, we can rewrite Eq. TE22-2 as

$$\frac{1}{4\pi\varepsilon_0} \frac{8|qq_p|}{x^2} = \frac{1}{4\pi\varepsilon_0} \frac{2|qq_p|}{(x-L)^2}. \tag{TE22-3}$$

(Note that only the charge magnitudes appear in Eq. TE22-3.) Rearranging Eq. TE22-3 gives us

$$\left(\frac{x-L}{x}\right)^2 = \frac{1}{4}.$$

After taking the square roots of both sides, we have

$$\frac{x-L}{x} = \frac{1}{2},$$

which gives us

$$x = 2L. \tag{Answer}$$

The equilibrium at $x = 2L$ is unstable; that is, if the proton is displaced leftward from point R, then $|\vec{F}_1|$ and $|\vec{F}_2|$ both increase but $|\vec{F}_2|$ increases more (because q_2 is closer than q_1), and a net force will drive the proton farther leftward. If the proton is displaced rightward, both $|\vec{F}_1|$ and $|\vec{F}_2|$ decrease but $|\vec{F}_2|$ decreases more, and a net force will then drive the proton farther rightward. In a stable equilibrium, each time the proton is displaced slightly, it would return to the equilibrium position.

Touchstone Example 22-8-3

The nucleus in an iron atom has a radius of about 4.0×10^{-15} m and contains 26 protons.

(a) What is the magnitude of the repulsive electrostatic force between two of the protons that are separated by 4.0×10^{-15} m?

SOLUTION: The **Key Idea** here is that the protons can be treated as charged particles, so the magnitude of the electrostatic force on one from the other is given by Coulomb's law. Table 22-1 tells us that their charge is $+ e$. Thus, Eq. 22-6 gives us

$$|\vec{F}| = \frac{1}{4\pi\varepsilon_0}\frac{e^2}{r^2}$$

$$= \frac{(8.99 \times 10^9 \text{ N} \cdot \text{m}^2/\text{C}^2)(1.60 \times 10^{-19} \text{ C})^2}{(4.0 \times 10^{-15} \text{ m})^2}$$

$$= 14 \text{ N.} \qquad \text{(Answer)}$$

This is a small force to be acting on a macroscopic object like a cantaloupe, but an enormous force to be acting on a proton. Such forces should blow apart the nucleus of any element but hydrogen (which has only one proton in its nucleus). However, they don't, not even in nuclei with a great many protons. Therefore, there must be some enormous attractive force to counter this enormous repulsive electrostatic force.

(b) What is the magnitude of the gravitational force between those same two protons?

SOLUTION: The **Key Idea** here is like that in part (a): Because the protons are particles, the magnitude of the gravitational force on one from the other is given by Newton's equation for the gravitational force (Eq. 14-1). With m_p ($= 1.67 \times 10^{-27}$ kg) representing the mass of a proton, Eq. 14-1 gives us

$$|\vec{F}| = G\frac{m_p^2}{r^2}$$

$$= \frac{(6.67 \times 10^{-11} \text{ N} \cdot \text{m}^2/\text{kg}^2)(1.67 \times 10^{-27} \text{ kg})^2}{(4.0 \times 10^{-15} \text{ m})^2}$$

$$= 1.2 \times 10^{-35} \text{ N.} \qquad \text{(Answer)}$$

This result tells us that the (attractive) gravitational force is far too weak to counter the repulsive electrostatic forces between protons in a nucleus. Instead, the protons are bound together by an enormous force called (aptly) the *strong nuclear force*—a force that acts between protons (and neutrons) when they are close together, as in a nucleus.

Although the gravitational force is many times weaker than the electrostatic force, it is more important in large-scale situations because it is always attractive. This means that it can collect many small bodies into huge bodies with huge masses, such as planets and stars, that then exert large gravitational forces. The electrostatic force, on the other hand, is repulsive for charges of the same sign, so it is unable to collect either positive charge or negative charge into large concentrations that would then exert large electrostatic forces.

23 Electric Fields

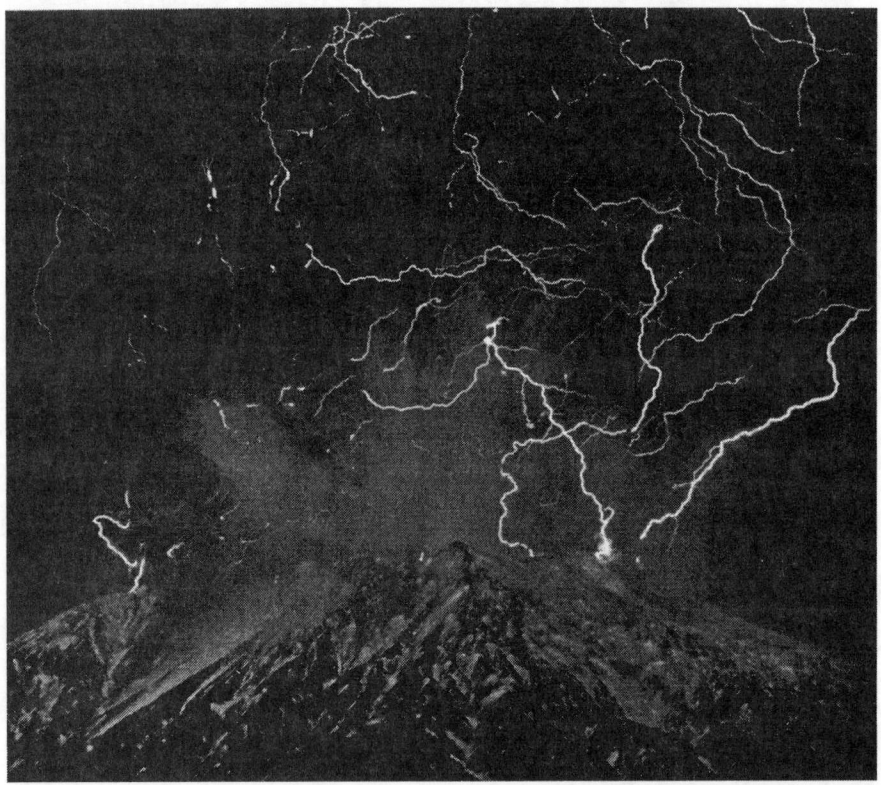

During the frequent eruptions of the Sakurajima volcano in Japan, multiple electrical discharges (sparks) flash over the volcano's crater, lighting up the sky and sending out sound waves that resemble thunder. However, this is not a lightning display in a thunderstorm, with electrified clouds of water drops discharging to the ground. This is something different.

How does the region above the volcano become electrified, and is there any way to tell whether the sparks travel up from the crater or down to it?

The answer is in this chapter.

23-1 Implications of Strong Electric Forces

We know how to handle problems in which only gravitational forces act on an object near Earth. In theory these problems are complicated because every object with mass will attract the object and so we have many force vectors to sum. However, the gravitational forces between most everyday objects (for example, between two cars) are not large enough to be of concern to us, so we can ignore them. Because the Earth is so much more massive than other objects found on or near it, the Earth is the only *significant* gravitational force that acts on objects near its surface.

Further, in Section 3-9 we discussed the fact experiments show the magnitude of the gravitational force on a given object is essentially constant and acts in a downward direction for all heights and locations near the surface of the Earth where people normally travel. This means it is easy to describe the gravitational force on any object relatively close to the Earth's surface. The force points straight down with a magnitude given by $\left|\vec{F}_{grav}\right| = mg$ (Eq. 3-6) where g denotes the local gravitational strength. In a local area where the curvature of the Earth is not noticeable, the direction of the force does not change as shown in Fig. 23-1a.

In contrast, the electric force is so strong even small objects like two Styrofoam cups rubbed with plastic bags exert observable forces on one another. We often have to consider electric forces exerted on a charged object by many different sources in our surroundings (from combs to electrons to Styrofoam cups). In addition, because these objects are relatively small, it is not hard to get "far" enough away from them so the magnitude of the force changes appreciably. Of course, a noticeable change in the magnitude of the gravitational force also occurs when an object moves far away from the Earth's surface as shown in Fig. 23-1b. Hence, we often find the electric forces on a charged object vary in direction and magnitude from point to point. The same can be said for gravitational forces in space travel where large distances are covered. We need to devise a method for keeping track of the influence of these various forces at different locations. The concept of *field* has been developed for this purpose.

23-2 Introduction to the Concept of a Field

Let's ask ourselves, "What force would an object feel if it were placed at a given point in space?" and then represent the result graphically. Consider the simple example of the gravitational force shown in Fig. 23-1b. We can calculate (or measure) the magnitude and direction of the gravitational force exerted by the Earth on another object of known mass. The more locations for which we do the calculation or make the measurement, the more detailed picture we will have of the distribution of relative forces experienced by our object. In general a **field** is defined as a three-dimensional representation of a physical quantity that can change with location.

Using Vectors to Map a Force Field

One way to represent a vector field is to draw a map with arrows representing the vectors at various points in space. For example, Fig. 23-1b shows the force at a few of the infinite number of points that make up a *gravitational vector force field* in the space surrounding the Earth. The magnitude of each force vector is directly proportional to both the mass of the Earth and the mass of the object of interest. We can also have fields associated with scalar quantities such as temperature and pressure. For example, the temperature at every point in a room can be measured. If the room contains both a good heater and a window open to cold winter air, the temperature at each point in the room will be different. We call the resulting distribution of temperatures a *temperature field.* In much the same way, you can imagine a *pressure field* in the atmosphere. Temperature *and* pressure are scalar quantities because they have no direction associated with them.

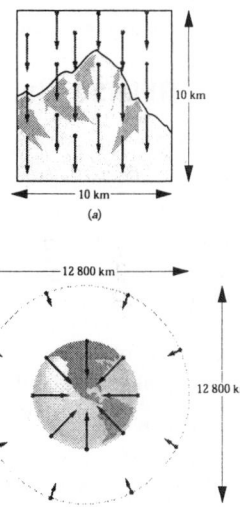

Fig. 23-1: Force field maps: (*a*) Force diagram showing the constant gravitational force on a near-Earth object placed at different locations within a vertical and horizontal distance of 10 km. Mt. Everest is in the background. The tail of each force vector has been placed at the location of the object. (*b*) If an object of a greater mass is placed near the Earth's surface, the magnitude of the force on it is greater but still constant at different locations near the Earth's surface. However, the direction defined as "downward" changes at points on the surface that are far apart from each other. Also, the gravitational force decreases to one-fourth of its former magnitude, a height equal to the Earth's radius.

Note that there are two separate aspects of the fields we just discussed. The first is a source of the quantity being measured. In the room discussed above, the heater and window can be thought of as sources of thermal energy that affect the temperature at different locations in the room. The second aspect is we have a measurement tool. In the case of the temperature field the tool is a thermometer. In the case of the pressure field, the tool is a pressure gauge. Without some kind of sensing tool, we could not "know" the fields. These are general characteristic of fields. There is a source (or sources) of the quantity being measured and there is some kind of a "test device" with which we measure the field.

In mapping *force fields*, the fixed object is called the **source object** because it is the object the force is exerted *by*. We will use the second object, the object that the force is exerted on as our measurement tool. It will be referred to as the **test object**.

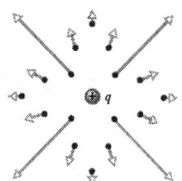

Fig. 23-2: Vectors representing the direction and relative magnitude of the force that would be exerted *by* a fixed positive source charge shown on a second positive test charge if the second charge was placed at the various locations of the vectors. The location of a vector is taken to be the location of the tail end of the vector. The magnitude of the force is given by the length of the vector.

Test Objects Should Not Change Field Values

Since all forces occur in pairs, it is possible for a test object to exert forces on the source objects, to change their locations, and hence to change the nature of the field. This makes the determination of the field values very complex unless the source objects have much greater magnitudes than the test object or are somehow fixed in space. The field concept is only useful in cases where the test object does not change the field. Here are several examples for which the field concept is useful:

1. In the Earth's gravitational force field, the location of a much smaller test mass has a negligible effect on the Earth's location and hence on its force field.

2. A conducting sphere with billions of excess electrons can act as source charges on a test charge consisting of a single proton. The electrons will stay put as they exert much stronger forces on each other than a single proton can exert on them.

3. A few electrons on the surface of an insulator such as a piece of Styrofoam can act as source charges that exert a net force on a single electron. The source charges are trapped on the insulator and will not reconfigure themselves.

▶The field concept is only useful when the test object is too small to change the nature of the field or the source objects are fixed.

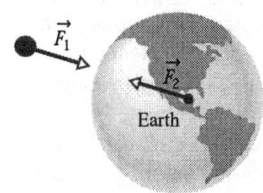

Fig. 23-3: An Earth-object system in which the interaction forces between a much less massive object and the Earth have the same magnitude. As the object falls, the movement of the Earth is negligible, so the gravitational force field surrounding the Earth does not change. The Earth does positive work on the falling object falling under the influence of the Earth's gravitational force, but the object does no work on the essentially stationary Earth.

Mapping the Force Field from a Single Charge

How can we map the electrostatic force field created by a source having a point-like charge represented as q acting on a test object with a charge represented by q_t? Instead of making force measurements to map the field, we can make theoretical predictions of what a map of the field would look like by using Coulomb's Law. We can calculate the magnitudes and directions of the forces on the test charge at many locations and then place an arrow at each location. The tail of the arrow (not its tip) is located over the point for which we have calculated (or measured) the force.

We are free to choose a convenient length for the first arrow we draw. But, the direction of the arrow must be the same as the force direction. Once the length of the first arrow to be drawn is chosen, a new arrow can be drawn at another point in space. However, the ratio of the length of the new arrow to that of the first arrow must be the same as the ratio of the new force to that of the first. For example, Coulomb's Law predicts if the distance between our test charge and the center of the source charge is doubled, the new force on it will be one-fourth of what the first force was. This is shown in Fig. 23-2 where the arrows at twice the distance are one-fourth the length. (Note that we are not putting down real objects at all of the possible locations surrounding the source charge. Instead we imagine or test what the force on our test charge would be if we placed it at one location at a time. If we put down lots of test objects at the same time they would exert forces on each other and mess things up.)

This type of *force field plot* is valuable in that it immediately tells us important information about the forces. For example, Figs. 23-2 and 23-4 show us the gravitational force exerted by an object is everywhere attractive and the force between two positively charged objects is everywhere repulsive. These two figures also show both of these forces act along the line connecting the centers of the two objects (they are "central forces"). They immediately remind us the forces are large close to the source charge (long arrows) and small farther from the source charge (short arrows).

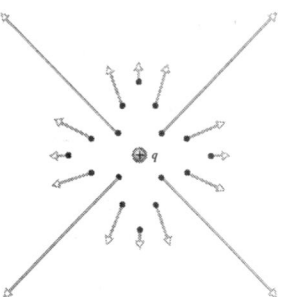

READING EXERCISE 23-1: In Fig. 23-4, suppose we chose an arrow that had a length of 36 mm to represent the force at a distance of 2 cm from the source charge. What would the length of the arrows representing the magnitude of the force on the same test charge be if it was:(a) 4 cm away from the center of the source charge? (b) 6 cm away from the center of the source charge?

READING EXERCISE 23-2: In measuring a field that has different values at each point in space, the "test" or measurement device must be small relative to the region of space over which the measurements are to be made. Why?

Fig. 23-4: Here the test charge is twice what it was in Fig. 23-2. Coulomb's Law tells us that if the test charge is doubled, the force it experiences at each location will be double that experienced by the original test charge shown in Fig. 23-2.

23-3 The Electric Field

Although the lengths of the arrows shown in Figs. 23-1 and 23-2 represent the magnitudes of the force experienced by a test object at various locations, the force magnitudes (and hence lengths of the arrows) would be different for different test objects. This is not very convenient since it means we will need an infinite number of field plots to represent all different test objects.

Figure 23-4 shows the force field of a test object with twice the charge as the one depicted in Fig. 23-2. Note every arrow shown in the figure doubles in length. That is, each arrow is scaled by the same factor based on how much larger or smaller the object's charge is. The force vectors scale as they do because for a given source charge, the electrostatic force is directly proportional to the magnitude of the test charge.

Figures 23-2 and 23-4 illustrate an important feature of force field mappings for electrostatic forces. Namely, the information we get from these force mappings about the direction and *relative* magnitude of the force at each location in space is independent of the characteristics of the test charge. In other words, all the test charges we might want to consider do not change the arrangement of the source charges interacting with them. In cases where the direction and *relative* magnitude of the force is independent of the characteristics of the test charge, we can avoid having to create a new map for every different test charge we might choose to consider.

This is exactly what we did in Section 14-2 when we noted that the magnitude of the gravitational force exerted by a source object such as the Earth of mass m_1 on a test object of mass m_2 is given by

$$\left| \vec{F}_{1\to 2} \right| = \left(\frac{Gm_1}{r^2} \right) m_2 \qquad (23\text{-}1)$$

where G is the gravitational constant and r is the distance between the objects. In order to facilitate calculation of the force *exerted on* various objects by the same source object we took advantage of this proportionality to define the magnitude of $\vec{g}_1$ as

$$|\vec{g}_1| = \left(\frac{Gm_1}{r^2} \right) \qquad (23\text{-}2)$$

where we called $|\vec{g}_1|$ the *local gravitational strength*. Using the field concept that we just developed we can now define the vector $\vec{g}_1$ as the **local gravitational field vector.** By combining Eqs. 23-1 and 23-2, we see that the force exerted by a source object of mass m_2 on a test object of mass m_1 can be determined using the simple expression

$$\vec{F}_{1 \to 2} = m_2 \vec{g}_1 . \qquad (23\text{-}3)$$

In other words, the gravitational force on an object at a certain point in space is the product of its mass and the gravitational field vector at that point in space.

Similarly, the magnitude of the electrostatic force is proportional to the charge of the object *being acted on* (as well as the charge of the object exerting the force). Hence, we will take the same approach as we did in defining a gravitational field vector. That is, we will define an electric field vector, $\vec{E}_1$ associated with a given source charge q_1 such that

$$\vec{F}_{1 \to 2} = q_2 \vec{E}_1 . \qquad (23\text{-}4)$$

Note because force is a vector and charge is a scalar, electric field must be a vector quantity. Rewriting this expression, we can define the electric field vector as:

$$\vec{E}_1 \equiv \frac{\vec{F}_{12}}{q_2} \qquad \text{(electric field)}. \qquad (23\text{-}5)$$

According to its definition, the **electric field** is the ratio of the electrostatic force and a test charge. That is, it is the force exerted by the source charge q_1 per unit of test charge q_2. We have defined the electric field associated with a source charge q_1 to be the force exerted by q_1 per coulomb of test charge q_2. This suggests the SI unit for electric field— the newton per coulomb (N/C). In comparison, recall the gravitational field was a measure of force per *mass* and the SI unit for the gravitational field g was the N/kg.

In order to give you some idea of how much force would be exerted on a one coulomb charge in various circumstances, Table 23-1 shows the electric fields that occur in a few physical situations. Remember though, one coulomb is a very large amount of charge. An object with 10,000 more protons than electrons would have only a charge on the order of 10^{-11} C.

TABLE 23-1 Some Electric Field Magnitudes

Field Location Or Situation	Value (N/C)
At the surface of a uranium nucleus	3×10^{21}
Within a hydrogen atom, at a radius of 5.29×10^{-11} m	5×10^{11}
Electric breakdown occurs in air	3×10^6
Near the charged drum of a photocopier	10^5
Near a charged plastic comb	10^3
In the lower atmosphere	10^2
Inside the copper wire of household circuits	10^{-2}

Although we use a test charge to define the electric field associated with a charged object, remember the electric field exists independently of the test charge just as the temperature in a room exists independent of whether or not there is a thermometer present to detect it. The test charge is simply the measurement device. The field at point P in Fig. 23-5 existed both before and after the test charge shown in the figure was put there. We must always assume the test charge does not alter the electric field we are defining, and so we should always imagine the test object to be point-like with a very small charge compared to the charge on the source object.

We asserted the concept of an electric field was useful because it simplified the determination of the forces on a test charge placed at different locations relative to a

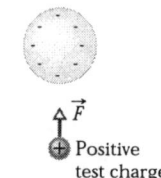

$\vec{F}$

⊕ Positive
test charge

Fig. 23-5: The electric field associated with a negatively charged source object points inward at all locations outside of the source. Since a legitimate test charge does not influence the source charges noticeably, we think of the electric field as existing whether or not a test charge is present to experience a force.

single source charge. But, what if our charged object experiences electrostatic forces from many source objects at the same time? As we mentioned in Section 22-8, experiments involving both gravitational and electrical forces have confirmed the net force exerted on a test object by a collection of source objects is the vector sum of the forces exerted by each individual source object. This principle of superposition of forces tells us the net electric field due to multiple sources can be calculated as the vector sum of the electric fields due to each of the source charges. The remainder of this chapter is primarily devoted to exploring how to use Coulomb's Law and the principle of superposition to find the electric fields associated with a single point charge and with relatively simple arrangements of charged objects. We will also explore the concept of *electric field lines* as an alternative to electric field vectors to represent electric fields visually.

23-4 The Electric Field Due to a Point Charge

The simplest of all possible charge distributions is a charge that can be approximated by a point with zero size. Protons, electrons, nuclei and ions can all be considered to be point charges. Understanding the interactions of these objects is vital to our understanding of the physical world. In fact, even large objects can be viewed as point charges when considered from afar.

In this section, we will determine the magnitude of the electric field due to a point-like charge q_1. We do this with the understanding if we know the form of the electric field of this charge, we know the form of the force exerted per unit charge on any other charge that we might bring into the region surrounding it. That is, once we know the form of the electric field, we qualitatively know the form of the force field. It is then a trivial calculation to determine the exact force exerted by charge q_1 on any other charge q_2.

Here we will calculate the electric field due to (or associated with) a positive charge q_1. We will also assume we use another positive charge q_2 to probe the region around q_1, testing the magnitude and direction of the force exerted by q_1. As we develop this expression for the electric field associated with a positive charge, we will consider how the situation would be different for a negative charge.

Let's consider what the force vector and electric field vector would look like for two positive point-like charges separated by a distance r as shown in Fig. 23-6. We know as like charges, q_2 experiences a repulsive electrical force caused by q_1. The magnitude of the force on our test charge q_2 can be found using Coulomb's Law:

$$\left|\vec{F}_{1\rightarrow 2}\right| = \frac{k|q_1||q_2|}{r^2}.$$ (23-6)

Since both q_1 and q_2 are positive charges, the absolute value signs are not needed in this case. However, they serve as a reminder the magnitude of the force is independent of the type (sign) of charge we have chosen to use in this development. *If* we place the test charge q_2 at some point in space, we can use Eq. 23-4 to express the magnitude of the force on the test charge due to the electric field created by q_1,

$$\left|\vec{F}_{1\rightarrow 2}\right| = q_2\left|\vec{E}_1\right|.$$ (23-7)

By combining Eq. 23-6 with Eq. 23-7, we can express the magnitude of the electric field at a distance r from a point charge of magnitude q_1 as

$$\left|\vec{E}_1\right| = k\frac{|q_1|}{r^2} \quad \text{(magnitude of the electric field due to a point charge).} \quad (23\text{-}8)$$

In agreement with our definition of the electric field as the force *per unit* charge, this expression is independent of the magnitude (and sign) of the test charge q_2 we use as the probe. This expression is valid everywhere around the point charge q_1.

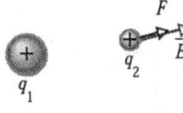

Fig. 23-6: When the charge on a source object and a point-like test object are both positive, the force between them is repulsive and both the vector representing the force on object 2 and the electric field vector at the location of object 2 point radially outward in the same direction.

The magnitude of the electric field due to a (positive or negative) point charge is given by the expression above. However, the electric field is a vector. Hence, we must still determine the direction of the electric field associated with our positive point charge q_1. From the definition of electric field,

$$\vec{E}_1 \equiv \frac{\vec{F}_{12}}{q_2} .$$
(23-9)

For a positive test charge q_2, we know that the direction of the field vector $\vec{E}_1$ is the same as the force vector $\vec{F}_{1\rightarrow2}$. This force points radially away from q_1 as shown in Fig. 23-6. Since the direction of the field is the same as the direction of the force for the positive charge q_2, we know the electric field created by the positive point charge q_1 must also point radially away from the charge q_1 as shown in Fig. 23-6.

Would the direction of the field change if the charge q_1 producing the field was negative rather than positive? The answer is yes. If the test charge q_2 is still positive, then the vector relationship above tells us the (sign) direction of the field is still the same as the direction of the force on q_2. However, if q_1 is negative and q_2 is positive, these unlike charges will attract one another. As shown in Fig. 23-7, the direction of the force on q_2 due to q_1 then points radially *toward* q_1 and so does the electric field.

▶**Force and Electric Field Directions:** (1) The direction of the force on a positive test charge is always the same as the electric field at the location of the positive test charge. (2) At any point in space, the direction of the electric field produced by a positive point charge is radially away from the charge. The direction of the electric field produced by a negative point charge is radially toward the charge. "Radially" means along the line connecting the charge and the point of evaluation.

READING EXERCISE 23-3: Rewrite the discussion of how we determine the direction of the field using a negative test charge rather than a positive test charge. Does it make any difference which type of charge we chose to use in determining the direction of the field?

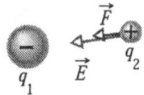

Fig. 23-7: When the charge on a source object is negative while a point-like test object is positive, the forces between them are attractive and both the vector representing the force on object 2 and the electric field vector at the location of object 2 point radially inward in the same direction.

23-5 Electric Field Plots and Lines

We will use several different ways to represent information about electric forces in a region of space pictorially. In the approach which we discussed in Section 23-2, we discussed the idea of an electric field plot that can be created. A systematic way to create a plot for a vector field associated with a point-like object is to: (1) choose a grid of points, (2) place the tail of an arrow representing the electric force exerted by a fixed source object on a particle-like test object. If the grid we chose is fine enough, we can get a good idea of what the force field will look like for those two objects at any point by looking at the force at nearby grid points. (However, we don't want to choose too fine a grid since then we will cover our page with arrows.) We call such a representation a *force vector field plot*. Such a force vector field plot is shown in Fig. 23-8a.

In a closely related representation, we consider only the source object by reverting to the concept of the electric field (force per unit test charge). We put the tail of an arrow representing the *electric field* associated with the source object at each point. As discussed above, this requires that we imagine the source is exerting a force on a positive test charge. This is shown in Fig. 23-8b. Aside from perhaps choosing the scaling arrows representing the vectors to have different lengths, this kind of a vector field plot looks similar to the force field plot since the relative lengths of the arrows at corresponding locations in the two vector plots will be the same.

In still another representation, we will use continuous lines to convey information about the direction of the field at different points. Since the magnitude and direction of the electric field usually changes smoothly, this turns out to be rather convenient. Michael Faraday, who introduced the idea of electric fields in the 19th century, thought of the space around a charged body as filled with *lines of force*. We now usually call these lines **electric field lines,** and although we attach no reality to them, they provide a nice way to visualize patterns of changing force.

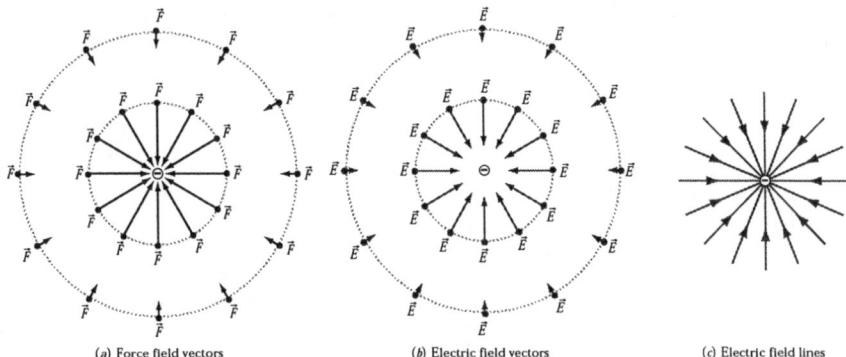

(a) Force field vectors   (b) Electric field vectors   (c) Electric field lines

Fig. 23-8: There are three methods commonly used to depict the pattern of forces a test charge might experience at different locations in space. (a) shows a vector force field, (b) shows a vector electric field, and (c) shows electric field lines with the arrowheads representing the direction of each line.

The field *lines* shown in Fig. 23-8c and Fig. 23-9 differ from the short straight field vectors because they always start or end on the source charge(s) and at any point, the direction of a straight field line or the direction of the tangent to a curved field line gives the direction of the electric field vector $\vec{E}$ at that point. Because the field lines point in the direction of the field, field lines must originate on positive charges and terminate on negative charges.

It is important to note we could draw field lines through every point in space. However, this would not be very helpful since our paper would be totally filled with field lines and we couldn't distinguish one from another. Instead, we will choose to draw a few field lines, with the number of lines leaving each positive charge (or ending on each negative charge) being proportional to the magnitude of each charge. If you choose to have 16 lines originating on a +4 uC charge, then you should have 8 lines ending on a –2 uC charge. This scaling of the number of field lines with magnitude of the charge turns out to be quite convenient since then the field lines are forced to be closely packed together where the field is strong and far apart where it is weak. We can see this in Fig. 23-8. In other words, the average density of field lines (the number of lines crossing through a small area perpendicular to their direction) is proportional to the strength of the field. We then have the following rule:

▶Electric field lines extend away from positive charge (where they originate) and toward negative charge (where they terminate). The density of field lines is proportional to the strength of the field.

Figure 23-10a shows a sphere of uniform negative charge. If we place a *positive* test charge anywhere near the sphere, an electrostatic force pointing *toward* the center of the sphere will act on the test charge as shown. In other words, the electric field vectors at all points near the sphere are directed radially toward the sphere. This pattern of vectors is neatly displayed by the field lines in Fig. 23-10b, which point in the same direction as the force and field vectors. Moreover, the spreading of the field lines with distance from the sphere tells us that the magnitude of the electric field decreases with distance from the sphere (as we expect). If the sphere of Fig. 23-10 were of uniform *positive* charge, the electric field vectors at all points near the sphere would be directed radially *away from* the sphere. Thus, the electric field lines would also extend radially away from the sphere.

Figure 23-11a shows part of an infinitely large, nonconducting *sheet* (or plane) with a uniform distribution of positive charge on one side. If we were to place a positive test charge at any point near the sheet of Fig. 23-11a, the net electrostatic force acting on the test charge would be perpendicular to the sheet, because forces acting in all other directions would cancel one another as a result of the symmetry. Moreover, the net force on the test charge would point away from the sheet as shown. Thus, the electric field

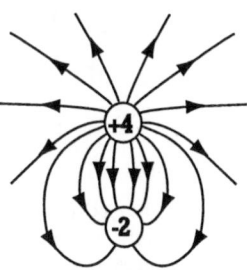

Fig. 23-9: Electric field lines for a $-2q$ and $+4q$ charge configuration.

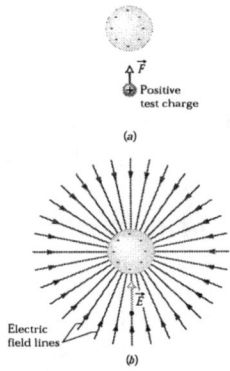

Fig. 23-10: (a) The electrostatic force $\vec{F}$ acting on a positive test charge near a sphere of uniform negative charge. (b) The electric field vector $\vec{E}$ at the location of the test charge, and the electric field lines in the space near the sphere. The field lines extend toward the negatively charged sphere. (They originate on distant positive charges.)

vector at any point in the space on either side of the sheet is also perpendicular to the sheet and directed away from it (Figs. 23-11b and c). Since the charge is uniformly distributed along the sheet, all the field vectors have the same magnitude. Such an electric field, with the same magnitude and direction at every point, is a uniform electric field.

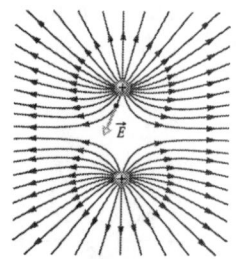

Fig. 23-12: Field lines for two equal positive point charges. The charges repel each other. (The lines terminate on distant negative charges.) To "see" the actual three-dimensional pattern of field lines, mentally rotate the pattern shown here about an axis passing through both charges in the plane of the page. The three-dimensional pattern and the electric field it represents are said to have rotational symmetry about that axis. The electric field vector at one point is shown; note that it is tangent to the field line through that point.

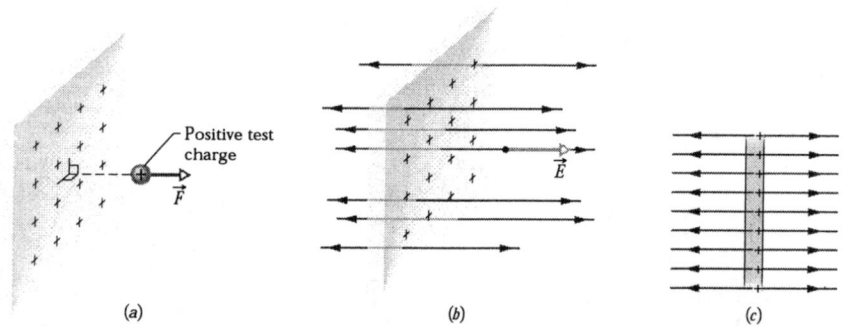

(a)   (b)   (c)

Fig. 23-11: (a) The electrostatic force $\vec{F}$ on a positive test charge near a very large, nonconducting sheet with uniformly distributed positive charge on one side. (b) The electric field vector $\vec{E}$ at the location of the test charge, and the electric field lines in the space near the sheet. The field lines extend away from the positively charged sheet. (c) Side view of (b).

Of course, no real nonconducting sheet (such as a flat expanse of plastic) is infinitely large, but if we consider a region near the middle of a real sheet and not near its edges, the field lines through that region are arranged as in Figs. 23-11b and c.

Figure 23-12 shows the field lines for two equal positive charges. Although we do not often use field lines quantitatively, they are very useful to visualize what is going on.

READING EXERCISE 23-4: Explain why electric field lines pointing in the direction of the electric field means electric field lines must originate on positive charges and terminate on negative charges.

23-6 The Electric Field Due to Multiple Charges

In the real world, problems are seldom as simple as one charged object exerting a force on another. A more common occurrence is several charges are present and the force exerted on the test charge is the net result of the forces due to each of the charges. We first discussed the issue of superposition of forces back in Chapter 3. We discussed it in specific reference to electrostatic forces in Chapter 22. We will use it again here to develop an understanding of how we can calculate the net electric field associated with a group of charges.

If we place a positive test charge q_0 near n point charges $q_1, q_2, ..., q_n$, as shown in Fig. 23-13, the forces exerted by each of the individual charges superimpose so that the net force $\vec{F}_0$ from the n point charges acting on the test charge is

$$\vec{F}_{0,\text{net}} = \vec{F}_{1\to0} + \vec{F}_{2\to0} + \cdots + \vec{F}_{n\to0}. \tag{23-10}$$

For each of the terms in the expression above, we can replace the individual forces with the equivalent expressions based on the definition of the electric field. For example,

$$\vec{F}_{1\to0} = q_0\vec{E}_1, \tag{23-11}$$

where $\vec{E}_1$ is the electric field associated with charge q_1. If we make such replacements for each force term in the expression above we have:

$$\vec{F}_{0,\text{net}} = q_0\vec{E}_{\text{net}} = q_0\vec{E}_1 + q_0\vec{E}_2 + \ldots q_0\vec{E}_n. \tag{23-12}$$

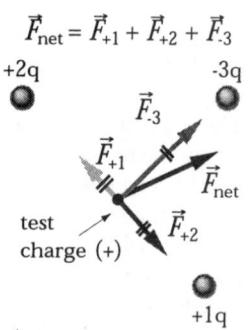

Fig. 23-13: Three point charges exert forces on a small test charge at a point in space. These force vectors superimpose to yield a net force.

Here $\vec{E}_{net}$ is the resultant electric field associated with the entire group of charges. If we divide both sides of the expression

$$q_0 \vec{E}_{net} = q_0 \vec{E}_1 + q_0 \vec{E}_2 + \dots q_0 \vec{E}_n \qquad (23\text{-}13)$$

by q_0, the result is an expression for the net electric field associated with a group of charges. Namely,

$$\vec{E}_{net} = \vec{E}_1 + \vec{E}_2 + \dots \vec{E}_n . \qquad (23\text{-}14)$$

This expression shows us that the principle of superposition applies to electric fields as well as to electrostatic forces. When doing calculations, however, it is important to remember that we are adding vectors here. Hence, the addition is more complex than simply adding numbers together.

READING EXERCISE 23-5: The figure here shows a proton p and an electron e on an x axis. What is the direction of the electric field due to the electron at (a) point S and (b) point R? What is the direction of the net electric field at (c) point R and (d) point S?

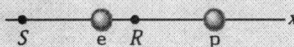

Touchstone Example 23-6-1, at the end of this chapter, illustrates how to use what you learned in this section.

TE

23-7 The Electric Field Due to an Electric Dipole

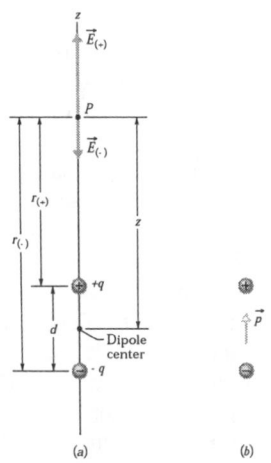

Figure 23-14a shows two charged particles of magnitude q but of opposite sign, separated by a distance d. As mentioned in the caption of Fig. 23-14, we call this configuration an electric dipole. Separation of positive and negative charge in an electrically neutral object occurs quite naturally. For example, recall the discussion of polarization in Chapter 22. As a result, true electric dipoles and approximations of electric dipoles are reasonably common. Hence, we will take some time to develop an expression for the electric field due to a dipole. We start with the idea of superposition of electric fields that we discussed in Section 23-6.

Let us find the electric field due to the dipole of Fig. 23-14a at a point P, a distance z from the midpoint of the dipole and on the axis through the particles, which is called the dipole axis. From symmetry, the electric field $\vec{E}$ at point P—and also the fields $\vec{E}_{(+)}$ and $\vec{E}_{(-)}$ due to the separate charges that make up the dipole—must lie along the dipole axis, which we have taken to be a z axis. Applying the superposition principle for electric fields, we find that the magnitude $|\vec{E}|$ of the electric field at P is

$$|\vec{E}| = |\vec{E}_{(+)}| - |\vec{E}_{(-)}|$$

$$= \frac{1}{4\pi\varepsilon_0} \frac{|q|}{r_{(+)}^2} - \frac{1}{4\pi\varepsilon_0} \frac{|q|}{r_{(-)}^2} \qquad (23\text{-}15)$$

$$= \frac{|q|}{4\pi\varepsilon_0 (z - \frac{1}{2}d)^2} - \frac{|q|}{4\pi\varepsilon_0 (z + \frac{1}{2}d)^2}$$

After a little algebra, we can rewrite this equation as

$$|\vec{E}| = \frac{|q|}{4\pi\varepsilon_0 z^2} \left[\left(1 - \frac{d}{2z}\right)^{-2} - \left(1 + \frac{d}{2z}\right)^{-2} \right]. \qquad (23\text{-}16)$$

Fig. 23-14: (a) An electric dipole. The electric field vectors $\vec{E}_{(+)}$ and $\vec{E}_{(-)}$ at point P on the dipole axis result from the dipole's two charges. P is at distances $r_{(+)}$ and $r_{(-)}$ from the individual charges that make up the dipole. (b) The dipole moment $\vec{p}$ of the dipole points from the negative charge to the positive charge.

We are usually interested in the electrical effect of a dipole only at distances that are large compared with the dimensions of the dipole—that is, at distances such that $z >> d$. At such large distances, we have $d/z << 1$ in the expression above. We can then expand the two quantities in the brackets in that equation by the binomial theorem (Appendix E), obtaining for those quantities

$$\left[\left(1+\frac{2d}{2z(1!)}+\cdots\right)-\left(1-\frac{2d}{2z(1!)}+\cdots\right)\right].$$

Thus,
$$|\vec{E}| = \frac{|q|}{4\pi\varepsilon_0 z^2}\left[\left(1+\frac{d}{z}+\cdots\right)-\left(1-\frac{d}{z}+\cdots\right)\right]. \tag{23-17}$$

The unwritten terms in these two expansions involve d/z raised to progressively higher powers. Since $d/z << 1$, the contributions of those terms are progressively less, and to approximate E at large distances, we can neglect them. Then, in our approximation, we can rewrite this expression as

$$|\vec{E}| = \frac{|q|}{4\pi\varepsilon_0 z^2}\frac{2d}{z} = \frac{1}{2\pi\varepsilon_0}\frac{|q|d}{z^3}. \tag{23-18}$$

The product qd, which involves the two intrinsic properties q and d of the dipole, is the magnitude $|\vec{p}|$ of a vector quantity known as the **electric dipole moment** $\vec{p}$ of the dipole. (The unit of $\vec{p}$ is the coulomb-meter.) Thus, we can rewrite the equation above as

$$|\vec{E}| = \frac{1}{2\pi\varepsilon_0}\frac{|\vec{p}|}{z^3}. \qquad \text{(electric dipole).} \tag{23-19}$$

The direction of $\vec{p}$ is taken to be from the negative to the positive end of the dipole, as indicated in Fig. 23-14b. We can use $\vec{p}$ to specify the orientation of a dipole.

The expression for the electric field due to a dipole shows if we measure the electric field of a dipole only at distant points, we can never find q and d separately, only their product. The field at distant points would be unchanged if, for example, q were doubled and d simultaneously halved. Thus, the dipole moment is a basic property of a dipole.

Although Eq. 23-19 holds only for distant points along the dipole axis, it turns out E for a dipole varies as $1/r^3$ for all distant points, regardless of whether they lie on the dipole axis; here r is the distance between the point in question and the dipole center.

Inspection of the field lines in Fig. 23-14 shows the direction of $\vec{E}$ for distant points on the dipole axis is always the direction of the dipole moment vector $\vec{p}$. This is true whether point P in Fig. 23-14a is on the upper or the lower part of the dipole axis.

Inspection of Eq. 23-19 shows if you double the distance of a point from a dipole, the electric field at the point drops by a factor of 8. If you double the distance from a single point charge, however (see Eq. 23-8), the electric field drops only by a factor of 4. Thus the electric field of a dipole decreases more rapidly with distance than does the electric field of a single charge. The physical reason for this rapid decrease in electric field for a dipole is that from distant points a dipole looks like two equal but opposite charges that almost—but not quite—coincide. Thus, their electric fields at distant points almost—but not quite—cancel each other.

Volcanic Lightning

When the Sakurajima volcano erupts, as seen in this chapter's opening photograph, it spews ash into the air. That ash results when liquid water within the volcano, suddenly converted to steam by the flow of hot lava, shatters rock. The liquid-to-steam conversion and the explosion of rock cause positive and negative charges to separate. Then, as the steam and ash are spewed into the air, they form a cloud that contains pockets of positive charge and pockets of negative charge.

As these pockets grow, the electric fields between adjacent pockets and between pockets and the volcano crater increase in magnitude. Whenever the magnitude of about 3×10^6 N/C is reached, the air undergoes electric breakdown and begins to conduct current. These momentary conducting paths appear in the air where the electric field has ionized air molecules, freeing some of their electrons. These electrons, propelled by the field, collide with air molecules in their way, which causes those molecules to emit light. We can see these brief paths, commonly called sparks, because of the light they emit. (A small-scale example of sparking can be seen around the charged metal cap in Fig. 23-15.)

The sparks above the volcano snake their way either down from a charge pocket to the crater wall or vice-versa. You can tell the direction of a spark by how any dead-end branches on it are forked. If the branches fork downward, then the spark snaked its way downward. (See the bright spark extending from the right side of the photograph to the crater wall.) If the branches fork upward, then the spark snaked its way upward. (See the lower part of the central bright spark on the crater wall.) Sometimes a downward-snaking spark and an upward-snaking spark meet each other. Can you find an example in the photograph?

23-8 The Electric Field Due to a Line of Charge

So far we have considered the electric field produced by one or, at most, a few point charges. We now consider charge distributions consisting of a great many closely spaced point charges (perhaps billions) spread along a line, over a surface, or within a volume. Such distributions can be treated as if they were **continuous** rather than discrete. Since these distributions can include an enormous number of point charges, we find the electric fields that they produce by means of calculus rather than by considering the point charges one by one. In this section we discuss the electric field caused by a line of charge. In the next chapter, we shall find the field inside a uniformly charged sphere.

When we deal with continuous charge distributions, it is most convenient to express the charge on an object as a *charge density* rather than as a total charge. For a line of charge, for example, we would report the linear charge density (or charge per unit length) λ, whose SI unit is the coulomb per meter. Table 23-2 shows the other charge densities we shall be using.

TABLE 23-2: Some Measures of Charge Densities

Name	Symbol	SI Unit
Charge	q	C
Linear charge density	λ	C/m
Surface charge density	σ	C/m^2
Volume charge density	ρ	C/m^3

Figure 23-16 shows a thin ring of radius R with a uniform positive linear charge density λ around its circumference. We may imagine the ring to be made of plastic or some other insulator, so the charges can be regarded as fixed in place. What is the electric field $\vec{E}$ at point P, a distance z from the plane of the ring along its central axis?

To answer, we cannot just use the expression for the electric field set up by a point charge, because the ring is obviously not a point charge. However, we can mentally divide the ring into differential elements of charge so small they are like point charges, and then we can apply Eq. 23-8,

$$\left|\vec{E}_q\right| = \frac{1}{4\pi\varepsilon_0}\frac{|q_1|}{r^2},$$

to each of them. Next, we can add the electric fields set up at P by all the differential elements. The vector sum of all those fields gives us the field set up at P by the ring.

Let $|d\vec{s}|$ be the (arc) length of any differential element of the ring. Since λ is the charge per unit length, the element has a charge of magnitude

$$|dq| = |\lambda \, d\vec{s}|. \qquad (23\text{-}20)$$

Fig. 23-15: The metal cap is so charged the electric field it produces in the surrounding space causes the air there to undergo electric breakdown. The visible sparks reveal where momentary conducting paths are set up in the air, along which the electric field has removed electrons from their molecules and then accelerated them into collisions with the molecules.

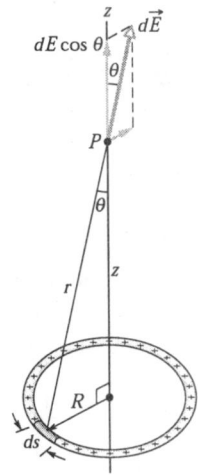

Fig. 23-16: A ring of uniform positive charge. A differential element of charge occupies a length ds (greatly exaggerated for clarity). This element sets up an electric field dE at point P. The component of dE along the central axis of the ring is $dE \cos \theta$.

This differential charge sets up a differential electric field $d\vec{E}$ at point P, which is a distance r from the element. Treating the element as a point charge, and using the equation above for the dq, we can express the magnitude of $d\vec{E}$ as

$$\left|d\vec{E}\right| = \frac{1}{4\pi\varepsilon_0}\frac{\left|d\vec{q}\right|}{r^2} = \frac{1}{4\pi\varepsilon_0}\frac{\left|\lambda\,d\vec{s}\right|}{r^2}. \qquad (23\text{-}21)$$

From Fig. 23-16, we see we can use the Pythagorean theorem to rewrite the equation above as

$$\left|d\vec{E}\right| = \frac{1}{4\pi\varepsilon_0}\frac{\left|\lambda\,d\vec{s}\right|}{(z^2 + R^2)}. \qquad (23\text{-}22)$$

Figure 23-16 shows us that the vector $d\vec{E}$ is at an angle θ to the central axis (which we have taken to be a z axis) and has components perpendicular to and parallel to that axis.

Every element of charge in the ring sets up a differential field $d\vec{E}$ at P, with magnitude given by the expression above. All the $d\vec{E}$ vectors have identical components parallel to the central axis, in both magnitude and direction. All these $d\vec{E}$ vectors have components perpendicular to the central axis as well; these perpendicular components are identical in magnitude but point in different directions. In fact, for any perpendicular component that points in a given direction, there is another one that points in the opposite direction. The sum of this pair of components, like the sum of all other pairs of oppositely-directed components, is zero.

Thus, the perpendicular components cancel and we need not consider them further. This leaves the parallel components; they all have the same direction, so the net electric field at P is their sum.

The parallel component of $d\vec{E}$ shown in Fig. 23-16 has magnitude $dE \cos\theta$. The figure also shows us that

$$\cos\theta = \frac{z}{r} = \frac{z}{(z^2 + R^2)^{1/2}}. \qquad (23\text{-}23)$$

Then multiplying our expressions for $d\vec{E}$ and $\cos\theta$ gives us the parallel component of $d\vec{E}$,

$$\left|d\vec{E}_\parallel\right| = \left|d\vec{E}\right|\cos\theta = \frac{\left|z\lambda\,d\vec{s}\right|}{4\pi\varepsilon_0(z^2 + R^2)^{3/2}}. \qquad (23\text{-}24)$$

To add the parallel components $dE \cos\theta$ produced by all the elements, we integrate this expression around the circumference of the ring, from $s = 0$ to $s = 2\pi R$. Since the only quantity that varies during the integration is s, the other quantities can be moved outside the integral sign. The integration then gives us an electric field magnitude of

$$\left|\vec{E}\right| = \int\left|d\vec{E}\right|\cos\theta = \left|\frac{z\lambda}{4\pi\varepsilon_0(z^2 + R^2)^{3/2}}\int_0^{2\pi R} d\vec{s}\right|$$
$$= \frac{\left|z\lambda\right|(2\pi R)}{4\pi\varepsilon_0(z^2 + R^2)^{3/2}}. \qquad (23\text{-}25)$$

Since λ is the charge per length of the ring, the term $\lambda(2\pi R)$ is q, the total charge on the ring. We can then rewrite this expression as

$$\left|\vec{E}\right| = \frac{\left|qz\right|}{4\pi\varepsilon_0(z^2 + R^2)^{3/2}} \qquad \text{(charged ring).} \qquad (23\text{-}26)$$

If the charge on the ring is negative, instead of positive as we have assumed, the magnitude of the field at P is still given by this expression. However, the electric field vector then points toward the ring instead of away from it.

Let us evaluate this equation for the electric field for a point on the central axis so far away that $z \gg R$. For such a point, the expression $z^2 + R^2$ can be approximated as z^2, and Eq. 23-26 becomes

$$|\vec{E}| = \frac{1}{4\pi\varepsilon_0} \frac{|q|}{z^2} \qquad \text{(charged ring at large distance).} \qquad (23\text{-}27)$$

This is a reasonable result, because from a large distance, the ring "looks" like a point charge. If we replace z with r we have the usual expression for the electric field due to a point charge.

Let us next check Eq. 23-26 for a point at the center of the ring—that is, for $z = 0$. At that point, this expression tells us that $E = 0$. This is a reasonable result, because if we were to place a test charge at the center of the ring, there would be no net electrostatic force acting on it; the force due to any element of the ring would be canceled by the force due to the element on the opposite side of the ring. If the force at the center of the ring were zero, the electric field there would also have to be zero.

Touchstone Example 23-8-1, at the end of this chapter, illustrates how to use what you learned in this section.

TE

23-9 A Point Charge in an Electric Field

In the preceding sections we worked at the first of our two tasks: given a charge distribution, find the electric field it produces in the surrounding space. Here we begin the second task: to determine what happens to a charged particle when it is in an electric field that is produced by other stationary or slowly moving charges.

What happens is an electrostatic force acts on the particle. This force, a vector quantity, is given by

$$\vec{F} = q\vec{E}, \qquad (23\text{-}28)$$

in which q is the charge of the particle (including its sign) and $\vec{E}$ is the electric field other charges have produced at the location of the particle. (The field is not the field set up by the particle itself; to distinguish the two fields, the field acting on the particle in the equation above is often called the external field. A charged particle (or object) is not affected by its own electric field.) Equation 23-28 tells us:

▶The electrostatic force $\vec{F}$ acting on a charged particle located in an external electric field $\vec{E}$ has the direction of $\vec{E}$ if the charge q of the particle is positive and has the opposite direction if q is negative.

READING EXERCISE 23-6: (a) In the figure, what is the direction of the electrostatic force on the electron due to the electric field shown? (b) In which direction will the electron accelerate if it is moving parallel to the y axis before it encounters the electric field? (c) If, instead, the electron is initially moving rightward, will its speed increase, decrease, or remain constant?

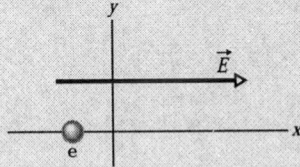

Touchstone Example 23-9-1, at the end of this chapter, illustrates how to use what you learned in this section.

TE

23-10 A Dipole in an Electric Field

We have defined the electric dipole moment $\vec{p}$ of an electric dipole to be a vector pointing from the negative to the positive end of the dipole. As you will see, the behavior of a dipole in a uniform external electric field $\vec{E}$ can be described completely in terms of the two vectors $\vec{E}$ and $\vec{p}$, with no need of any details about the dipole's structure.

A molecule of water (H_2O) is an electric dipole; Fig. 23-17 shows why. There the black dots represent the oxygen nucleus (having eight protons) and the two hydrogen nuclei (having one proton each). The colored enclosed areas represent the region in which electrons can be located around the nuclei.

In a water molecule, the two hydrogen atoms and the oxygen atom do not lie on a straight line but form an angle of about 105°, as shown in Fig. 23-17. As a result, the molecule has a definite "oxygen side" and "hydrogen side." Moreover, the 10 electrons of the molecule tend to remain closer to the oxygen nucleus than to the hydrogen nuclei. This makes the oxygen side of the molecule slightly more negative than the hydrogen side and creates an electric dipole moment $\vec{p}$ that points along the symmetry axis of the molecule as shown. If the water molecule is placed in an external electric field, it behaves as would be expected of the more abstract electric dipole of Fig. 23-9.

To examine this behavior, we now consider such an abstract dipole in a uniform external electric field $\vec{E}$, as shown in Fig. 23-18a. We assume that the dipole is a rigid structure that consists of two centers of opposite charge, each of magnitude q, separated by a distance d. The dipole moment $\vec{p}$ makes an angle θ with field $\vec{E}$. Since effects far away from the molecule only depend on $\vec{p}$, it doesn't matter how we choose to think about it, this is, as long as the forces aren't strong enough to change the shape of the molecule. We can imagine it as being a ball-and-stick rigid dipole.

Electrostatic forces act on the charged ends of the dipole. Because the electric field is uniform, those forces act in opposite directions (as shown in Fig. 23-18) and with the same magnitude $|\vec{F}| = |q\vec{E}|$. Thus, because the field is uniform, the net force on the dipole from the field is zero and the center of mass of the dipole does not move. However, the forces on the charged ends do produce a net torque τ on the dipole about its center of mass. The center of mass lies on the line connecting the charged ends, at some distance x from one end and thus a distance $d - x$ from the other end. From Eq. 11-28 ($|\vec{\tau}| = |\vec{r}||\vec{F}|\sin\phi$), we can write the magnitude of the net torque τ as

$$|\vec{\tau}| = |\vec{F}||\vec{x}|\sin\phi + |\vec{F}||d - x|\sin\phi = |\vec{F}|d\sin\phi. \qquad (23\text{-}29)$$

We can also write the magnitude of $\vec{\tau}$ in terms of the magnitudes of the electric field $|\vec{E}|$ and the dipole moment $|\vec{p}| = |q|d$. To do so, we substitute $|q\vec{E}|$ for $|\vec{F}|$ and $|\vec{p}|/|\vec{q}|$ for d, finding the magnitude of τ is

$$|\vec{\tau}| = |\vec{p}||\vec{E}|\sin\phi. \qquad (23\text{-}30)$$

We know the direction of the vector $\vec{\tau}$ is given by the right-hand rule. We see the result for both the magnitude and direction is given by the cross product as

$$\vec{\tau} = \vec{p} \times \vec{E} \qquad \text{(torque on a dipole).} \qquad (23\text{-}31)$$

Vectors $\vec{p}$ and $\vec{E}$ are shown in Fig. 23-18b. The torque acting on a dipole tends to rotate $\vec{p}$ (hence the dipole) into the direction of $\vec{E}$, thereby reducing ϕ. In Fig. 23-18, such rotation is clockwise. As we discussed in Chapter 11, we can represent a torque component τ that gives rise to a clockwise rotation by including a minus sign with the magnitude of the torque. With that notation, the torque of Fig. 23-18 is

$$\vec{\tau} = \tau\,\hat{k} = \left(-|\vec{\tau}| - |p||\vec{E}|\sin\phi\right)\hat{k}. \qquad (23\text{-}32)$$

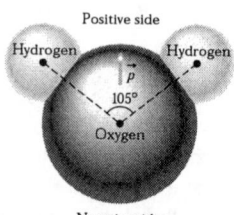

Fig. 23-17: A molecule of H_2O, showing the three nuclei (represented by dots) and the regions in which the electrons orbit the nuclei. The electric dipole moment $\vec{p}$ points from the (negative) oxygen side to the (positive) hydrogen side of the molecule.

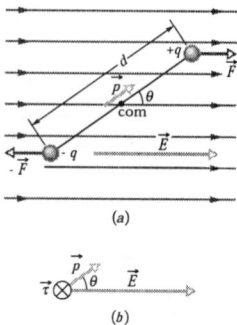

Fig. 23-18: (a) An electric dipole in a uniform electric field $\vec{E}$. Two centers of equal but opposite charge are separated by distance d. The line between them represents their rigid connection. (b) Field $\vec{E}$ causes a torque $\vec{\tau}$ on the dipole. The direction of $\vec{\tau}$ is into the plane of the page, as represented by the symbol $\otimes$.

READING EXERCISE 23-7: The figure shows four orientations of an electric dipole in an external electric field. Rank the orientations according to (a) the magnitude of the torque on the dipole and (b) the potential energy of the dipole, greatest first.

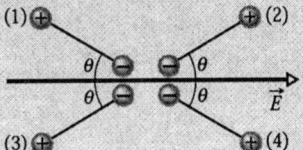

Touchstone Example 23-6-1

Figure TE23-1a shows three particles with charges $q_1 = +2Q$, $q_2 = -2Q$, and $q_3 = -4Q$, each a distance d from the origin. We assume Q is positive. What net electric field E is produced at the origin?

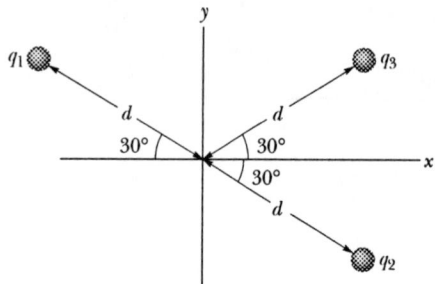

Fig. TE23-1 Three particles with charges q_1, q_2, and q_3 are at the same distance d from the origin.

SOLUTION: We need to find the electric field vectors E_1, E_2, and E_3 that act at the origin. The **Key Idea** is that we can pick a more convenient coordinate system to describe these electric field vectors. A $x' - y'$ coordinate system that is rotated by 30° in a clockwise direction has q_1 and q_2 lying along its x axis, as shown in Fig. TE23-2a.

Another **Key Idea** is that charges q_1, q_2, and q_3 produce electric field vectors E_1, E_2, and E_3, respectively, at the origin, and the net electric field is the vector sum $E = E_1 + E_2 + E_3$. To find this sum, we first must find the magnitudes and orientations of the three field vectors. To find the magnitude of E_1, which is due to q_1, we use Eq. 23-8, substituting d for r and $2Q$ for $|q|$ and obtaining

$$|E_1| = \frac{1}{4\pi\varepsilon_0} \frac{2Q}{d^2}.$$

Similarly, we find the magnitudes of the fields E_2 and E_3 to be

$$|E_2| = \frac{1}{4\pi\varepsilon_0} \frac{2Q}{d^2} \quad \text{and} \quad |E_3| = \frac{1}{4\pi\varepsilon_0} \frac{4Q}{d^2}.$$

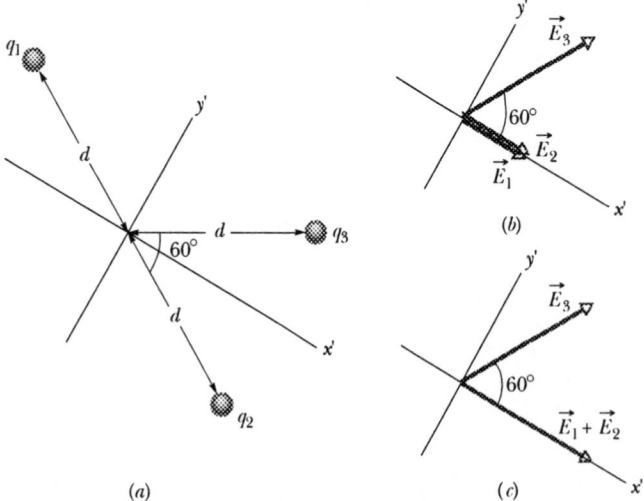

Fig. TE23-2 (a) The electric field vectors E_1, E_2, and E_3 at the origin due to the three particles. (b) The electric field vector E_3 and the vector sum $E_1 + E_2$ at the origin.

We next must find the orientations of the three electric field vectors at the origin. Because q_1 is a positive charge, the field vector it produces points directly *away* from it, and because q_2 and q_3 are both negative, the field vectors they produce point directly *toward* each of them. Thus, the three electric fields produced at the origin by the three charged particles are oriented as in Fig. 23-2b. (*Caution:* Note that we have placed the tails of the vectors at the point where the fields are to be evaluated; doing so decreases the chance of error.)

We can now add the fields vectorially as outlined for forces in TE22-7-1c. However, here we can use symmetry to simplify the procedure. From Fig. TE23-2b, we see that E_1 and E_2 have the same direction. Hence, their vector sum points along the positive x' axis and has the magnitude

$$|E_1 + E_2| = \frac{1}{4\pi\varepsilon_0}\frac{2Q}{d^2} + \frac{1}{4\pi\varepsilon_0}\frac{2Q}{d^2} = \frac{1}{4\pi\varepsilon_0}\frac{4Q}{d^2},$$

or

$$E_{1x'} + E_{2x'} = \frac{1}{4\pi\varepsilon_0}\frac{4Q}{d^2}.$$

This sum happens to equal the magnitude of E_3.

We must now combine two vectors, E_3 and the vector sum $E_1 + E_2$, that have the same magnitude. We do this by resolving E_3 into its x' and y' components.

$$E_{3,x'} = |E_3|\cos 60° = \tfrac{1}{2}|E_3|$$

$$E_{3,y'} = |E_3|\sin 60° = \tfrac{\sqrt{3}}{2}|E_3|.$$

Then we find $E_{x'}$ and $E_{y'}$ components.

$$E_{x'} = E_{1x'} + E_{2x'} + E_{3x'}$$

$$= \frac{1}{4\pi\varepsilon_0}\frac{4Q}{d^2} + \frac{1}{2}\frac{1}{4\pi\varepsilon_0}\frac{4Q}{d^2}$$

$$= \frac{3}{2}\left(\frac{1}{4\pi\varepsilon_0}\frac{4Q}{d^2}\right).$$

$$E_{y'} = \frac{\sqrt{3}}{2}\left(\frac{1}{4\pi\varepsilon_0}\frac{4Q}{d^2}\right).$$

Using vector notation we get

$$E = \left(\frac{1}{4\pi\varepsilon_0}\frac{4Q}{d^2}\right)\left(\tfrac{3}{2}\,\hat{i}' + \tfrac{\sqrt{3}}{2}\,\hat{j}'\right),\qquad\text{(Answer)}$$

where $\hat{i}'$ and $\hat{j}'$ are unit vectors in the x'-y' coordinate system. The magnitude of E is given by

$$|E| = \sqrt{E_{x'}^2 + E_{y'}^2}$$

$$= \frac{6.39Q}{4\pi\varepsilon_0 d^2}.\qquad\text{(Answer)}$$

Touchstone Example 23-8-1

Figure TE23-3a shows a plastic rod having a uniformly distributed charge $-Q$. We assume Q is positive, so $-Q$ is negative. The rod has been bent in a 120° circular arc of radius r. We place coordinate axes such that the axis of symmetry of the rod lies along the x axis and the origin is at the center of curvature P of the rod. In terms of Q and r, what is the electric field $\vec{E}$ due to the rod at point P?

SOLUTION The **Key Idea** here is that, because the rod has a continuous charge distribution, we must find an expression for the electric fields due to differential elements of the rod and then sum those fields via calculus. Consider a differential element having arc length ds and located at an angle θ above the x axis (Fig. TE23-3b). If we let λ represent the linear charge density of the rod, our element ds has a differential charge of magnitude

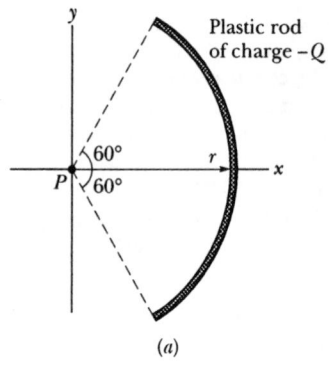

(a)

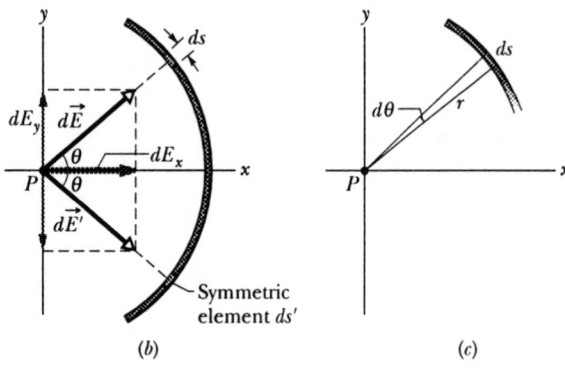

(b)   (c)

Fig. TE23-3 (a) A plastic rod of charge $-Q$ in a circular section of radius r and central angle 120°; point P is the center of curvature of the rod. (b) A differential element in the top half of the rod, at an angle θ to the x axis and of arc length ds, sets up a differential electric field $\vec{dE}$ at P. An element ds', symmetric to ds about the x axis, sets up a field $\vec{dE}'$ at P with the same magnitude. (c) Arc length ds makes an angle $d\theta$ about point P.

$$dq = \lambda \, ds. \qquad \text{(TE23-1)}$$

Our element produces a differential electric field $\vec{dE}$ at point P, which is a distance r from the element. Treating the element as a point charge, we can rewrite Eq. 23-8 to express the magnitude of $\vec{dE}$ as

$$|\vec{dE}| = \frac{1}{4\pi\varepsilon_0} \frac{|dq|}{r^2} = \frac{1}{4\pi\varepsilon_0} \frac{|\lambda| \, ds}{r^2}. \qquad \text{(TE23-2)}$$

The direction of $\vec{dE}$ is towards ds, because charge dq is negative.

Our element has a symmetrically located (mirror image) element ds' in the bottom half of the rod. The electric field $\vec{dE}'$ set up at P by ds' also has the magnitude given by Eq. 23-31, but the field vector points toward ds' as shown in Fig. TE23-2b. If we resolve the electric field vectors of ds and ds' into x and y components as shown in Fig. TE23-3b, we see that their y components cancel (because they have equal magnitudes and are in opposite directions). We also see that their x components have equal magnitudes and are in the same direction.

Thus, to find the electric field set up by the rod, we need sum (via integration) only the x components of the differential electric fields set up by all the differential elements of the rod. From Fig. TE23-3b and Eq. TE23-2, we can write the component dE_x set up by ds as

$$|dE_x| = |\vec{dE}| \cos\theta = \frac{1}{4\pi\varepsilon_0} \frac{|\lambda|}{r^2} \cos\theta \, ds. \qquad \text{(TE23-3)}$$

Equation TE23-3 has two variables, θ and s. Before we can integrate it, we must eliminate one variable. We do so by replacing ds, using the relation

$$ds = r \, d\theta,$$

in which $d\theta$ is the angle at P that includes arc length ds (Fig. TE23-2c). With this replacement, we can integrate Eq. TE23-3 over the angle made by the rod at P, from $\theta = -60°$ to $\theta = 60°$; that will give us the magnitude of the electric field at P due to the rod:

$$|\vec{E}| = \int |dE_x| = \int_{-60°}^{60°} \frac{1}{4\pi\varepsilon_0} \frac{|\lambda|}{r^2} \cos\theta \, r \, d\theta$$

$$= \frac{|\lambda|}{4\pi\varepsilon_0 r} \int_{-60°}^{60°} \cos\theta \, d\theta = \frac{|\lambda|}{4\pi\varepsilon_0 r} \left[\sin\theta \right]_{-60°}^{60°}$$

$$= \frac{|\lambda|}{4\pi\varepsilon_0 r} [\sin 60° - \sin(-60°)]$$

$$= \frac{1.73|\lambda|}{4\pi\varepsilon_0 r}. \tag{TE23-4}$$

(If we had reversed the limits on the integration, we would have gotten the same result but with a minus sign. Since the integration gives only the magnitude of $\vec{E}$, we would then have discarded the minus sign.)

To evaluate the magnitude of λ, we note that the rod has an angle of 120° and so is one-third of a full circle. Its arc length is then $2\pi r/3$, and its linear charge density must be

$$|\lambda| = \frac{\text{charge}}{\text{length}} = \frac{Q}{2\pi r/3} = \frac{0.477Q}{r}.$$

Substituting this into Eq. TE23-4 and simplifying give us

$$E = \frac{(1.73)(0.477Q)}{4\pi\varepsilon_0 r^2}$$

$$= \frac{0.83Q}{4\pi\varepsilon_0 r^2}. \tag{Answer}$$

The direction of $\vec{E}$ is toward the rod, along the axis of symmetry of the charge distribution. We can write $\vec{E}$ in unit-vector notation as

$$\vec{E} = \frac{0.83Q}{4\pi\varepsilon_0 r^2} \, \hat{i}.$$

Touchstone Example 23-9-1

Figure TE23-3 shows the deflecting plates of an ink-jet printer, with superimposed coordinate axes. An ink drop with a mass m of 1.3×10^{-10} kg and a negative charge of magnitude $Q = 1.5 \times 10^{-13}$ C enters the region between the plates, initially moving along the x axis with speed $v_x = 18$ m/s. The length L of the plates is 1.6 cm. The plates are charged and thus produce an electric field at all points between them. Assume that field $\vec{E}$ is downward directed, uniform, and has a magnitude of 1.4×10^6 N/C. What is the vertical deflection of the drop at the far edge of the plates? (The gravitational force on the drop is small relative to the electrostatic force acting on the drop and can be neglected.)

SOLUTION: The drop is negatively charged and the electric field is directed *downward*. The **Key Idea** here is that, from Eq. 23-4, a constant electrostatic force of magnitude $|Q\vec{E}|$ acts *upward* on the charged drop. Thus, as the drop travels parallel to the x axis at constant speed v_x, it accelerates upward with some constant acceleration a_y where a_y is positive. Applying Newton's Second Law ($F_y = ma_y$) for components along the y axis, we find that

$$a_y = \frac{F_y}{m} = \frac{|Q\vec{E}|}{m}. \tag{TE23-5}$$

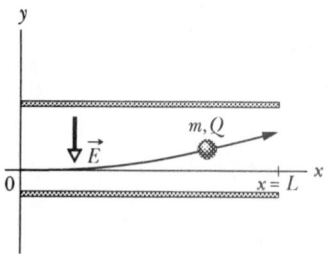

Fig. TE23-4 An ink drop of mass m and charge magnitude $|Q|$ is deflected in the electric field of an ink-jet printer.

Let t represent the time required for the drop to pass through the region between the plates. During t the vertical and horizontal displacements of the drop are

$$y = \tfrac{1}{2}a_y t^2 \quad \text{and} \quad L = v_x t, \tag{TE23-6}$$

respectively. Eliminating t between these two equations and substituting Eq. TE23-5 for a_y, we find

$$
\begin{aligned}
y &= \frac{QEL^2}{2mv_x^2} \\
&= \frac{(1.5 \times 10^{-13}\,\text{C})(1.4 \times 10^6\,\text{N/C})(1.6 \times 10^{-2}\,\text{m})^2}{(2)(1.3 \times 10^{-10}\,\text{kg})(18\,\text{m/s})^2} \\
&= 6.4 \times 10^{-4}\,\text{m} \\
&= 0.64\,\text{mm}. \hspace{4cm} \text{(Answer)}
\end{aligned}
$$

24 Gauss' Law

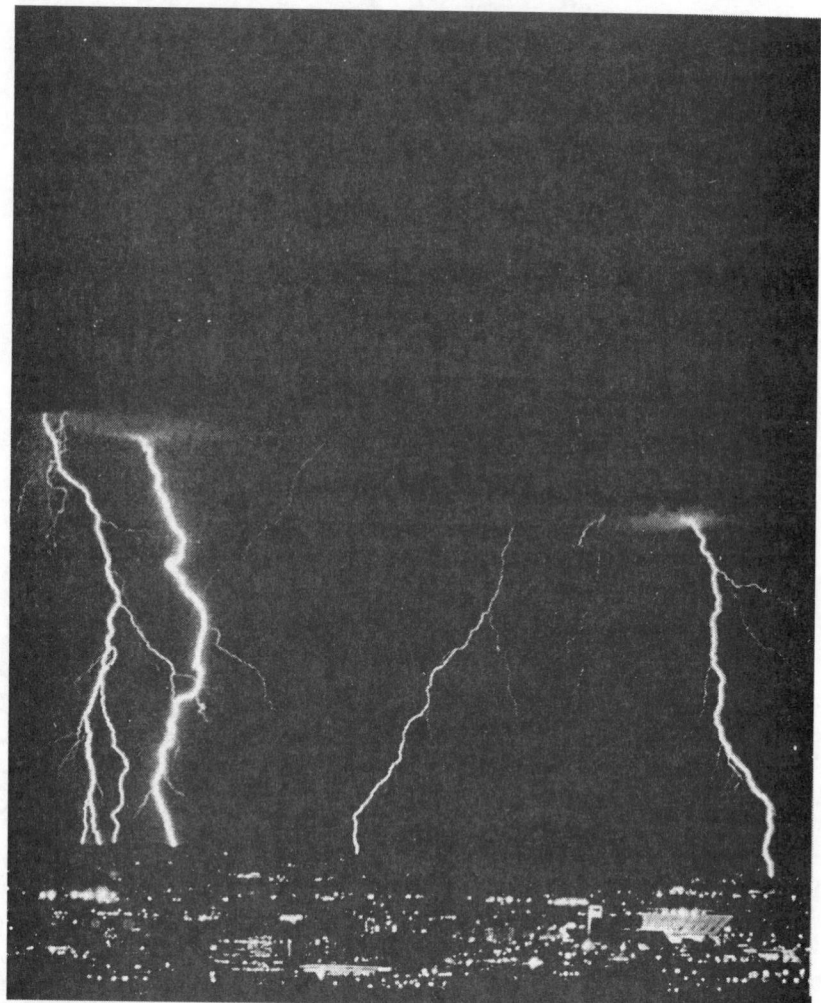

Lightning strikes Tucson in a brilliant display, each strike delivering about 10^{20} electrons from the cloud base to the ground.

How wide is a lightning strike? Since a strike can be seen from kilometers away, is it as wide as, say, a car?

The answer is in this chapter.

24-1 An Alternative to Coulomb's law

We associate a vector electric field with a distribution of charges. The electric field has a vector at every location in space telling us what force a test charge q_t will experience at a given location. In Sections 23-6 through 23-10 in the last chapter, we used Coulomb's law and the principle of superposition to calculate the electric field vectors at various points in space due to charges that were distributed in different ways. Although Coulomb's law can be used to calculate the electric force (and hence electric field) exerted on a test charge by any possible arrangement of charges we could imagine, this is usually a very difficult task. For example, calculating the electric field outside the surface of a hollow conducting sphere with charges arranged symmetrically as shown in Fig. 24-1 would require us to do a triple integration.

In Chapter 23 we used Coulomb's law to find electric fields from charge distributions, but what if we want to turn our calculation around and determine a distribution of charges from an electric field pattern? Unless our distribution of charges is very simple, this reverse calculation is also difficult to perform using Coulomb's law. Thus Coulomb's law appears to be valid but difficult to use in many circumstances. In this chapter we introduce another method for relating a known electric field to the charge distribution generating it, and, conversely for relating a known charge distribution to its associated electric field.

To explore how we might find a general relationship between a collection of charges and electric field, let's consider the electric field associated with the simplest possible charge distribution—a point charge. By applying Coulomb's law we already found that the magnitude of the charge's electric field *decreases* as the inverse square of the distance r, as expressed in Eq. 23-8,

$$\left|\vec{E}\right| = k\frac{|q|}{r^2}.$$

But, if we construct an imaginary spherical surface around our source charge we find that the surface area of the sphere *increases* as the square of the distance of the spherical surface from the source charge. The equation for the surface area is given by $A = 4\pi r^2$. Thus, we see that the product of the electric field magnitude and the surface area of any imaginary spherical boundary is constant no matter how large or small the distance from the charge is, as shown in Eq. 24-1,

$$\left|\vec{E}\right|A = k\frac{|q|}{r^2}(4\pi r^2) = \frac{1}{4\pi\varepsilon_0}\frac{|q|}{r^2}(4\pi r^2) = \frac{|q|}{\varepsilon_0}. \qquad (24\text{-}1)$$

This is a remarkable result for two reasons. First, as the electric field magnitude gets smaller, the area over which it can act gets larger by exactly the same amount. Second, the product of the electric field magnitude anywhere on a spherical surface and the area of the spherical surface is proportional to the charge enclosed by that surface. Notice also, mathematically this product looks rather like a volume flux described in Chapter 15 if we simply replace the fluid velocity vectors with electric field vectors.

Is it possible flux is a useful concept in helping us relate charge distributions to electric fields? Is there the proportionality between enclosed charge and the "flux" at a closed spherical surface for more complex charge distributions? Does the proportionality exist when the closed surface takes on other shapes? These questions were addressed by German mathematician and physicist Carl Friedrich Gauss (1777-1855). Thus, we begin our study of Gauss' approach to relating charge distributions and electric fields by defining a new quantity called electric flux.

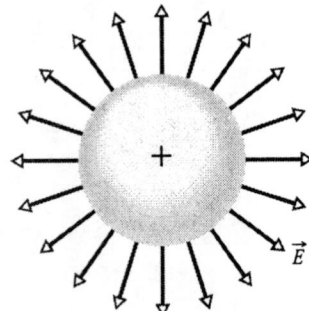

Fig. 24-1: If a single charge is located at the center of an imaginary sphere, Coulomb's law tells us the magnitude of the electric field vectors is the same at all points on the surface of the sphere and the direction of each electric field vector is normal (perpendicular) to the surface. Only the field vectors that lie in the plane of the page are shown in this drawing.

24-2 Electric Flux

The volume flux defined in Chapter 15 allows us to calculate the amount of fluid flowing through any very small element of a larger surface area, provided we know what the nature of the velocity vector field is characterizing the motion of the fluid. If we look at the ith element of the larger area, the *volume flux* element for that small area is defined as the scalar or dot product of the normal vector representing an area element and the velocity vector at the location of the area element as shown in Eq. 15-27,

$$\Phi_i \equiv \vec{v}_i \cdot \Delta \vec{A}_i \qquad \text{(volume flux definition for a small area element).}$$

> ▶ *What is a normal vector?* Recall we defined the normal vector representing a small flat element of area. It's a vector that points at right angles, or normal, to the plane of the area and has a magnitude equal to the area. If the element of area is part of a closed surface completely surrounding a space, we define the normal vector to be pointing *out* of the surface. (See Section 15-10 for details.)

Although *electric flux* does not involve the flow of anything, we define it in a way mathematically analogous to volume flux introduced in Chapter 15. An **electric flux element** is defined as the dot product of the normal vector representing an area element and the electric field vector at the location of the area element as shown in Eq. 24-2,

$$\Phi_i \equiv \left|\vec{E}_i\right|\left|\Delta \vec{A}_i\right|\cos\theta = \vec{E}_i \cdot \Delta \vec{A}_i \qquad \text{(electric flux definition for a small area),} \quad (24\text{-}2)$$

where θ is the angle between the two vectors. Some possible orientations for area elements and electric field vectors needed to calculate electric flux elements are shown in Fig. 24-2.

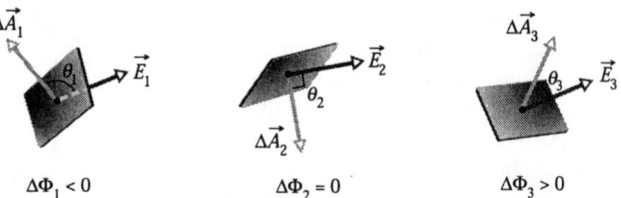

$$\Delta\Phi_1 < 0 \qquad\qquad \Delta\Phi_2 = 0 \qquad\qquad \Delta\Phi_3 > 0$$

Fig. 24-2: Three small areas that subtend different angles with respect to various electric field vectors. The first flux element is negative, the second zero, and the third positive. Note that nothing is "flowing" in order for electric flux to exist.

As is the case for volume flux, if our area is not small enough to be considered as flat or if the electric field vectors are not uniform over the area we choose, then we must break the area into small elements that are essentially flat. We can then determine the net electric flux as the sum of individual flux elements. This is given by

$$\begin{aligned}\Phi_{net} &= \Phi_1 + \Phi_2 + ... \Phi_N \\ &= \vec{E}_1 \cdot \Delta \vec{A}_1 + \vec{E}_2 \cdot \Delta \vec{A}_2 + ... \vec{E}_N \cdot \Delta \vec{A}_N \qquad \text{(net electric flux),} \quad (24\text{-}3) \\ &= \sum_{n=1}^{N} \vec{E}_n \cdot \Delta \vec{A}_n \end{aligned}$$

where $\vec{E}_1$, $\vec{E}_2$, $\vec{E}_3$, and so on represent the electric field vectors at the location of each of the N area elements. The flux associated with an electric field is a scalar, and its SI unit is the newton-meter-squared per coulomb or $\left[N \cdot m^2/C\right]$.

In everyday language the term flux is often used to represent flow or change. This is suggested by expressions such as "the economy is in a state of flux." Whenever either the magnitude of the normal electric field component or the area increases, the electric flux "at the area" increases even though nothing is flowing. Although the electric field may be changing over time due to a redistribution of electric charges, electric flux can exist whenever an electric field exists, even when the vector field is not changing.

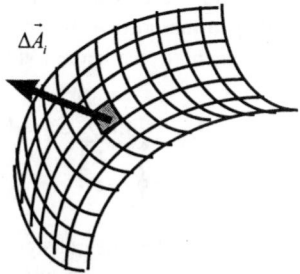

Fig. 24-3: In order to make net flux calculations, a curved surface area must be divided into N small area elements. Each element must be small enough so it is essentially flat and has electric field vectors that have the same magnitude and direction at every location on the surface. The ith area element and its normal vector are shown.

➤Instead of representing change or flow, **electric flux** at an area represents the summation over a surface of flux elements. Each flux element represents the product of an area element on the surface and the normal component of the electric field vector at a point on that area element.

24-3 Net Flux at a Closed Surface

In the introductory section we posed the question of whether there is a proportionality between an enclosed charge distribution and the "flux" at a surface that encloses it. To answer this question we need to carefully examine the procedures for determining net electric flux at an imaginary surface that encloses charges. The word "enclose" is important here. In the discussion that follows, we will not be discussing calculations of electric flux at any arbitrary surface. We will limit our discussion to the electric flux at closed surfaces that are continuous and connected. That is, a closed surface must be without holes. Nothing can get into or out of such surfaces without passing through the surface itself.

In order to define the net electric flux at any closed surface, consider Fig. 24-4, which shows an arbitrary (*asymmetric*) imaginary surface immersed in a *nonuniform* electric field. For historical reasons, any imaginary closed surface used in the calculation of a net electric flux is called a **Gaussian surface**. Since the electric field vector might be different at each location on our Gaussian surface, we must divide the entire surface into small area elements and take the sum as shown in Eq. 24-3.

Let's consider the arbitrary closed surface shown in Fig. 24-4. The vectors $\Delta\vec{A}$ and $\vec{E}$ for each square have some angle θ between each other. Figure 24-4 shows an enlarged view of three small squares (1, 2, and 3) on the Gaussian surface, and the angle θ between the two vectors. Our net flux equation (Eq. 24-3) instructs us to visit each square on the Gaussian surface, to evaluate the scalar product $\vec{E} \cdot \Delta\vec{A}$ for the two vectors $\vec{E}$ and $\Delta\vec{A}$ we find there, and to sum the results algebraically (that is, with signs included) for all the squares that make up the surface. The sign or a zero resulting from each scalar product determines whether the flux at a square is positive, negative, or zero. Squares like 1, in which $\vec{E}$ points inward, make a negative contribution to the sum. Squares like 2, in which $\vec{E}$ lies in the surface, make zero contribution. Squares like 3, in which $\vec{E}$ points outward, make a positive contribution.

The exact definition of the flux of the electric field at a surface is found by allowing the area of the squares shown in Fig. 24-4 to become smaller and smaller, approaching a differential limit dA. The area vectors then approach a differential limit $d\vec{A}$. Thus, the electric flux at a closed surface is given by the integral of the electric field components parallel to the normal of each surface area element over the magnitude of each surface area element. In mathematical notation the equation for electric flux becomes

$$\Phi_{\text{net}} \equiv \lim_{\Delta A \to 0} \sum_{n=1}^{N} \vec{E}_n \cdot \Delta\vec{A}_n$$

(net electric flux at a Gaussian surface). (24-4)

$$= \oint \vec{E} \cdot d\vec{A}$$

The circle on the integral sign indicates that the integration is to be taken over the entire closed surface (Gaussian surface).

Touchstone Examples 24-3-1 and 24-3-2, at the end of this chapter, illustrate how to use what you learned in this section.

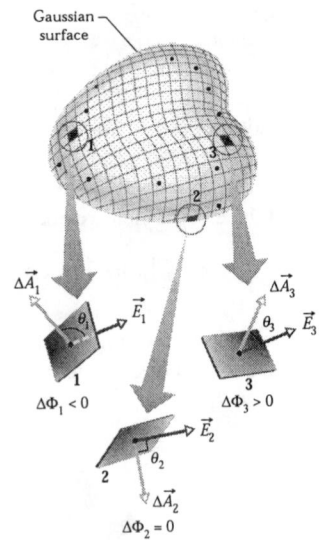

Gaussian surface

Fig. 24-4: A Gaussian surface of arbitrary shape is immersed in an electric field. The surface is divided into small area elements. The electric field vectors and the area vectors are shown for three representative area elements marked 1, 2, and 3. The other electric field vectors are not

TE

24-4 Gauss' Law

Let's return for a moment to the consequence of Coulomb's law we presented in the first section where we surrounded a single charge with a spherical Gaussian surface. We found that a flux-like quantity (namely the product of the magnitude of the electric field at the sphere's surface multiplied by the area of the sphere's surface) is equal to a constant times the enclosed charge. The surprising thing is this is true no matter what the radius of the sphere is, because the amount by which the surface area of the sphere increases just compensates for the amount by which the electric field magnitude decreases. This suggests the net flux though a Gaussian surface of any shape enclosing a single charge will be proportional to the amount of charge enclosed, shown in Fig. 24-5.

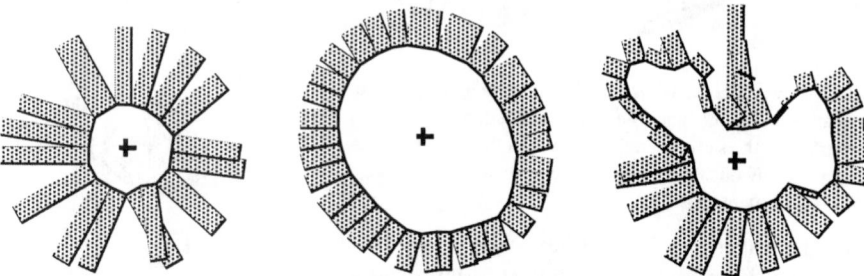

Fig. 24-5: Three different imaginary Gaussian surfaces surround the same point charge. The drawings show only a two-dimensional cross section of three dimensional surfaces. The contribution of small flux elements is represented by rectangles. Note whenever part of a surface is close to the charge, the flux elements are bigger but there are fewer of them. It is plausible to expect the net flux consisting of the sum of all the flux elements for each of the shapes will be the same.

A single point charge yields the same net electric flux at a Gaussian surface no matter what its shape. By extension, if there are two charges inside a Gaussian surface, each charge contributes its proportional share to the net flux no matter where the charges are located, provided they are *inside* the Gaussian surface. This leads us to a statement of Gauss' law that describes a plausible general relationship between the net flux on a Gaussian surface of any shape and the total enclosed charge no matter how it is distributed. The mathematical expression of Gauss' law is then

$$\Phi_{net} = \frac{q_{enc}}{\varepsilon_0} \qquad \text{(Gauss' law).} \qquad (24\text{-}5)$$

By substituting the definition of electric flux at a Gaussian surface, $\Phi_{net} \equiv \oint \vec{E} \cdot d\vec{A}$ we can also write Gauss' law as

$$\Phi_{net} = \oint \vec{E} \cdot d\vec{A} = \frac{q_{enc}}{\varepsilon_0} \qquad \text{(Gauss' law).} \qquad (24\text{-}6)$$

The use of the permittivity constant for a vacuum, ε_0, in Eqs. 24-5 and 24-6, indicates this form of Gauss' law only holds when the net charge is located in a vacuum or, for most practical purposes, in air. In Section 28-6, we modify Gauss' law to include situations in which so-called dielectric materials such as paper, oil, or water are present. In Fig. 24-6 we show how the net flux can have the same value for three different charge distributions involving the same amount of enclosed charge.

Interpreting Gauss' Law

If you know Gauss' law, you can calculate just how much net positive charge is enclosed. To make the calculation, you need know only "how much" electric field is intercepted by any surface enclosing a collection of charges. This "how much" involves the *flux* of the electric field at the surface.

In the equations above, the net charge q_{enc} is the algebraic sum of all the *enclosed* positive and negative charges, and it can be positive, negative, or zero. We include the sign, rather than just use the magnitude of the enclosed charge, because the sign tells us something about the net flux at the Gaussian surface: If q_{enc} is positive, the net flux is *outward*; if q_{enc} is negative, the net flux is *inward*, as shown in Fig. 24-7.

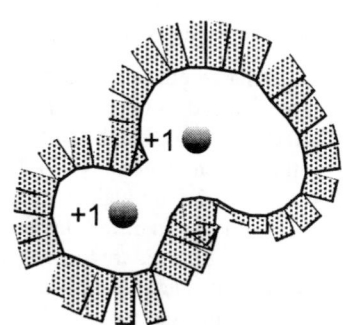

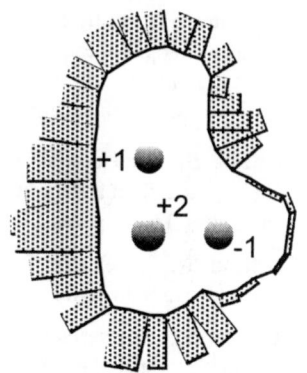

Fig. 24-6: Each one of the Gaussian surfaces encloses a different charge distribution. The total enclosed charge is the same in all three cases. When the electric flux is calculated at each area element on a Gaussian surface using Coulomb's law its value is represented by a grey rectangle. The total space covered by all of the grey rectangles turns out to be the same, which is compatible with the predictions of Gauss' law.

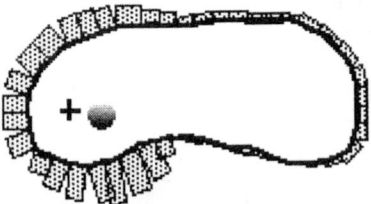 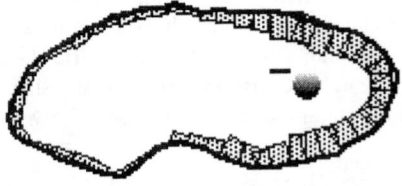 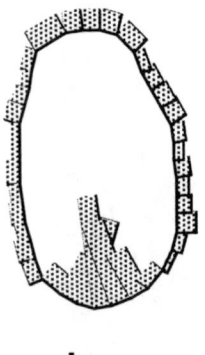

Fig. 24-7: Each one of the Gaussian surfaces shown in cross section encloses the same magnitude of charge. But in one case the charge is negative and in the other the charge is positive. The magnitude of the net flux at the Gaussian surface is the same in each case. When the net charge is positive, the net flux is positive, which is depicted by outward flux elements. When it is negative, so is the net flux, which is depicted by inward flux elements.

Charge outside a Gaussian surface, no matter how large or how close it may be, is not included in the term q_{enc} in Gauss' law. We expect this since there is no source of electric field inside the surface, and negative and positive flux elements will cancel each other, as shown in Fig. 24-8. This is mathematically analogous to the case of stream water flowing through an imaginary closed box as discussed in Section 15-11. The exact form or location of the charges inside the Gaussian surface is also of no concern; the only things that matter are the magnitude and sign of the net enclosed charge. The quantity $\vec{E}$ on the left side of Eq. 24-6, however, is the electric field resulting from *all* charges, both those inside and those outside the Gaussian surface. This may seem to be inconsistent, but keep in mind the electric field due to a charge outside the Gaussian surface contributes zero net flux on the surface.

Let us apply these ideas to Fig. 24-9, which shows two point charges, equal in magnitude but opposite in sign. Four Gaussian surfaces are also shown, in cross section. Let us consider each in turn.

Surface S_1. The electric field is outward for all points on this surface. Thus, the flux of the electric field at this surface is positive, and so is the net charge within the surface, as Gauss' law requires. (That is, if Φ is positive, q_{enc} must be also.)

Surface S_2. The electric field is inward for all points on this surface. Thus, the flux of the electric field is negative and so is the enclosed charge, as Gauss' law requires.

Surface S_3. This surface encloses no charge, and thus $q_{enc} = 0$. Gauss' law (Eq. 24-6) requires the net flux of the electric field at this surface to be zero. That is reasonable because the field vectors point inward for the portion of the surface nearest to the positive charge and outward for the portion of the surface near the negative charge. In calculating the net flux, the inward and outward flux elements cancel each other.

Surface S_4. This surface encloses no *net* charge, because the enclosed positive and negative charges have equal magnitudes. Gauss' law requires the net flux of the electric field at this surface be zero. That is reasonable because in this case the field vectors point outward for the portion of the surface nearest to the positive charge and inward for the portion of the surface near the negative charge. In calculating the net flux, the outward and inward flux elements cancel each other.

What would happen if we were to bring an enormous charge Q up close to surface S_4 in Fig. 24-9? The pattern of the electric field would certainly change, but the net flux for the four Gaussian surfaces would not change. We can understand this because the inward and outward flux elements associated with the added Q at any of the four surfaces would cancel each other, making no contribution to the net flux at any of them. The value of Q would not enter Gauss' law in any way, because Q lies outside all four of the Gaussian surfaces that we are considering.

Fig. 24-8: A charge is located outside a Gaussian surface. When the electric flux is calculated at each area element using Coulomb's law, its value is represented by a grey rectangle. The net flux is zero as the negative inward flux at the portion of the surface near the charge just cancels the positive outward flux at the location of the portions of the surface far away from the charge.

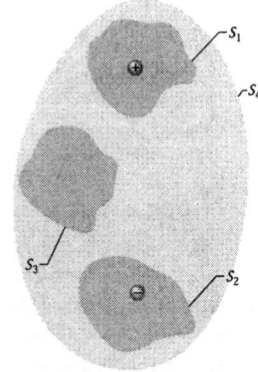

Fig. 24-9: Two point charges, equal in magnitude but opposite in sign, and the field lines that represent their net electric field. Four Gaussian surfaces are shown in cross section. Surface S_1 encloses the positive charge. Surface S_2 encloses the negative charge. Surface S_3 encloses no charge. Surface S_4 encloses both charges, and thus no net charge.

READING EXERCISE 24-1: The figure shows three situations in which a Gaussian cube sits in an electric field. The arrows indicate the directions of the electric field vectors for the top, front and right faces of each cube. The flux at the six sides of each cube is listed in the table below. In which situations does the cube enclose (a) a positive net charge, (b) a negative net charge, and (c) zero net charge?

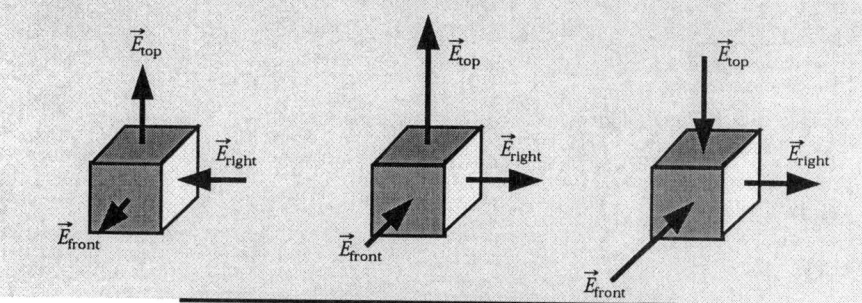

Flux [N·M²/C]

Face	Cube 1	Cube 2	Cube 3
Front	+2	−4	−7
Back	−3	−3	+8
Left	+7	+3	+2
Right	−4	+5	+5
Top	+5	+10	−6
Bottom	−7	−6	−5

24-5 Symmetry in Charge Distributions

Why go through all this trouble to develop a method of calculating electric fields that is equivalent to Coulomb's law? As we have mentioned, it is because Gauss' law makes it much simpler to calculate the field for highly symmetric charge distributions. What we mean by symmetric charge distributions are arrangements of charges that can be rotated about an axis or reflected in a mirror and still look the same. For example, Fig 24-10 shows several symmetric objects.

However, Gauss' law only offers us an advantage in dealing with highly symmetric charge distributions. For example, objects like an electric dipole, a line of charge or a disk of charge are arguably symmetric. Unfortunately, they are not symmetric enough to make Gauss' law an advantage over Coulomb's law. There are only a few charge distributions with sufficient symmetry for Gauss' law to be useful. These include single point charges and spherically symmetric, cylindrical and planar charge distributions. However, there are many physical situations for which these geometries are important. Hence, Gauss' law is an important and useful tool.

Why do charge distributions need to be so symmetric in order for Gauss' law to be helpful? It is because, in most situations, the charge distribution does not possess enough symmetry to enable us to find a convenient Gaussian surface over which the field has a constant magnitude. If we can find a surface over which the electric field is constant, we can take the electric field outside of the integral sign and Gauss' law goes from

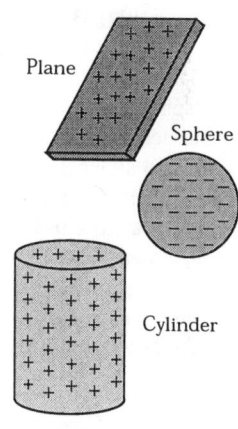

Fig. 24-10: Some symmetrically charged objects—a plane, a sphere, and a cylinder.

$$\varepsilon_0 \int \vec{E} \cdot d\vec{A} = q_{enc} \qquad \text{to} \qquad \varepsilon_0 \left| \vec{E} \right| \int \cos\theta \left| d\vec{A} \right| = \left| q_{enc} \right|.$$

Better still is to be able to find a Gaussian surface over which both the electric field and the angle between the field and area vectors θ are constant. In that case, both the electric field and the cosine functions can be moved outside the integral and Gauss' law becomes:

$$\varepsilon_0 \left| \vec{E} \right| \cos\theta \int \left| d\vec{A} \right| = \left| q_{enc} \right|.$$

This expression is very easy to evaluate because the integral of $d\vec{A}$ is simply the area of the Gaussian surface. Hence, if we can find a Gaussian surface over which the field and angle θ are constant, Gauss' law allows us to calculate the electric field of an extended charge distribution without doing an integral. In those cases, Gauss' law tells us that

$$|\vec{E}| = \frac{|q_{enc}|}{\varepsilon_0 A \cos\theta} \qquad (24\text{-}7)$$

where A is the area of the Gaussian surface, θ is the angle between the field and the area vectors and q_{enc} is the net charge *enclosed by the Gaussian surface.*

However, for most objects, we cannot easily use Gauss' law to find the field because the flux integral on the left hand side of the expression

$$\varepsilon_0 \int \vec{E} \cdot d\vec{A} = q_{enc}$$

is too complicated to evaluate. In these cases, Gauss' law is still valid. It is no simpler to use than Coulomb's law.

The issue of symmetry is important in using Gauss' law to determine the electric field for two reasons. The first was discussed above—Gauss' law only has an advantage over Coulomb's law for highly symmetric situations. The second is the symmetry of the charge distribution suggests to us what shape Gaussian surface to use. For example, if the charge distribution is a sphere, the Gaussian surface should be a sphere. Some examples of using Gauss' law on highly symmetric distributions are discussed below.

24-6 Application of Gauss' Law to Highly Symmetric Charge Distributions

Spherical Symmetry

Figure 24-11 shows a charged spherical shell of total charge q and radius R and two concentric spherical Gaussian surfaces, S_1 and S_2. (Note we chose the shape of the Gaussian surface to mirror the symmetry of the charge distribution). Because the charge distribution is so symmetric and the Gaussian surface mirrors that symmetry, the electric field is the same at every point of the surface S_2. Further, the angle between the electric field $\vec{E}$ and the area $\vec{A}$ is constant. The area vector points radially outward at all points on S_2. By symmetry and an understanding of point charges, we know the electric field also points radially outward at all points on S_2. Hence, the angle θ is constant and equal to $0°$ at all points on the surface. Applying Gauss' law to surface S_2 then comes down to evaluating the expression Eq. 24-7,

$$|\vec{E}| = \frac{|q_{enc}|}{\varepsilon_0 A \cos\theta}.$$

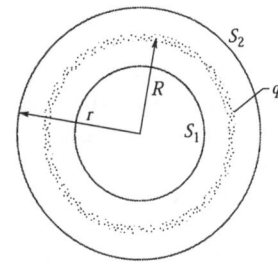

Fig. 24-11: A thin, uniformly charged, spherical shell with total charge q, in cross section. Two Gaussian surfaces S_1 and S_2 are also shown in cross section. Surface S_2 encloses the shell, and S_1 encloses only the empty interior of the shell.

The cosine of $0°$ is one and the area of a sphere (the Gaussian sphere) of radius r is $4\pi r^2$. Hence, for any $r \geq R$, we would find that

$$|\vec{E}| = \frac{1}{4\pi\varepsilon_0} \frac{|q_{enc}|}{r^2} \qquad \text{(spherical shell, field at } r \geq R \text{).} \qquad (24\text{-}8)$$

This is the same field that would be set up by a point charge q at the center of the shell of charge. Thus, a shell of charge q would produce the same force on a charged particle placed outside the shell as would a point charge q located at the center of the shell.

➤A shell of uniform charge attracts or repels a charged particle that is outside the shell as if all the shell's charge were concentrated at the center of the shell.

Applying Gauss' law to surface S_1, for which $r < R$, leads directly to

$$|\vec{E}| = 0 \qquad \text{(spherical shell, field at } r < R), \qquad (24\text{-}9)$$

because this Gaussian surface encloses no charge. Thus, if a charged particle were enclosed by the shell, the shell would exert no net electrostatic force on it.

▶A shell of uniform charge exerts no electrostatic force on a charged particle that is located inside the shell. This proves the first shell theorem.

Any spherically symmetric charge distribution, such as that of Fig. 24-12, can be constructed with a nest of concentric spherical shells. For purposes of applying the two shell theorems stated above, the volume charge density ρ should have a single value for each shell but need not be the same from shell to shell. Thus, for the charge distribution as a whole, ρ, can vary, but only with r, the radial distance from the center. We can then examine the effect of the charge distribution "shell by shell."

In Fig. 24-12a the entire charge lies within a Gaussian surface with $r > R$. The charge produces an electric field on the Gaussian surface as if the charge were a point charge located at the center, and Eq. 24-8 holds.

Figure 24-12b shows a Gaussian surface with $r < R$. To find the electric field at points on this Gaussian surface, we consider two sets of charged shells—one set inside the Gaussian surface and one set outside. The charge lying *outside* the Gaussian surface does not set up a net electric field on the Gaussian surface. Gauss' law tells us that the charge *enclosed* by the surface sets up an electric field as if that enclosed charge were concentrated at the center. Letting q' represent that enclosed charge, we can then rewrite

$$|\vec{E}| = \frac{1}{4\pi\varepsilon_0} \frac{|q|}{r^2}$$

as

$$|\vec{E}| = \frac{1}{4\pi\varepsilon_0} \frac{|q'|}{r^2} \qquad \text{(spherical distribution, field at } r < R).$$

If the full charge q enclosed within radius R is uniform, then q' enclosed within radius r in Fig. 24-12b is proportional to q:

$$\frac{\text{charge enclosed by } r}{\text{volume enclosed by } r} = \frac{\text{full charge}}{\text{full volume}},$$

or

$$\frac{q'}{\frac{4}{3}\pi r^3} = \frac{q}{\frac{4}{3}\pi R^3}. \qquad (24\text{-}10)$$

This gives us

$$q' = q\frac{r^3}{R^3}. \qquad (24\text{-}11)$$

Substituting this into

$$|\vec{E}| = \frac{1}{4\pi\varepsilon_0} \frac{|q'|}{r^2}$$

yields

$$|\vec{E}| = \left(\frac{|q|}{4\pi\varepsilon_0 R^3}\right) r \qquad \text{(uniform charge, field at } r \le R). \qquad (24\text{-}12)$$

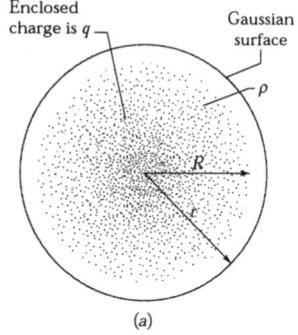

(a)

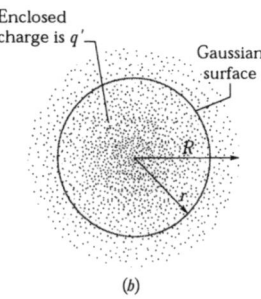

(b)

Fig. 24-12: The scalars represent a spherically symmetric distribution of charge of radius R, whose volume charge density ρ is a function only of distance from the center. The charged object is not a conductor, and the charge is assumed to be fixed in position. A concentric spherical Gaussian surface with $r > R$ is shown in (a). A similar Gaussian surface with $r < R$ is shown in (b).

Cylindrical Symmetry

Figure 24-13 shows a section of an infinitely long cylindrical plastic rod with a uniform positive linear charge density λ. Let us find an expression for the magnitude of the electric field $\vec{E}$ at a distance r from the axis of the rod.

Our Gaussian surface should match the symmetry of the problem, which is cylindrical. We choose a circular cylinder of radius r and length h, coaxial with the rod. The Gaussian surface must be closed, so we include two end caps as part of the surface.

Imagine now, while you are not watching, someone rotates the plastic rod around its longitudinal axis or turns it end for end. When you look again at the rod, you will not be able to detect any change. We conclude from this symmetry the only uniquely specified direction in this problem is along a radial line. Thus, at every point on the cylindrical part of the Gaussian surface, $\vec{E}$ must have the same magnitude $|\vec{E}|$ and (for a positively charged rod) must be directed radially outward.

Since $2\pi r$ is the circumference of the cylinder and h is its height, the area A of the cylindrical surface is $2\pi rh$. The flux of $\vec{E}$ at this cylindrical surface is then

$$\Phi = |\vec{E}|A\cos\theta = |\vec{E}|(2\pi rh)\cos 0 = |\vec{E}|(2\pi rh).$$

There is no flux at the end caps because $\vec{E}$, being radially directed, is parallel to the end caps at every point.

The charge enclosed by the surface is λh so that Gauss' law,

$$\varepsilon_0\Phi_{net} = q_{enc},$$

the magnitudes of the terms reduce to $\qquad \varepsilon_0|\vec{E}|(2\pi rh) = |\lambda|h,$

yielding $\qquad\qquad \boxed{|\vec{E}| = \dfrac{|\lambda|}{2\pi\varepsilon_0 r}} \qquad$ (line of charge). (24-13)

This is the electric field due to an infinitely long, straight line of charge, at a point that is a radial distance r from the line. The direction of $\vec{E}$ is radially outward from the line of charge if the charge is positive, and radially inward if it is negative. Equation 24-13 also approximates the field of a *finite* line of charge, at points that are not too near the ends (compared with the distance from the line).

Nonconducting Sheet

Figure 24-14 shows a portion of a thin, infinite, nonconducting sheet with a uniform (positive) surface charge density σ. A sheet of thin plastic wrap, uniformly charged on one side, can serve as a simple model. Let us find the electric field $\vec{E}$ a distance r in front of the sheet.

A useful Gaussian surface is a closed cylinder with end caps of area A, arranged to pierce the sheet perpendicularly as shown. From symmetry, $\vec{E}$ must be perpendicular to the sheet and hence to the end caps. Furthermore, since the charge is positive, $\vec{E}$ is directed *away* from the sheet, and thus the electric field vectors point in an outward direction from the two Gaussian end caps. Because the electric field vectors are perpendicular to the normal vectors on the curved surface, there is no flux at this portion of the Gaussian surface. Thus $\vec{E}\cdot d\vec{A}$ is simply $|\vec{E}|dA$; then Gauss' law,

$$\varepsilon_0\oint\vec{E}\cdot d\vec{A} = q_{enc},$$

becomes $\qquad\qquad \varepsilon_0(|\vec{E}|A + |\vec{E}|A) = |\sigma|A,$

where σA is the charge enclosed by the Gaussian surface. This gives

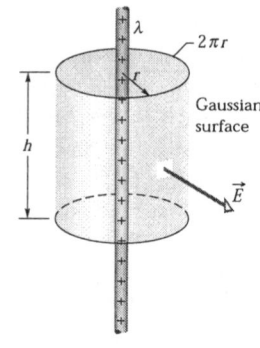

Fig. 24-13: A Gaussian surface in the form of a closed cylinder surrounds a section of a very long, uniformly charged, cylindrical plastic rod.

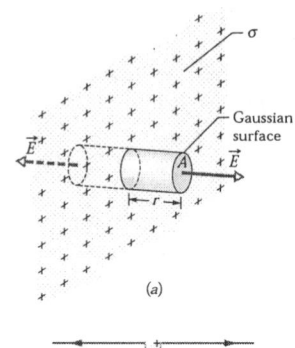

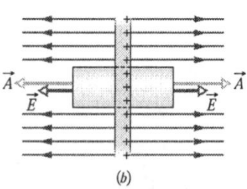

Fig. 24-14: Perspective view (*a*) and side view (*b*) of a portion of a very large, thin plastic sheet, uniformly charged on one side to surface charge density σ. A closed cylindrical Gaussian surface passes through the sheet and is perpendicular to it.

$$|\vec{E}| = \frac{|\sigma|}{2\varepsilon_0} \qquad \text{(sheet of charge).} \qquad (24\text{-}14)$$

Since we are considering an infinite sheet with uniform charge density, this result holds for any point at a finite distance from the sheet. The equation above agrees with what we would have found by integration of the electric field components that are produced by individual charges. (That would be a very time-consuming and challenging integration, and note how much more easily we obtain the result with Gauss' law. That is one reason for devoting a whole chapter to that law: for certain symmetric arrangements of charge, it is much easier to use than integration of field components.)

Touchstone Example 24-6-1, at the end of this chapter, illustrates how to use what you learned in this section.

TE

24-7 Gauss' Law and Coulomb's Law

If Gauss' law and Coulomb's law are equivalent, we should be able to derive each from the other. Here we derive Coulomb's law from Gauss' law and some symmetry considerations.

Figure 24-15 shows a positive point charge q, around which we have drawn a concentric spherical Gaussian surface of radius r. Let us divide this surface into differential areas dA. By definition, the area vector $d\vec{A}$ at any point is perpendicular to the surface and directed outward from the interior. From the symmetry of the situation, we know at any point the electric field $\vec{E}$ is also perpendicular to the surface and directed outward from the interior. Thus, since the angle θ between $\vec{E}$ and $d\vec{A}$ is zero, we can rewrite Gauss' law as

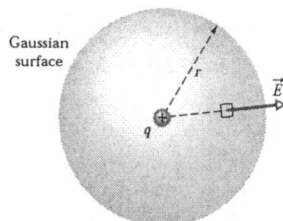

Fig. 24-15: A spherical Gaussian surface centered on a point charge q.

$$\varepsilon_0 \oint \vec{E} \cdot d\vec{A} = \varepsilon_0 \oint |\vec{E}| \, |d\vec{A}| = q_{enc}. \qquad (24\text{-}15)$$

Here $q_{enc} = q$. Although the magnitude of the vector $\vec{E}$ varies radially with the distance from q, it has the same value everywhere on the spherical surface. Since the integral in this equation is taken over that surface, $|\vec{E}|$ is a constant in the integration and can be brought out in front of the integral sign. That gives us

$$\varepsilon_0 |\vec{E}| \oint |d\vec{A}| = |q_{enc}|. \qquad (24\text{-}16)$$

The integral is now merely the sum of all the differential areas dA on the sphere and thus is just the surface area, $4\pi r^2$. Substituting this, we have

$$\varepsilon_0 |\vec{E}|(4\pi r^2) = |q_{enc}|.$$

or

$$|\vec{E}| = \frac{1}{4\pi\varepsilon_0} \frac{|q_{enc}|}{r^2}. \qquad (24\text{-}17)$$

This is exactly the electric field due to a point charge (Eq. 23-8), which we found using Coulomb's law. Thus, Gauss' law is equivalent to Coulomb's law.

READING EXERCISE 24-2: There is a certain net flux Φ_{net} at a Gaussian sphere of radius r enclosing an isolated charged particle. Suppose the enclosing Gaussian surface is changed to (a) a larger Gaussian sphere, (b) a Gaussian cube with edge length equal to r, and (c) a Gaussian cube with edge length equal to $2r$. In each case, is the net flux at the new Gaussian surface greater than, less than, or equal to Φ_{net}?

Touchstone Example 24-7-1, at the end of this chapter, illustrates how to use what you learned in this section.

TE

24-8 A Charged Isolated Conductor

Gauss' law permits us to prove an important theorem about isolated conductors:

▶If an excess charge is placed on an isolated conductor, that amount of charge will move entirely to the surface of the conductor. None of the excess charge will be found within the body of the conductor.

This might seem reasonable, considering charges with the same sign repel each other. You might imagine that, by moving to the surface, the added charges are getting as far away from each other as they can. We turn to Gauss' law for verification of this speculation.

Figure 24-16a shows, in cross section, an isolated lump of copper hanging from an insulating thread and having an excess charge q. We place a Gaussian surface just inside the actual surface of the conductor.

The electric field inside this conductor must be zero. If this were not so, the field would exert forces on the conduction (free) electrons, which are always present in a conductor, and thus current would always exist within a conductor. (That is, charge would flow from place to place within the conductor.) Of course, there are no such perpetual currents in an isolated conductor, and so the internal electric field is zero.

(An internal electric field *does* appear as a conductor is being charged. However, the added charge quickly distributes itself in such a way that the net internal electric field—the vector sum of the electric fields due to all the charges, both inside and outside—is zero. The movement of charge then ceases, because the net force on each charge is zero; the charges are then in *electrostatic equilibrium.*)

If $\vec{E}$ is zero everywhere inside our copper conductor, it must be zero for all points on the Gaussian surface because that surface, though close to the surface of the conductor, is definitely inside the conductor. This means the flux at the Gaussian surface must be zero. Gauss' law then tells us the net charge inside the Gaussian surface must also be zero. Then because the excess charge is not inside the Gaussian surface, it must be outside that surface, which means it must lie on the actual surface of the conductor.

An Isolated Conductor with a Cavity

Figure 24-16b shows the same hanging conductor, but now with a cavity totally within the conductor. It is perhaps reasonable to suppose when we scoop out the electrically neutral material to form the cavity, we do not change the distribution of charge or the pattern of the electric field that exists in Fig. 24-16a. Again, we must turn to Gauss' law for a quantitative proof.

We draw a Gaussian surface surrounding the cavity, close to its surface but inside the conducting body. Because $\vec{E} = 0$ inside the conductor, there can be no flux at this new Gaussian surface. Therefore, from Gauss' law, that surface can enclose no net charge. We conclude there is no net charge on the cavity walls; all the excess charge remains on the outer surface of the conductor, as in Fig. 24-16a.

The Conductor Removed

Suppose, by some magic, the excess charges could be "frozen" into position on the conductor's surface, perhaps by embedding them in a thin plastic coating, and suppose then the conductor could be removed completely. This is equivalent to enlarging the cavity of Fig. 24-16b until it consumes the entire conductor, leaving only the charges. The electric field would not change at all; it would remain zero inside the thin shell of charge and would remain unchanged for all external points. This shows us the electric field is set up by the charges and not by the conductor. The conductor simply provides an initial pathway for the charges to take up their positions.

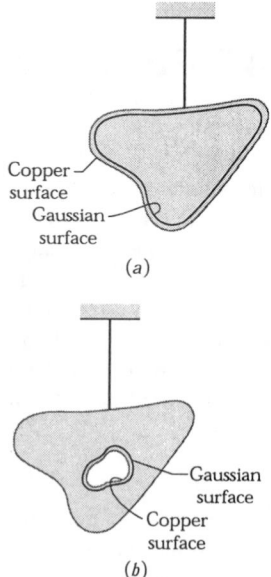

Copper
surface
Gaussian
surface

(a)

Gaussian
surface
Copper
surface

(b)

Fig. 24-16: (a) A lump of copper with a charge q hangs from an insulating thread. A Gaussian surface is placed within the metal, just inside the actual surface. (b) The lump of copper now has a cavity within it. A Gaussian surface lies within the metal, close to the cavity surface.

The External Electric Field

You have seen that the excess charge on an isolated conductor moves entirely to the conductor's surface. However, unless the conductor is spherical, the charge does not distribute itself uniformly. Put another way, the surface charge density σ (charge per unit area) varies over the surface of any nonspherical conductor. Generally, this variation makes the determination of the electric field set up by the surface charges very difficult.

However, the electric field just outside the surface of a conductor is easy to determine using Gauss' law. To do this, we consider a section of the surface small enough to permit us to neglect any curvature and thus to take the section to be flat. We then imagine a tiny cylindrical Gaussian surface to be embedded in the section as in Fig. 24-17: One end cap is fully inside the conductor, the other is fully outside, and the cylinder is perpendicular to the conductor's surface.

The electric field $\vec{E}$ at and just outside the conductor's surface must also be perpendicular to that surface. If it were not, then it would have a component along the conductor's surface exerting forces on the surface charges, causing them to move. However, such motion would violate our implicit assumption that we are dealing with electrostatic equilibrium. Therefore, $\vec{E}$ is perpendicular to the conductor's surface.

We now sum the flux at the Gaussian surface. There is no flux at the internal end cap, because the electric field within the conductor is zero. There is no flux at the curved surface of the cylinder, because internally (in the conductor) there is no electric field and externally the electric field is parallel to the curved portion of the Gaussian surface. The only flux at the Gaussian surface is at the external end cap, where $\vec{E}$ is perpendicular to the plane of the cap. We assume the cap area A is small enough that the field magnitude $|\vec{E}|$ is constant over the cap. Then the magnitude of the flux at the cap is $|\vec{E}|A$, and that is the net flux magnitude $|\Phi|$ at the Gaussian surface.

The charge q_{enc} enclosed by the Gaussian surface lies on the conductor's surface in an area A. If σ is the charge per unit area, then q_{enc} is equal to σA. When we substitute σA for q_{enc} and $|\vec{E}|A$ for $|\Phi|$, Gauss' law

$$\varepsilon_0|\Phi| = |q_{enc}|,$$

becomes

$$\varepsilon_0|\vec{E}|A = |\sigma|A,$$

from which we find
$$|\vec{E}| = \frac{|\sigma|}{\varepsilon_0} \quad \text{(conducting surface).} \quad (24\text{-}18)$$

Thus, the magnitude of the electric field at a location just outside a conductor is proportional to the surface charge density at that location on the conductor. If the charge on the conductor is positive, the electric field is directed away from the conductor as in Fig. 24-17. It is directed toward the conductor if the charge is negative.

The field vectors in Fig. 24-17 point toward negative charges somewhere in the environment. If we bring those charges near the conductor, the charge density at any given location on the conductor's surface changes, and so does the magnitude of the electric field. However, the relation between σ and $|\vec{E}|$ is still given by Eq. 24-18

$$|\vec{E}| = \frac{|\sigma|}{\varepsilon_0}.$$

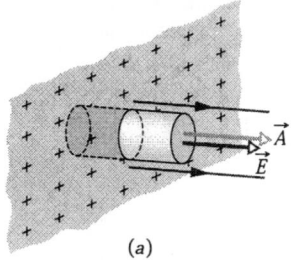

(a)

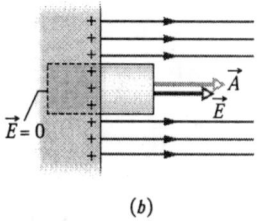

(b)

Fig. 24-17: Perspective view (a) and side view (b) of a tiny portion of a large, isolated conductor with excess positive charge on its surface. A (closed) cylindrical Gaussian surface, embedded perpendicularly in the conductor, encloses some of the charge. Electric field lines pierce the external end cap of the cylinder, but not the internal end cap. The external end cap has area A and area vector $\vec{A}$.

The Faraday Cage

The fact an isolated conductor with a cavity has no electric field inside of it has led to the construction of a very valuable electrical device. Many research environments today involve the measurement of very low power electrical signals. This might occur when measuring the electrical signals from the neuron of a live mouse running a maze or while trying to measure the electrical properties of a microscopic device meant as part of a micro-miniaturized computer chip. In our modern world there are numerous electrical signals running through space, arising from everything from the 60 Hz power running in our walls to the radio signals from TV stations and cellular phones. These ubiquitous signals can interfere with sensitive electrical measurements. To prevent these stray electric fields from messing up their measurements, researchers conduct their experiment inside a metal cage of wire mesh known as a Faraday Cage. This cage can prevent even strong electrical signals from producing electric fields inside the cage. The principle of the Faraday Cage is what makes it reasonably safe to be inside an automobile in a lightning storm. Even if lightning strikes your car, the effects inside the conductor will be substantially reduced.

Touchstone Example 24-8-1, at the end of this chapter, illustrates how to use what you learned in this section.

TE

Touchstone Example 24-3-1

Figure TE24-1 shows a Gaussian surface in the form of a cylinder of radius R immersed in a uniform electric field $\vec{E}$, with the cylinder axis parallel to the field. What is the flux Φ of the electric field through this closed surface?

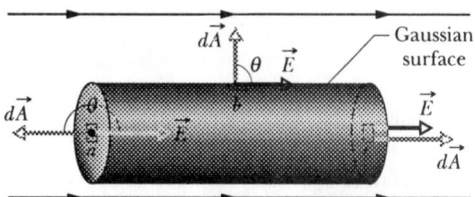

Fig. TE24-1 A cylindrical Gaussian surface, closed by end caps, is immersed in a uniform electric field. The cylinder axis is parallel to the field direction.

SOLUTION: The **Key Idea** here is that we can find the flux Φ through the surface by integrating the scalar product $\vec{E} \cdot d\vec{A}$ over the Gaussian surface. We can do this by writing the flux as the sum of three terms: integrals over the left cylinder cap a, the cylindrical surface b, and the right cap c. Thus, from Eq. 24-4,

$$\Phi = \oint \vec{E} \cdot d\vec{A}$$

$$= \int_a \vec{E} \cdot d\vec{A} + \int_b \vec{E} \cdot d\vec{A} + \int_c \vec{E} \cdot d\vec{A}. \qquad \text{(TE24-1)}$$

For all points on the left cap, the angle θ between $\vec{E}$ and $d\vec{A}$ is 180° and the magnitude E of the field is constant. Thus,

$$\int_a \vec{E} \cdot d\vec{A} = \int E(\cos 180°)\, dA = -E \int dA = -EA,$$

where $\int dA$ gives the cap's area, A ($= \pi R^2$). Similarly, for the right cap, where $\theta = 0$ for all points,

$$\int \vec{E} \cdot d\vec{A} = \int |\vec{E}(\cos 0)|\, dA = |\vec{E}|A.$$

Finally, for the cylindrical surface, where the angle θ is 90° at all points,

$$\int_b \vec{E} \cdot d\vec{A} = \int |\vec{E}(\cos 90°)|\, dA = 0.$$

Substituting these results into Eq. TE24-1 leads us to

$$\Phi = -|\vec{E}|A + 0 + |\vec{E}|A = 0. \qquad \text{(Answer)}$$

This result is perhaps not surprising because the field lines that represent the electric field all pass entirely through the Gaussian surface, entering through the left end cap, leaving through the right end cap, and giving a net flux of zero.

Touchstone Example 24-3-2

A *nonuniform* electric field given by $\vec{E} = (3.0\ \text{N/C} \cdot \text{m})\hat{i} + (4.0\ \text{N/C})\hat{j}$ pierces the Gaussian cube shown in Fig. TE24-2. (E is in newtons per coulomb and x is in meters.) What is the electric flux through the right face, the left face, and the top face?

SOLUTION: The **Key Idea** here is that we can find the flux Φ through the surface by integrating the scalar product $\vec{E} \cdot d\vec{A}$ over each face.

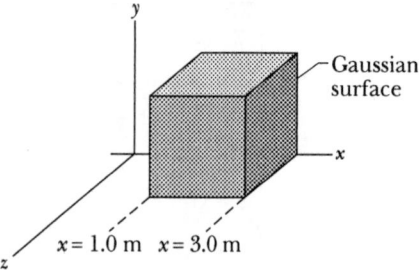

Fig. TE24-2 A Gaussian cube with one edge on the x axis lies within a nonuniform electric field.

Right face: An area vector $\vec{A}$ is always perpendicular to its surface and always points away from the interior of a Gaussian surface. Thus, the vector $d\vec{A}$ for the right face of the cube must point in the positive x direction. In unit vector notation, then,

$$d\vec{A} = dA\hat{\imath}.$$

From Eq. 24-4, the flux Φ_r through the right face is then

$$\Phi_r = \int \vec{E} \cdot d\vec{A} = \int [(3.0\ \text{N/C} \cdot \text{m})x\hat{\imath} + (4.0\ \text{N/C})\hat{\jmath}] \cdot (dA\hat{\imath})$$

$$= \int [(3.0\ \text{N/C} \cdot \text{m})(x)(dA)\hat{\imath} \cdot \hat{\imath} + (4.0\ \text{N/C})(dA)\hat{\jmath} \cdot \hat{\imath}]$$

$$= \int (3.0\ \text{N/C} \cdot \text{m})x\, dA + (0.0\ \text{N} \cdot \text{m}^2/\text{C}) = (3.0\ \text{N/C} \cdot \text{m}) \int x\, dA.$$

We are about to integrate over the right face, but we note that x has the same value everywhere on that face—namely, $x = 3.0$ m. This means we can substitute that constant value for x. Then

$$\Phi_r = (3.0\ \text{m}) \int (3.0\ \text{N/C} \cdot \text{m})\, dA = (9.0\ \text{N/C}) \int dA.$$

Now the integral merely gives us the area $A = 4.0\ \text{m}^2$ of the right face, so

$$\Phi_r = (9.0\ \text{N/C})(4.0\ \text{m}^2) = 36\ \text{N} \cdot \text{m}^2/\text{C}. \qquad \text{(Answer)}$$

Left face: The procedure for finding the flux through the left face is the same as that for the right face. However, two factors change. (1) The differential area vector $d\vec{A}$ points in the negative x direction and thus $d\vec{A} = -dA\hat{\imath}$. $\hat{\imath}$(2) The term x again appears in our integration, and it is again constant over the face being considered. However, on the left face, $x = 1.0$ m. With these two changes, we find that the flux Φ_l through the left face is

$$\Phi_l = -12\text{N} \cdot \text{m}^2/\text{C}. \qquad \text{(Answer)}$$

Top face: The differential area vector $d\vec{A}$ points in the positive y direction and thus $d\vec{A} = dA\hat{\jmath}$. The flux Φ_t through the top face is then

$$\Phi_t = \int [(3.0\ \text{N/C} \cdot \text{m})x\hat{\imath} + (4.0\ \text{N/C})\hat{\jmath}] \cdot (dA\hat{\jmath})$$

$$= \int [(3.0\ \text{N/C} \cdot \text{m})(xdA)\hat{\imath} \cdot \hat{\jmath} + (4.0\ \text{N/C})(dA)\hat{\jmath} \cdot \hat{\jmath}]$$

$$= \int [(0.0\ \text{N} \cdot \text{m}^2/\text{C}) + (4.0\ \text{N/C})\, dA] = (4.0\ \text{N/C}) \int dA$$

$$= 16\ \text{N} \cdot \text{m}^2/\text{C}. \qquad \text{(Answer)}$$

Touchstone Example 24-6-1

Figure TE24-3a shows portions of two large, parallel, nonconducting sheets, each with a fixed uniform charge on one side. The magnitudes of the surface charge densities are

$\sigma_{(+)} = 6.8$ μC/m^2 for the positively charged sheet and $\sigma_{(-)} = 4.3$ μC/m^2 for the negatively charged sheet.

Find the electric field $\vec{E}$ (a) to the left of the sheets, (b) between the sheets, and (c) to the right of the sheets.

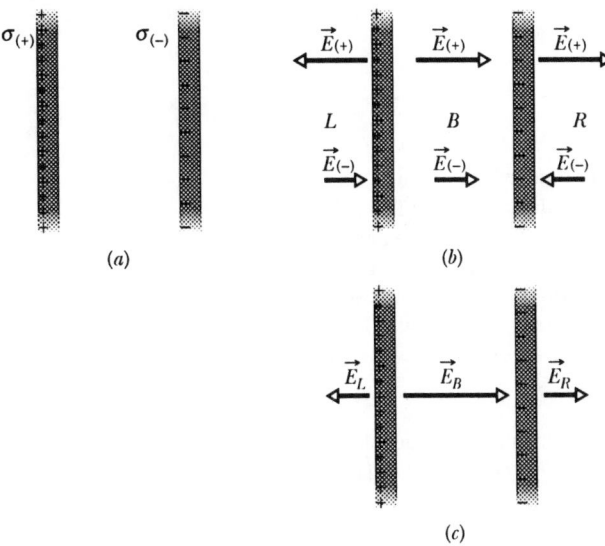

Fig. TE24-3 (a) Two large, parallel sheets, uniformly charged on one side. (b) The individual electric fields resulting from the two charged sheets. (c) The net field due to both charged sheets, found by superposition.

SOLUTION: The **Key Idea** here is that with the charges fixed in place, we can find the electric field of the sheets in Fig. TE24-3a by (1) finding the field of each sheet as if that sheet were isolated and (2) algebraically adding the fields of the isolated sheets via the superposition principle. (We can add the fields algebraically because they are parallel to each other.) From Eq. 24-14, the magnitude $E_{(+)}$ of the electric field due to the positive sheet at any point is

$$|\vec{E}_{(+)}| = \frac{|\sigma_{(+)}|}{2\varepsilon_0} = \frac{6.8 \times 10^{-6} \text{ C/m}^2}{(2)(8.85 \times 10^{-12} \text{ C}^2/\text{N} \cdot \text{m}^2)}$$
$$= 3.84 \times 10^5 \text{ N/C}.$$

Similarly, the magnitude $|\vec{E}_{(-)}|$ of the electric field at any point due to the negative sheet is

$$|\vec{E}_{(-)}| = \frac{|\sigma_{(-)}|}{2\varepsilon_0} = \frac{4.3 \times 10^{-6} \text{ C/m}^2}{(2)(8.85 \times 10^{-12} \text{ C}^2/\text{N} \cdot \text{m}^2)}$$
$$= 2.43 \times 10^5 \text{ N/C}.$$

Figure TE24-3b shows the fields set up by the sheets to the left of the sheets (L), between them (B), and to their right (R).

The resultant fields in these three regions follow from the superposition principle. To the left, the field magnitude is

$$|\vec{E}_L| = |\vec{E}_{(+)}| - |\vec{E}_{(-)}|$$
$$= 3.84 \times 10^5 \text{ N/C} - 2.43 \times 10^5 \text{ N/C}$$
$$= 1.4 \times 10^5 \text{ N/C}. \qquad \text{(Answer)}$$

Because $|\vec{E}_{(+)}|$ is larger than $|\vec{E}_{(-)}|$, the net electric field $\vec{E}_L$ in this region is directed to the left, as Fig. TE24-3c shows. To the right of the sheets, the electric field $\vec{E}_R$ has the same magnitude but is directed to the right, as Fig. TE24-3c shows.

Between the sheets, the two fields add and we have

$$|\vec{E}_B| = |\vec{E}_{(+)}| + |\vec{E}_{(-)}|$$
$$= 3.84 \times 10^5 \text{ N/C} + 2.43 \times 10^5 \text{ N/C}$$
$$= 6.3 \times 10^5 \text{ N/C}. \qquad \text{(Answer)}$$

The electric field $\vec{E}_B$ is directed to the right.

Touchstone Example 24-7-1

The visible portion of a lightning strike is preceded by an invisible stage in which a column of electrons extends downward from a cloud to the ground. These electrons come from the cloud and from air molecules that are ionized within the column. The linear charge density λ along the column is typically -1×10^{-3} C/m. Once the column reaches the ground, electrons within it are rapidly dumped to the ground. During the dumping, collisions between the moving electrons and the air within the column result in a brilliant flash of light. If air molecules break down (ionize) in an electric field exceeding 3×10^6 N/C, what is the radius of the column?

SOLUTION: One Key Idea here is that, although the column is not straight or infinitely long, we can approximate it as being a line of charge as in Fig. 24-13. (Since it contains a net negative charge, its electric field $\vec{E}$ points radially inward.) Then, according to Eq. 24-13, the field's magnitude E decreases with distance from the axis of the column of charge.

A second Key Idea is that the surface of the column of charge must be at the radius r where the magnitude of $|\vec{E}|$ is 3×10^6 N/C, because air molecules within the radius ionize

Fig. TE24-4 Lightning strikes a 20-m-high sycamore. Because the tree was wet, most of the charge traveled through the water on it and the tree was unharmed.

Fig. TE24-5 Ground currents from a lightning strike have burned grass off this golf green, exposing the soil.

while those farther out do not. Solving Eq. 24-13 for r and inserting the known data, we find the radius of the column to be

$$r = \frac{\lambda}{2\pi\varepsilon_0 E}$$

$$= \frac{1 \times 10^{-3} \text{ C/m}}{(2\pi)8.85 \times 10^{-12} \text{ C}^2/\text{N} \cdot \text{m}^2)(3 \times 10^6 \text{ N/C})}$$

$$= 6 \text{ m}.$$

(The radius of the luminous portion of a lightning strike is smaller, perhaps only 0.5 m. You can get an idea of the width from Fig. TE24-4.) Although the radius of the column may be only 6 m, do not assume that you are safe if you are at a somewhat greater distance from the strike point, because the electrons dumped by the strike travel along the ground. Such *ground currents* are lethal. Figure TE24-5 shows evidence of ground currents.

Touchstone Example 24-8-1

Figure TE24-6a shows a cross section of a spherical metal shell of inner radius R. A point charge of -5.0 μC is located at a distance $R/2$ from the center of the shell. If the shell is electrically neutral, what are the (induced) charges on its inner and outer surfaces? Are those charges uniformly distributed? What is the field pattern inside and outside the shell?

SOLUTION: Figure TE24-6b shows a cross section of a spherical Gaussian surface within the metal, just outside the inner wall of the shell. One **Key Idea** here is that the electric field must be zero inside the metal (and thus on the Gaussian surface inside the metal). This means that the electric flux through the Gaussian surface must also be zero. Gauss' law then tells us that the *net* charge enclosed by the Gaussian surface must be zero. With a point charge of -5.0 μC within the shell, a charge of $+5.0$ μC must lie on the inner wall of the shell.

If the point charge were centered, this positive charge would be uniformly distributed along the inner wall. However, since the point charge is off-center, the distribution of positive charge is skewed, as suggested by Fig. TE24-6b, because the positive charge tends to collect on the section of the inner wall nearest the (negative) point charge.

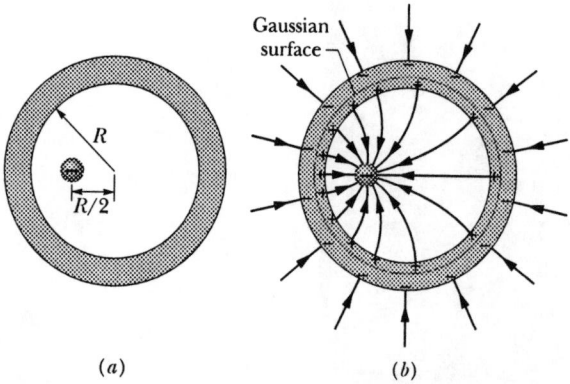

(a) (b)

Fig. TE24-6 (a) A negative point charge is located within a spherical metal shell that is electrically neutral. (b) As a result, positive charge is nonuniformly distributed on the inner wall of the shell, and an equal amount of negative charge is uniformly distributed on the outer wall.

A second **Key Idea** is that because the shell is electrically neutral, its inner wall can have a charge of $+5.0~\mu$C only if electrons, with a total charge of $-5.0~\mu$C, leave the inner wall and move to the outer wall. There they spread out uniformly, as is also suggested by Fig. TE24-6b. This distribution of negative charge is uniform because the shell is spherical and because the skewed distribution of positive charge on the inner wall cannot produce an electric field in the shell to affect the distribution of charge on the outer wall.

The field lines inside and outside the shell are shown approximately in Fig. TE24-6b. All the field lines intersect the shell and the point charge perpendicularly. Inside the shell the pattern of field lines is skewed owing to the skew of the positive charge distribution. Outside the shell the pattern is the same as if the point charge were centered and the shell were missing. In fact, this would be true no matter where inside the shell the point charge happened to be located.

25 Electric Potential

While enjoying the Sequoia National Park from a lookout platform, this woman found her hair rising from her head. Amused, her brother took her photograph. Five minutes after they left, lightning struck the platform, killing one person and injuring seven.

What had caused the woman's hair to rise?

The answer is in this chapter.

25-1 Introduction

In the last few chapters we have explored the nature of interaction forces between charged particles. We have developed the concept of electric field as a way to represent the forces a point charge could experience at any point in space surrounding a collection of charges. In certain situations it is difficult to describe the interactions of charges in terms of an electric field. This is analogous to problems encountered in describing the motion of an object in the presence of gravitational forces. We developed the concepts of work and energy in Chapters 9 and 10 to deal with these problems.

We will investigate the application of the concepts of work and energy to situations where the forces involved are electrostatic forces. We will find because work and energy are scalar quantities, an analysis using the concept of electric potential is simpler than using the electric field concept. In this chapter we will develop the concept of electric potential—commonly referred to as voltage. We will then explore some of its properties, including how charges are distributed on a metal conductor placed in an electric field.

Since the concept of potential or voltage is essential to an understanding of the energy transformations in simple electric circuits, we will use the concept of electric potential in the next chapter to help us understand the role that batteries play in maintaining currents in circuits.

25-2 Electric Potential Energy

Newton's law for the gravitational force and Coulomb's law for the electrostatic force are mathematically similar. In Section 22-7 we saw that the magnitude of the gravitational force between two particle-like masses depends directly on the product of the masses and inversely with the square of the distance between them (Eq. 14-1). In like manner, the electrostatic forces between two point charges depend directly on the product of the charge magnitudes and inversely with the square of the distance between them (Eq. 22-4). This similarity gives us a starting point in our search for concepts related to the interactions between charged objects that will be useful. In this chapter, we consider whether some of the general features we have established for the gravitational force apply to the electrostatic force as well.

For example, the gravitational force is a *conservative force*. The work done by it is independent of the path along which an object moves. In experimental tests the work done by the electrostatic force has also been found to be *path independent*. If a charged particle moves from point i to point f while an electrostatic force is acting on it, the work W done by the force is the same for all paths between points i and f. Hence, we can infer that the electrostatic force is a conservative force as well.

Definition of Electric Potential Energy

In Chapter 10 we defined potential energy as the energy associated with the configuration of a system of objects that interact and hence exert forces on each other. We then proceeded to define gravitational potential energy as the negative of the amount of gravitational work objects in the system do on each other when they are reconfigured or $\Delta U = -W_g$ (Eq. 10-4). Clearly, this general definition can be applied to a system of charges that interact by means of electrostatic forces.

Thus, if electrostatic forces, like gravitational forces, are conservative, then it makes sense to assign an **electric potential energy change** ΔU to a system of interacting charges in a similar manner. If we cause or allow a system to change its configuration from an initial potential energy state i to a different final state f, the electrostatic interaction forces do a total amount of work W_e on the particles in the system. In order to reconfigure a system of interacting charges, an external agent must do an amount of work $-W_e$ on the system. We then define a system's potential energy change ΔU as the amount of work an external agent must do to reconfigure it. This can be expressed symbolically as

$$\Delta U \equiv U_f - U_i = W_{\text{external}} = -W_e. \qquad (25\text{-}1)$$

Figure 25-1 shows a system of charges losing electrostatic potential energy as a result of a natural reconfiguration. Figure 25-2 shows the same system gaining electrostatic potential energy when an external agent causes the system to reconfigure by doing positive work. This causes the system to do negative work on itself.

(a) Initial configuration   (b) Final configuration

Fig. 25-1: (a) A system of three charges is in an initial configuration in which the charges are separated and have an electric potential energy U_i associated with them. (b) If q_2 is much greater than q_1 and q_3, these smaller charges will be naturally attracted to the larger one. Thus, the charges will coalesce into a final configuration. The net electrostatic work the charges do on each other is positive so the system loses potential energy. Thus $U_f < U_i$ so that $\Delta U < 0$.

(a) Initial configuration   (b) Final configuration

Fig. 25-2: (a) A system of three charges is in an initial configuration in which the charges are close together and have an electric potential energy U_i associated with them. (b) If q_2 is much greater than q_1 and q_3, these smaller charges will be naturally attracted to the larger charge and it will take positive external work, W_{ext}, to pull the charges apart. The net electrostatic work the charges do on each other is negative so the system gains potential energy. Thus $U_f > U_i$ so that $\Delta U > 0$.

As you may recall, we determined only differences in gravitational potential energy were physically significant. In most cases the system of masses we considered consisted of a single object near the Earth's surface and the Earth. We typically choose a convenient height at which to set the gravitational potential energy to zero. For example, we might define an Earth-object system as having zero potential energy when an object is at floor level or, another time, at the level of a tabletop. In so doing, we set the absolute scale for gravitational potential energy differently in different situations. This is legitimate only if the differences in potential energy are meaningful.

Although potential energy difference is also of primary importance in keeping track of electric potential energy, it is *customary to define the electric potential energy of a system of charges to be zero when the particles are all infinitely separated from each other*. Using this reference for electric potential energy makes sense because the charges making up such a system have no interaction forces in that configuration. Because we use a standard reference potential instead of moving it around as we do for typical Earth-object systems allows us to find unique values of U_i and U_f. Suppose several charged particles come together from initially infinite separations (state *i*) to form a system of nearby particles (state *f*). Then using the conventional reference configuration, the initial potential energy U_i is zero. If W_∞ represents the work done by the electrostatic forces between particles during the move in from infinity, then from Equation 25-1,

$$\Delta U = U_f - U_i = U_f = -W_e. \qquad (25\text{-}2)$$

Since U_i is zero, the final potential energy U_f of the system can simply be denoted as U so the electrical potential energy of a system is the negative of the total work done on all the charges in the system by the electrostatic forces they exert on each other. In terms of symbols

$$U \equiv -W_{e,\infty}. \qquad (25\text{-}3)$$

As usual, the use of the symbol "$\equiv$" signifies that the expression is a definition.

By analogy to the gravitational case, we might assume if only conservative forces act within a closed system experiencing no external forces, the total energy of the system is conserved and hence maintains a constant value. Extensive experimentation has never provided evidence to contradict this assumption. Thus, we assume electrostatic forces are conservative and use the idea of conservation of energy extensively throughout this chapter.

External Forces and Energy Conservation

Since opposite charges attract, they will come together naturally if they are free to move closer to each other. Similarly, like charges that are free will move apart. In these cases the charges "fall together" and the potential energy of the system of charges will be reduced. We can raise the potential energy of a system of charges by using energy from another external system. Two common examples of external agents that can raise the potential energy of a system of charges are the Van de Graaff generator and the battery. Van de Graaff generators use mechanical energy to force charges of like sign onto metal conductors. Batteries use chemical potential energy (which is actually a microscopic form of electrical potential energy) to force charges unnaturally onto an electrode having the same sign charges.

Suppose an *external force* outside of the system under consideration causes a test particle of charge q to move from an initial location to a final location in the presence of the electric field generated by the source charges in the system. As the test charge moves, our outside force does work $W_{external}$ on the charge. At the same time, the electric field does work W_e on it. By the work-kinetic energy theorem, the change ΔK in the kinetic energy of the particle is

$$\Delta K = K_f - K_i = W_{external} + W_e \qquad (25\text{-}4)$$

Now suppose the particle is stationary before and after the move. Then K_f and K_i are both zero, and this reduces to

$$W_{external} = -W_e \qquad (25\text{-}5)$$

That is, the work, $W_{external}$, done by our external force during the move is equal to the negative of the work W_e done by the electric field—provided there is no change in kinetic energy.

By using Equation 25-5 to substitute $W_{external}$ into $\Delta U = U_f - U_i = -W_e$, we can relate the work done by our external force to the change in the potential energy of the particle during the move. We find

$$\Delta U = U_f - U_i = W_{external}. \qquad (25\text{-}6)$$

So in what direction will a positive or negative charge move if released in a given potential? Will the charge move to higher or lower potential? The expression above can be used to determine this. For example, let the external force (perhaps the push or pull of your hand) do positive work. Then recalling the sign convention associated with work in general, we know there must be energy added to the system in order for the motion to occur. That is, we must force the motion to occur. If $W_{external}$ is positive, then ΔU must also be positive (by the equation above) and so we know that $U_f > U_i$. In other words, the motion of a positive charge from a lower potential energy to a higher potential energy requires positive work to be done on the system. This motion would not happen if the particle were simply released. This is very similar to the situation encountered in Section 9-8 where it takes external work to lift an object in the presence of the Earth's attractive gravitational force.

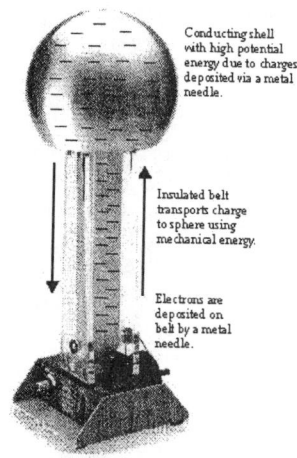

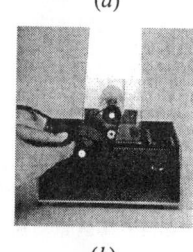

(a)

(b)

Fig. 25-3: A Van de Graaff generator uses mechanical energy from either (a) a motor or (b) a hand crank to transport charge to a conducting sphere, raising its potential energy. (Photo courtesy of PASCO scientific.)

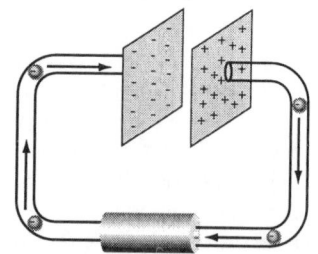

Fig. 25-4: A 1.5 V D-cell can act as an external agent that does the work needed to move electrons through a wire from a metal plate with excess positive charges to one with excess negative charges.

READING EXERCISE 25-1: Why is a configuration with charges separated by an infinite distance a good choice for our reference (zero) potential energy? Would a zero separation be equally good? Why or why not?

READING EXERCISE 25-2: In the figure, a proton moves from point i to point f in a uniform electric field directed as shown. (a) Does the electric field do positive or negative work on the proton? (b) Does the electric potential energy of the proton increase or decrease?

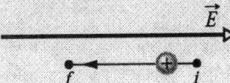

Touchstone Example 25-2-1, at the end of this chapter, illustrates how to use what you learned in this section.

TE

25-3 Electric Potential

When considering gravitational potential energy we dealt primarily with a system consisting of the Earth and a single object much smaller than the Earth. If the object were to fall toward the Earth the interaction forces between them would be equal in magnitude, but as the object moves toward the Earth, the Earth's motion would be negligibly small. Thus the change in the system's gravitational potential energy would simply be the change in potential energy of the falling object. Similarly, as we did in Section 23-2, we can consider systems in which a small "test" charge moves in the presence of an electric field, but does not change the electric field significantly. In these systems, the electric potential energy of the system can be calculated as the negative of work done by the electric field on a single test charge as we bring it to a location of interest from infinity.

➤In the next several chapters we will focus primarily on systems in which the change in potential energy of a single test charge moving in an electric field is for all practical purposes the same as the change in potential energy of the entire system of charges.

Defining Electric Potential

Recall we defined and used the concept of electric field as the *electric force per unit charge* so we could easily analyze the forces experienced by a charge of any sign or magnitude. It is advantageous to develop an analogous concept for the determination of the electric potential energy of a system associated with the change in location of a test charge of any reasonable sign or magnitude. We will now take the same approach with electric potential energy by defining **electric potential** as a potential energy *per unit charge*. Once we have chosen a reference configuration with zero energy, our electric potential (potential energy per unit charge) has a unique value at any point in space. For example, suppose we place a test particle of positive charge 1.60×10^{-19} C at a point in an electric field where the particle has an electric potential energy of 2.40×10^{-17} J. Then the electric potential, ΔV, of the system associated with the change in location of a test charge can be calculated as

$$\Delta V = \frac{2.40 \times 10^{-17} \text{ J}}{1.60 \times 10^{-19} \text{ C}} = 150 \text{ J/C}.$$

Next, suppose we replace that test particle with one having twice as much positive charge, 3.20×10^{-19} C. We would find the second particle has an electric potential energy of 4.80×10^{-17} J, twice that of the first particle. However, the potential energy per unit charge would be the same, still 150 J/C.

Thus, the system potential energy per unit charge, which can be symbolized as U/q, is independent of the charge q of the test particle we happen to be considering. It is *characteristic only of the electric field* that is present. The potential energy per unit charge at a point in an electric field is defined as the **electric potential** ΔV (or simply the **potential**) at that point. Thus ΔV is defined as

$$\Delta V \equiv \frac{U}{q}. \tag{25-7}$$

Note potential energy and charge are both scalar quantities, so the electric potential is also a scalar, not a vector.

The electric potential difference ΔV associated with moving a charge q between any two points i and f in an electric field is equal to the difference in potential energy per unit charge between the two points:

$$\Delta V = V_f - V_i = \frac{U_f}{q} - \frac{U_i}{q} = \frac{\Delta U}{q}. \tag{25-8}$$

Using $\Delta U = U_f - U_i = -W_e$ (Eq. 25-1) to substitute the work done by electrostatic forces $-W_e$ for ΔU in the equation above, we can define the potential difference between points i and f as

$$\Delta V = V_f - V_i \equiv -\frac{W_e}{q} \qquad \text{(potential difference defined).} \tag{25-9}$$

The potential difference between two points is the negative of the work done by the electrostatic force to move a unit charge from one point to the other. A potential difference can be positive, negative, or zero, depending on the signs and magnitudes of the charge q and the electrostatic work W_e.

If we set $U_i = 0$ at infinity as our reference potential energy, then according to Eq. 25-7, the electric potential must also be zero there. Then using Equation 25-9, we can define the electric potential ΔV at any point in an electric field to be

$$\Delta V = -\frac{W_{e,\infty}}{q} \qquad \text{(potential defined),} \tag{25-10}$$

where $W_{e,\infty}$ is the work done by the electrostatic force on a charged particle as that particle moves in from infinity to point f. A potential ΔV can be positive, negative, or zero, depending on the signs and magnitudes of q and $W_{e,\infty}$.

The SI unit for electric potential that follows from Eq. 25-10 is the joule per coulomb. This combination occurs so often a special unit, the *volt* (abbreviated V) is used to represent it. Thus,

$$1 \text{ volt} \equiv 1 \text{ joule per coulomb.} \tag{25-11}$$

Unfortunately, the terms electric potential energy and electric potential are very similar. They are not the same thing. This is probably one of the reasons why it is so common to refer to electric potential as voltage after its unit—the volt.

This new unit called the volt allows us to adopt a more conventional unit for the electric field $\vec{E}$, which we have measured up to now in newtons per coulomb. With two unit conversions, we obtain

$$1 \text{N/C} = \left[1 \frac{\text{N}}{\text{C}} \right] \left[\frac{1 \text{ V} \cdot \text{C}}{1 \text{ J}} \right] \left[\frac{1 \text{ J}}{1 \text{ N} \cdot \text{m}} \right] = 1 \text{ V/m.} \tag{25-12}$$

The conversion factor in the second set of parentheses comes from Eq. 25-11; that in the third set of parentheses is derived from the definition of the joule. From now on, we shall express values of the electric field in volts per meter rather than in newtons per coulomb.

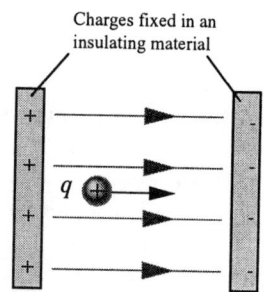

Fig. 25-5: A test charge moves in an electric field created by a stable configuration of source charges. If the test charge doesn't affect the electric field significantly as it changes location, the change in electric potential, ΔV, of the system (consisting of the source charges and the test charge) is due entirely to the work per unit charge done on the text charge by the electric field.

The Electron Volt

Because we often have situations in which the charges involved are very small (a few times the charge of an electron), we define an energy unit that is a convenient one for energy measurements in the atomic and subatomic domain. One *electron-volt* (eV) is the energy equal to the work required to move a single elementary charge e, such as that of the electron or the proton, through a potential difference of exactly one volt. Equation 25-9,

$$\Delta V = V_f - V_i = -\frac{W_e}{q},$$

tells us that the magnitude of this work is $q\,\Delta V$, so

$$1 \text{ eV} = e(1 \text{ V})$$
$$= (1.60 \times 10^{-19} \text{ C})(1 \text{ J/C}) = 1.60 \times 10^{-19} \text{ J}. \qquad (25\text{-}13)$$

READING EXERCISE 25-3: In the figure shown in Reading Exercise 25-2, we moved a proton from point i to point f in a uniform electric field directed as shown. (a) Does our force do positive or negative work? (b) Does the proton move to a point of higher or lower potential?

25-4 Equipotential Surfaces

We are interested in what our knowledge of electric potential can tell us about how small test charges might move. We can infer from the discussion above that charged particles will not spontaneously move from one point to another point of equal potential. This is quite analogous to movement of mass in a gravitational field. A skier on a flat surface with no kinetic energy will not spontaneously move from one part of the surface to another. On the other hand, if the skier is on a slope and is free to move, the skier will spontaneously start moving down the slope. Thus, it would be useful to know where all the points of equal potential energy are in a given region of space. That way, we can predict the motion of charges. An **equipotential surface** is defined as a surface having the same potential at all points on it.

Let's consider the electric field associated with a source consisting of a single fixed point charge we designate as the source charge. What happens if we place a test charge at a distance r from the source charge and move it around? If we move the charge anywhere on the surface of a sphere of radius r, no electrostatic work is done on the test charge as it is always moving perpendicular to the electric field vectors. However, we cannot move our test charge from one distance from the source charge to another distance without the electric field doing work on it. This is illustrated in Fig. 25-6. Thus, any sphere centered on the source charge is an equipotential surface. If our source charge is positive, then the potential decreases as the distance from the source charge increases. Thus the equipotential surfaces associated with a positive point charge consist of a family of concentric spheres centered on the source charge. Each sphere has a different potential.

An equipotential surface can be either imaginary, such as a mathematical sphere or a real, physical surface such as the outside of a wire. The set of all equipotential surfaces fills all of space, since every point in space has some value of electric potential associated with it. We could draw an equipotential surface through any one of these points, just like we can draw a field line through every point in space. However, in order to simplify illustrations and diagrams, we typically show just a few of the surfaces.

No net work W is done on a charged particle by an electric field when the particle moves between two points i and f on the same equipotential surface. This follows from Eq. 25-9,

$$\Delta V = V_f - V_i = -\frac{W_e}{q},$$

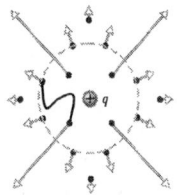

Fig. 25-6: All of the electric field vectors created by the presence of a single point charge point radially outward in three dimensions. If a test charge moves around on a sphere that is centered on the charge (dotted circle), no work is done on it by the electric field since all the electric field vectors on surface elements of the sphere are normal to the sphere. If the charge is moved from one radius to the other (black squiggly line) it has to move parallel to the field vectors some of the time, and work is done on it.

which tells us W must be zero if $V_f = V_i$. Because of the path independence of work (and thus of potential energy and potential), $W = 0$ for *any* path connecting points *i* and *f*, regardless of whether that path lies entirely on the equipotential surface. In other words, if the charge moves away from the equal potential surface during the motion, the work done (positive or negative) is exactly canceled by the work done (negative or positive) in moving back onto the surface.

Figure 25-7 shows a *family* of equipotential surfaces, associated with the electric field due to some distribution of charges. The work done by the electrostatic force on a charged particle as the particle moves from one end to the other of paths I and II is zero because each of these paths begins and ends on the same equipotential surface. The work done as the charged particle moves from one end to the other of paths III and IV is not zero but has the same value for both these paths because the initial and final potentials are identical for the two paths; that is, paths III and IV connect the same pair of equipotential surfaces.

As we already noted, the equipotential surfaces produced by a point charge or a spherically symmetrical charge distribution are a family of concentric spheres. For a uniform electric field it is not difficult to see the equipotential surfaces are a family of planes perpendicular to the field lines.

The fact that the value of the potential is constant along an equipotential surface implies that the electric field must always be perpendicular to the equipotential surfaces. Why? Because, if $\vec{E}$ were *not* perpendicular to an equipotential surface, it would have a component lying along that surface. This component would then do work on a charged particle as it moved along the surface. However, once again we use Eq. 25-9,

$$\Delta V = V_f - V_i = -\frac{W_e}{q},$$

to prove work cannot be done if the surface is truly an equipotential surface. So, the only possible conclusion is $\vec{E}$ must be everywhere perpendicular to the surface.

If the electric field lines must be perpendicular to the equipotential surface, then conversely the equipotential surfaces must be perpendicular to the field lines. Thus, equipotential surfaces are always perpendicular to $\vec{E}$, which is tangent to the field lines. Figure 25-8 shows electric field lines and cross sections of the equipotential surfaces for a uniform electric field and for the field associated with a point charge and with an electric dipole.

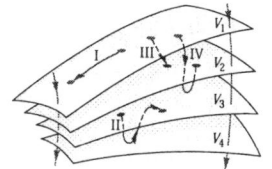

Fig. 25-7: Portions of four equipotential surfaces at electric potentials $V_1 = 100$ V, $V_2 = 80$ V, $V_3 = 60$ V, and $V_4 = 40$ V. Four paths along which a test charge may move are shown. Two electric field lines are also indicated.

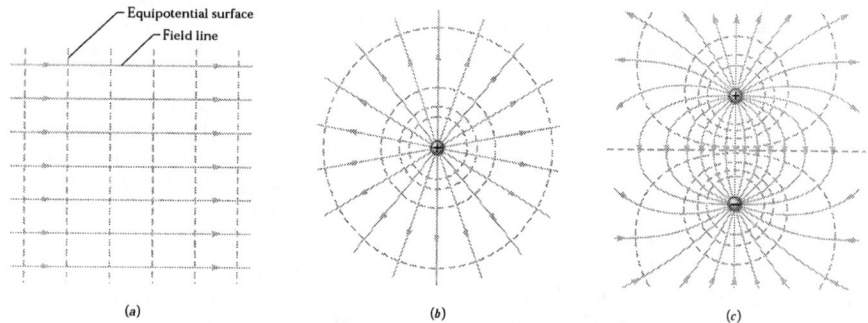

Fig. 25-8: Electric field lines (solid lines with arrows) and cross sections of equipotential surfaces (dashed lines) for (a) a uniform field, (b) the field of a point charge, and (c) the field of an electric dipole.

25-5 Calculating Potential from an E-Field

Can we calculate the potential difference between any two points *i* and *f* in an electric field if we know the electric field vector $\vec{E}$ all along any path connecting those points?

We can if we can find the work done on a charge by the field as the charge moves from i to f, and then use Eq. 25-9 again,

$$\Delta V = V_f - V_i = -\frac{W_e}{q}.$$

For example, consider an arbitrary electric field, represented by the field lines in Fig. 25-9, and a positive test charge q_t moving along the path shown from point i to point f. At any point on the path, an electrostatic force $q_t\vec{E}$ acts on the charge as it moves through an infinitesimally small differential displacement $\vec{ds}$. From Chapter 9, we know the differential work dW done on a particle by a force $\vec{F}$ during a displacement $\vec{ds}$ is

$$dW = \vec{F} \cdot \vec{ds}. \qquad (25\text{-}14)$$

For the situation of Fig. 25-9, $\vec{F} = q_t\vec{E}$, and Eq. 25-14 above becomes

$$dW = q_t\vec{E} \cdot \vec{ds}. \qquad (25\text{-}15)$$

To find the total work W done on the particle by the field as the particle moves from point i to point f, we sum—via integration—the differential works done on the charge as it moves through all the differential displacements $\vec{ds}$ along the path:

$$W = q_t \int_i^f \vec{E} \cdot \vec{ds}. \qquad (25\text{-}16)$$

If we substitute the total work W from Eq. 25-16 into Eq. 25-9, $\Delta V = V_f - V_i = -W/q$, we find

$$V_f - V_i = -\int_i^f \vec{E} \cdot \vec{ds}. \qquad (25\text{-}17)$$

Thus, the potential difference $V_f - V_i$ between any two points i and f in an electric field is equal to the negative of the *line integral* (meaning the integral along a particular path) of $\vec{E} \cdot \vec{ds}$ from i to f. However, because the electrostatic force is conservative, all paths (whether easy or difficult to use) yield the same result.

If the electric field is known throughout a certain region, Eq. 25-17 allows us to calculate the difference in potential between any two points in the field. If we choose the potential ΔV_i at point i to be zero, then Eq. 25-17 becomes

$$\Delta V = -\int_i^f \vec{E} \cdot \vec{ds}, \qquad (25\text{-}18)$$

in which we have dropped the subscript f on V_f. Equation 25-18 gives us the potential ΔV at any point f in the electric field *relative to the zero potential* at point i. If we let point i be at infinity, then Eq. 25-18 gives us the potential ΔV at any point f relative to the zero potential at infinity.

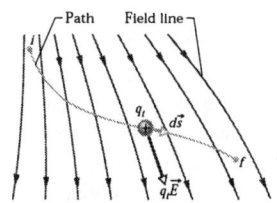

Fig. 25-9: A test charge q_t moves from point i to point f along the path shown in a nonuniform electric field represented by curved electric field lines. During a displacement $\vec{ds}$, an electrostatic force q_tE acts on the test charge. This force points in the direction of the field line at the location of the test charge.

READING EXERCISE 25-4: The figure shows a family of parallel equipotential surfaces (in cross section) and five paths along which we shall move an electron from one surface to another. (a) What is the direction of the electric field associated with the surfaces? (b) For each path, is the work we do positive, negative, or zero? (c) Rank the paths according to the work we do, greatest first.

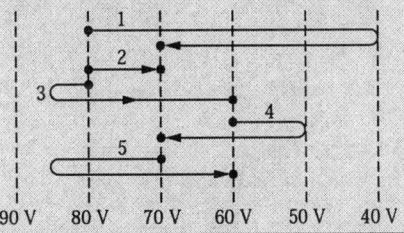

Touchstone Example 25-5-1, at the end of this chapter, illustrates how to use what you learned in this section.

TE

25-6 Potential Due to a Point Charge

Imagine a single point charge in space. What would the value of the potential be at a distance of 3 meters away from the charge? Consider a point P at a distance R from a fixed particle of positive charge q as in Fig. 25-10. To use Equation 25-17

$$V_f - V_i = -\int_i^f \vec{E} \cdot d\vec{s},$$

we imagine that we move a positive test charge q_t from some initial position to infinity. Mathematically, the simplest path between point P and infinity involves traveling in the same direction as the electric field vectors point at all times so no vector components of the electric field have to be considered. Because the path we take does not matter, let us choose a line extending radially outward from the source charge through the initial position of the test charge and then on to infinity.

We must then evaluate the dot product

$$\vec{E} \cdot d\vec{s} = |\vec{E}| \cos\theta |d\vec{s}|. \qquad (25\text{-}19)$$

The electric field $\vec{E}$ in Fig. 25-10 is directed radially outward from the fixed particle. Thus, the differential displacement $d\vec{s}$ of the test particle along our chosen path (also radially outward) has the same direction as $\vec{E}$. That means that the angle $\theta = 0$ and $\cos\theta = 1$. Because the path is radial, let us write ds as dr. Then, substituting the limits R and ∞, we can write Eq. 25-17

$$V_f - V_i = -\int_i^f \vec{E} \cdot d\vec{s},$$

as

$$V_f - V_i = -\int_R^\infty |\vec{E}| \, dr. \qquad (25\text{-}20)$$

Next, we set $V_f = 0$ (at ∞) and $V_i = \Delta V$ (at R). Then, for the magnitude of the electric field at the site of the test charge, we substitute from Chapter 23:

$$|\vec{E}| = \frac{1}{4\pi\varepsilon_0} \frac{|q|}{r^2}. \qquad (25\text{-}21)$$

With these changes, Eq. 25-20 then gives us

$$0 - \Delta V = -\frac{|q|}{4\pi\varepsilon_0} \int_R^\infty \frac{1}{r^2} dr = \frac{|q|}{4\pi\varepsilon_0} \left[\frac{1}{r}\right]_R^\infty \qquad (25\text{-}22)$$

$$= -\frac{1}{4\pi\varepsilon_0} \frac{|q|}{R}.$$

Solving for ΔV and switching R to r, we then have

$$\Delta V = \frac{1}{4\pi\varepsilon_0} \frac{|q|}{r}, \qquad (25\text{-}23)$$

as the electric potential ΔV due to a particle of charge q, at any radial distance r from the particle.

Although we have derived this expression for a positively charged particle, the derivation holds also for a negatively charged particle, in which case, q is a negative quantity. Note that the sign of ΔV is the same as the sign of q:

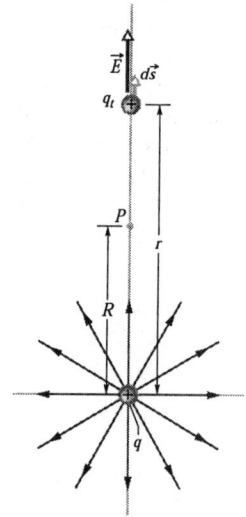

Fig. 25-10: The positive point charge q produces an electric field $\vec{E}$ and an electric potential ΔV at point P. We find the potential by moving a test charge q_t from its initial location at point P to infinity. The test charge is shown at distance r from the point charge, undergoing differential displacement $d\vec{s}$.

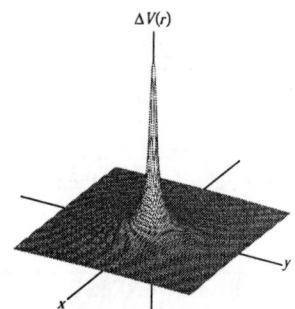

Fig. 25-11: A computer-generated plot of the electric potential $\Delta V(r)$ due to a positive point charge located at the origin of an xy plane. The potentials at points in that plane are plotted vertically. (Curved lines have been added to help you visualize the plot.) The infinite value of ΔV predicted by Eq. 25-23 for $r = 0$ is not plotted.

►A positively charged particle produces a positive electric potential. A negatively charged particle produces a negative electric potential.

Figure 25-11 shows a computer-generated plot of Eq. 25-23 for a positively charged particle; the magnitude of ΔV is plotted vertically. Note the magnitude increases as $r \to 0$. In fact, according to the expression above, ΔV is infinite at $r = 0$, although Fig. 25-11 shows a finite, smoothed-off value there.

Equation 25-23 also gives the electric potential *outside or on the external surface of* a spherically symmetric charge distribution. We can prove this by using an electrostatic analogy to the shell theorem we found so useful in our study of gravitation (Section 14-2). This theorem allows us to replace the actual spherical charge distribution with an equal charge concentrated at its center. Then the derivation leading to Eq. 25-23 follows, provided we do not consider a point within the actual distribution.

25-7 Potential Due to a Group of Point Charges

We found in Chapter 23 the electric field arising from a group of point charges satisfies a superposition principle. That is, the total electric field is the sum of the individual electric fields arising from each individual point charge. Since the potential V is the line integral of the electric field and the integral of a sum of terms is the sum of the integrals, we also get a superposition principle for electrostatic potential.

Hence, we can find the net potential at a point due to a group of point charges with the help of superposition. We separately calculate the potential resulting from each charge at the given point separately, using Eq. 25-23 with the sign of the charge included. Then we sum the potentials. For n charges, the net potential is

$$\Delta V = \sum_{i=1}^{n} V_i = \frac{1}{4\pi\varepsilon_0} \sum_{i=1}^{n} \frac{q_i}{r_i} \qquad (n \text{ point charges}). \qquad (25\text{-}24)$$

Here q_i is the value of the ith charge, and r_i is the radial distance of the given point from the ith charge. The sum in Equation 25-24 is an *algebraic sum*, not a vector sum like the sum used to calculate the electric field resulting from a group of point charges. Herein lies an important computational advantage of potential over electric field: it is a lot easier to sum several scalar quantities than to sum several vector quantities whose directions and components must be considered.

READING EXERCISE 25-5: The figure shows three arrangements of two protons. Rank the arrangements according to the net electric potential produced at point P by the protons, greatest first.

Touchstone Examples 25-7-1 and 25-7-2, at the end of this chapter, illustrate how to use what you learned in this section.

TE

25-8 Potential Due to an Electric Dipole

Electrically neutral matter is made of equal amounts of positive and negative charges. Electric forces pull in opposite directions on those charges. Thus, an electric field can cause a small separation of the positive and negative charges in matter (called polarization). In addition, many molecules distribute their electrons throughout their volume in a non-uniform way. This results in their having more positive charge on one end and one negative charge on the other end.

A small separation produces an electric field very similar to that of a pair of equal and opposite charges separated by a small distance. If the charges were right on top of each other, their electric fields would cancel and they would appear neutral. But if they are a bit separated, their fields don't cancel perfectly, leaving a field pattern known as an electric dipole. The electric dipole fields produced by molecules play an essential role in a large number of processes in chemistry and biology, as well as in determining the electrical properties of matter such as color and transparency.

Now let us apply Eq. 25-24,

$$\Delta V = \sum_{i=1}^{n} V_i = \frac{1}{4\pi\varepsilon_0} \sum_{i=1}^{n} \frac{q_i}{r_i},$$

to an electric dipole to find the potential at an arbitrary point P in Fig. 25-12a. At P, the positive point charge (at distance $r_{(+)}$) sets up potential $V_{(+)}$ and the negative point charge (at distance $r_{(-)}$) sets up potential $V_{(-)}$. Then the net potential at P is given by Eq. 25-24 as

$$\Delta V = \sum_{i=1}^{2} V_i = V_{(+)} + V_{(-)} = \frac{1}{4\pi\varepsilon_0} \left(\frac{q}{r_{(+)}} + \frac{-q}{r_{(-)}} \right)$$

$$= \frac{q}{4\pi\varepsilon_0} \frac{r_{(-)} - r_{(+)}}{r_{(-)} r_{(+)}}. \qquad (25\text{-}25)$$

Naturally occurring dipoles—such as those possessed by many molecules—are quite small, so we are usually interested only in points that are relatively far from the dipole, such that $r \gg d$, where d is the distance between the charges. Under those conditions, the approximations that follow from Fig. 25-12b are

$$r_{(-)} - r_{(+)} \approx d \cos\theta \quad \text{and} \quad r_{(-)} r_{(+)} \approx r^2.$$

If we substitute these quantities into Eq. 25-25, we can approximate ΔV to be

$$\Delta V = \frac{q}{4\pi\varepsilon_0} \frac{d \cos\theta}{r^2},$$

where θ is measured from the dipole axis as shown in Fig. 25-12a. We can now write ΔV as

$$V = \frac{1}{4\pi\varepsilon_0} \frac{p \cos\theta}{r^2} \qquad \text{(electric dipole)}, \qquad (25\text{-}26)$$

in which $p(= qd)$ is the magnitude of the electric dipole moment $\vec{p}$ defined in Section 23-7. The vector $\vec{p}$ is directed along the dipole axis, from the negative to the positive charge. (Thus, θ is measured from the direction of $\vec{p}$.)

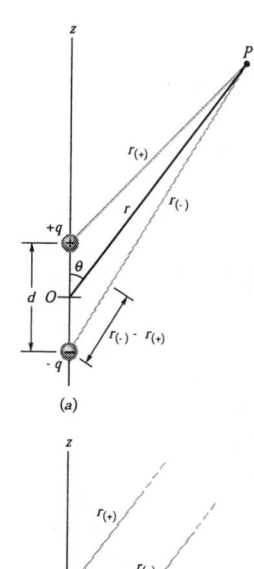
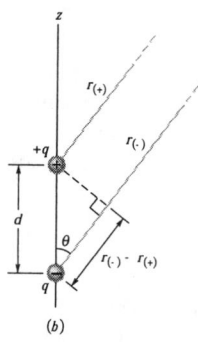

Fig. 25-12: (a) Point P is a distance r from the midpoint O of a dipole. The line OP makes an angle θ with the dipole axis. (b) If P is far from the dipole, the lines of lengths $r_{(+)}$ and $r_{(-)}$ are approximately parallel to the line of length r, and the dashed black line is approximately perpendicular to the line of length $r_{(-)}$.

READING EXERCISE 25-6: Suppose three points are set at equal (large) distances r from the center of the dipole in Fig. 25-12: Point a is on the dipole axis above the positive charge, point b is on the axis below the negative charge, and point c is on a perpendicular bisector through the line connecting the two charges. Rank the points according to the electric potential of the dipole there, greatest (most positive) first.

Induced Dipole Moment

Many molecules such as water have *permanent* electric dipole moments. In other molecules (called non-polar molecules) and in every isolated atom, the centers of the positive and negative charges coincide (Fig. 25-13a) and thus no dipole moment is set up. However, if we place an atom or a non-polar molecule in an external electric field, the field distorts the electron orbits and separates the centers of positive and negative charge

(Fig. 25-13b). Because the electrons are negatively charged, they tend to be shifted in a direction opposite the field. This shift sets up a dipole moment $\vec{p}$ pointing in the direction of the field. This dipole moment is said to be induced by the field, and the atom or molecule is then said to be polarized by the field (it has a positive side and a negative side). When the field is removed, the induced dipole moment and the polarization disappear.

25-9 Potential Due to a Continuous Charge Distribution

When a charge distribution q is continuous (as on a uniformly charged thin rod or disk), we cannot use a summation to find the potential ΔV at a point P. Instead, we must choose a differential element of charge dq. A differential element of charge is a very small bit of charge, small enough so we can treat it as if it were a point charge. We can then determine the potential dV at P due to dq, and then integrate over the entire charge distribution.

Let us again take the zero of potential to be at infinity. If we treat the element of charge dq as a point charge, then we can use Eq. 25-23

$$\Delta V = \frac{1}{4\pi\varepsilon_0}\frac{q}{r},$$

to express the potential dV at point P due to dq:

$$dV = \frac{1}{4\pi\varepsilon_0}\frac{dq}{r} \qquad \text{(positive or negative } dq\text{).} \qquad (25\text{-}27)$$

Here r is the distance between P and dq. To find the total potential ΔV at P, we integrate to sum the potentials due to all the charge elements:

$$\Delta V = \int dV = \frac{1}{4\pi\varepsilon_0}\int\frac{dq}{r}. \qquad (25\text{-}28)$$

The integral must be taken over the entire charge distribution. Note because the electric potential is a scalar, there are *no vector components* to consider in the equation above.

We now examine a continuous charge distribution, a line of charge.

Line of Charge

In Figure 25-14a, a thin nonconducting rod of length L has a positive charge of uniform linear density λ. Let us determine the electric potential ΔV due to the rod at point P, a perpendicular distance d from the left end of the rod.

We consider a differential element dx of the rod as shown in Fig. 25-14b. This (or any other) element of the rod has a differential charge of

$$dq = \lambda\,dx. \qquad (25\text{-}29)$$

This element produces a potential dV at point P, which is a distance $r = (x^2 + d^2)^{1/2}$ from the element. Treating the element as a point charge, we can use Eq. 25-27,

$$dV = \frac{1}{4\pi\varepsilon_0}\frac{dq}{r},$$

to write the potential dV as

$$dV = \frac{1}{4\pi\varepsilon_0}\frac{dq}{r} = \frac{1}{4\pi\varepsilon_0}\frac{\lambda\,dx}{(x^2+d^2)^{1/2}}. \qquad (25\text{-}30)$$

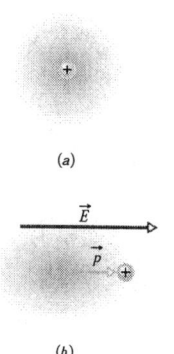

(a)

(b)

Fig. 25-13: (a) An atom, showing the positively charged nucleus and the negatively charged electrons (shading). The centers of positive and negative charge coincide. (b) If the atom is placed in an external electric field $\vec{E}$, the electron orbits are distorted so that the centers of positive and negative charge no longer coincide. An induced dipole moment $\vec{p}$ appears. The distortion is greatly exaggerated here.

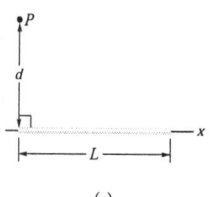

(a)

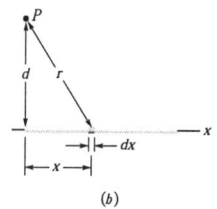

(b)

Fig. 25-14: (a) A thin, uniformly charged rod produces an electric potential ΔV at point P. (b) An element of charge produces a differential potential dV at P.

Since the charge on the rod is positive and we have taken $\Delta V = 0$ at infinity, we know dV in this expression must be positive.

We now find the total potential ΔV produced by the rod at point P by integrating along the length of the rod, from $x = 0$ to $x = L$. We evaluate the integral using an integral table or a symbolic manipulation program like Mathcad or Maple. We then find

$$\Delta V = \int dV = \int_0^L \frac{1}{4\pi\varepsilon_0} \frac{\lambda}{(x^2 + d^2)^{1/2}} dx.$$

$$= \frac{\lambda}{4\pi\varepsilon_0} \int_0^L \frac{dx}{(x^2 + d^2)^{1/2}}$$

$$= \frac{\lambda}{4\pi\varepsilon_0} \left[\ln\left(x + (x^2 + d^2)^{1/2}\right) \right]_0^L$$

$$= \frac{\lambda}{4\pi\varepsilon_0} \left[\ln\left(L + (L^2 + d^2)^{1/2}\right) - \ln d \right].$$

We can simplify this result by using the general relation $\ln A - \ln B = \ln(A/B)$. We then find

$$\Delta V = \frac{\lambda}{4\pi\varepsilon_0} \ln\left[\frac{L + (L^2 + d^2)^{1/2}}{d} \right]. \tag{25-31}$$

Because ΔV is the sum of positive values of dV, it should be positive—but does this expression give a positive ΔV? Since the argument of the logarithm is greater than one, the logarithm is a positive number and ΔV is indeed positive.

25-10 Calculating the Field from the Potential

In Section 25-5, you saw how to find the potential at a point f if you know the electric field along a path from a reference point to point f. In this section, we propose to go the other way—that is, to find the electric field when we know the potential. As Figure 25-8 shows, graphically finding the direction of the field is easy: If we know the potential ΔV at all points near an assembly of charges, we can draw in a family of equipotential surfaces. The electric field lines, sketched perpendicular to those surfaces, reveal the direction of $\vec{E}$. What we are seeking here is the mathematical equivalent of this graphical procedure.

Figure 25-15 shows cross sections of a family of closely spaced equipotential surfaces, the potential difference between each pair of adjacent surfaces being dV. As the figure suggests, the field $\vec{E}$ at any point P is perpendicular to the equipotential surface through P.

Suppose a positive test charge q_0 moves through a displacement $d\vec{s}$ from one equipotential surface to the adjacent surface. From Equation 25-9, we can relate the change in electric potential to the work done by the electric field on our test charge

$$\Delta V = V_f - V_i = -\frac{W_e}{q}.$$

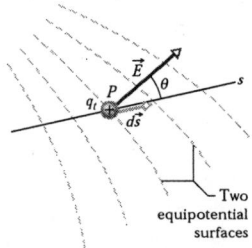

Fig. 25-15: A test charge q_0 moves a distance $d\vec{s}$ from one equipotential surface to another. (The separation between the surfaces has been exaggerated for clarity.) The displacement $d\vec{s}$ makes an angle θ with the direction of the electric field $\vec{E}$.

Let's consider the potential difference associated with an infinitesimally small displacement denoted by $d\vec{s}$. We see that the electric field does an infinitesimal amount of work on the test charge during the move. Using Equation 25-9, we can denote this as $-q_0 dV$. From Equation 25-15, $dW = q_0 \vec{E} \cdot d\vec{s}$, and Fig. 25-15, we see that the infinitesimal work done by the force may also be written as $(q_0 \vec{E}) \cdot d\vec{s}$, or $q_0 |\vec{E}|(\cos\theta)|d\vec{s}|$ where θ is the angle between the electric field and displacement vectors as shown in Fig. 25-15. Equating these two expressions for the work yields

$$-q_0 dV = q_0 |\vec{E}|(\cos\theta)|d\vec{s}|, \tag{25-32}$$

or
$$|\vec{E}|\cos\theta = -\frac{dV}{ds}. \tag{25-33}$$

Since $E = |\vec{E}|\cos\theta$ is the component of $\vec{E}$ in the direction of $d\vec{s}$, the equation above becomes

$$E_s = -\frac{\partial V}{\partial s}. \tag{25-34}$$

We have added a subscript to the component E and switched to the partial derivative symbols to emphasize that this expression involves only the variation of ΔV along a specified axis (here called the s axis) and only the component of $\vec{E}$ along that axis. In words, Equation 25-34 is essentially the inverse of Eq. 25-17,

$$V_f - V_i = -\int_i^f \vec{E} \cdot d\vec{s},$$

and states:

➤ The component of $\vec{E}$ in any direction is the negative of the rate of change of the electric potential with distance in that direction.

If we take the s axis to be, in turn, the x, y, and z axes, we find that the x, y, and z components of $\vec{E}$ at any point are

$$E_x = -\frac{\partial V}{\partial x}; \quad E_y = -\frac{\partial V}{\partial y}; \quad E_z = -\frac{\partial V}{\partial z}. \tag{25-35}$$

Thus, if we know ΔV for all points in the region around a charge distribution—that is, if we know the function $\Delta V(x, y, z)$—we can find the components of $\vec{E}$, and thus $\vec{E}$ itself, at any point by taking partial derivatives.

For the simple situation in which the electric field $\vec{E}$ is uniform, the equipotential surfaces are a set of parallel planes that lie perpendicularly to the direction of the electric field. In addition, for a given potential difference, the distance between any two equipotential planes is the same. So, when the electric field is uniform, we can rewrite Eq. 25-34,

$$E_s = -\frac{\partial V}{\partial s},$$

as
$$\vec{E} = -\frac{\Delta V}{\Delta s}, \tag{25-36}$$

where s is perpendicular to the equipotential surfaces. The component of the electric field is zero in any direction parallel to the equipotential surfaces. Thus, for a given potential difference ΔV, the magnitude of the electric field is given by the potential difference divided by the distance between any two equipotential surfaces.

READING EXERCISE 25-7: The figure shows three pairs of parallel plates with the same separation, and the electric potential of each plate. The electric field between the plates is uniform and perpendicular to the plates. (a) Rank the pairs according to the magnitude of the electric field between the plates, greatest first. (b) For which pair is the electric field pointing rightward? (c) If an electron is released midway between the third pair of plates, does it remain there, move rightward at constant speed, move leftward at constant speed, accelerate rightward, or accelerate leftward?

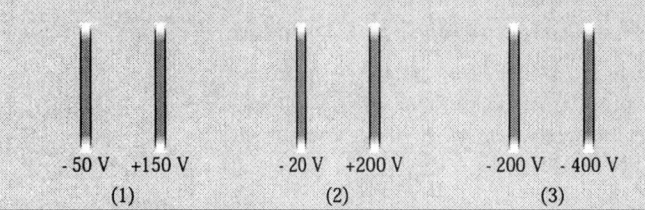

| | - 50 V +150 V | - 20 V +200 V | - 200 V - 400 V |
| | (1) | (2) | (3) |

READING EXERCISE 25-8: In what ways is the superposition principle for energy discussed above the same as, and different from, the superposition principle for electric field?

Touchstone Example 25-10-1, at the end of this chapter, illustrates how to use what you learned in this section.

TE

25-11 Potential of a Charged Isolated Conductor

In Section 24-8, we concluded $\vec{E} = 0$ for all points inside an electrically isolated conductor. We then used Gauss' law to prove that an excess charge placed on an isolated conductor lies entirely on its surface. (This is true even if the conductor has an empty internal cavity.) Here we use the first of these facts to prove an extension of the second:

▶An excess charge placed on an isolated conductor will distribute itself on the surface of that conductor so that all points of the conductor—whether on the surface or inside—come to the same potential. This is true even if the conductor has an internal cavity and even if that cavity contains a net charge.

This fact is rather obvious since any potential difference inside a conductor requires an electric field inside it. The non-zero electric field would, in turn, cause the free conduction electrons to redistribute themselves until the potential difference disappears.

The mathematical proof an electrically isolated conductor is an equipotential region follows directly from Eq. 25-17,

$$V_f - V_i = -\int_i^f \vec{E} \cdot d\vec{s}.$$

Since $\vec{E} = 0$ for all points within a conductor, it follows directly that $V_f = V_i$ for all possible pairs of points i and f in the conductor.

A Spherical Shell with No External Electric Field

Figure 25-16a shows a plot of potential against radial distance r from the center for an isolated spherical conducting shell of 1.0 m radius, having a net excess charge of 1.0 μC. In the absence of an external field, we know by symmetry the surface charges will be uniformly distributed over the surface of the shell. For points outside the shell, we can calculate $\Delta V(r)$, the electric potential. Obviously this potential also has a spherical symmetry and can be given by Eq. 25-23,

$$\Delta V = \frac{1}{4\pi\varepsilon_0} \frac{q}{r},$$

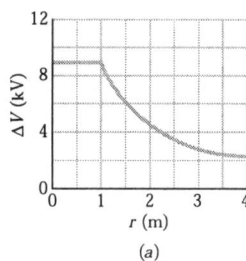

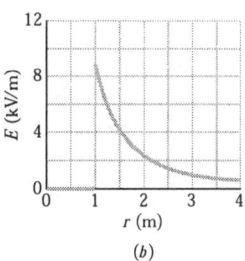

Fig. 25-16: (*a*) A plot of $\Delta V(r)$ both inside and outside a charged spherical shell of radius 1.0 m. (*b*) A plot of the electric field magnitude, $E(r)$, for the same shell.

because the total charge on the shell, denoted as q, behaves for external points as if it were concentrated at the center of the shell. That equation holds right up to the surface of the shell. Now let us push a small test charge through the shell—assuming a small hole exists—to its center. No extra work is needed to do this because no net electric force acts on the test charge once it is inside the shell. Thus, the potential at all points inside the shell has the same value as on the surface, as Fig. 25-16a shows.

Figure 25-16b shows the variation of electric field with radial distance for the same shell. Note $\vec{E} = 0$ everywhere inside the shell. The curves of Fig. 25-16b can be derived from the curve of Fig. 25-16a by differentiating with respect to r, using Eq. 25-34 (the derivative of a constant, recall, is zero). The curve of Fig. 25-16a can be derived from the curves of Fig. 25-16b by integrating

$$E_s = -\frac{\partial V}{\partial s}.$$

The Charge Distribution on a Non Spherical Conductor

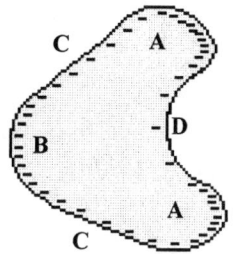

Consider a non-spherical charged conductor. Assume the conductor is electrically isolated and there is no external electric field in its vicinity. It turns out its surface charges do not distribute themselves uniformly. When compared to the uniform density of excess charge on a spherical conductor, the charges redistribute themselves so there is a higher charge density when the radius of curvature is convex and small and a lower charge density where the radius of curvature is concave and small (Fig. 25-17). Why? We can use the characteristics of equipotential surfaces to develop a qualitative explanation for this phenomenon.

The explanation is as follows: there is no electric field inside the conductor and the electric field at each point on the surface of the conductor must be normal (in other words perpendicular) to the surface. This requirement is obvious since any component of electric field parallel to the surface would cause free electrons to reconfigure themselves until all tangential components along the surface disappear. This also means the entire surface of our conductor is an equipotential surface no matter what its shape is. However, if we are far away from our charged conductor, the equipotential surfaces look more and more like those of a point charge. Thus, the family of equipotential surfaces that are each some chosen ΔV apart from the previous one become more and more spherical in shape. As the successive equipotential surfaces morph (change shape) slowly from that of our odd shaped surface to that of a sphere, it is obvious the parts of the equipotential surfaces near small radius convex surface elements must be closer together than those elements having large radii of curvature. This is shown in Fig. 25-18. Now, equipotential surfaces more closely spaced occur where the electric field is the strongest and can do the most work on test charges. But the electric field is largest where the charge density that is its source is largest.

Fig. 25-17: The charge density on a conductor is greatest on a convex surface with a small radius of curvature (A) and least on a concave surface having small radius of curvature (D). The ranking of charge density is $A > B \gg C > D$.

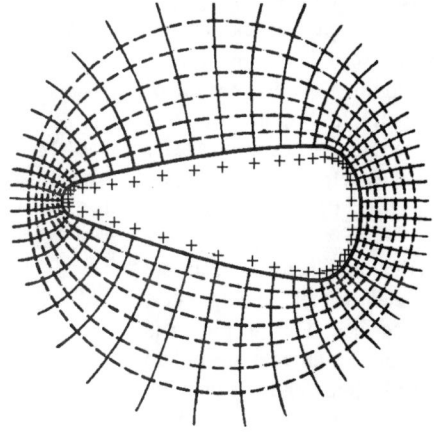

Fig. 25-18: The net positive charge on an odd-shaped isolated conductor distributes itself on the conductor's surface so the electric field generated by it is zero inside and normal to the surface elements of the conductor. This requires the equipotential surfaces (shown with dotted lines) to be closest together on the left where the conductor's convex radius of curvature is smallest. The electric field lines (shown with solid lines) and the excess charges also have the greatest density on the left where the curvature of the conductor's surface is smallest.

An Isolated Conductor in an External Electric Field

Suppose an *uncharged* isolated conductor is placed in an *external electric field*, as in Fig. 25-18. The electric field at the conductor's surface must have the same characteristics as it does when no external field is present. However, this doesn't mean its charges will be distributed in the same way as if no external electric field were present. All points of the conductor still come to a single potential regardless of whether the conductor is electrically neutral or has an excess charge. The free conduction electrons distribute themselves on the surface in such a way the electric field they produce at interior points cancels the external electric field that would otherwise be there. Furthermore, the electron distribution causes the net electric field at all points on the surface to be normal to the surface. If the conductor in Fig. 25-18 could somehow be removed, leaving the surface charges frozen in place, the pattern of the electric field would remain absolutely unchanged, for both exterior and interior points.

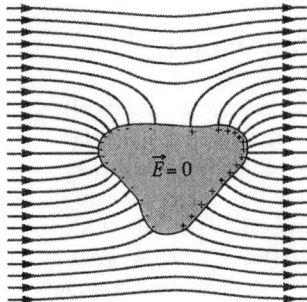

Fig. 25-19: An uncharged conductor is suspended in an external electric field. The free electrons in the conductor distribute themselves on the surface as shown, so as to reduce the net electric field inside the conductor to zero and make the net field normal to each surface element.

One common natural source of an external electric field that can affect isolated metal objects are excess negative charges at the bases of clouds contributing to the onset of thunderstorms. Such an external electric field can cause charge separation in metal objects such as golf clubs and rock hammers. Since these objects have points where the curvature is high, the surface charge density—and thus the external electric field, which is proportional to it—may reach very high values. The air around sharp points may become ionized, producing the corona discharge that golfers and mountaineers see on their tools when thunderstorms threaten. Such corona discharges, like hair that stands on end, are often the precursors of lightning strikes.

Fig: 25-20: This enhancement of the chapter's opening photograph shows the result of an overhead cloud system creating a strong electric field $\vec{E}$ near a woman's head. Many of the hair strands extended along the field, which was perpendicular to the equipotential surfaces and greatest where those surfaces were closest, near the top of her head.

   The cells and blood inside a human body contain salt water that acts as a conductor. The natural oil found on hair is also conductive. A person placed in a strong electric field can act like an uncharged conductor as shown in Fig. 25-20. We can now explain what happened to the woman in the opening photograph for this chapter. Because she was standing on a platform connected to the mountainside, she was at about the same potential as the mountainside. Overhead, a cloud system that had a high degree of charge separation with excess negative charges at its base moved in and created a strong electric field around her and the mountainside. Electrostatic forces due to this field drove some of the conduction electrons in the woman downward through her body, leaving her head and strands of her hair positively charged. The magnitude of this electric field was apparently large, but less than the value of about 3×10^6 V/m needed to cause electrical breakdown of the air molecules. (That value was exceeded when lightning struck the platform shortly after the picture was taken.)

   As we just discussed, the surface charges on a non-spherical conductor concentrate in regions where the curvature is greatest. Thus, we expect the electric field to be greatest near the top of the woman's head which is an equipotential surface. This suspicion is confirmed because the strands of her hair, containing excess positive charge, are pulled out most strongly where her head has the most curvature. And, the strands of hair are extended along the direction of $\vec{E}$ perpendicular to her head. Since the magnitude of $\vec{E}$

was greatest just above her head, this is where the equipotential surfaces were most closely spaced. A sketch showing this close spacing is shown in Fig. 25-20.

The lesson here is simple. If an electric field causes the hairs on your head to stand up, you'd better run for shelter rather than pose for a snapshot.

What if Lightning Might Strike?

Speaking of lightning, what is the best way to protect yourself in case of lightning? There are two ways to protect yourself, both based on a knowledge of how conductors behave in electric fields. First, it is wise to enclose oneself in a cavity inside a relatively spherical conducting shell, where the electric field is guaranteed to be zero. A car (unless it is a convertible) is almost ideal (Fig. 25-21) as it protects the passengers from the effects of lightning for the same reason that the Faraday Cage shown in Chapter 24 protects the young girl from the high voltage caused by the transfer of charge to it by a Tesla coil. Second, if you live in an area where thunderstorms are common, you can embed the base of a tall metal lightning rod in the ground. Recall that the bottoms of thunderclouds have an excess of negative charge which can create strong electric fields at the Earth's surface. The free electrons in the rod will move toward the ground leaving a large accumulation of positive metal ions at the sharp point at the top of the rod. This will attract electrons from the atmosphere to the rod and down to the ground in a corona discharge process that can serve to prevent a major discharge or lightning strike in your vicinity from happening all at once.

Fig. 25-21: A large spark jumps to a car's body and then exits by moving across the insulating left front tire (note the flash there), leaving the person inside unharmed because the electric potential difference remains zero inside the car.

READING EXERCISE 25-9: The figure below shows the region in the neighborhood of a negatively-charged conducting sphere and a large positively-charged conducting plate extending far beyond the region shown. Someone claims lines A through F are possible field lines describing the electric field lying in the region between the two conductors. (a) Examine each of the lines and indicate whether it is a correctly drawn field line. If a line is not correct, explain why. (b) Redraw the diagram with a pattern of field lines which is more nearly correct.

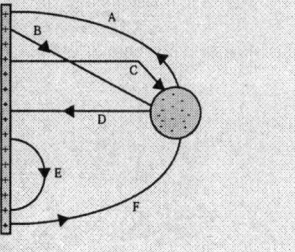

READING EXERCISE 25-10: Why are the equipotential surfaces shown in Fig. 25-20 closer together just above the woman's head than they are at the side of her head?

Touchstone Example 25-2-1

Electrons are continually being knocked out of air molecules in the atmosphere by cosmic-ray particles coming in from space. Once released, each electron experiences an electrostatic force $\vec{F}$ due to the electric field $\vec{E}$ that is produced in the atmosphere by charged particles already on Earth. Near Earth's surface the electric field has the magnitude $E = 150$ N/C and is directed downward. What is the change ΔU in the electric potential energy of a released electron when the electrostatic force causes it to move vertically upward through a distance $d = 520$ m (Fig. TE25-1)?

Fig. TE25-1 An electron in the atmosphere is moved upward through displacement $\vec{d}$ by an electrostatic force $\vec{F}$ due to an electric field $\vec{E}$.

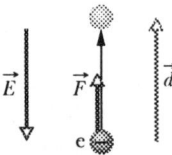

SOLUTION: We need three **Key Ideas** here. One is that the change ΔU in the electric potential energy of the electron is related to the work W done on the electron by the electric field. Equation 25-1 ($\Delta U = -W$) gives the relation. A second **Key Idea** is that the work done by a constant force $\vec{F}$ on a particle undergoing a displacement $\vec{d}$ is

$$W = \vec{F} \cdot \vec{d}. \tag{TE25-1}$$

Finally, the third **Key Idea** is that the electrostatic force and the electric field are related by $\vec{F} = q\vec{E}$, where here q is the charge of an electron ($= -1.6 \times 10^{-19}$ C). Substituting for $\vec{F}$ in Eq. TE25-1 and taking the dot produce yield

$$W = q\vec{E} \cdot \vec{d} = q|\vec{E}||\vec{d}| \cos \theta, \tag{TE25-2}$$

where θ is the angle between the directions of $\vec{E}$ and $\vec{d}$. The field $\vec{E}$ is directed downward and the displacement $\vec{d}$ is directed upward, so $\theta = 180°$. Substituting this and other data into Eq. TE25-2, we find

$$W = (-1.6 \times 10^{-19}\,\text{C})(150\,\text{N/C})(520\,\text{m}) \cos 180°$$
$$= 1.2 \times 10^{-14}\,\text{J}.$$

Equation 25-1 then yields

$$\Delta U = -W = -1.2 \times 10^{-14}\,\text{J}. \tag{Answer}$$

This result tells us that during the 520 m ascent, the electric potential energy of the electron *decreases* by 1.2×10^{-14} J.

Touchstone Example 25-5-1

(a) Figure TE25-2a shows two points i and f in a uniform electric field $\vec{E}$. The points lie on the same electric field line (not shown) and are separated by a distance d. Find the potential difference $V_f - V_i$ by moving a positive test charge q_0 from i to f along the path shown, which is parallel to the field direction.

SOLUTION: The **Key Idea** here is that we can find the potential difference between any two points in an electric field by integrating $\vec{E} \cdot d\vec{s}$ along a path connecting those two points according to Eq. 25-17. We do this by mentally moving a test charge q_0 along that path, from initial point i to final point f. As we move such a test charge along the path in Fig. TE25-2a, its differential displacement $d\vec{s}$ always has the same direction as $\vec{E}$. Thus, the angle θ between $\vec{E}$ and $d\vec{s}$ is zero and the dot product in Eq. 25-17 is

$$\vec{E} \cdot d\vec{s} = |\vec{E}||d\vec{s}| \cos \theta = |\vec{E}||d\vec{s}|. \tag{TE25-3}$$

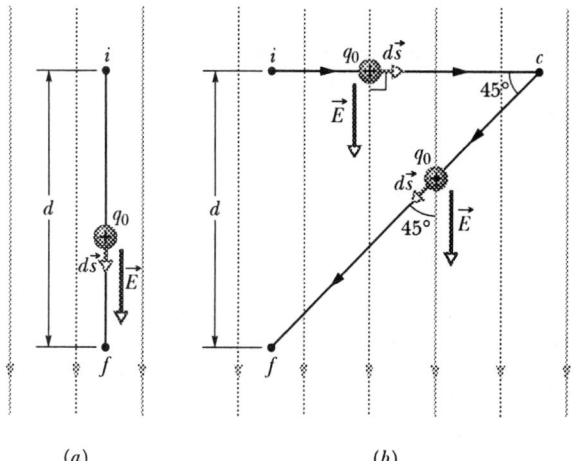

(a)   (b)

Fig. TE25-2 (a) A test charge q_0 moves in a straight line from point i to point f, along the direction of a uniform electric field. (b) Charge q_0 moves along path icf in the same electric field.

Equations 25-17 and TE25-3 then give us

$$V_f - V_i = -\int_i^f \vec{E} \cdot d\vec{s} = -\int_i^f |\vec{E}| |d\vec{s}|. \qquad \text{(TE25-4)}$$

Since the field is uniform, E is constant over the path and can be moved outside the integral, giving us

$$V_f - V_i = -|\vec{E}| \int_i^f |d\vec{s}| = -|\vec{E}| |\vec{d}|, \qquad \text{(Answer)}$$

in which the integral is simply the length d of the path. The minus sign in the result shows that the potential at point f in Fig. TE25-2a is lower than the potential at point i. This is a general result: The potential always decreases along a path that extends in the direction of the electric field lines.

(b) Now find the potential difference $V_f - V_i$ by moving the positive test charge q_0 from i to f along the path icf shown in Fig. TE25-2b.

SOLUTION: The **Key Idea** of (a) applies here too, except now we move the test charge along a path that consists of two lines: ic and cf. At all points along line ic, the displacement $d\vec{s}$ of the test charge is perpendicular to $\vec{E}$. Thus, the angle ϕ between $\vec{E}$ and $d\vec{s}$ is 90°, and the dot product $\vec{E} \cdot d\vec{s}$ is 0. Equation 25-17 then tells us that points i and c are at the same potential: $\Delta V_c - \Delta V_i = 0$.

For line cf we have $\phi = 45°$ and, from Eq. 25-17,

$$V_f - V_i = -\int_c^f \vec{E} \cdot d\vec{s} = -\int_c^f |\vec{E}| (\cos 45°) |d\vec{s}|$$

$$= -|\vec{E}| (\cos 45°) \int_c^f |d\vec{s}|.$$

The integral in this equation is just the length of line cf; from Fig. TE25-2b, that length is $d/\sin 45°$. Thus,

$$V_f - V_i = -|\vec{E}| (\cos 45°) \frac{|\vec{d}|}{\sin 45°} = -|\vec{E}| |\vec{d}|. \qquad \text{(Answer)}$$

25-21

This is the same result we obtained in (a), as it must be; the potential difference between two points does not depend on the path connecting them. Moral: When you want to find the potential difference between two points by moving a test charge between them, you can save time and work by choosing a path that simplifies the use of Eq. 25-17.

Touchstone Example 25-7-1

What is the electric potential at point P, located at the center of the square of point charges shown in Fig. TE25-3a? The distance d is 1.3 m, and the charges are

$$q_1 = +12 \text{ nC}, \qquad q_3 = -31 \text{ nC},$$
$$q_2 = -24 \text{ nC}, \qquad q_4 = +17 \text{ nC}.$$

SOLUTION: The **Key Idea** here is that the electric potential ΔV at P is the algebraic sum of the electric potentials contributed by the four point charges. (Because electric potential is a scalar, the orientations of the point charges do not matter.) Thus, from Eq. 25-24, we have

$$\Delta V = \sum_{i=1}^{4} \Delta V_i = \frac{1}{4\pi\varepsilon_0} \left(\frac{q_1}{r} + \frac{q_2}{r} + \frac{q_3}{r} + \frac{q_4}{r} \right).$$

The distance r is $d/\sqrt{2}$, which is 0.919 m, and the sum of the charges is

$$q_1 + q_2 + q_3 + q_4 = (12 - 24 + 31 + 17) \times 10^{-9} \text{ C}$$
$$= 36 \times 10^{-9} \text{ C}.$$

Thus, $$\Delta V = \frac{(8.99 \times 10^9 \text{ N} \cdot \text{m}^2/\text{C}^2)(36 \times 10^{-9} \text{ C})}{0.919 \text{ m}}$$

$$\approx 350 \text{ V}. \qquad \text{(Answer)}$$

Close to any of the three positive charges in Fig. TE25-3a, the potential has very large positive values. Close to the single negative charge, the potential has very large negative values. Therefore, there must be points within the square that have the same intermediate potential as that at point P. The curve in Fig. TE25-3b shows the intersection of the plane of the figure with the equipotential surface that contains point P. Any point along that curve has the same potential as point P.

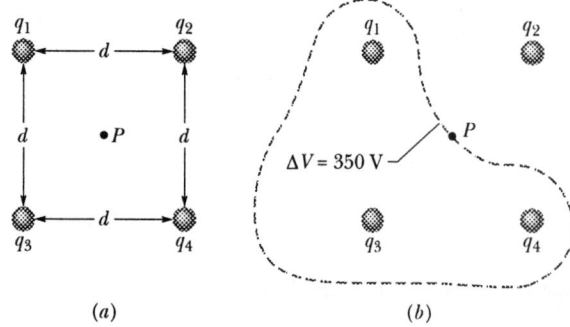

(a)   (b)

Fig. TE25-3 (a) Four point charges are held fixed at the corners of a square. (b) The closed curve is a cross section, in the plane of the figure, of the equipotential surface that contains point P. (The curve is only roughly drawn.)

Touchstone Example 25-7-2

(a) In Fig. TE25-4a, 12 electrons (of charge $-e$) are equally spaced and fixed around a circle of radius R. Relative to $\Delta V = 0$ at infinity, what are the electric potential and electric field at the center C of the circle due to these electrons?

SOLUTION: The **Key Idea** here is that the electric potential ΔV at C is the algebraic sum of the electric potentials contributed by all the electrons. (Because electric potential is a scalar, the orientations of the electrons do not matter.) Because the electrons all have the same negative charge $-e$ and are all the same distance R from C, Eq. 25-24 gives us

$$\Delta V = -12 \frac{1}{4\pi\varepsilon_0} \frac{e}{R}. \qquad \text{(Answer)} \quad \text{(TE25-4)}$$

For the electric field at C, the **Key Idea** is that electric field is a vector quantity and thus the orientation of the electrons *is* important. Because of the symmetry of the arrangement in Fig. TE25-4a, the electric field vector at C due to any given electron is canceled by the field vector due to the electron that is diametrically opposite it. Thus, at C,

$$\vec{E} = 0. \qquad \text{(Answer)}$$

(b) If the electrons are moved along the circle until they are nonuniformly spaced over a 120° arc (Fig. TE25-4b), what then is the potential at C? How does the electric field at C change (if at all)?

SOLUTION: The potential is still given by Eq. TE25-4, because the distance between C and each electron is unchanged and orientation is irrelevant. The electric field is no longer zero, because the arrangement is no longer symmetric. There is now a net field that is directed toward the charge distribution.

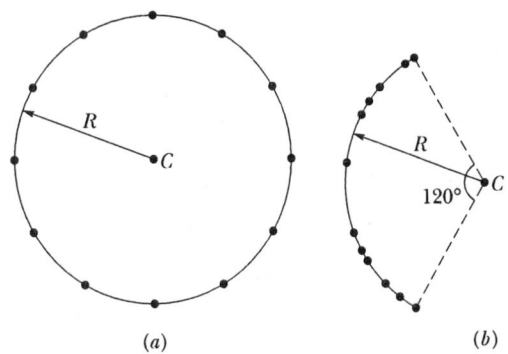

(a)   (b)

Fig. TE25-4 (a) Twelve electrons uniformly spaced around a circle. (b) Those electrons are now nonuniformly spaced along an arc of the original circle.

Touchstone Example 25-10-1

The electric potential at any point on the central axis of a uniformly charged disk is given by

$$\Delta V = \frac{\sigma}{2\varepsilon_0}\left(\sqrt{z^2 + R^2} - z\right).$$

Starting with this expression, derive an expression for the electric field at any point on the axis of the disk.

SOLUTION: We want the electric field $\vec{E}$ as a function of distance z along the axis of the disk. For any value of z, the direction of $\vec{E}$ must be along that axis because the disk has circular symmetry about that axis. Thus, we want the component E_z of $\vec{E}$ in the direction of z. Then the **Key Idea** is that this component is the negative of the rate of change of the electric potential with distance z. Thus, from the last of Eqs. 25-35, we can write

$$E_z = -\frac{\partial V}{\partial z} = -\frac{\sigma}{2\varepsilon_0}\frac{d}{dz}\left(\sqrt{z^2 + R^2} - z\right)$$

$$= \frac{\sigma}{2\varepsilon_0}\left(1 - \frac{z}{\sqrt{z^2 + R^2}}\right). \qquad \text{(Answer)}$$

26 Current and Resistance

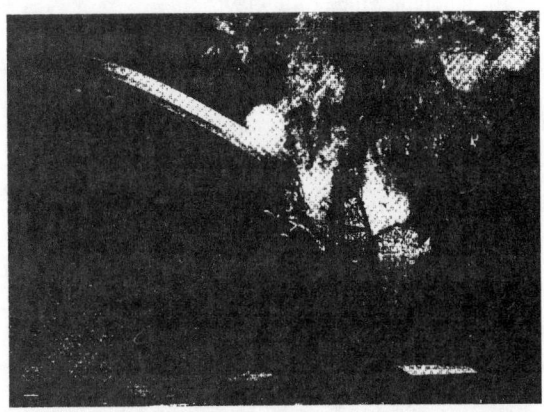

The pride of Germany and a wonder of its time, the zeppelin *Hindenburg*—almost the length of three football fields—was the largest flying machine that had ever been built. Although it was kept aloft by 16 cells of highly flammable hydrogen gas, it made many trans-Atlantic trips without incident. In fact, German zeppelins, which all depended on hydrogen, had never suffered an accident due to the hydrogen. However, shortly after 7:21 p.m. on May 6, 1937, as the *Hindenburg* was ready to land at the U.S. Naval Air Station at Lakehurst, New Jersey, the ship burst into flames. Its crew had been waiting for a rainstorm to diminish, and handling ropes had just been let down to a navy ground crew, when ripples were sighted on the outer fabric of the ship about one-third of the way forward from the stern. Seconds later a flame erupted from that region, and a red glow illuminated the interior of the ship. Within 32 seconds the burning ship fell to the ground.

After so many successful flights of hydrogen-floated zeppelins, why did this zeppelin burst into flames?

The answer is in this chapter.

26-1 Batteries, Current and Charge Flow

Electric currents are both important and common. Examples abound, ranging from the large currents constituting lightning strikes to the tiny nerve currents regulating our muscular activity. The currents in household wiring, in light bulbs and electrical appliances are familiar. A beam of electrons—which is also a continuous transport of charge or current—moves through an evacuated space in the picture tube of a television set. Charged particles flow in the ionized gases of fluorescent lamps, in the semiconductors used in pocket calculators and in the chips that control microwave ovens and electric dishwashers.

On a global scale, charged particles trapped in the Van Allen radiation belts surge back and forth above the atmosphere between Earth's north and south magnetic poles. On the scale of the solar system, enormous currents of protons, electrons, and ions fly radially outward from the Sun as the *solar wind*. On the galactic scale, cosmic ray currents, which consist largely of energetic protons, stream through our Milky Way galaxy, some reaching the Earth.

Chapters 22 through 25 deal largely with *electrostatics*—that is, with charges that do not move continuously. The phenomena and concepts we discussed in those chapters included the nature and origin of charge, the interaction force between charged objects, electric field, electric potential energy and potential difference. How are these electrostatic concepts and phenomenon related to our experiences with electric circuits and currents? We will investigate this question in this chapter.

Let's begin by discussing the origin of the electric battery used to power many small circuits. By the end of the eighteenth century, Alessandro Volta had discovered when two metal plates were placed in contact with a moist conductor, they seemed to have electrical properties. In order to magnify this effect Volta piled up pairs of unlike metals. By grasping the plates (terminal) at each end of the pile with his hands, he claimed to feel the conduction of electricity through his body on a continuous basis. What evidence supports there is a similarity between the continuous electricity Volta associated with his voltaic pile (or that we associate with a modern battery) and the electrostatic charges we have been studying?

Electrostatics, Batteries and Electricity in Motion

In order to investigate the relationship between a battery and electrostatic charges, let's do several experiments using a hanging ball that can hold electric charge. We will make use of an electroscope, like the one shown in Fig. 26-1a, which consists of a housing electrically insulated from a central conductor. Suppose the electroscope housing is connected to one metal plate while its central conductor is connected to a second metal plate as shown in Fig. 26-1b. If the two plates are charged electrostatically, the deflection of the electroscope leaf can be used to measure the charge on the plates. When a light uncharged ball coated with a conductor is suspended between the charged plates, a fascinating sequence of events occurs. At first the ball is attracted by induction to the most strongly charged plate. As soon as it touches the plate, electrons are exchanged, and the excess charge on the ball is the same as the plate's. The ball is then repelled from the plate it is in contact with, and attracted to the opposite plate. This process continues as the ball oscillates back and forth transporting charges between the plates more and more slowly until there is no more charge to transport. Thus, the rate of swing slows down and stops as the electroscope leaf deflection decreases to zero. If we connect a wire between the plates, we can produce the same transition from a deflected electroscope leaf to a non-deflected leaf. Except in this case, the discharge takes place very rapidly.

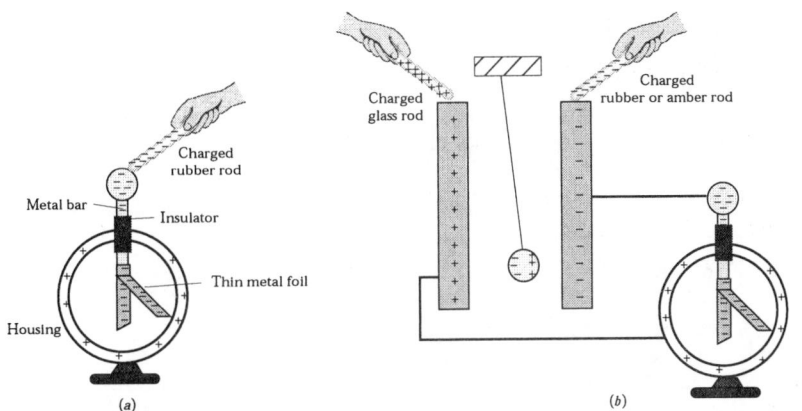

Fig. 26-1: (*a*) An electroscope (like that depicted previously in Fig. 22-7) can be used to measure the amount of excess charge on metal plates. (*b*) A light conducting ball will oscillate between two charged plates until all the net charge on one plate is transferred to the oppositely charged plate.

Now let's attempt to transport "charge" in two additional ways: 1) By attaching a positive battery terminal to one plate while attaching the negative terminal to the other plate; and 2) by connecting the two plates to a generator. After we charge the objects in these two new ways, we will compare the behavior of the hanging charged ball and the electroscope leaf and note similarities and differences.

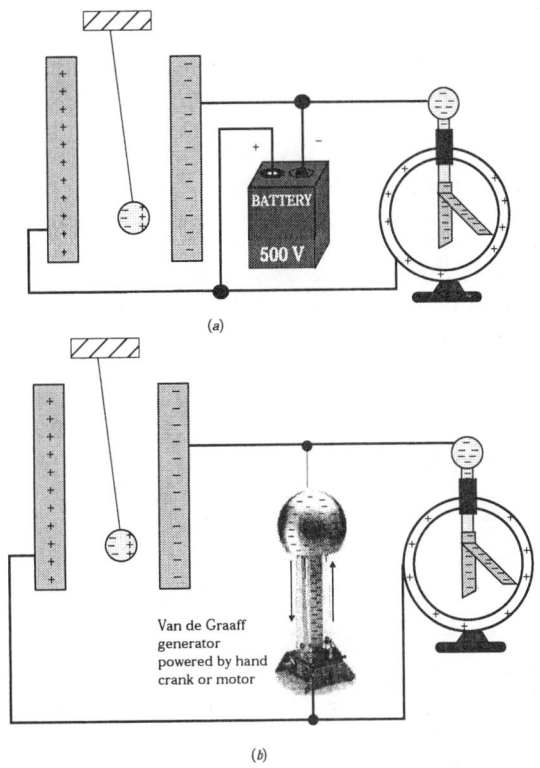

Fig. 26-2: (*a*) When the terminals of a relatively high voltage battery are attached to the metal plates, the light conducting ball oscillates between two charged plates continuously. (*b*) When the base and charged sphere of a Van de Graaff generator are attached to the metal plates, the light conducting ball also oscillates between two charged plates continuously.

The results of the two new experiments are depicted in Fig. 26-2. What we observe in both of the new experiments is the ball swings back and forth in the same way as it did when the plates were charged electrostatically, with one exception—the swinging does not "run down" or stop.

Since the battery and the generator produce the same visible effects (deflection of the electroscope and swinging of the ball) as electrostatic charging with rubber and glass rods, we infer the underlying electric effects are the same in all three cases. The observations we have discussed, as well as many other observations, indicate that static electrical charge and its transport (an electric current) in seemingly unrelated circumstances are in fact the same basic phenomena.

However, the persistence of the swinging in the presence of the battery and the generator implies a more continuous transport of charge between the plates via the ball.

In addition, if the ball is removed and a wire is connected between the plates, the electroscope leaf also maintains its deflection. The implication of these observations is transport of charge between the plates must be taking place continuously. Also, the continuous nature of the effects associated with the battery and the generator cease just as rapidly as they do for electrostatic charging when their connection to the experimental apparatus is broken.

These observations are reinforced by the fact a conducting flashlight bulb will not light unless the battery, bulb, and the wires connecting the battery to the bulb make a complete loop or circuit as shown in Fig. 26-3. We conclude:

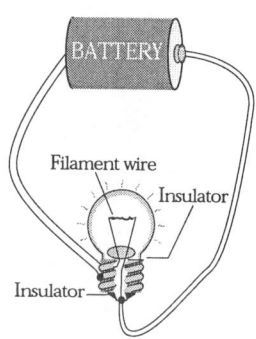

> There is continuous transport of charge when a battery or generator is connected to a closed conducting loop. This transport or movement of charge is called an **electric current**; if the loop is broken, the transport stops. Such closed loops are called **electric circuits**, or just **circuits**.

Fig. 26-3: It takes a complete loop between a battery, wires, and a conducting bulb to light a bulb. But, then the bulb is continuously lighted until the battery runs out of chemical potential energy.

We now understand batteries are devices that do this by means of chemical reactions. This leads us to the following definition of an ideal battery:

> An ideal battery is a device maintaining a constant potential difference between its terminals by means of chemical reactions.

The rearrangements of chemical bonds in a battery provide the energy needed to cause the separation of charges that maintains a potential difference between terminals— at least until all the chemical reactions have taken place. When they are fresh, alkaline batteries that are readily available in stores behave like ideal batteries.

It is commonly believed batteries store excess charge and a battery dies when this excess charge is used up. The fact people often refer to "charging" and "discharging" batteries is evidence of this belief. The excess charge a fresh alkaline flashlight battery, or a 1.5 volt D-cell would have to store to keep a flashlight bulb lit as long as it does is more than 20 000 coulombs. This is a hundred million times the amount of charge we can typically place on a light metal coated ball on a string. *Yet, there appear to be no forces between a charged hanging ball and a D-cell battery, so overall the battery seems to be electrically neutral.*

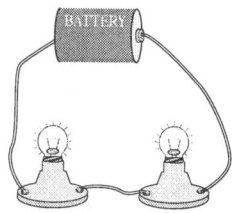

Fig. 26-4: When two identical bulbs are placed in series with a battery, they both glow with the same brightness. We conclude the same current is passing through both bulbs. This indicates that the battery is not a source of excess charge used up in the circuit.

What other evidence is there batteries store chemical energy rather than charge? If excess charge is used up when a battery is attached to a circuit, then where does it go? It is often suggested this charge is transformed into "heat" energy or "light" energy. In this case we would predict if we placed two flashlight bulbs in series with our flashlight battery, some or all of the current passing through the first bulb would be used up so little or no current would pass through the second bulb.

Based on the observation shown in Fig. 26-4, we conclude:

> When wires and other conducting elements such as bulbs and other types of resistors are placed between the battery terminals to make a single loop or circuit, the battery acts as a pump that pushes charge carriers already available in the wires around the loop.

Batteries are not the only devices that can pump charges through electrical circuits. Another very common way to establish a potential difference in a circuit is to establish a closed connection to an *electric generator*. Through electrical connections (wires) from a generating plant, a potential difference is established between connection points in the electrical outlets of our homes and workplaces. Other devices used to establish potential differences in circuits include *solar cells* (winglike panels on spacecrafts and buildings), *fuel cells* (which power the space shuttles) and *thermopiles* (which use thermal energy to provide onboard electrical power for spacecraft and remote stations in Antarctica). A source of electric potential does not have to be an instrument—living systems, ranging from electric eels to human beings to plants, have physiological devices for production of potential differences.

Although the devices we have listed differ widely in their modes of operation, they all perform the same basic function—they maintain a potential difference between their

The world's largest battery, housed in Chino, California, has a power capability of 10 MW, which is put to use during peak power demands on the electric system served by Southern California Edison.

terminals and thus they can do work on charge carriers and produce an electric current in a closed circuit.

READING EXERCISE 26-1: Describe the behavior one would expect of the swinging ball if there is not a continuous transport of charge through the wires to the plate when the battery or outlet are connected.

READING EXERCISE 26-2: Explain why the word "circuit," as defined in its non-technical sense, is an appropriate term for application to electrical situations.

26-2 Circuit Diagrams

As we move into the remaining sections in this chapter and the next, we will be drawing more circuits with elements such as batteries, bulbs, wires, and switches. We will also be introducing some new elements such as ohmic resistors. For convenience we will introduce some of the symbols scientists and engineers have created to represent circuits. Here are the symbols for some of the circuit elements we have mentioned or depicted so far.

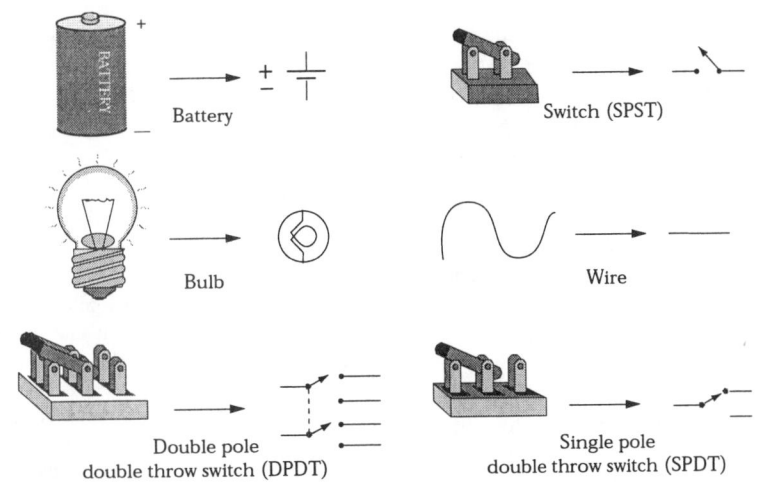

Battery

Switch (SPST)

Bulb

Wire

Double pole double throw switch (DPDT)

Single pole double throw switch (SPDT)

Fig. 26-5: Some circuit symbols.

Using these symbols, the circuit shown in Fig. 26-3 with a switch added can be represented as shown in Fig. 26-6.

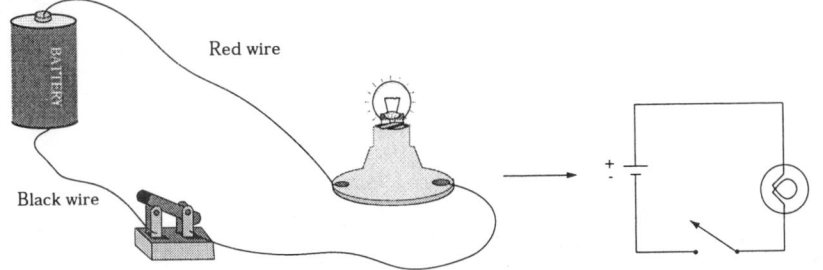

Red wire

Black wire

Fig. 26-6: A circuit sketch and corresponding diagram.

26-3 Electric Current

If a battery is not a source of charged particles, then how can it pump current through wires and other conductors? What exactly is a current? Do both positive and negative charges move? If not, which type does move? And how do we know? These are the questions we address is this section of the chapter. Let's start out by discussing how a battery can produce a current in a wire or other conductors that bend around or are oddly shaped.

The Electric Field inside a Current Carrying Wire

Since the electric field vectors between two plates are uniform and point directly from the positively charged plate to the negatively charged plate, it is easy to imagine placing a conducting wire directly between two charged plates and having charge carriers flow until the plates are neutralized as shown in Fig. 26-7. Note that it is impossible to tell whether the charge carriers are positive or negative because the end result is the same. If the charge carriers are positive, they are attracted to the negative plate and neutralize the negative charges. Since they came from the plate initially having positive charge, that plate also becomes neutral. Alternatively, if the charge carriers are negative, they are attracted to the positive plate and neutralize it. The plate they flowed from now has no excess negative charges and also becomes neutral. If one checks for the original charge separation and then later for neutrality using a familiar device such as an electroscope or a hanging charged ball, the outcome of the observations will be identical whether the carriers are positive or negative. This is shown in Fig. 26-7. Thus, if we analyze circuits assuming that the charge carriers are positive, we will predict exactly the same behaviors as if the carriers are negative.

What should happen if the two oppositely charged plates are replaced by the terminals of a battery? We can use the same type of reasoning used in the study of electrostatics to explain these observations that a steady state current flows through a battery. Since there is a potential difference between the terminals, an electric field is established. But the initial electric fields in the neighborhood of the terminals are not represented by nice straight uniform vectors. Yet, when wires and conducting bulbs in series with each other bend around between the two terminals as shown in Figs. 26-3, 26-4 and 26-6, we can place a current detector, such as an ammeter or small compass, at different points along the wire. We discover *the electric current or rate of flow of charge carriers is the same in all points along the circuit.* This implies two things. First, the electric field at every point in the connecting wires and bulb wire points along the wires no matter how they twist and turn. How can this be? Second, the charge carriers don't accelerate and flow at unequal rates.

To explain how the electric field lines can align with a wire with bends, we can turn to the reasoning used in electrostatics to determine a net electric field resulting from the flow of conventional current. If there is a component of electric field not parallel to the direction of a wire, charge carriers in the conductor will move toward the surface of the wire so rapidly that the process is, for all practical purposes, instantaneous. This is shown in Fig. 26-8.

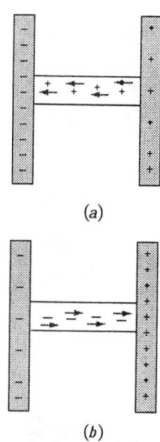

(a)

(b)

Fig. 26-7: No ordinary experiment will allow us to detect whether the charge carriers in a conducting wire transfering charge between two oppositely charged conducting plates are: (*a*) positive or (*b*) negative. In either case the two plates will become neutral very rapidly.

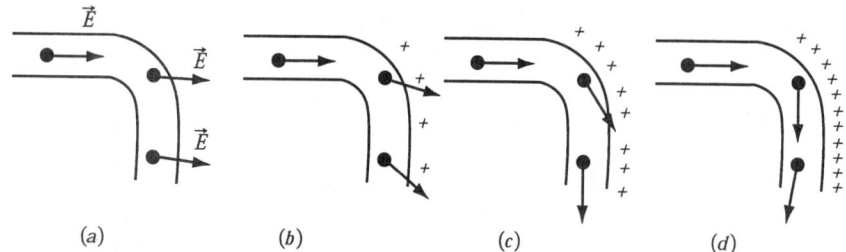

(a)   (b)   (c)   (d)

Fig. 26-8: Depiction of what happens to the electric field inside a wire connected to the terminals of a battery at its first bend immediately after a circuit is completed. In this diagram the left end of the wire is attached to the positive terminal and after several bends the other end (not shown) is attached to the negative terminal. We assume conventional current with positive charge carriers. (*a*) The initial electric field causes positive charges to move toward the bend. (*b*) As the charges accumulate at the bend the net field vectors that are a superposition of charges at the terminal and charges at the bend become directed along the wire. (*c*) The charges stop building up as soon as the net electric field vectors are directed along the wire. At this point, the rest of the charges start flowing along the wire. (d) If too many charges build up at a bend, the electric field acts to remove the excess charges from the bend. (This diagram is adapted from R.W. Chabay and B.A Sherwood, *Electric and Magnetic Interactions*, New York, Wiley, 1995, pg. 216.)

It can be readily shown if the surface charges are uniform along the surface of a straight length of wire that the net electric field inside the wire will be zero. Thus, in a whole circuit we assume the electric field needed to sustain current flow is maintained by a continuous variation in the surface charge density along the wire with some additional charges at the bends. A rough diagram depicting this is shown in Fig. 26-9.

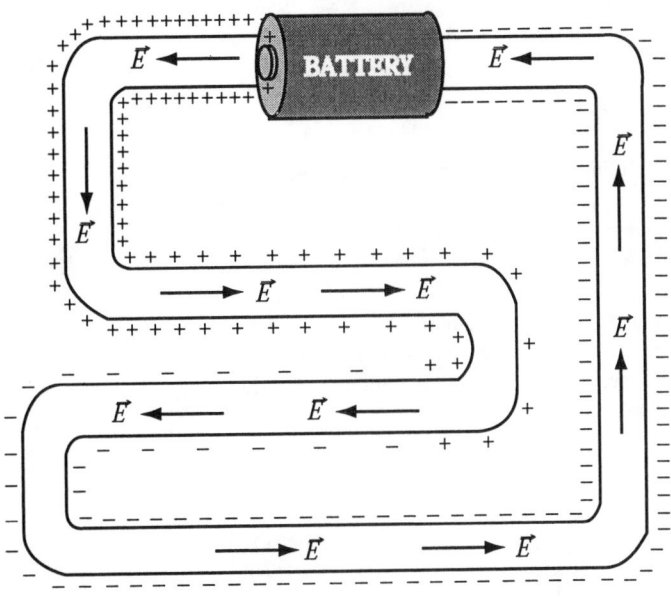

Fig. 26-9: A metal wire with a circular cross section is attached to a battery. This is a crude approximation of what the sign and relative densities of surface charges on the wire might look like within about a few nanoseconds (1 ns = 10^{-9} s) of the time the circuit is completed. This surface charge density causes the electric field vectors to line up with the wire, even as it bends. (This diagram is adapted from R.W. Chabay and B.A Sherwood, *Electric and Magnetic Interactions*, New York, Wiley, 1995, p. 210).

Defining Current Mathematically

Note Figure 26-3 shows a complete circuit where current pushed through the circuit by the battery flows through a conducting wire, which probably has a circular cross section, and then through the filament of a bulb which is usually a very thin wire that may have a rectangular cross section. When we say the current or rate of flow of charge is constant, we need to develop a mathematical definition for current taking these differences into account.

Figure 26-10 shows a section of a conducting loop with different cross-sectional areas in which current has been established. If net charge dq passes through a hypothetical plane (such as aa') in time dt, then the current through that plane is defined as

$$i \equiv \frac{dq}{dt} \qquad \text{(definition of current). (26-1)}$$

Regardless of the details of the geometry of the charge flow, we can find the net charge passing through the plane in a time interval extending from 0 to t by integration:

$$q = \int dq = \int_0^t i \, dt, \qquad (26\text{-}2)$$

in which the current i may be a function of time. Referring back to our knowledge of flux, we can envision the result of a flow of both positive and negative charge. Although an electric current is a stream of moving charges, we do not refer to all moving charges as an electric current. Specifically, we only say there is an electric current through a given surface when there is a *net* flow of charge through that surface.

As we mentioned already, observations show under steady-state conditions, the current is the same in all parts of a series circuit where there are no junctions or alternate paths for the current to take. The current or rate of charge flow is the same passing through the planes aa', bb', and cc' shown in Fig. 26-10. Indeed, the current is a flux and

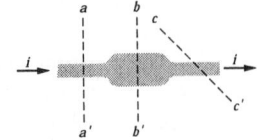

Fig. 26-10: The current i or charge per unit time through the conductor has the same value at planes aa', bb', and cc' as long as the planes cut through the entire conductor at the points of intersection.

is the same for all planes that pass completely through the conducting elements in a series circuit, no matter what their location or orientation. This leads us to assume charge is conserved. Under the steady-state conditions assumed here, a charge carrier must pass through plane aa' for every charge carrier that passes through plane cc'. In the same way, if we have a steady flow of water through a garden hose, a drop of water must leave the nozzle for every drop that enters the hose at the other end. The amount of water in the hose is a conserved quantity.

The unit for current is called the *ampere* (A) and it can be related to the coulomb by the expression

$$1 \text{ ampere} = 1 \text{ A} = 1 \text{ coulomb} / \text{second} = 1 \text{ C/s}.$$

However, this expression is not the official SI definition of the ampere. Since the currents have magnetic forces associated with them, the ampere is actually defined in terms of the forces between parallel current carrying wires. Since we introduced a temporary electrostatic definition of the coulomb in Section 22-7, we will continue to use it for now. When the formal definition of the ampere is introduced in Section 30-2 as part of our study of currents and magnetic effects, we can redefine the coulomb as the net charge that passes through a surface in 1 second when 1 ampere of current flows. In the end, everything has been adjusted so we will be dealing with the same numerical values whether we define the ampere in terms of the coulomb or vice versa!

The Directions of Currents

Let us return to the question we started to address earlier in this section—whether there are positive or negative charges moving when a current is established in electrically neutral conductors. In the situation depicted in Fig. 26-7, we concluded unless we do experiments at the atomic level, a flow of positive charge in one direction is indistinguishable from a flow of negative charge in the opposite direction.

Early experimenters with electricity had no knowledge of atomic structure and could only use macroscopic observations of electrical effects to guide them. They assumed charge carriers are positive. This assumption is still commonly used. For example, if an electrician tells you current flows from the "positive" terminal of a battery toward the "negative" terminal of a battery, he or she is applying this assumption. Even though we now believe negatively charged electrons are the charge carriers in conductors, for historical reasons we will stick with the assumption the charge carriers are positive. This historical assumption makes it easier to use traditional references on electricity, and all the characteristics of circuits we will study on a macroscopic level will be exactly the same. This early assumption would even be correct, if Benjamin Franklin had decided to designate the excess charges on rubber rods as positive and those on glass to be negative!

Although we now believe charge carriers in conductors are negative, other currents are not necessarily made up of negative charge carriers. For example, protons streaming out of our Sun create positive currents. Also charge carriers in fluids can be either positive ions (atoms with missing electrons) or negative electrons or ions (atoms with extra electrons). In fact, the movement of charge in most batteries is due to the migration of positive ions that undergo chemical reactions.

Current arrows show only a direction (or sense) of flow of charge carriers along the connected conductors as they bend and turn between battery terminals, not a fixed direction in space. Since current is actually a flux which is a scalar quantity, *these current arrows do not represent vectors with magnitude and direction.*

▶A current arrow, although not a mathematical vector, is drawn in the direction in which positive charge carriers would move through wires and circuit elements from a higher potential to a lower (more negative) potential, even if the actual charge carriers are negative and move in the opposite direction.

Charge Conservation at Junctions

So far we have only considered *series* circuits for which there is only one path for charge carriers to follow. Many times there are parallel circuits in which charge carriers encounter a junction where they can take two or more paths. To work with more complicated circuits involving the flow of current through resistive circuit elements such as bulbs, we need to introduce more precise definitions of *series* and *parallel*. These definitions are summarized in Fig. 26-11.

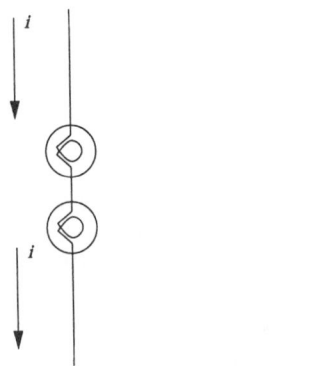

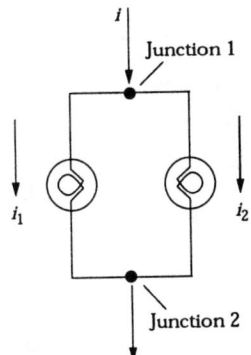

Fig. 26-11: A light bulb is used as an example of a simple conductive circuit element. A single *series* path is shown on the left. A parallel path is shown on the right and the current can split at Junction 1 and come together at Junction 2.

Series connection:
Two resistors are in series if they are connected so that the same current that passes through one bulb passes through the other.

Parallel connection:
Two resistors are in parallel if their terminals are connected so that at each junction one terminal of one bulb is directly connected to one terminal of the other.

The right side of Fig, 26-11 shows the moving charge carriers splitting up at a junction and then moving in parallel. If the bulbs are identical how do the currents split at Junction 1? What happens when they come back together at Junction 2? If charge is conserved, we expect the current coming into Junction 1 will divide equally so half takes the left path and half the right. When the currents combine at Junction 2, they should add up to the original current. We can verify our predictions by making a very simple observation with a couple of flashlight batteries in series and 4 bulbs as shown in Fig. 26-12. Because bulbs A and D have the same brightness as each other, this fact verifies if positive current carriers come from the positive terminal of the battery and flow through bulb A, then the amount of current flowing through bulb D and back through the battery is the same. The fact bulbs B and C have the same brightness as each other but are dimmer than A and D suggests the current is splitting in half at Junction 1. We conclude

$$i = i_1 + i_2 = \frac{i}{2} + \frac{i}{2} \qquad \text{(special case where parallel elements are identical)} \qquad (26\text{-}3)$$

If charge is conserved, the magnitudes of the currents in two parallel branches must add to yield the magnitude of the current in the original conductor even when the branches have different circuit elements. Thus, we predict

$$i = i_1 + i_2. \qquad \text{(general case)} \qquad (26\text{-}4)$$

We can verify this more general equation by putting another bulb, C2, in series with C. Then we will find B is brighter than the C and C2 bulbs, but even though all bulbs are dimmer than before (for reasons we will explore later), the brightness of A and D still match each other.

As Fig. 26-13*b* suggests, bending or reorienting the wires in space does not change the validity of this equation.

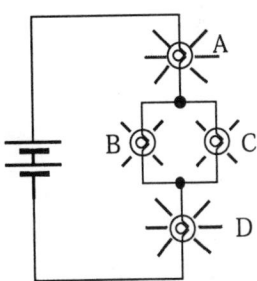

Fig. 26-12: In order to verify what we think would happen to current and the junctions, four identical light bulbs and a battery are connected into a circuit having both series and parallel elements. Observations tell us the brightness of bulb A is the same as that in bulb D. Bulbs B and C have the same brightness as each other but are much dimmer than A and D.

READING EXERCISE 26-3: Suppose a switch is closed allowing a battery to act as a pump and set up a flow of charges through wires and a bulb. (a) Will the overall circuit, consisting of the battery, bulb, and wires, remain electrically neutral, become positive, or become negative? Explain. (b) Will the wires remain electrically neutral, become positive, or become negative? Explain.

READING EXERCISE 26-4: Apply your understanding of the concept of flux, the nature of electrical charge and the definition of current presented to explain why a flow of equal amounts of opposite charge in opposite directions would not be considered a current. That is, explain why there must be a net flow of charge through a surface in order for there to be a current.

READING EXERCISE 26-5: The figure below shows a portion of a circuit. What are the magnitude and direction of the current i in the lower right-hand wire?

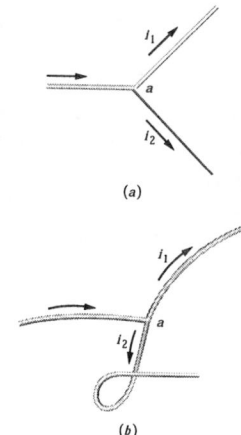

(a)

(b)

Fig. 26-13: The relation $i = i_1 + i_2$ is true at junction a no matter what the orientation in space of the three wires.

Touchstone Example 26-3-1, at the end of this chapter, illustrates how to use what you learned in this section.

TE

26-4 Potential Difference, Current and Resistance

Current is a very real and important part of modern life, and we have well-established convenient ways to measure it. The device with which one measures current is called an **ammeter**. Potential difference is measured with a device called a **voltmeter**. An ammeter and voltmeter along with their circuit symbols are depicted in Fig. 26-14.

Current measurements

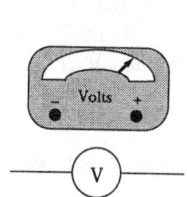

Ammeter symbol

Voltmeter symbol

Fig. 26-14: An analog ammeter for measuring current and an analog voltmeter for measuring potential difference (or "voltage"), along with their circuit symbols.

Often ammeters and voltmeters are combined in a device used to measure either potential difference or current. When the two or more meters are combined, the meter is typically called a **multimeter**. A digital multimeter is shown in Fig. 26-15. Many modern digital multimeters are also capable of measuring other quantities which we will also be studying, such as resistance and capacitance.

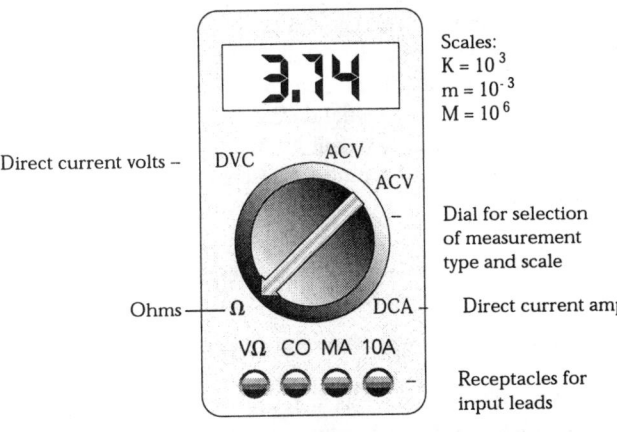

Scales:
K = 10^3
m = 10^{-3}
M = 10^6

Direct current volts –

Dial for selection
of measurement
type and scale

Ohms

Direct current amps

Receptacles for
input leads

Fig. 26-15: The digital
multimeter pictured can be
configured to act as an
ammeter to measure current
through a given part of a
circuit, a voltmeter to
measure potential difference
across any two points in a
circuit, or the resistance of
any circuit element.

In everyday applications of physics, like designing electronic devices, we often need to know what effect adding more circuit elements will have on the amount of current. Given devices like ammeters and voltmeters, with which we can measure current and potential difference, we can do quantitative studies of the relationship between current and potential difference. For example, what will happen to the current flowing through a circuit element such as a bulb that is part of a circuit if we add more batteries in series with our original battery? What will happen to the current in a conducting wire as voltage increases? The experimental set-up for this investigation is shown in Fig. 26-16. The results are shown in Fig. 26-17 which shows graphs of applied potential difference, V and the resulting current, i, that flows through two different circuit elements.

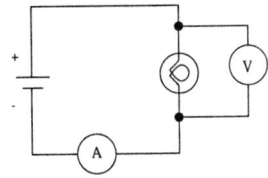

Fig. 26-16: A basic circuit
for measuring the current
flowing through a circuit
element as a function of
potential difference across it.

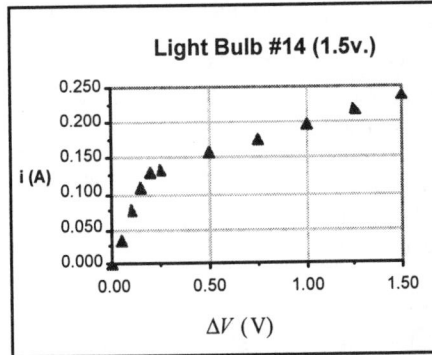

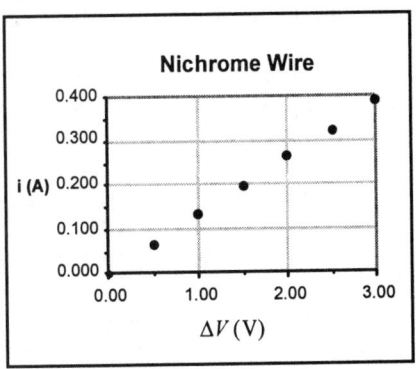

Fig. 26-17: The graph on the left shows the current flowing through a #14 lightbulb as a function of potential difference between the terminals of the lightbulb. The graph on the right shows the current flowing through a length of cylindrical nichrome wire as a function of potential difference between the ends of the wire.

There are several interesting conclusions we can draw from looking at the two graphs in Fig. 26-17. First, we see from the graphs as the potential difference increases, the current flowing through a given device increases. Second, it is not possible to tell how much current flows by just knowing the potential difference across part of a circuit. For instance, when 1.0 V is placed across the light bulb, the current through it is 0.20 A, but the same potential difference only causes 0.13 A to flow through the nichrome wire. Thus, we must know about the characteristics of our device. Third, for some circuit elements (such as the length of nicrome wire), the current is proportional to the potential difference, ΔV, across it. For others (such as the light bulb) there is not a direct proportionality.

For reasons we will reveal in Section 26-5, *any circuit element*, such as the nichrome wire, *for which there is a proportional relationship between potential difference and current* is known as an **ohmic device**. In contrast, *any circuit element*, such as the #14 light bulb, *for which there is not a proportional relationship between potential difference and current* is known as an **non-ohmic device**.

Regardless of whether a circuit element is ohmic or non-ohmic we know for a given potential difference across it a certain current will flow. It is obvious when a small potential difference causes a relatively large current to flow, the circuit element has a small resistance to the flow of current. Conversely, when the same potential difference is impressed across another circuit element and the current flow is small, then we say the resistance is large. Mathematically we can define resistance as the ratio of potential difference, ΔV, and current at that potential difference,

$$R = \frac{\Delta V}{i} \qquad \text{(definition of } R\text{)}. \qquad (26\text{-}5)$$

Here we use the notation ΔV to emphasize we are dealing with the *difference* in potential at two locations in a circuit.

The SI unit for resistance that follows from the equation above is the volt per ampere. This combination occurs so often that we give it a special name, the **ohm** (symbol Ω); that is,

$$1\,\text{ohm} = 1\,\Omega = 1\ \text{volt/ampere}$$
$$= 1\,\text{V/A}. \qquad (26\text{-}6)$$

For an ohmic device like the nicrome wire we can rewrite the equation as

$$\Delta V = iR, \qquad (26\text{-}7)$$

and note we will get the same value for R no matter what potential difference we impress across the device. We must be *careful* in the case of a non-ohmic device to specify at what potential difference we are measuring the current, i, in order to determine its resistance.

A conductor whose function in a circuit is to provide a specified ohmic resistance independent of the potential difference impressed across it is called a **resistor** (see Fig. 26-18). Carbon resistors are the most standard sources of resistance used in electrical circuits for several reasons. A light bulb has a resistance that increases with temperature and current and thus doesn't make a good circuit element when quantitative attributes are important. The resistance of carbon resistors doesn't vary with the amount of current passing through them. Carbon resistors are inexpensive to manufacture and can be produced with low or high resistances.

A typical carbon resistor contains a form of carbon, known as graphite, suspended in a hard glue binder. It usually is surrounded by a plastic case with a color code painted on it. It is instructive to look at samples of carbon resistors that have been cut down the middle as shown in Figure 26-20.

Fig. 26-18: An assortment of carbon resistors. The circular bands are color-coding marks that identify the value of the resistance.

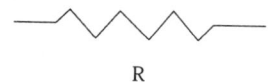

R

Fig. 26-19: Circuit diagram symbol for an ohmic resistor.

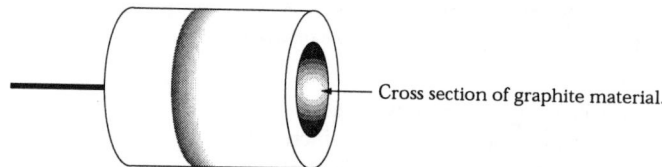

Cross section of graphite material.

Fig. 26-20: A cutaway view of a carbon resistor showing a cross-sectional area of the graphite material.

26-5 Resistance and Resistivity

The next question we will investigate is the way in which the resistance of ohmic circuit elements such as metal wires or carbon resistors depend on their geometries. That is, how does the resistance of a short, broad object change if we stretch it so it is long and thin? In order to determine this, we fix our investigation on a single material. For example, we might experiment with copper wire. Copper wire is commonly used in electric circuits because it has a very low resistance compared to other circuit elements. Thus, it can be used to connect circuit elements without adding much resistance to a circuit.

To start with, we will keep the thickness of the object (a copper wire) fixed and just increase or decrease the length of the wire. If we apply a potential difference across the ends of the wire and use current and potential difference measurements, we can determine its resistance as a function of length. We find that there is a proportionality between the length L of the wire and the resistance of the wire R. Thus, we can write

$$R = kL .$$

If instead we fix the length of the copper wire and vary the thickness or cross-sectional area A then we find the resistance of the copper decreases as the cross-sectional area A of the wire increases. In fact we get an inverse relationship so that

$$R = k' \frac{1}{A} .$$

Combining these two expressions gives us

$$R = kk' \frac{L}{A} .$$

To simplify this expression, we can replace the product of the two constants with a single new constant and write

$$R = \rho \frac{L}{A}. \qquad (26\text{-}8)$$

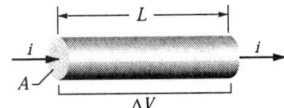

Fig. 26-21: A potential difference ΔV is applied between the ends of a wire of length L and cross section A, establishing a current i.

These investigations and the resulting relationship above are important for a number of reasons. First, they imply that current flows through the volume of the conductor, not along the surface. This knowledge will come in handy as we continue to think about current flow through wires and other objects.

Next, we might ask ourselves on what (if anything) does the constant of proportionality ρ depend? Is it the same for all materials? The answer is no. What we observe is if we apply the same potential difference between the ends of geometrically similar (same L and same A) rods of copper and of glass, very different currents result. This investigation implies ρ varies with material. That is, it is a property of the *material* from which the object is fashioned. We call this property of a material the material's **resistivity**.

We have just made an important distinction:

➤ Resistance is a property of an object. Resistivity is a property of a material.

However, the two are related. If we know the resistivity of a substance such as copper, we can calculate the resistance of a length of wire with a given cross-sectional area made of that substance using the relationship between resistance and resistivity given above.

TABLE 26-1: Resistivities Of Some Materials at Room Temperature (20°C)

Material	Resistivity, ρ ($\Omega \cdot m$)	Temperature Coefficient of Resistivity, α (K^{-1})
Typical Metals		
Silver	1.62×10^{-8}	4.1×10^{-3}
Copper	1.69×10^{-8}	4.3×10^{-3}
Aluminum	2.75×10^{-8}	4.4×10^{-3}
Tungsten	5.25×10^{-8}	4.5×10^{-3}
Iron	9.68×10^{-8}	6.5×10^{-3}
Platinum	10.6×10^{-8}	3.9×10^{-3}
Manganin[a]	48.2×10^{-8}	0.002×10^{-3}
Typical Semiconductors		
Silicon, pure	2.5×10^{3}	-70×10^{-3}
Silicon, n-type[b]	8.7×10^{-4}	
Silicon, p-type[c]	2.8×10^{-3}	
Typical Insulators		
Glass	$10^{10} - 10^{14}$	
Fused quartz	$\sim 10^{16}$	

[a]An alloy specifically designed to have a small value of α.
[b]Pure silicon doped with phosphorus impurities to a charge carrier density of 10^{23} m^{-3}.
[c]Pure silicon doped with aluminum impurities to a charge carrier density of 10^{23} m^{-3}.

READING EXERCISE 26-6: In the section above, we cited the fact the resistance of a wire to current flow was inversely proportional to the cross-sectional area of the wire as evidence the current flows through the volume of the wire rather than along the surface of the wire. (a) Justify this assertion. (b) What expression would you expect to replace Eq. 26-7 if the current flow was along the surface of the wire instead?

READING EXERCISE 26-7: The figure shows three cylindrical copper conductors along with their face areas and lengths. Rank them according to the current through them, greatest first, when the same potential difference ΔV is placed across their lengths.

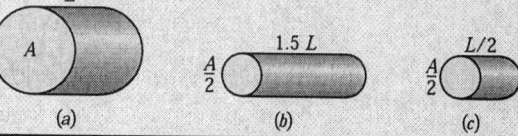

Variation with Temperature

The values of most physical properties vary with temperature, and resistivity is no exception. Fig. 26-22, for example, shows the variation of this property for copper over a wide temperature range. The relation between temperature and resistivity for copper—and for metals in general—is fairly linear over a rather broad temperature range. For such linear relations we can write an empirical approximation good enough for most engineering purposes,

$$\rho - \rho_0 = \rho_0 \alpha (T - T_0). \tag{26-9}$$

Here T_0 is a selected reference temperature and ρ_0 is the resistivity at that temperature. Usually $T_0 = 293$ K (room temperature), for which $\rho_0 = 1.69 \times 10^{-8} \, \Omega \cdot m$ for copper.

Because temperature enters into this expression only as a difference, it does not matter whether you use the Celsius or Kelvin scale in that equation because the sizes of degrees on these scales are identical. The quantity α, called the *temperature coefficient of resistivity*, is chosen so that the equation gives good agreement with experiment for temperatures in the chosen range. Some values of α for metals are listed in Table 26-1.

The *Hindenburg*

When the zeppelin *Hindenburg* was preparing to land, the handling ropes were let down to the ground crew. Exposed to the rain, the ropes became wet (and thus were able to conduct a current). In this condition, the ropes "grounded" the metal framework of the zeppelin to which they were attached; that is, the wet ropes formed a conducting path between the framework and the ground, making the electric potential of the framework the same as the ground's. This should have also grounded the outer fabric of the zeppelin. The *Hindenburg*, however, had been the first zeppelin to have its outer fabric painted with a sealant of large electrical resistivity. The fabric remained at the electric potential of the atmosphere at the zeppelin's altitude of about 43 m. Due to the rainstorm, that potential was large relative to the potential at ground level.

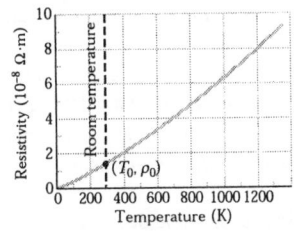

Fig. 26-22: The resistivity of copper as a function of temperature. The dot on the curve marks a convenient reference point ($T_0 = 293$ K and $\rho_0 = 1.69 \times 10^{-8}\ \Omega\cdot m$).

The handling of the ropes apparently ruptured one of the hydrogen cells and released hydrogen between that cell and the zeppelin's outer fabric, causing the reported rippling of the fabric. There was then a dangerous situation: the fabric was wet with conducting rainwater and was at a potential much different from the framework of the zeppelin's. Apparently, charge flowed along the wet fabric and then sparked through the released hydrogen to reach the metal framework of the zeppelin, igniting the hydrogen in the process. The burning rapidly ignited the cells of hydrogen in the zeppelin and brought the ship down. If the sealant on the outer fabric of the *Hindenburg* had been of less resistivity (like that of earlier and later zeppelins), the *Hindenburg* disaster probably would not have occurred.

26-6 Ohm's Law

As we discussed in Section 26-4, a resistor is a conductor with a specified resistance. It has that same resistance no matter what the magnitude and direction (*polarity*) of the applied potential difference. Other conducting devices, however, might have resistances that change with the applied potential difference.

Fig. 26-23*a* shows how to distinguish such devices. A potential difference ΔV is applied across the device being tested, and the resulting current i through the device which is measured as ΔV is varied in both magnitude and polarity. The polarity of ΔV is arbitrarily taken to be positive when the left terminal of the device is at a higher potential than the right terminal. The direction of the resulting current (from left to right) is arbitrarily assigned a plus sign. The reverse polarity of ΔV (with the right terminal at a higher potential) is then negative; the current it causes is assigned a minus sign.

Fig. 26-23*b* is a plot of i versus ΔV for one device. This plot is a straight line passing through the origin, so the ratio i/V (which is the slope of the straight line) is the same for all values of ΔV. This means the resistance $R = \Delta V/i$ of the device is independent of the magnitude and polarity of the applied potential difference ΔV.

Fig. 26-23*c* is a plot for another conducting device. Current can exist in this device only when the polarity of ΔV is positive and the applied potential difference is more than about 1.5 V. When current does exist, the relation between i and ΔV is not linear; it depends on the value of the applied potential difference ΔV.

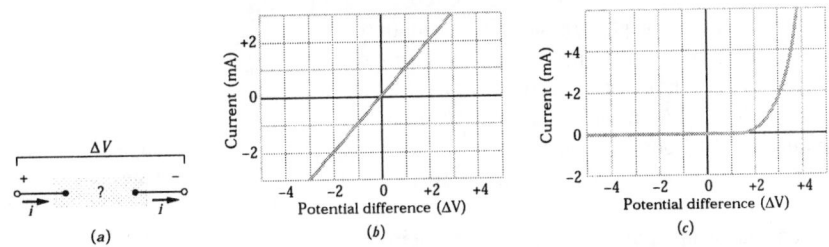

Fig. 26-23: (*a*) A potential difference ΔV is applied to the terminals of a device, establishing a current i. (*b*) A plot of current i versus applied potential difference ΔV when the device is a 1000Ω resistor. (*c*) A plot when the device is a semiconducting *pn* junction diode.

We distinguish between the two types of device by saying one obeys Ohm's law and the other does not.

➤**Ohm's law** is an assertion the current through a device is *always* directly proportional to the potential difference applied to the device.

(This assertion is correct only in certain situations; still, for historical reasons, the term "law" is used.) The device of Fig. 26-23b—which turns out to be a $1000\,\Omega$ resistor—obeys Ohm's law. The device of Fig. 26-23c—which turns out to be a so-called *pn* junction diode—does not.

➤A conducting device obeys Ohm's law when the resistance of the device is independent of the magnitude and polarity of the applied potential difference.

Modern microelectronics—and therefore much of the character of our present technological civilization—depends almost totally on devices that do *not* obey Ohm's law. Your calculator, for example, is full of them.

It is often contended that $\Delta V = iR$ is a statement of Ohm's law. That is not true! This equation is the defining equation for resistance, and it applies to all conducting devices, whether they obey Ohm's law or not. If we measure the potential difference ΔV across, and the current i through, any device, even a *pn* junction diode, we can find its resistance *at that value of* ΔV as $R = \Delta V/i$. The essence of Ohm's law, however, is a plot of i versus ΔV is linear; that is, the value of R is independent of the value of ΔV.

READING EXERCISE 26-8: The following table gives the current i (in amperes) through two devices for several values of potential difference ΔV (in volts). From these data, determine which device does not obey Ohm's law.

Device 1		Device 2	
ΔV	i	ΔV	i
2.00	4.50	2.00	1.50
3.00	6.75	3.00	2.20
4.00	9.00	4.00	2.80

26-7 A Microscopic View of Current, Resistance and Ohm's Law

To find out *why* particular materials obey Ohm's law, we must look into the details of the conduction process at the atomic level. Here we consider only conduction in metals, such as copper. We base our analysis on the *free-electron model*, in which we assume the conduction electrons in the metal are free to move throughout the volume of a sample, like the molecules of a gas in a closed container. We also assume the electrons collide not with one another but only with atoms of the metal.

According to classical physics, the electrons should have a Maxwellian speed distribution somewhat like that of the molecules in a gas. In such a distribution (see Section 20-7), the average electron speed would be proportional to the square root of the absolute temperature. The motions of electrons are, however, governed not by the laws of classical physics but by those of quantum physics. As it turns out, an assumption much closer to the quantum reality is conduction electrons in a metal move with a single effective speed $|\vec{v}_{eff}|$, and this speed is essentially independent of the temperature. For copper, $|\vec{v}_{eff}| \approx 1.6 \times 10^6$ m/s.

When we apply an electric field to a metal sample, the electrons modify their random motions slightly and drift very slowly—in a direction opposite that of the field—with an average drift speed $|\vec{v}_d|$. The drift speed in a typical metallic conductor is about 5×10^{-7} m/s, less than the effective speed $(1.6 \times 10^6$ m/s) by many orders of magnitude. Figure 26-24 suggests the relation between these two speeds. The dark lines

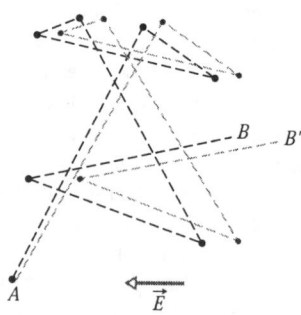

Fig. 26-24: The dark lines show an electron moving from *A* to *B*, making six collisions en route. The lighter lines show what its path might be in the presence of an applied electric field $\vec{E}$. Note the steady drift in the direction of $-\vec{E}$. (Actually, the lighterlines should be slightly curved, to represent the parabolic paths followed by the electrons between collisions, under the influence of an electric field.)

show a possible random path for an electron in the absence of an applied field; the electron proceeds from A to B, making six collisions along the way. The lighterlines show how the same events might occur when an electric field $\vec{E}$ is applied. We see that the electron drifts steadily to the right, ending at B' rather than at B. Fig. 26-24 was drawn with the assumption that $|\vec{v}_d| \approx 0.02 v_{eff}$. However, because the actual value is more like $|\vec{v}_d| \approx (10^{-13})|\vec{v}_{eff}|$, the drift displayed in the figure is greatly exaggerated.

The motion of the conduction electrons in an electric field $\vec{E}$ is thus a combination of the motion due to random collisions and that due to $\vec{E}$. When we consider all the free electrons, their random motions average to zero and make no contribution to the drift speed. Thus, the drift speed is due only to the effect of the electric field on the electrons.

26-8 Current Density

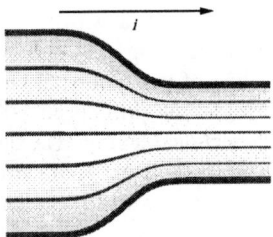

Sometimes we are interested in the current i in a particular conductor. At other times we take a localized view and study the flow of charge through a cross section of the conductor at a particular point. To describe this flow, we can use the **current density** $\vec{J}$, which has the same direction as the velocity of the moving charges if they are positive and the opposite direction if they are negative. For each element of the cross section, the magnitude $|\vec{J}|$ is equal to the current per unit area through that element. We can write the amount of current through the element as $\vec{J} \cdot d\vec{A}$, where $d\vec{A}$ is the area vector of the element, perpendicular to the element. The total current through the surface is then

$$i = \int \vec{J} \cdot d\vec{A}. \qquad (26\text{-}10)$$

Fig. 26-25: Streamlines representing current density in the flow of charge through a constricted conductor.

If the current is uniform across the surface and parallel to $d\vec{A}$, then $\vec{J}$ is also uniform and parallel to $d\vec{A}$. Then Equation 26-10 becomes

$$|i| = \int |\vec{J}||d\vec{A}| = |\vec{J}| \int |d\vec{A}| = |\vec{J}|A,$$

so
$$|\vec{J}| = \frac{i}{A}, \qquad (26\text{-}11)$$

where A is the total area of the surface. From the equations above, we see that the SI unit for current density is the ampere per square meter (A/m^2).

In Chapter 23 we saw we can represent an electric field with electric field lines. Fig. 26-25 shows how current density can be represented with a similar set of lines, which we can call *streamlines*. The current, which is toward the right in Fig. 26-25, makes a transition from the wider conductor at the left to the narrower conductor at the right. Because charge is conserved during the transition, the amount of charge and thus the amount of current cannot change. However, the current density does change—it is greater in the narrower conductor. The spacing of the streamlines suggests this increase in current density; streamlines that are closer together imply greater current density.

Drift Speed

In the last century, careful experimentation has shown when a conductor, like a piece of metal wire, is not attached to a battery or other source of potential difference, its conduction electrons are not at rest, bound to the positively charged nucleus. Instead, they have some kinetic energy and move randomly around in the wire, with no net motion in any direction. The free electrons (conduction electrons) in an isolated length of copper wire are in random motion at speeds of the order of 10^6 m/s.

However, we know from our experiments with the relationship between potential difference and current if we do not apply a potential difference to the wire, there can be no current. How do we reconcile these ideas? If you pass a hypothetical plane through such a wire, conduction electrons pass through it *in both directions* at the rate of many billions per second—but, there is *no net transport* of charge and thus *no current* through the wire.

Does this random motion of the charges disappear if we apply a potential difference to the conductor? There is no reason to assume it does, and in fact it does not disappear. Instead, when you connect the ends of the wire to a battery, you slightly bias the flow in one direction, with the result there now is a net transport of charge, and thus an electric current through the wire. The electrons still move randomly, but now they tend to *drift* with a **drift speed** $|\vec{v}_d|$ in the direction opposite of the applied electric field causing the current. The drift speed is tiny compared to the speeds in the random motion. For example, in the copper conductors of household wiring, electron drift speeds are perhaps 10^{-5} or 10^{-4} m/s, compared to random-motion speeds of around 10^6 m/s.

We can use Fig. 26-26 to relate the drift speed $|\vec{v}_d|$ of the conduction electrons in a current through a wire to the magnitude $|\vec{J}|$ of the current density in the wire. For convenience, Fig. 26-26 shows the equivalent drift of *positive* charge carriers in the direction of the applied electric field $\vec{E}$. Let us assume these charge carriers all move with the same drift speed $|\vec{v}_d|$ and the current density $\vec{J}$ is uniform across the wire's cross-sectional area A. The number of charge carriers in a length L of the wire is nAL, where n is the number of carriers per unit volume. The total charge of the carriers in the length L, each with a charge magnitude of $\pm e$, is then

$$q = \pm(nAL)e.$$

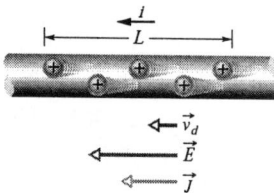

Fig. 26-26: Positive charge carriers drift at speed $|\vec{v}_d|$ in the direction of the applied electric field $\vec{E}$. By convention, the direction of the current density J and the sense of the current arrow are drawn in that same direction.

where the sign is determined by whether the carriers are negative or positive. Because the carriers all move along the wire with speed $|\vec{v}_d|$, this total charge moves through any cross section of the wire in the time interval

$$t = \frac{L}{|\vec{v}_d|}.$$

Equation 26-1 tells us the current i is the time rate of transfer of charge across a cross section of wire, so here we have

$$i = \frac{q}{t} = \frac{\pm nALe}{L/|\vec{v}_d|} = \pm nAe|\vec{v}_d|. \qquad (26\text{-}12)$$

Solving for the drift speed and recalling Eq. 26-11 $(|\vec{J}| = i/A)$, we obtain

$$|\vec{v}_d| = \frac{|i|}{nAe} = \frac{|\vec{J}|}{ne},$$

or, extended to vector form,

$$\vec{J} = \pm(ne)\vec{v}_d. \qquad (26\text{-}13)$$

Here the product *ne*, whose SI unit is the coulomb per cubic meter (C/m^3), is the *carrier charge density*. For positive carriers, we use the plus sign so expression above predicts $\vec{J}$ and $\vec{v}_d$ have the same direction. For negative carriers, we use the minus sign so $\vec{J}$ and $\vec{v}_d$ have opposite directions.

READING EXERCISE 26-9: The figure shows conduction electrons moving leftward through a wire. Are the following leftward or rightward: (*a*) the current *i*, (*b*) the current density $\vec{J}$, (*c*) the electric field $\vec{E}$ in the wire?

Touchstone Example 26-8-1, at the end of this chapter, illustrates how to use what you learned in this section.

TE

26-9 A More General View of Resistivity, Ohm's Law and Drift Velocity

As we have done several times before, we will now try to take a more general view of things. We did this when we invented the idea of electric field to allow us to describe the interaction forces between charged objects in a more general way than picking two specific objects to consider. We did this again when we invented the idea of electric potential to allow us to describe the electric potential energy in a more general way.

The macroscopic quantities V, i, and R are of greatest interest when we are making electrical measurements on specific conductors. They are the quantities we read directly on meters. We now turn to the microscopic quantities $\vec{E}$, $\vec{J}$, and ρ so we can consider the fundamental electrical properties of materials.

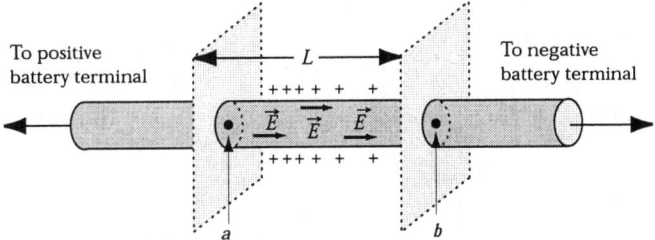

Fig. 26-27: A length L between points a and b along a current-carrying conductor.

Resistivity Revisited

As you should recall, the resistance of an object R is related to the potential difference across the object ΔV and the resulting current through the object i according to the relation:

$$R = \left| \frac{\Delta V}{i} \right|.$$

We can write this expression in a more general form if we replace the potential difference ΔV with an expression involving the more general quantity of electric field $\vec{E}$. From Chapter 25, we know that the relationship between the electric field and the potential difference between two locations a and b is

$$\Delta V_{ab} = \int_{a}^{b} \vec{E} \cdot d\vec{s}.$$

For a wire of length L with one end at location a and the other at location b, (Fig. 26-27), the electric field $\vec{E}$ set up within the wire is constant. As a result, the expression above becomes

$$\Delta V_{ab} = \pm |\vec{E}| L.$$

where we use the plus sign if $\vec{E}$ and $d\vec{s}$ are in the same direction and the minus sign if $\vec{E}$ and $d\vec{s}$ point in opposite directions. Combining this expression with

$$R = \left| \frac{\Delta V_{ab}}{i} \right|$$

gives us

$$R = \frac{|\vec{E}| L}{|i|} = \frac{|\vec{E}| L}{|\vec{J}| A}.$$

The substitution for i comes from the relationship between current i, current density $\vec{J}$ and the cross section of the wire A. Comparing this relation with

$$R = \rho \frac{L}{A},$$

leads us to the more general expression for resistivity that we were seeking:

$$\rho = \frac{|\vec{E}|}{|\vec{J}|} \qquad \text{(definition of } \rho \text{)}. \qquad (26\text{-}14)$$

If we combine the SI units of $\vec{E}$ and $\vec{J}$ we get, for the unit of ρ, the ohm-meter $(\Omega \cdot \text{m})^{-1}$:

$$\frac{\text{unit}\,(|\vec{E}|)}{\text{unit}\,(|\vec{J}|)} = \frac{\text{V/m}}{\text{A/m}^2} = \frac{\text{V}}{\text{A}}\,\text{m} = \Omega \cdot \text{m}.$$

(Do not confuse the *ohm-meter*, the unit of resistivity, with the *ohmmeter*, which is an instrument that measures resistance.)

We can rewrite this expression in vector form as

$$\vec{E} = \rho \vec{J}. \qquad (26\text{-}15)$$

However, be aware that these two relations hold only for *isotropic* materials—materials whose electrical properties are the same in all directions.

If an electron of mass m is placed in an electric field of magnitude $|\vec{E}|$, the electron will experience an acceleration given by Newton's Second Law:

$$\vec{a} = \frac{\vec{F}}{m} = \frac{e\vec{E}}{m}. \qquad (26\text{-}16)$$

The nature of the collisions experienced by conduction electrons is such that, after a typical collision, each electron will—so to speak—completely lose its memory of its previous drift velocity. Each electron will then start off fresh after every encounter, moving off in a random direction. In the average time τ between collisions, the average electron will acquire a drift speed of $|\vec{v}_d| = a\tau$. Moreover, if we measure the drift speeds of all the electrons at any instant, we will find their average drift speed is also $a\tau$. Thus at any instant, on average, the electrons will have drift speed $|\vec{v}_d| = a\tau$. Then the equation above gives us

$$|\vec{v}_d| = a\tau = \frac{e|\vec{E}|\tau}{m}. \qquad (26\text{-}17)$$

Combining this result with $\vec{J} = ne\vec{v}_d$ yields the drift velocity of

$$\vec{v}_d = \frac{\vec{J}}{ne} = \frac{e\vec{E}\tau}{m},$$

which we can write as

$$\vec{E} = \left(\frac{m}{e^2 n \tau}\right)\vec{J}.$$

Comparing this with Eq. 26-15 ($\vec{E} = \rho \vec{J}$) leads to

$$\rho = \frac{m}{e^2 n \tau}. \qquad (26\text{-}18)$$

Conductivity

As well as referring to the resistivity of a material, we often speak of the conductivity σ of a material. This is simply the reciprocal of its resistivity, so

$$\sigma = \frac{1}{\rho} \qquad \text{(definition of } \sigma \text{).} \qquad (26\text{-}19)$$

The SI unit of conductivity is the reciprocal ohm-meter, $(\Omega \cdot \text{m})^{-1}$. The unit name mhos per meter is sometimes used (mho is ohm backward). The definition of σ allows us to write Eq. 26-14 in the alternative form

$$\vec{J} = \sigma \vec{E}. \qquad (26\text{-}20)$$

Ohm's Law Revisited

Equation 26-17 may be taken as a statement metals obey Ohm's law if we can show for metals, their resistivity ρ is a constant, independent of the strength of the applied electric field $\vec{E}$. Because n, m, and e are constant, this reduces to convincing ourselves τ, the average time (or *mean free time*) between collisions, is a constant, independent of the strength of the applied electric field. Indeed, τ can be considered to be a constant because the drift speed $|\vec{v}_d|$ caused by the field is about a billion times smaller than the effective speed $|\vec{v}_{\text{eff}}|$, so the time between collisions is hardly affected by the field.

We can express Ohm's law in a more general way if we focus on conducting *materials* rather than on conducting *devices*. The relevant relation is then $\vec{E} = \rho \vec{J}$, which is the analog of $|\Delta V| = |i|R$.

➤A conducting material obeys Ohm's law when the resistivity of the material is independent of the magnitude and direction of the applied electric field.

All homogeneous materials, whether they are conductors like copper or semiconductors like pure silicon or silicon containing special impurities, obey Ohm's law within some range of values of the electric field. If the field is too strong, however, there are departures from Ohm's law in all cases.

Touchstone Example 26-9-1, at the end of this chapter, illustrates how to use what you learned in this section.

TE

26-10 Power in Electric Circuits

Figure 26-28 shows a circuit consisting of a battery B that is connected by wires, which we assume have negligible resistance, to an unspecified conducting device. The device might be a resistor, a storage battery (a rechargeable battery), a motor, or some other electrical device. The battery maintains a potential difference of magnitude ΔV across its own terminals, and thus (because of the wires) across the terminals of the unspecified device, with a greater potential at terminal a of the device than at terminal b.

Since there is an external conducting path between the two terminals of the battery, and since the potential differences set up by the battery are maintained, a steady current i is produced in the circuit, directed from terminal a to terminal b. The amount of charge dq moving between those terminals in time interval dt is equal to $i\,dt$. This charge dq moves through a decrease in potential of magnitude ΔV, and thus its electric potential energy decreases in magnitude by the amount

$$dU = dq\,\Delta V = i\,dt\Delta V.$$

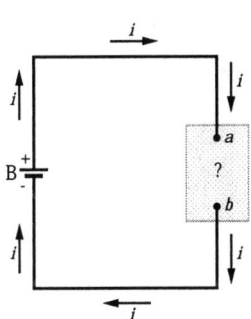

Fig. 26-28: A battery B sets up a current i in a circuit containing an unspecified conducting device.

The principle of conservation of energy tells us the decrease in electric potential energy from a to b is accompanied by a transfer of energy to some other form. The power P associated with that transfer is the rate of transfer dU/dt, which is

$$P = i\Delta V \qquad \text{(rate of electric energy transfer)}. \qquad (26\text{-}21)$$

Moreover, this power P is also the rate energy is transferred from the battery to the unspecified device. If that device is a motor connected to a mechanical load, the energy is transferred as work done on the load. If the device is a storage battery being charged, the energy is transferred to stored chemical energy in the storage battery. If the device is a resistor, the energy is transferred to internal thermal energy, tending to increase the resistor's temperature.

The unit of power following from the equation above is the volt-ampere $(V \cdot A)$. We can write it as

$$1 \text{ V} \cdot \text{A} = \left(1\frac{J}{C}\right)\left(1\frac{C}{s}\right) = 1\frac{J}{s} = 1 \text{ W}.$$

The course of an electron moving through a resistor at constant drift speed is much like that of a stone falling through water at constant terminal speed. The average kinetic energy of the electron remains constant, and its lost electric potential energy appears as thermal energy in the resistor and the surroundings. On a microscopic scale this energy transfer is due to collisions between the electron and the molecules of the resistor, which leads to an increase in the temperature of the resistor lattice. The mechanical energy thus transferred to thermal energy is *dissipated* (lost), because the transfer cannot be reversed.

For a resistor or some other device with resistance R, we can combine Eqs. 26-5 $(R = V/i)$ and 26-21 to obtain, for the rate of electric energy dissipation due to a resistance, either

$$P = i^2 R \qquad \text{(resistive dissipation)} \qquad (26\text{-}22)$$

or

$$P = \frac{(\Delta V)^2}{R} \qquad \text{(resistive dissipation)}. \qquad (26\text{-}23)$$

The wire coils within a toaster have appreciable resistance. When there is a current through them, electrical energy is transferred to thermal energy of the coils, increasing their temperature. The coils then emit infrared radiation and visible light that will toast (or burn) bread.

Caution: We must be careful to distinguish these two new equations from Eq. 26-21: $P = i\Delta V$ applies to electric energy transfers of all kinds; $P = i^2 R$ and $P = (\Delta V)^2/R$ apply only to the transfer of electric potential energy to thermal energy in a device with resistance.

READING EXERCISE 26-10: A potential difference ΔV is connected across a device with resistance R, causing current i through the device. Rank the following variations according to the change in the rate at which electrical energy is converted to thermal energy due to the resistance, greatest change first: (a) ΔV is doubled with R unchanged, (b) i is doubled with R unchanged, (c) R is doubled with ΔV unchanged, (d) R is doubled with i unchanged.

Touchstone Example 26-10-1, at the end of this chapter, illustrates how to use what you learned in this section.

26-11 Semiconductors

Semiconducting devices are at the heart of the microelectronic revolution that ushered in the information age. Table 26-2 compares the properties of silicon—a typical semiconductor—and copper—a typical metallic conductor. We see silicon has significantly fewer charge carriers, a much higher resistivity, and a temperature coefficient of resistivity that is both large and negative. Thus, although the resistivity of copper increases with temperature, that of pure silicon decreases.

Pure silicon has such a high resistivity it is effectively an insulator and not of much direct use in microelectronic circuits. However, its resistivity can be greatly reduced in a controlled way by adding minute amounts of specific "impurity" atoms in a process called *doping*. Table 26-1 gives typical values of resistivity for silicon before and after doping with two different impurities.

We can roughly explain the differences in resistivity (and thus in conductivity) between semiconductors, insulators, and metallic conductors in terms of the energies of their electrons. In a metallic conductor such as copper wire, most of the electrons are firmly locked into place within the molecules; much energy would be required to free them so they could move and participate in an electric current. However, in good conductors, the outermost electrons of one atom nearly overlap with the next atom. Thanks to the laws of quantum mechanics, particles can pass through small energy barriers and so the electrons can move from one atom to the next without any additional energy. In this situation each individual electron can be shared by lots of atoms and we can picture the electron as moving freely among those atoms. Thus, the electric field set up in the wire when a potential difference is applied drives a current through the conductor.

In an insulator, energy is required to free electrons so they can move through the material. Thermal energy cannot supply enough energy, and neither can any reasonable electric field applied to the insulator. Thus, no electrons are available to move through the insulator, and hence no current occurs even with an applied electric field.

A semiconductor is like an insulator *except* the energy required to free some electrons is not quite so great. More important, doping can supply electrons or positive charge carriers held very loosely within the material and easy to get moving. Moreover, by controlling the doping of a semiconductor, we can control the density of charge carriers participating in a current, and thereby can control some of its electrical properties. Most semiconducting devices, such as transistors and junction diodes, are fabricated by the selective doping of different regions of the silicon with impurity atoms of different kinds.

Table 26-2: Some Electrical Properties Of Copper And Silicon[a]

Property	Copper	Silicon
Type of material	Metal	Semiconductor
Charge carrier density, m^{-3}	9×10^{28}	1×10^{16}
Resistivity, $\Omega \cdot$ m	2×10^{-8}	3×10^{3}
Temperature coefficient of resistivity, K^{-1}	$+4 \times 10^{-3}$	-70×10^{-3}

[a]Rounded to one significant figure for easy comparison.

Let us now look again at Eq. 26-18 for the resistivity of a conductor:

$$\rho = \frac{m}{e^2 n \tau},$$ (Eq. 26-18)

where n is the number of charge carriers per unit volume and τ is the mean time between collisions of the charge carriers. (We derived this equation for conductors, but it also applies to semiconductors.) Let us consider how the variables n and τ change as the temperature is increased.

In a conductor, n is large but very nearly constant; that is, its value does not change appreciably with temperature. The increase of resistivity with temperature for metals (Fig. 26-29) is caused by an increase in the collision rate of the charge carriers, which shows up in Eq. 26-18 above as a decrease in τ, the mean time between collisions.

In a semiconductor, n is small but increases very rapidly with temperature as the increased thermal agitation makes more charge carriers available. This causes a *decrease* of resistivity with increasing temperature, as indicated by the negative temperature coefficient of resistivity for silicon in Table 26-2. The same increase in collision rate we noted for metals also occurs for semiconductors, but its effect is swamped by the rapid increase in the number of charge carriers.

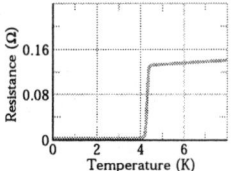

Fig. 26-29: The resistance of mercury drops to zero at a temperature of about 4 K.

26-23

26-12 Superconductors

In 1911, Dutch physicist Kamerlingh Onnes discovered the resistivity of mercury absolutely disappears at temperatures below about 4 K (Fig. 26-29). This phenomenon of **superconductivity** is of vast potential importance in technology because it means charge can flow through a superconducting conductor without producing thermal energy losses. Currents created in a superconducting ring, for example, have persisted for several years without any measurable decrease; the electrons making up the current require a force and a source of energy at start-up time, but not thereafter.

A disk-shaped magnet is levitated above a superconducting material that has been cooled by liquid nitrogen. The goldfish is along for the ride.

Prior to 1986, the technological development of superconductivity was throttled by the cost of producing the extremely low temperatures that were required to achieve the effect. In 1986, however, new ceramic materials were discovered that become superconducting at considerably higher (and thus cheaper to produce) temperatures. Practical application of superconducting devices at room temperature may eventually become feasible.

Superconductivity is a much different phenomenon from conductivity. In fact, the best of the normal conductors, such as silver and copper, cannot become superconducting at any temperature, and the new ceramic superconductors are actually insulators when they are not at low enough temperatures to be in a superconducting state.

One explanation for superconductivity is the electrons making up the current move in coordinated pairs. One of the electrons in a pair may electrically distort the molecular structure of the superconducting material as it moves through, creating a short-lived concentration of positive charge nearby. The other electron in the pair may then be attracted toward this positive charge. According to the theory, such coordination between electrons would prevent them from colliding with the molecules of the material and thus would eliminate electrical resistance. The theory worked well to explain the pre-1986, lower temperature superconductors, but new theories appear to be needed for the newer, higher temperature superconductors.

In the next chapter we restrict ourselves to the study—within the framework of classical physics—of *steady* currents of *conduction electrons* moving through *metallic conductors* such as copper wires.

Touchstone Example 26-3-1

Water flows through a garden hose at a volume flow rate dV/dt of 450 cm³/s. What is the current of negative charge?

SOLUTION: The current i of negative charge is due to the electrons in the water molecules moving through the hose. The current is the rate at which that negative charge passes through any plane that cuts completely across the hose. Thus, the Key Idea here is that we can write the current in terms of the number of molecules that pass through such a plane per second as

$$i = (\text{charge/electron})(\text{electrons/molecule})(\text{molecules/second})$$

or
$$i = (e)(10)\frac{dN}{dt}.$$

We substitute 10 electrons per molecule because a water (H_2O) molecule contains 8 electrons in the single oxygen atom and 1 electron in each of the two hydrogen atoms.

We can express the rate dN/dt in terms of the given volume flow rate dV/dt by first writing

$$\text{molecules/second} = (\text{molecules/mole})(\text{moles/unit mass})(\text{mass/unit volume})(\text{volume/second})$$

"Molecules/mole" is Avogadro's number N_A. "Moles/unit mass" is the inverse of the mass/mole, which is the molar mass M of water. "Mass/unit volume" is the (mass) density ρ_{mass} of water. The volume per second is the volume flow rate dV/dt. Thus, we have

$$\frac{dN}{dt} = N_A\left(\frac{1}{M}\right)\rho_{mass}\left(\frac{dV}{dt}\right) = \frac{N_A\rho_{mass}}{M}\frac{dV}{dt}.$$

Substituting this into the equation for i, we find

$$i = 10eN_AM^{-1}\rho_{mass}\frac{dV}{dt}.$$

N_A is 6.02×10^{23} molecules/mol, or 6.02×10^{23} mol⁻¹, and ρ_{mass} is 1000 kg/m³. We can get the molar mass of water from the molar masses listed in Appendix F: We add the molar mass of oxygen (16 g/mol) to twice the molar mass of hydrogen (1 g/mol), obtaining 18 g/mol = 0.018 kg/mol. Then

$$i = (10)(1.6 \times 10^{-19}\text{ C})(6.02 \times 10^{23}\text{ mol}^{-1})$$
$$\times (0.018\text{ kg/mol})^{-1}(1000\text{ kg/m}^3)(450 \times 10^{-6}\text{ m}^3\text{/s})$$
$$= 2.41 \times 10^7\text{ C/s} = 2.41 \times 10^7\text{ A}$$
$$= 24.1\text{ MA.} \hspace{3cm} \text{(Answer)}$$

This current of negative charge is exactly compensated by a current of positive charge associated with the nuclei of the three atoms that make up the water molecule. Thus, there is no net flow of charge through the hose.

Touchstone Example 26-8-1

What is the drift speed of the conduction electrons in a copper wire with radius $r = 900$ μm when it has a uniform current $i = 17$ mA? Assume that each copper atom contributes one conduction electron to the current and the current density is uniform across the wire's cross section.

SOLUTION: We need three Key Ideas here:

1. The drift speed $|\vec{v}_d|$ is related to the current density $\vec{J}$ and the number n of conduction electrons per unit volume according to Eq. 26-12, which we can write in magnitude form as $J = ne|\vec{v}_d|$.

2. Because the current density is uniform, its magnitude $|\vec{J}|$ is related to the given current i and wire size by Eq. 26-10 ($|\vec{J}| = i/A$, where A is the cross-sectional area of the wire).

3. Because we assume one conduction electron per atom, the number n of conduction electrons per unit volume is the same as the number of atoms per unit volume.

Let us start with the third idea by writing

$$n = \text{(atoms/unit volume)} = \text{(atoms/mole)(moles/unit mass)(mass/unit volume)}.$$

The number of atoms per mole is just Avogadro's number N_A ($= 6.02 \times 10^{23}$ mol^{-1}). Moles per unit mass is the inverse of the mass per mole, which here is the molar mass M of copper. The mass per unit volume is the (mass) density ρ_{mass} of copper. Thus,

$$n = N_A \left(\frac{1}{M} \right) \rho_{mass} = \frac{N_A \rho_{mass}}{M}.$$

Taking copper's molar mass M and density ρ_{mass} from Appendix F, we then have (with some conversions of units)

$$n = \frac{(6.02 \times 10^{23} \text{ mol}^{-1})(8.96 \times 10^3 \text{ kg/m}^3)}{63.54 \times 10^{-3} \text{ kg/mol}}$$

$$= 8.49 \times 10^{28} \text{ electrons/m}^3$$

or

$$n = 8.49 \times 10^{28} \text{ m}^{-3}.$$

Next let us combine the first two key ideas by writing

$$\frac{i}{A} = nev_d.$$

Substituting for A with πr^2 ($= 2.54 \times 10^{-6}$ m^2), and solving for the component of drift velocity, v_d, in the direction of the current, we then find

$$v_d = \frac{i}{ne(\pi r^2)}$$

$$= \frac{17 \times 10^{-3} \text{ A}}{(8.49 \times 10^{28} \text{ m}^{-3})(1.6 \times 10^{-19} \text{ C})(2.54 \times 10^{-6} \text{ m}^2)}$$

$$= 4.9 \times 10^{-7} \text{ m/s},$$ (Answer)

which is only 1.8 mm/h, slower than a sluggish snail.

You may well ask: "If the electrons drift so slowly, why do the room lights turn on so quickly when I throw the switch?" Confusion on this point results from not distinguishing between the drift speed of the electrons and the speed at which *changes* in the electric field configuration travel along wires. This latter speed is nearly that of light; electrons everywhere in the wire begin drifting almost at once, including into the lightbulbs. Similarly, when you open the valve on your garden hose, with the hose full of water, a pressure wave travels along the hose at the speed of sound in water. The speed at which the water itself moves through the hose—measured perhaps with a dye marker—is much slower.

Touchstone Example 26-9-1

(a) What is the mean free time τ between collisions for the conduction electrons in copper?

SOLUTION: The **Key Idea** here is that the mean free time τ of copper is approximately constant, and in particular does not depend on any electric field that might be applied to a sample of the copper. Thus, we need not consider any particular value of applied electric field. However, because the resistivity ρ displayed by copper under an electric field depends on τ, we can find the mean free time τ from Eq. 26-17 ($\rho = m/e^2 n\tau$). That equation gives us

$$\tau = \frac{m}{ne^2\rho}.$$

We take the value of n, the number of conduction electrons per unit volume in copper, from Touchstone Example 26-8-1. We take the value of ρ from Table 26-1. The denominator then becomes

$(8.49 \times 10^{28}\ \text{m}^{-3})(1.6 \times 10^{-19}\ \text{C})^2(1.69 \times 10^{-8}\ \Omega \cdot \text{m})$

$$= 3.67 \times 10^{-17}\ \text{C}^2 \cdot \Omega/\text{m}^2 = 3.67 \times 10^{-17}\ \text{kg/s},$$

where we converted units as

$$\frac{\text{C}^2 \cdot \Omega}{\text{m}^2} = \frac{\text{C}^2 \cdot \text{V}}{\text{m}^2 \cdot \text{A}} = \frac{\text{C}^2 \cdot \text{J/C}}{\text{m}^2 \cdot \text{C/s}} = \frac{\text{kg} \cdot \text{m}^2/\text{s}^2}{\text{m}^2/\text{s}} = \frac{\text{kg}}{\text{s}}.$$

Using these results and substituting for the electron mass m, we then have

$$\tau = \frac{9.1 \times 10^{-31}\ \text{kg}}{3.67 \times 10^{-17}\ \text{kg/s}} = 2.5 \times 10^{-14}\ \text{s}. \qquad \text{(Answer)}$$

(b) The mean free path λ of the conduction electrons in a conductor is the average distance traveled by an electron between collisions. (This definition parallels that in Section 20-6 for the mean free path of molecules in a gas.) What is λ for the conduction electrons in copper, assuming that their effective speed v_{eff} is 1.6×10^6 m/s?

SOLUTION: The **Key Idea** here is that the distance d any particle travels in a certain time t at a constant speed v is $d = vt$. For the electrons in copper, this gives us

$$\lambda = v_{\text{eff}}\tau = (1.6 \times 10^6\ \text{m/s})(2.5 \times 10^{-14}\ \text{s})$$
$$= 4.0 \times 10^{-8}\ \text{m} = 40\ \text{nm}. \qquad \text{(Answer)}$$

This is about 150 times the distance between nearest-neighbor atoms in a copper lattice. Thus, on the average, each conduction electron passes many copper atoms before finally hitting one.

Touchstone Example 26-10-1

You are given a length of uniform heating wire made of a nickel–chromium–iron alloy called Nichrome; it has a resistance R of 72 Ω. At what rate is energy dissipated in each of the following situations? (1) A potential difference of 120 V is applied across the full length of the wire. (2) The wire is cut in half, and a potential difference of 120 V is applied across the length of each half.

SOLUTION: The **Key Idea** is that a current in a resistive material produces a transfer of mechanical energy to thermal energy; the rate of transfer (dissipation) is given by Eqs. 26-20 to 26-22. Because we know the potential ΔV and resistance R, we use Eq. 26-22, which yields, for situation 1,

$$P = \frac{\Delta V^2}{R} = \frac{(120\ \text{V})^2}{72\ \Omega} = 200\ \text{W}. \qquad \text{(Answer)}$$

In situation 2, the resistance of each half of the wire is (72 Ω)/2, or 36 Ω. Thus, the dissipation rate for each half is

$$P' = \frac{(120\ \text{V})^2}{36\ \Omega} = 400\ \text{W},$$

and that for the two halves is

$$P = 2P' = 800\ \text{W}. \qquad \text{(Answer)}$$

This is four times the dissipation rate of the full length of wire. Thus, you might conclude that you could buy a heating coil, cut it in half, and reconnect it to obtain four times the heat output. Why is this unwise? (What would happen to the amount of current in the coil?)

27 Circuits

The electric eel (*Electrophorus*) lurks in rivers of South America, killing the fish on which it preys with pulses of current. It does so by producing a potential difference of several hundred volts along its length; the resulting current in the surrounding water, from near the eel's head to the tail region, can be as much as one ampere. If you were to brush up against this eel while swimming, you might wonder (after recovering from the very painful stun):

How can the creature manage to produce a current that large without shocking itself?

The answer is in this chapter.

27-1 Electric Currents and Circuits

Knowing how to analyze circuits by predicting the currents through their elements and the potential differences across them is valuable. Such knowledge enables engineers and scientists to design electrical devices and helps them make productive use of existing devices. Our goal in this chapter is to understand the behavior of relatively simple electric circuits by applying the concepts such as current, potential difference, and resistors developed in the last chapter. We will start by considering very simple ideal circuits and then go on to consider ideal circuits with multiple loops and batteries such as those shown in Fig. 27-1. Toward the end of the chapter we will introduce the concept of emf or electromotive force associated with batteries and other power sources. In particular, we will consider how to extend our analysis to the behavior of circuits powered by non-ideal batteries that have internal resistors.

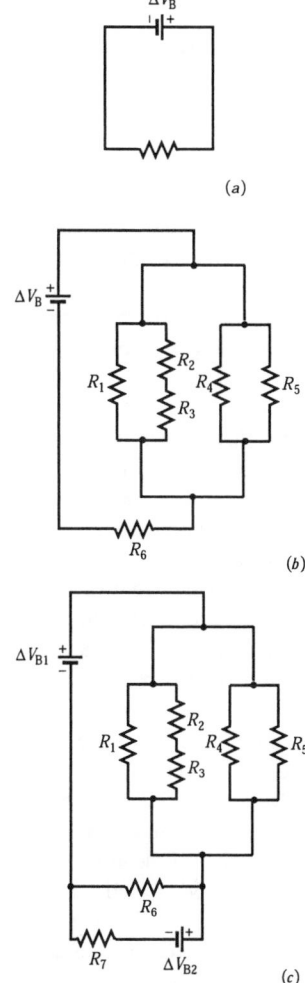

Fig. 27-1: Several types of ideal circuits we will learn to analyze in this chapter consist of ideal batteries, conducting wires with negligible resistance and ohmic resistors. (*a*) A single loop circuit. (*b*) A single battery multiple loop circuit. (*c*) A multiple loop circuit with multiple batteries.

Ideal Circuits

As we so often do in developing physical ideas, we start by analyzing how a system behaves under ideal conditions. Only then do we introduce real world complexities that require us to modify our methods of analysis. The ideal circuits we consider first have three characteristics:

1. **They are powered by ideal batteries.** As stated in the first section of Chapter 26, an ideal battery "maintains a constant potential difference across its terminals." This means there is a negligible amount of "electric friction" and the potential difference, ΔV_B, across the terminals of an ideal battery stays the same regardless of the amount of current flowing through it. But as the chemical potential energy of an ideal battery decreases, it develops some *internal resistance* and the potential difference across its terminals decreases when more current flows through it.

2. **All circuit elements, other than the battery and connecting wires, are ohmic devices having a significant resistor.** In general, carbon resistors obey Ohm's law and have resistance of $1\,\Omega$ or more. As discussed in Section 26-4, an *ohmic device has a constant value of resistance, R, that is not a function of the amount of current passing through it.* We will make use of the fact the potential difference across the terminals of an ohmic device is directly proportional to the current, i, flowing through it and is given by $\Delta V = iR$ (Eq. 26-7).

3. **Ideal conducting wires connect the battery to circuit elements.** Copper wiring is used in most circuits found in consumer devices, households and industries. We can use Eq. 26-7 and data from Table 26-2 to determine the resistance of a 30 cm length of common 22 gauge copper wire is less than $0.02\,\Omega$. Thus the potential difference between the ends of even a relatively long continuous connecting wire is for all practical purposes negligible when compared to $1\,\Omega$. We assume there is essentially no potential drop across connecting wires.

READING EXERCISE 27-1: Show that the resistance of a 30 cm (≈ 12 inch) length of 22 gauge copper wire of diameter 0.024 cm has a resistance of about $0.02\,\Omega$. **Hint:** You will need to use information from Table 26-2 along with Eq. 26-7.

READING EXERCISE 27-2: Explain how a small length of ohmic conductor with negligible resistance that can carry a current, i, can also have a negligible potential drop between its ends.

27-2 Current and Potential Difference in Single-Loop Circuits

Suppose we want to design or operate an electrical device such as a CD player or refrigerator. The operation of the given device will require a certain minimum current or potential difference. How would we calculate the amount of current in a circuit or the potential difference between two points within the device? That is the topic of this section.

We will start out our discussion of current flow in circuits by focusing on the part of the circuit outside of the battery. That is, we will focus on current flow from one battery terminal, through the circuit and back to the other terminal. We start the chapter by focusing on circuit elements that are external to the battery. At the end of the chapter we will review and extend our previous discussions about what goes on inside devices like batteries and generators.

Consider the simple *single-loop* circuit of Fig. 27-2. The circuit consists of an ideal battery, a resistor, R, and two ideal connecting wires. Unless otherwise indicated, we assume that wires in circuits have negligible resistance. Their function, then, is merely to provide pathways along which charge carriers can move. Through use of stored chemical energy (a form of internal potential energy), the battery keeps one of its terminals (called the positive terminal and often labeled +) at a higher electric potential than the other terminal (called the negative terminal and labeled –).

The mobile negative charge carriers in the circuit wires move preferentially toward the positive terminal and away from the negative terminal. As a result, for the circuit shown in Fig. 27-2, we have a net flow of negative charge in a counterclockwise direction. In Chapter 26, we discussed the fact a flow of negative electrons in one direction is macroscopically indistinguishable from a flow of positive charges in the other direction. For historical reasons we continue the practice established in the last chapter of working with current as if the charge carriers are positive.

Recall the direction of the conventional current through external circuit elements is always away from the positive terminal and toward the negative terminal of a battery. The direction of the conventional current in the circuit shown in Fig. 27-2 is noted with arrows that are labeled *i*. Unless otherwise noted, we will continue the practice of using conventional (positive) current in our analysis of electric circuits. We will reach the same conclusions about the fundamental behavior of circuits as we would if we had used electron currents.

To begin learning to calculate currents in circuits, let's start with the ideal circuit depicted in Fig. 27-2. We have marked the points just before and after each element with the letters *a*, *b*, *c*, and *d*. Let's start at point *a* and proceed around the circuit in either direction, adding any changes in potential we encounter. Once we return to our starting point, we must also have returned to our starting potential. In words, the potential energy change per unit of charge traveling through the battery plus the potential energy change of the charge traveling through the wires and the resistors must be zero. This can be denoted as

$$\Delta V_{a \to b} + \Delta V_{b \to c} + \Delta V_{c \to d} + \Delta V_{d \to a} = \Delta V_{a \to a} = 0.$$

For our simple circuit in Fig. 27-2 the charges gain potential while traveling from *a* to *b* due to the energy boost from the battery so that $\Delta V_{a \to b} = V_b - V_a = \Delta V_B$. The charges then flow freely from *b* to *c* through the first segment of the ideal conductor with no potential loss since the wire has a negligible resistance. Then the charges flow through the resistor, *R*. Finally, they flow back to point *a*, through another length of ideal wire.

$$\Delta V_B + \Delta V_{c \to d} = 0,$$

where ΔV_B represents a positive change in potential per unit charge as charges proceed from point *a* to point *b* by moving through the battery. Recall if our ohmic resistor has a fixed value R, then we noted in Eq. 26-7 $\Delta V = iR$ where *i* is the current flowing through

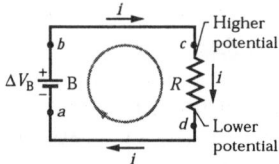

Fig. 27-2: A single-loop circuit in which a resistance R is connected across an ideal battery B with potential difference ΔV_B. The resulting current *i* is the same throughout the circuit.

the circuit. However, Eq. 26-7 didn't specify whether the ΔV refers to $\Delta V_{c \to d}$ or $\Delta V_{d \to c} = -\Delta V_{c \to d}$. It is clear from the context if we proceed through the loop from c to d, $\Delta V_{c \to d}$ must be negative so it will cancel the ΔV_B which we know is positive. This tells us the following about the mathematics of finding the potential difference across a resistor:

$$\Delta V_{d \to c} = iR \text{ and } \Delta V_{c \to d} = -iR.$$

In other words, charges lose potential as they travel through a resistor. This makes sense physically because resistors give off energy in the form of heat and light. So our battery acts as a pump to increase the potential energy of a charge and the charge loses potential energy in passing through a resistive device.

This can be summarized as the loop rule.

►**LOOP RULE:** The algebraic sum of the changes in potential encountered in a complete traversal of any loop of a circuit must be zero.

This is often referred to as *Kirchhoff's loop rule* (or *Kirchhoff's voltage law*), after German physicist Gustav Robert Kirchhoff. This rule is analogous to what happens when you hike around a mountain. If you start from any point on a mountain and return to the same point after walking around it, the algebraic sum of the changes in elevation you encounter must be zero. Thus, you end up at the same gravitational potential as you had before you started. Although we developed this rule through consideration of a single loop circuit, it also holds for any complete loop in a *multi-loop* circuit, no matter how complicated. Right now we will focus on single loop circuits, but we will return to the discussion of more complicated circuits in Section 27-4.

In Fig. 27-2, we will start at point a, whose potential is V_a, and mentally walk clockwise around the circuit until we are back at a, keeping track of potential changes as we move. (Our starting point is at the low-potential terminal of the battery—the negative terminal). The potential difference between the battery terminals is equal to ΔV_B. When we pass through the battery to the high-potential terminal, the change in potential is positive.

As we walk along the top wire to the top end of the resistor, there is no potential change because the wire has negligible resistance; it is at the same potential as the high-potential terminal of the battery. So too is the top end of the resistor. When we pass through the resistor in the direction of the current flow, the potential decreases by an amount equal to $-iR$. We know the potential decreases because we are moving from the higher potential terminal of the resistor to the lower potential terminal.

For a walk around a single loop circuit of total resistance R in *the direction of the current flow* our loop rule gives us

$$\Delta V_B - iR = 0.$$

Solving this equation for i gives us

$$i = \frac{\Delta V_B}{R} \qquad \text{(single loop circuit).} \qquad (27\text{-}1)$$

If we apply the loop rule to a complete walk around a single loop circuit of total resistance R *against the direction of current flow*, the rule gives us

$$-\Delta V_B + iR = 0,$$

and we again find that

$$i = \frac{\Delta V_B}{R} \qquad \text{(single loop circuit).}$$

Thus, you may mentally circle a loop in either direction to apply the loop rule.

To prepare for circuits more complex than Fig. 27-2, let us summarize two rules for finding potential differences as we move around a chosen loop:

▶**RESISTANCE RULE:** For a move through a resistor in the direction of the conventional current, the change in potential is $-iR$; in the opposite direction of current flow it is $+iR$.

▶**POTENTIAL RULE:** For a move through a source of potential difference from low potential (for example, the negative terminal on a battery denoted a) to high potential (for example, the positive terminal on a battery denoted b) the change in potential is positive and given by $V_b - V_a = \Delta V_B$; in the opposite direction it is negative and given by $V_a - V_b = -\Delta V_B$.

What happens to the amount of current as it passes through a resistor? Is current going into the resistor the same as the current coming out of the resistor? Or, does a resistor (for example, a light bulb) "use up" current? Recall in Fig. 26-4 we depicted observations involving batteries and bulbs that clearly showed current is constant throughout a single loop circuit when resistors are connected in series. You can easily replicate these observations using fresh flashlight batteries, copper wires, and 1.5 V bulbs.

READING EXERCISE 27-3: It is asserted above we can infer the current flow into and out of a resistor is the same because two light bulbs connected in series glow equally brightly. Describe the brightness of the second bulb in Fig 26-4 relative to the first bulb under the following assumptions: (a) All the current is used up by the first bulb; (b) Most of the current is used up by the first bulb; (c) A small amount of the current was used up by the first bulb.

READING EXERCISE 27-4: The figure shows the conventional current I in a single-loop circuit with a battery B and a resistor R (and wires of negligible resistance). At points a, b, and c, rank (a) the magnitude of the current and (b) the electric potential, greatest first.

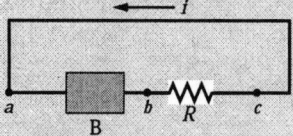

READING EXERCISE 27-5: Consider a circuit consisting of a battery with a potential difference of 12 volts across its terminals, zero resistance wire and a resistor of resistance R like the one shown in Reading Exercise 27-4. How (if at all) does the current in the circuit change if we increase the value of the resistance R? How (if at all) does the current in the circuit change if we decrease the value of the resistance R?

READING EXERCISE 27-6: Justify the following statement: "The potential rule, discussed above, is fundamentally a statement of conservation of energy."

27-3 Series Resistance

We now turn our attention to more complicated single loop circuits. Figure 27-3a shows three resistors connected in series to an ideal battery with potential difference ΔV_B between its terminals. Note the three resistors are connected one after another between b & c, c & d, and d & a. Also an ideal battery maintains a potential difference across the series of resistors (between points a and b). If we apply the loop rule for charges moving in the direction of conventional current from point a at the negative terminal of the battery and proceeding through the loop until we encounter point a again we get:

$$\Delta V_{a \to b} + \Delta V_{b \to c} + \Delta V_{c \to d} + \Delta V_{d \to a} = 0. \qquad (27\text{-}2)$$

Because we know that current is not used up by a resistor, we know the current flowing through the loop is the same everywhere, and so the current through each resistor must be the same. We also assume there is not potential difference along any segment of wire. If

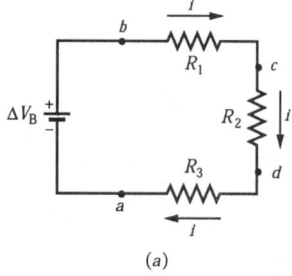

(a)

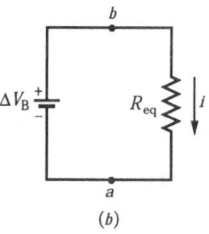

(b)

Fig. 27-3: (a) Three resistors are connected in series between points a and b. (b) An equivalent circuit, with the three resistors replaced with their equivalent resistance R_{eq}.

we consider the three resistors separately, applying the loop rule in the same manner (starting at the positive terminal of the battery and proceeding through the loop in the direction of conventional current) gives:

$$\Delta V_B + (-iR_1) + (-iR_2) + (-iR_3) = 0 .$$

By rearranging terms in the equation above we get

$$\Delta V_B - i(R_1 + R_2 + R_3) = 0 \qquad (27\text{-}3)$$

and defining an equivalent resistance as $R_{eq} = R_1 + R_2 + R_3$ we find Eq. 27-3 reduces to the same form as Eq. 27-1 with the equivalent resistance playing the role of the resistance in a circuit that has only one resistance. This is illustrated in Fig. 27-3.

Equating these two expressions tells us two things. First, the potential difference across the whole series of resistors is equal to the sum of the potential differences across each resistor. Second, the potential difference across the whole series of resistors is equal to the potential difference across our ideal battery. Figure 27-3*b* shows the equivalent resistance, with a new resistor R_{eq}, that can replace the three resistors of Fig. 27-3*a*.

The result $R_{eq} = R_1 + R_2 + R_3$ is not surprising because it is compatible with the experimental findings we presented in Section 26-5: the resistance of a length of wire is directly proportional to its length (Eq. 26-8). Imagine three different carbon resistors like those depicted in the last chapter (Fig. 26-20). Suppose these resistors are connected by ideal conductors (with almost no resistivity) having the same graphite material in their centers each with the same cross-sectional area. Giving the resistors different values of resistance would involve having the centers of the resistors have three different lengths. We would then expect the total resistance to be proportional to the sum of the three lengths of the resistors' graphite centers.

Obviously, we can extend our method of finding the equivalent resistance from 3 to *N* resistors by expanding Eq. 27-3 into the equation

$$R_{eq} = R_1 + R_2 + R_3 + \dots R_N = \sum_{j=1}^{N} R_j \quad (N \text{ resistors in series}). \qquad (27\text{-}4)$$

Note when resistors are in series, their equivalent resistance is always *greater* than any of the individual resistors. Also, the current moving through resistors wired in series can move along only a single route. If there are additional routes, so the currents in different resistors are different, the resistors are not connected in series.

In general:

▶ *N* resistors connected in *series* can be replaced with an equivalent resistor of value $R_{eq} = R_1 + R_2 + R_3 + \dots R_N$. The equivalent resistor has the same current flowing through it as through the actual resistors and the same *total* potential difference across it ΔV as that across all *N* of the actual resistors.

In short, we conclude if we replace a series of resistors, with a single equivalent resistor, the new circuit will have the same overall potential differences and currents as the original one did.

More on Ammeters

Analog ammeters work by measuring the torque exerted by magnetic forces on a current carrying wire. We discuss more about their operation in Chapter 29 on magnetic fields. However, we continue our discussion of these devices from the last chapter and consider some important attributes the ammeter must have.

Recall from Chapter 26 to measure the current in a wire, you are to break or cut the wire and insert the ammeter in series with an arm of the circuit so the current to be measured passes through the meter. (In Fig. 27-4, ammeter A is set up to measure current

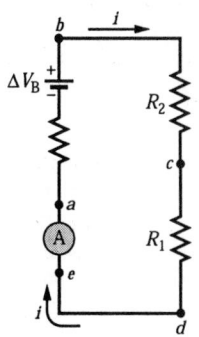

Fig. 27-4: This depicts how an ammeter can be inserted into a series circuit to meaure the current that flows through the circuit.

i.). If the ammeter has an appreciable resistance, it will increase the equivalent resistance of the array of resistors in series and serve to reduce the current that is being measured.

When measuring the current in a circuit (or anything else for that matter) it is imperative the measurement tool does not change the quantity you are trying to measure. Hence, it is essential the resistance R_A of the ammeter be very small compared to other resistances in the circuit. Otherwise, the presence of the meter will significantly change the current flow in the circuit and measured current will be an inaccurate representation of the true current.

READING EXERCISE 27-7: In Fig. 27-3*a*, if $R_1 > R_2 > R_3$, rank the three resistances according to (a) the current through them and (b) the potential difference across them, greatest first.

READING EXERCISE 27-8: Consider an ammeter inserted into the circuit shown in Fig. 27-4. Assume the ammeter is inserted above the resistor R_1 (closer to the negative terminal). Compare the amount of current flowing through R_1 under the following three conditions: (a) without the ammeter inserted, (b) if the ammeter has a resistance $r \ll R_1$, and (c) if the ammeter has a resistance $r = R_1$. Explain your reasoning. Discuss the implications of your result on designing an ammeter.

27-4 Multiloop Circuits

Figure 27-5 shows a circuit containing two loops. There are two points (*b* and *d*) at which the current branches split off or come together. We call such branching points **junctions**. For the circuit shown in Fig. 27-5, we would say there are two junctions, at *b* and *d*, and there are three *branches* connecting these junctions. The branches are the left branch (*bad*), the right branch (*bcd*), and the central branch (*bd*).

What are the currents in the three branches? We arbitrarily label the currents, using a different subscript for each branch. Because current is not used up and there are no additional branching points, current i_1 has the same value everywhere in branch *bad*, i_2 has the same value everywhere in branch *bcd*, and i_3 is the current through branch *bd*. The directions of the currents are assumed arbitrarily.

Consider junction *d* for a moment: Charge comes into that junction via incoming currents i_1 and i_3, and it leaves via outgoing current i_2. Because charge particles are neither created nor destroyed at the junction, the total incoming charge must be equal to the total outgoing charge. Hence, through conservation of charge arguments, we conclude the total incoming current must equal the total outgoing current:

$$i_{in} = i_{out},$$

or
$$i_1 + i_3 = i_2. \tag{27-5}$$

You can easily check applying this condition to junction *b* leads to exactly the same equation. This expression for the current in branch 2 thus suggests a general principle:

▶JUNCTION RULE: The sum of the currents entering any junction must be equal to the sum of the currents leaving that junction.

This rule is often called *Kirchhoff's junction rule* (or *Kirchhoff's current law*). It is simply a statement of the conservation of charge for a steady flow of charge—there is neither a build-up nor a depletion of charge at a junction. Thus, our basic tools for solving complex circuits are the *loop rule* (based on the conservation of energy) and the *junction rule* (based on the conservation of charge).

The relationship between i_1, i_2, and i_3 above is a single equation involving three unknowns. To solve the circuit completely (that is, to find all three currents), we need two more equations involving those same unknowns. We obtain them by applying the loop rule twice. In the circuit of Fig. 27-5, we have three loops from which to choose: the

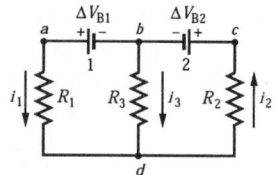

Fig. 27-5: A multiloop circuit consisting of three branches: left-hand branch *bad*, right-hand branch *bcd*, and central branch *bd*. The circuit also consists of three loops: left-hand loop *badb*, right-hand loop *bcdb*, and big loop *badcb*.

left-hand loop (*badb*), the right-hand loop (*bcdb*), and the big loop (*badcb*). Which two loops we choose does not matter—let's choose the left-hand loop and the right-hand loop.

If we traverse the left-hand loop in a counterclockwise direction from point *b*, the loop rule gives us

$$\Delta V_{B1} - i_1 R_1 + i_3 R_3 = 0, \qquad (27-6)$$

where ΔV_{B1} is the difference in potential between the terminals of Battery 1. If we traverse the right-hand loop in a counterclockwise direction from point *b*, the loop rule gives us an equation involving Battery 2,

$$-i_3 R_3 - i_2 R_2 - \Delta V_{B2} = 0. \qquad (27-7)$$

We now have three equations (Eqs. 27-5, 27-6, and 27-7) containing the three unknown currents, and they can be solved by a variety of mathematical techniques.

If we had applied the loop rule to the big loop, we would have obtained (moving counterclockwise from *b*) the equation

$$\Delta V_{B1} - i_1 R_1 - i_2 R_2 - \Delta V_{B2} = 0.$$

This equation may look like fresh information, but in fact it is only the sum of Eqs. 27-6 and 27-7. (It would, however, yield the proper results when used with Eq. 27-5 and either 27-6 or 27-7.)

It is important to note the assumed direction of the currents in the branch of the circuit do not have to be correct in order to get a correct solution. We must only keep track of the assumptions we have made. If in solving the resulting algebraic expressions, we find one of our currents turns out to have a negative value, then (because of the negative value), we know we made a wrong assumption about the direction of the current in that branch of the circuit.

27-5 Parallel Resistance

Figure 27-6*a* shows three resistances connected by branching junctions. Resistances that are parts of separate loops like those in Fig. 27-6*a* are said to be connected *in parallel* to the battery. Resistors connected "in parallel" are directly wired together on one side and directly wired together on the other side and a potential difference ΔV is applied across the pair of connected sides. Thus, the resistances have the same potential difference ΔV across them, producing a current through each. In general,

▶When a potential difference ΔV is applied across resistances connected in parallel, the resistances all have that same potential difference ΔV.

Notice we have again labeled the currents in each of the branches i_1, i_2, and i_3. We have discussed the way in which the current into a junction is equal to the current out of the junction. We have not yet discussed in what proportions currents divide when there is a branch (a choice of path) in a circuit. Are all three currents i_1, i_2, and i_3 equal? If not, which of these currents is largest? The answer to this question becomes more clear when we write out the expressions for current through each of the resistors in Fig. 27-6 using the potential rule for loops. For the case pictured here, we have

$$i_1 = \frac{\Delta V}{R_1}, \; i_2 = \frac{\Delta V}{R_2} \text{ and } i_3 = \frac{\Delta V}{R_3}. \qquad (27-8)$$

Since each resistor is connected so it has the same potential difference across it, it is straightforward to see how the sizes of the currents compare to each other. If the resistances are all equal, the current through each is the same. However, if the three resistances are not equal, more current flows through the smaller resistances. This

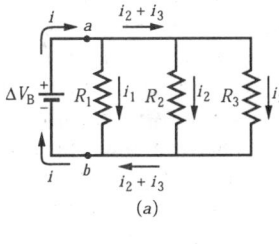

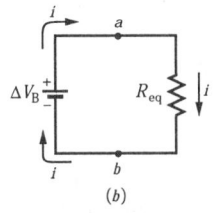

Fig. 27-6: (*a*) Three resistors connected in parallel across points *a* and *b*. (*b*) An equivalent circuit, with the three resistors replaced with their equivalent resistance R_{eq}.

outcome is consistent with what we might predict based solely on an understanding a resistor is just that—a device that resists the flow of current.

> ➤For resistors connected in parallel to the same source of potential difference, the least current flows in the branch with the largest resistance.

Returning to the circuit shown in Fig. 27-6a, suppose we wanted to save some money on the purchase of resistors and wanted to replace the three resistors in parallel with an equivalent resistor R_{eq}. The equivalent resistor R_{eq} would need to have all of the same characteristics as the three resistors in parallel. Figure 27-6b shows the three parallel resistances replaced with an equivalent resistance R_{eq}. The applied potential difference ΔV_B is maintained by a battery. We can see from this figure the potential difference across the equivalent resistance would have to be the same as the potential difference applied across each of the original resistors. Furthermore, the equivalent resistor would have to have the same total current $(i_1 + i_2 + i_3)$ through it as the original three resistors.

> ➤Resistances connected in parallel can be replaced with an equivalent resistance R_{eq}. If the equivalent resistance has the same potential difference applied across it, then the current through it will equal the sum of currents flowing th.rough the original resistors.

To derive an expression for R_{eq} in Fig. 27-6b, we first write the current in each of the resistors in Fig. 27-6a as

$$i_1 = \frac{\Delta V}{R_1}, \ i_2 = \frac{\Delta V}{R_2} \text{ and } i_3 = \frac{\Delta V}{R_3},$$ (Eq. 27-8)

where ΔV is the potential difference between a and b. If we apply the junction rule at point a in Fig. 27-6a and then substitute these values, we find

$$i = i_1 + i_2 + i_3 = \Delta V\left(\frac{1}{R_1} + \frac{1}{R_2} + \frac{1}{R_3}\right).$$ (27-9)

If we instead consider the parallel combination with the equivalent resistance R_{eq} (Fig. 27-6b), we would have

$$i = \frac{\Delta V}{R_{eq}} = \Delta V\left(\frac{1}{R_{eq}}\right).$$ (27-10)

Comparing the two equations above leads to

$$\frac{1}{R_{eq}} = \frac{1}{R_1} + \frac{1}{R_2} + \frac{1}{R_3}.$$ (27-11)

The result $1/R_{eq} = 1/R_1 + 1/R_2 + 1/R_3$ is not surprising because it is compatible with the experimental findings we presented in Section 26-5: the resistance of a length of wire is inversely proportional to its cross-sectional area (Eq. 26-8). Imagine three different carbon resistors like those depicted in the last chapter (Fig. 26-20) connected in parallel. Then giving them different values of resistance would involve having the centers of the resistors have three different cross-sectional areas. We would then expect the total cross-sectional area to be the sum of the three cross-sectional areas of the resistors' graphite centers so $A_{eq} = A_1 + A_2 + A_3$. But since the cross-sectional area and resistance are inversely proportional, we get $1/R_{eq} = 1/R_1 + 1/R_2 + 1/R_3$.

Extending Eq. 27-11 to the case of n resistors, we have

$$\frac{1}{R_{eq}} = \sum_{j=1}^{n}\frac{1}{R_j}$$ (*n* resistors in parallel). (27-12)

Since we often deal with the case of two resistors in parallel, it is worth it for us to consider this case a bit more. For the case of two resistors, the equivalent resistance is

$$\frac{1}{R_{eq}} = \frac{1}{R_1} + \frac{1}{R_2} .$$

With a bit of algebra, this becomes

$$R_{eq} = \frac{R_1 R_2}{R_1 + R_2} . \qquad\qquad (27\text{-}13)$$

If you accidentally took the equivalent resistance to be the sum divided by the product, you would notice at once this result would be dimensionally incorrect.

Note when two or more resistors are connected in parallel, the equivalent resistance is smaller than any of the combining resistances.

More on the Voltmeter

Recall from our discussion in Chapter 26 a meter used to measure potential differences is called a *voltmeter*. In order to measure the potential difference between any two points in the circuit, the voltmeter terminals are connected across those points, without breaking or cutting the wire. In Fig. 27-7, voltmeter V is set up to measure the potential difference across a resistor R_1. As you can see, the voltmeter is inserted in parallel with points d and e in the circuit.

In order for the presence of the voltmeter to not affect the measurement, it is essential the resistance R_V of a voltmeter be *very large* compared to the resistance of the circuit element across which the voltmeter is connected. Otherwise, the meter becomes an important circuit element by drawing a significant current through itself. This change in current flow can alter the potential difference to be measured.

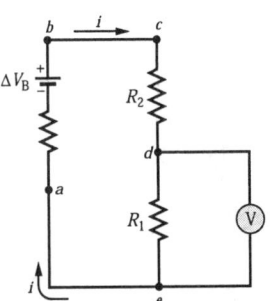

Fig. 27-7: A single-loop circuit, showing how to connect a voltmeter (V).

READING EXERCISE 27-9: A battery, with potential ΔV_B across it, is connected to a combination of two identical resistors and then has current i through it. What are the potential difference across and the current through either resistor if the resistors are (a) in series, and (b) in parallel?

READING EXERCISE 27-10: Consider the voltmeter inserted into the circuit shown in Fig. 27-7. Describe what would happen if the voltmeter has a resistance $r << R_{eq}$. How would this affect the potential difference measured across the resistor R_1? Describe what would happen if the voltmeter has a resistance $r >> R_{eq}$. How would this affect the potential difference measured across the resistor R_1? Which case would give the most "accurate" measure of the potential difference across the resistor when the voltmeter is not a part of the circuit?

READING EXERCISE 27-11: Suppose the resistors in Fig. 27-6 are all identical light bulbs. Rank the brightness of the three bulbs. Compare the brightness of each of the bulbs to the brightness of one of the bulbs alone connected to the same battery.

Touchstone Examples 27-5-1 and 27-5-2, at the end of this chapter, illustrate how to use what you learned in the previous two sections.

TE

27-6 Energy Considerations and Emf Devices

So far we have discussed ideal batteries that can be characterized as maintaining a constant potential difference between their terminals no matter what current is flowing through them. Also, we have concentrated on analyzing what happens in the part of the circuit that lies outside the battery. In this section we consider a bit more about what goes on inside batteries and energizing devices and more about how real, not so ideal, batteries behave.

The amazing thing about a battery and other energizing devices is, inside such a device, charge carriers enter at a low potential energy and emerge from it at a higher electric potential. Energy transformations inside an energizing device enable charges to overcome the forces exerted on them by the electric field inside the device. The effect is inside a device, positive carriers seem to move opposite from the electric field, and negative charge carriers with the electric field. There must be some other force present inside an energizing device enabling charges to swim upstream against electrical forces. The outdated term given to this "force" is electromotive force. Its abbreviation which we still use today, is emf. How is this force defined? Where does it come from in a typical battery?

We define the emf, ε, of an energizing device in terms of the work done per unit charge on charges flowing into it:

$$\varepsilon = \frac{dW}{dq} \qquad \text{(definition of } \varepsilon \text{).} \qquad (27\text{-}14)$$

In words, the emf of an energizing device is the work per unit charge the device does in moving charge from its low-potential terminal to its high-potential terminal. The SI unit for emf is the joule/coulomb. In Chapter 25 we defined one joule/coulomb as the *volt*. There must be some source of energy within an energizing device, enabling it to do work on the charges by forcing them to move as they do. The energy source may be chemical, as in a battery or a fuel cell. Temperature differences may supply the energy, as in a thermopile; or the Sun may supply it, as in a solar cell.

When an energizing device is connected to a circuit, the device transfers energy to the charge carriers passing through it. Let's take a look at one example of how chemical action can do this. For this purpose we will consider the chemical reactions that take place inside one cell of a lead acid battery used in most automobiles. A lead acid battery consists of several cells wired together in series. A cell is two metal plates surrounded by a liquid bath of chemicals used to transfer charges between the plates. A lead-acid cell consists of a negative plate made of pure lead and a positive plate made of lead-oxide. These plates are immersed in sulfuric acid mixed with water. The acid dissociates in the water into hydronium ions (H_3O^+) and bisulfate ions (HSO_4^-). This is shown in Fig. 27-8. Both the lead and lead oxide can react with the bisulfate ions as follows:

$$Pb + HSO_4^- + H_2O \rightarrow PbSO_4 + H_3O^+ + 2e^-$$
$$PbO_2 + HSO_4^- + 3H_3O^+ + 2e^- \rightarrow PbSO_4 + 4H_2O$$

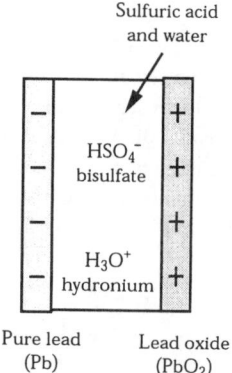

Fig. 27-8: The chemical constituents of the lead acid battery.

The two electrons produced on the pure lead plate pile up on it. The second reaction removes the two electrons it needs from the lead oxide plate. Thus, each time the pair of reactions occur, electrons are added to the negative plate and removed from the positive plate. If the cell were not connected to a circuit, the reactions would stop when the charge difference gets so large the energy needed to put more charges on the plates is greater than the energy released by the reactions. If the battery is connected to an external circuit, then as the charges are externally removed from one plate and put back on the other, the process can keep going continuously.

Note when we talk about a battery as a charge pump, this is somewhat misleading as *the electrons removed by the chemical reaction at one battery terminal (plate) are not the same electrons released at the other terminal.*

There are hundreds of different types of chemical batteries. The lead-acid battery action described here, simply serves as an example of how chemical reactions can cause charge separation in a battery.

READING EXERCISE 27-12: Consider the assumption we made above regarding the continuous nature of current in the circuit as it flows through the battery. Using examples from some of the many electric circuits connected to batteries with which you have experience, cite evidence this assumption is correct.

27-7 Internal Resistance

In our evaluation of circuits up to this point, we have assumed the current passes through the battery (or other emf source) without encountering any resistance within it. We call such a battery or other emf device "ideal."

An **ideal emf device** is one that lacks any resistance to the movement of charge through it. The potential difference between the terminals of an *ideal* emf device is equal to the emf of the device. For example, an ideal (perfectly fresh alkaline) battery with an emf of 12.0 V has a potential difference of 12.0 V between its terminals.

A **real emf device** has internal resistance to the movement of charge through it. For a real emf device (for example, a real battery), the only situation for which the potential difference between its terminals is equal to its emf is when the device is not connected to a circuit, and thus does not have current through it. However, when the device has current through it, the potential difference between its terminals differs from its emf.

Figure 27-9a shows a real battery, with internal resistance r, wired to an external resistor of resistance R. The internal resistance of the battery is the electrical resistance of the conducting materials of the battery and thus is an irremovable feature of the battery. However, as an illustration, a real battery is depicted in Fig. 27-9b as if it could be separated into an ideal battery with potential difference (and emf) ΔV_{B} between its terminals and a resistor of resistance r. The order in which the symbols for these separated parts are drawn does not matter.

If we apply the potential (loop) rule, proceeding clockwise and beginning at point a, the *changes* in potential give us

$$\Delta V_{B} + \Delta V_{R} = 0,$$

or
$$\varepsilon + \Delta V_{\text{Internal Resistance}} + \Delta V_{R} = 0. \tag{27-15}$$

Noting that we go through both resistances in the direction of the conventional current:

$$\varepsilon - ir - iR = 0. \tag{27-16}$$

Solving for the current, we find

$$i = \frac{\varepsilon}{R+r}. \tag{27-17}$$

Note this equation reduces to Eq. 27-1 if the battery is ideal. That is, for an ideal battery, $r = 0$.

Figure 27-9b shows graphically the changes in electric potential around the circuit. (To better link Fig. 27-9b with the *closed circuit* in Fig. 27-9a, imagine curling the graph into a cylinder with point a at the left overlapping point a at the right.) Note how traversing the circuit is like walking around a (potential) mountain and returning to your starting point—you also return to the starting elevation.

In this book, if a battery is not described as real or if no internal resistance is indicated, you can assume that it is ideal—but, of course, in the real world batteries are always real and have internal resistance.

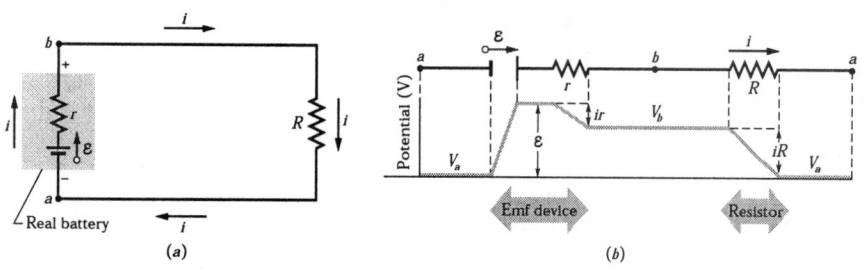

Fig. 27-9: (*a*) A single-loop circuit containing a real battery having internal resistance r and emf ε. (*b*) The same circuit, now spread out in a line. The potentials encountered in traversing the circuit clockwise from *a* are also shown. The potential V_a is arbitrarily assigned a value of zero, and other potentials in the circuit are graphed relative to V_a.

Implications of Internal Resistance in Real EMF Devices

In order to understand the implications of internal resistance in emf devices for real circuits, let's try to make our understanding a bit more quantitative. To start with, let's see how $\Delta V_B = \Delta V_{a \to b} = V_b - V_a$, the potential difference across the battery terminals in Fig. 27-9, is affected by the existence of an internal resistance in the battery. In order to calculate $V_b - V_a$, we start at point *a* and follow the shorter path around to *b* which takes us clockwise through the battery. We then have

$$V_a + \varepsilon - ir = V_b,$$

or

$$V_b - V_a = \Delta V_B = \varepsilon - ir, \qquad (27\text{-}18)$$

where r is the internal resistance of the battery and ε is the emf of the battery. This expression tells us the potential difference of the battery is equal to the emf minus the drop in potential associated with internal resistance.

Furthermore, if we refer back to Eq. 27-17,

$$i = \frac{\varepsilon}{R+r},$$

and substitute this expression for current (in the circuit shown in Fig. 27-9) into our expression for the potential difference across the battery terminals, we get

$$\Delta V_B = \varepsilon - \left(\frac{\varepsilon r}{R+r} \right).$$

With some algebra, we get the following generally applicable expression:

$$\Delta V_B = \varepsilon \frac{R}{R+r}. \qquad (27\text{-}19)$$

For example, suppose in Fig. 27-9, $\varepsilon = 12\,\text{V}$, $R = 10\,\Omega$, and $r = 2.0\,\Omega$. Then the equation above tells us the potential across the battery's terminals is

$$\Delta V_B = (12\,\text{V}) \frac{10\,\Omega}{10\,\Omega + 2.0\,\Omega} = 10\,\text{V}.$$

In "pumping" charge through itself, the battery (via electrochemical reactions) does work per unit charge of $\varepsilon = 12\,\text{J/C}$, or 12 V. However, because of the internal resistance of the battery, it produces a potential difference of only 10 J/C, or 10 V, across its terminals.

Note if the internal resistance becomes large as compared to the resistance in the circuit, the available potential difference of the battery, electrical generator or other emf device will drop significantly. The result of this drop in available potential difference is a reduction in the amount of current in the circuit. This is especially important to consider when designing circuits with many resistors in parallel (or even just a few resistors in parallel if they are small).

For example, consider the circuit shown in Fig. 27-6 (three resistors in parallel with a battery) and let $R = 3\,\Omega$ for each resistor. The equivalent resistance in the circuit would be $R_{eq} = 1\,\Omega$. If the potential difference source is taken to be an ideal battery (internal resistance $r = 0$), the current in the circuit is

$$i = \frac{\Delta V_B}{R_{eq}} = \frac{12\text{ V}}{1\,\Omega} = 12\text{ A}.$$

The 12 amps are split evenly between each branch (because the resistances are all equal), so each resistor has 4 amps of current flowing through it.

If however, the potential difference source is a real battery with $\varepsilon = 12$ V and internal resistance $r = 2.0\,\Omega$, then the available potential difference from the battery is

$$\Delta V_B = (12\text{ V})\frac{1\,\Omega}{1\,\Omega + 2.0\,\Omega} = 4\text{ V}.$$

The total current in the circuit is then

$$i = \frac{\Delta V_B}{R_{eq}} = \frac{4\text{ Volts}}{1\,\Omega} = 4\text{ A}.$$

This current is still split between each of the branches of the circuit. So for the case of the real battery, the current flowing through each resistor is now only 4/3 amp. In comparison to the 4 amps produced by the ideal battery, one can see how the internal resistance of an emf device can play a significant role in the functioning of real circuits.

Power

When a battery or some other type of emf device does work on the charge carriers to establish a current i, it transfers energy from its source of energy (such as the chemical source in a battery) to the charge carriers. Because a real emf device has an internal resistance r, it also transfers energy to internal thermal energy via resistive dissipation, discussed in Chapter 26. Let us relate these transfers.

The net rate P of energy transfer from the emf device to the charge carriers is given by:

$$P = i\Delta V, \tag{27-20}$$

where ΔV is the potential across the terminals of the emf device. (Note this is the power associated with the transfer). If we apply this expression to the circuit shown in Fig. 27-9, (from Eq. 27-20 above) we can substitute $\Delta V_B = \varepsilon - ir$ into Eq. 27-20 to find

$$P = i(\varepsilon - ir) = i\varepsilon - i^2 r. \tag{27-21}$$

We see the term $i^2 r$ in Eq. 27-21 is the rate P_r of energy transfer to thermal energy within the emf device:

$$P_r = i^2 r \qquad \text{(internal dissipation rate)}. \tag{27-22}$$

Then the term $i\varepsilon$ in Eq. 27-21 must be the rate P_{emf} at which the emf device transfers energy to *both* the charge carriers and to internal thermal energy. Thus,

$$P_{emf} = i\varepsilon \qquad \text{(power of emf device)}. \tag{27-23}$$

If a battery is being *recharged*, with a "wrong way" current through it, the energy transfer is then from the charge carriers to the battery—both to the battery's chemical energy and to the energy dissipated in the internal resistance r. The rate of change of the chemical energy is given by Eq. 27-23, the rate of dissipation is given by Eq. 27-22, and the rate at which the carriers supply energy is given by Eq. 27-20.

Touchstone Examples 27-7-1 and 27-7-2 at the end of this chapter, illustrate how to use what you learned in this section.

TE

Touchstone Example 27-5-1

Figure TE27-1a shows a multiloop circuit containing one ideal battery and four resistances with the following values:

$$R_1 = 20\ \Omega, \quad R_2 = 20\ \Omega, \quad \Delta V_B = 12\ \text{V},$$
$$R_3 = 30\ \Omega, \quad R_4 = 8.0\ \Omega.$$

(a) What is the current through the battery?

SOLUTION: First note that the current through the battery must also be the current through R_1. Thus, one **Key Idea** here is that we might find that current by applying the loop rule to a loop that includes R_1 because the current would be included in the potential difference across R_1. Either the left-hand loop or the big loop will do. Noting that the potential difference arrow of the battery points upward, so the current the battery supplies is clockwise, we might apply the loop rule to the left-hand loop, clockwise from point a. With i being the current through the battery, we would get

$$+\Delta V_B - iR_1 - iR_2 - iR_4 = 0 \qquad \text{(incorrect)}.$$

However, this equation is incorrect because it assumes that R_1, R_2, and R_4 all have the same current i. Resistances R_1 and R_4 do have the same current, because the current passing through R_4 must pass through the battery and then through R_1 with no change in value. However, that current splits at junction point b—only part passes through R_2, the rest through R_3.

To distinguish the several currents in the circuit, we must label them individually as in Fig. TE27-1b. Then, circling clockwise from a, we can write the loop rule for the left-hand loop as

$$+\Delta V_B - i_1 R_1 - i_2 R_2 - i_1 R_4 = 0.$$

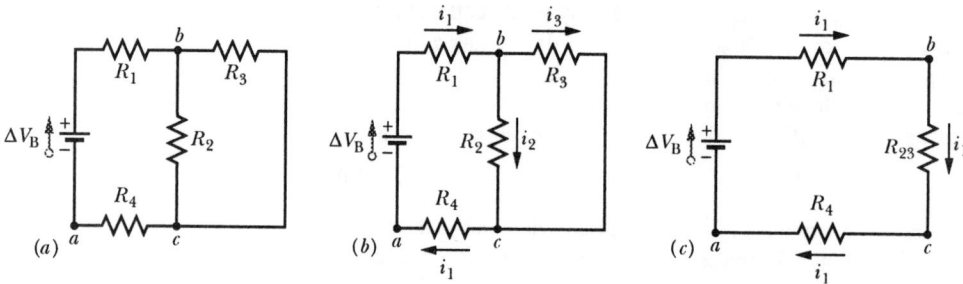

Fig. TE27-1 (a) A multiloop circuit with an ideal battery of terminal potential difference ΔV_B and four resistances. (b) Assumed currents through the resistances. (c) A simplification of the circuit, with resistances R_2 and R_3 replaced with their equivalent resistance R_{23}. The current through R_{23} is equal to that through R_1 and R_4.

Unfortunately, this equation contains two unknowns, i_1 and i_2; we would need at least one more equation to find them. We could do this by using the right hand loop.

A second **Key Idea** is that a much easier option is to simplify the circuit of Fig. TE27-1b by finding equivalent resistances. Note carefully that R_1 and R_2 are *not* in series and thus cannot be replaced with an equivalent resistance. However, since the ΔV across them is the same, R_2 and R_3 are in parallel, so we can use either Eq. 27-12 or Eq. 27-13 to find their equivalent resistance R_{23} since the ΔV across them is the same. From the latter,

$$R_{23} = \frac{R_2 R_3}{R_2 + R_3} = \frac{(20\ \Omega)(30\ \Omega)}{50\ \Omega} = 12\ \Omega.$$

We can now redraw the circuit as in Fig. TE27-1c; note that the current through R_{23} must be i_1 because charge that moves through R_1 and R_4 must also move through R_{23}. For this simple one-loop circuit, the loop rule (applied clockwise from point a) yields

$$+ \Delta V_B - i_1 R_1 - i_1 R_{23} - i_1 R_4 = 0.$$

Substituting the given data, we find

$$12 \text{ V} - i_1(20 \text{ }\Omega) - i_1(12 \text{ }\Omega) - i_1(8.0 \text{ }\Omega) = 0,$$

which gives us

$$i_1 = \frac{12 \text{ V}}{40 \text{ }\Omega} = 0.30 \text{ A}. \qquad \text{(Answer)}$$

(b) What is the current i_2 through R_2?

SOLUTION: One **Key Idea** here is that we must work backward from the equivalent circuit of Fig. TE27-1c, where R_{23} has replaced the parallel resistances R_2 and R_3. A second **Key Idea** is that, because R_2 and R_3 are in parallel, they both have the same potential difference across them as their equivalent R_{23}. We know the current through R_{23} is $i_1 = 0.30$ A. Thus, we can use Eq. 26-4 ($R = \Delta V/i$) to find the potential difference ΔV_{23} across R_{23}:

$$\Delta V_{23} = i_1 R_{23} = (0.30 \text{ A})(12 \text{ }\Omega) = 3.6 \text{ V}.$$

The potential difference across R_2 is thus 3.6 V, so the current i_2 in R_2 must be, by Eq. 26-4,

$$i_2 = \frac{\Delta V_2}{R_2} = \frac{3.6 \text{ V}}{20 \text{ }\Omega} = 0.18 \text{ A}. \qquad \text{(Answer)}$$

(a) What is the current i_3 through R_3?

SOLUTION: We can answer by using the same technique as in (b), or we can use this **Key Idea**: The junction rule tells us that at point b in Fig. TE27-1b, the incoming current i_1 and the outgoing currents i_2 and i_3 are related by

$$i_1 = i_2 + i_3.$$

This gives us

$$i_3 = i_1 - i_2 = 0.30 \text{ A} - 0.18 \text{ A} = 0.12 \text{ A}. \qquad \text{(Answer)}$$

Touchstone Example 27-5-2

Figure TE27-2 shows a circuit with three ideal batteries in it. The circuit elements have the following values:

$$\Delta V_{B_1} = 3.0 \text{ V}, \quad \Delta V_{B_2} = 6.0 \text{ V},$$
$$R_1 = 2.0 \text{ }\Omega, \quad R_2 = 4.0 \text{ }\Omega.$$

Find the magnitude and direction of the current in each of the three branches.

SOLUTION: It is not worthwhile to try to simplify this circuit, because no two resistors are in parallel, and the resistors that are in series (those in the right branch or those in the left branch) present no problem. So, our **Key Idea** is to apply the junction and loop rules.

Using arbitrarily chosen directions for the currents as shown in Fig. TE27-2, we apply the junction rule at point a by writing

$$i_3 = i_1 + i_2. \qquad \text{(TE27-1)}$$

An application of the junction rule at junction b gives only the same equation, so we next apply the loop rule to any two of the three loops of the circuit. We first arbitrarily choose the left-hand

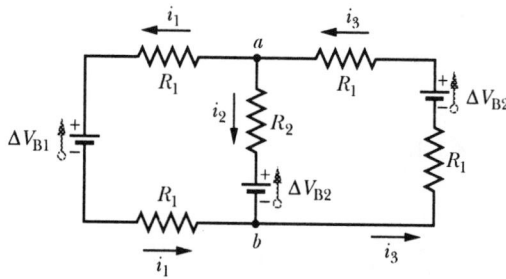

Fig. TE27-2 A multiloop circuit with three ideal batteries and five resistances.

loop, arbitrarily start at point a, and arbitrarily traverse the loop in the counterclockwise direction, obtaining

$$-i_1 R_1 - \Delta V_{B_1} - i_1 R_1 + \Delta V_{B_2} + i_2 R_2 = 0.$$

Substituting the given data and simplifying yield

$$i_1(4.0\ \Omega) - i_2(4.0\ \Omega) = 3.0\ \text{V}. \qquad \text{(TE27-2)}$$

For our second application of the loop rule, we arbitrarily choose to traverse the right-hand loop clockwise from point a, finding

$$+i_3 R_1 - \Delta V_{B_2} + i_3 R_1 + \Delta V_{B_2} + i_2 R_2 = 0.$$

Substituting the given data and simplifying yield

$$i_2(4.0\ \Omega) + i_3(4.0\ \Omega) = 0. \qquad \text{(TE27-3)}$$

Using Eq. TE27-1 to eliminate i_3 from Eq. TE27-3 and simplifying give us

$$i_1(4.0\ \Omega) + i_2(8.0\ \Omega) = 0. \qquad \text{(TE27-4)}$$

We now have a system of two equations (Eqs. TE27-2 and TE27-4) in two unknowns (i_1 and i_2) to solve either "by hand" (which is easy enough here) or with a "math package." (One solution technique is Cramer's rule, given in Appendix E.) We find

$$i_2 = -0.25\ \text{A}.$$

(The minus sign signals that our arbitrary choice of direction for i_2 in Fig. TE27-2 is wrong; i_2 should point up through ΔV_{B_2} and R_2.) Substituting $i_2 = -0.25$ A into Eq. TE27-4 and solving for i_1 then give us

$$i_1 = 0.50\ \text{A}. \qquad \text{(Answer)}$$

With Eq. TE27-1 we then find that

$$i_3 = i_1 + i_2 = 0.25\ \text{A}. \qquad \text{(Answer)}$$

The positive answers we obtained for i_1 and i_3 signal that our choices of directions for these currents are correct. We can now correct the direction for i_2 and write its magnitude as

$$i_2 = 0.25\ \text{A}. \qquad \text{(Answer)}$$

Touchstone Example 27-7-1

Let's consider a circuit with two *non-ideal* batteries that have internal resistances. Since the potential differences across the terminals of these batteries are not constant, we characterize these batteries in terms of their emf ($\varepsilon_1 \varepsilon_2$) and internal resistances ($r_1 r_2$). The

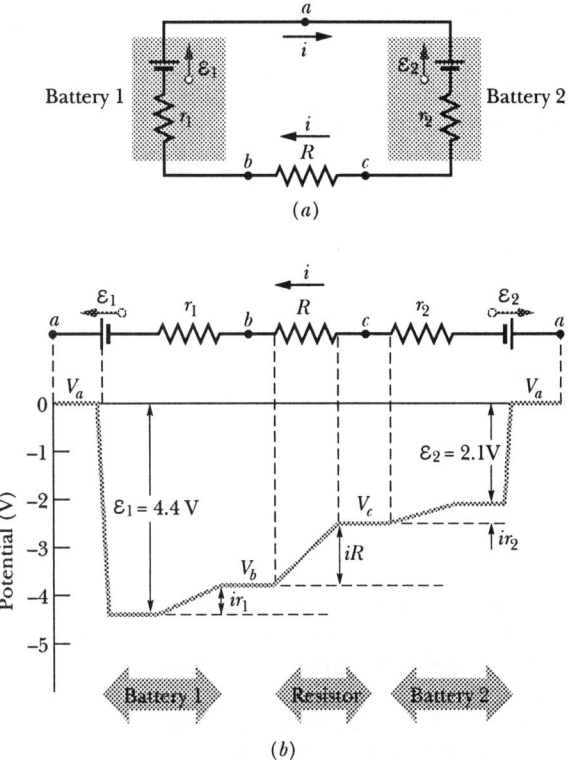

Fig. TE27-3 (*a*) A single-loop circuit containing two real batteries and a resistor. The batteries oppose each other; that is, they tend to send current in opposite directions through the resistor. (*b*) A graph of the potentials, counterclockwise from point *a*, with the potential at *a* arbitrarily taken to be zero. (To better link the circuit with the graph, mentally cut the circuit at *a* and then unfold the left side of the circuit toward the left and the right side of the circuit toward the right.)

emfs and resistances in the circuit of Fig. TE27-3*a* have the following values:

$$\varepsilon_1 = 4.4 \text{ V}, \quad \varepsilon_2 = 2.1 \text{ V},$$
$$r_1 = 2.3 \ \Omega, \quad r_2 = 1.8 \ \Omega, \quad R = 5.5 \ \Omega.$$

(a) What is the current *i* in the circuit?

SOLUTION: The **Key Idea** here is that we can get an expression involving the current *i* in this single-loop circuit by applying the loop rule. Although knowing the direction of *i* is not necessary, we can easily determine it from the emfs of the two batteries. Because ε_1 is greater than ε_2, battery 1 controls the direction of *i*, so the direction is clockwise. Let us then apply the loop rule by going counterclockwise—against the current—and starting at point *a*. We find

$$-\varepsilon_1 + ir_1 + iR + ir_2 + \varepsilon_2 = 0.$$

Check that this equation also results if we apply the loop rule clockwise or start at some point other than *a*. Also, take the time to compare this equation term by term with Fig. TE27-3*b*, which shows the potential changes graphically (with the potential at point *a* arbitrarily taken to be zero).

Solving the above loop equation for the current *i*, we obtain

$$i = \frac{\varepsilon_1 - \varepsilon_2}{R + r_1 + r_2} = \frac{4.4 \text{ V} - 2.1 \text{ V}}{5.5 \ \Omega + 2.3 \ \Omega + 1.8 \ \Omega}$$
$$= 0.2396 \text{ A} \approx 240 \text{ mA}. \qquad \text{(Answer)}$$

(b) What is the potential difference between the terminals of battery 1 in Fig. TE27-3a?

SOLUTION: The **Key Idea** is to sum the potential differences between points a and b. Let us start at point b (effectively the negative terminal of battery 1) and travel clockwise through battery 1 to point a (effectively the positive terminal), keeping track of potential changes. We find that

$$V_b - ir_1 + \varepsilon_1 = V_a,$$

which gives us

$$\begin{aligned} V_a - V_b &= -ir_1 + \varepsilon_1 \\ &= -(0.2396 \text{ A})(2.3 \text{ } \Omega) + 4.4 \text{ V} \\ &= +3.84 \text{ V} \approx 3.8 \text{ V}, \end{aligned} \qquad \text{(Answer)}$$

which is less than the emf of the battery. You can verify this result by starting at point b in Fig. TE27-3a and traversing the circuit counterclockwise to point a.

Touchstone Example 27-7-2

Electric fish are able to generate current with biological cells called *electroplaques,* which are physiological emf devices. The electroplaques in the South American eel shown in the photograph that opens this chapter are arranged in 140 rows, each row stretching horizontally along the body and each containing 5000 electroplaques. The arrangement is suggested in Fig. 28-11a; each electroplaque has an emf ε of 0.15 V and an internal resistance r of 0.25 Ω. The water surrounding the eel completes a circuit between the two ends of the electroplaque array, one end at the animal's head and the other near its tail.

(a) If the water surrounding the eel has resistance $R_w = 800$ Ω, how much current can the eel produce in the water?

SOLUTION: The **Key Idea** here is that we can simplify the circuit of Fig. 28-11a by replacing combinations of emfs and internal resistances with equivalent emfs and resistances. We first consider a single row. The total emf ε_{row} along a row of 5000 electroplaques is the sum of the emfs:

$$\varepsilon_{row} = 5000\varepsilon = (5000)(0.15 \text{ V}) = 750 \text{ V}.$$

The total resistance R_{row} along a row is the sum of the internal resistances of the 5000 electroplaques:

$$R_{row} = 5000r = (5000)(0.25 \text{ } \Omega) = 1250 \text{ } \Omega.$$

We can now represent each of the 140 identical rows as having a single emf ε_{row} and a single resistance R_{row}, as shown in Fig. 28-11b.

In Fig. 28-11b, the emf between point a and point b on any row is $\varepsilon_{row} = 750$ V. Because the rows are identical and because they are all connected together at the left in Fig. 28-11b all points b in that figure are at the same electric potential. Thus, we can consider them to be connected so that there is only a single point b. The emf between point a and this single point b is $\varepsilon_{row} = 750$ V, so we can draw the circuit as shown in Fig. 28-11c.

Between points b and c in Fig. 28-11c are 140 resistances $R_{row} = 1250$ Ω, all in parallel. The equivalent resistance R_{eq} of this combination is given by Eq. 28-21 as

$$\frac{1}{R_{eq}} = \sum_{j=1}^{140} \frac{1}{R_j} = 140 \frac{1}{R_{row}},$$

or

$$R_{eq} = \frac{R_{row}}{140} = \frac{1250 \text{ } \Omega}{140} = 8.93 \text{ } \Omega.$$

Replacing the parallel combination with R_{eq}, we obtain the simplified circuit of Fig. 28-11d.

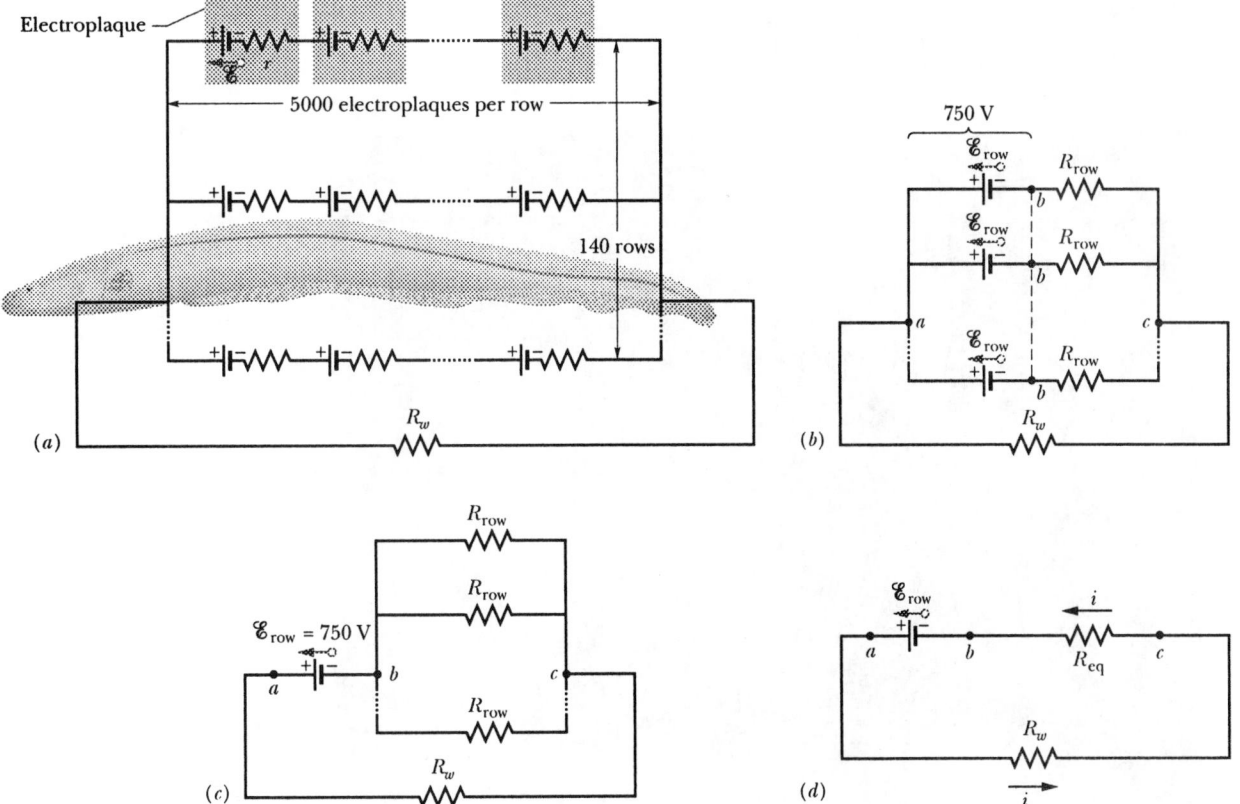

Fig. TE27-4 (a) A model of the electric circuit of an eel in water. Each electroplaque of the eel has an emf ε and internal resistance r. Along each of 140 rows extending from the head to the tail of the eel, there are 5000 electroplaques. The surrounding water has resistance R_w. (b) The emf ε_{row} and resistance R_{row} of each row. (c) The emf between points a and b is ε_{row}. Between points b and c are 140 parallel resistances R_{row}. (d) The simplifies circuit, with R_{eq} replacing the parallel combination.

Applying the loop rule to this circuit counterclockwise from point b, we have

$$\varepsilon_{\text{row}} - iR_w - iR_{\text{eq}} = 0.$$

Solving for i and substituting the known data, we find

$$i = \frac{\varepsilon_{\text{row}}}{R_w + R_{\text{eq}}} = \frac{750 \text{ V}}{800 \ \Omega + 8.93 \ \Omega}$$

$$= 0.927 \text{ A} \approx 0.93 \text{ A}. \qquad \text{(Answer)}$$

If the head or tail of the eel is near a fish, some of this current could pass along a narrow path through the fish, stunning or killing it.

(b) How much current i_{row} travels through each row of Fig. 28-11a?

SOLUTION: The **Key Idea** here is that since the rows are identical, the current into and out of the eel is evenly divided among them:

$$i_{\text{row}} = \frac{i}{140} = \frac{0.927 \text{ A}}{140} = 6.6 \times 10^{-3} \text{ A}. \qquad \text{(Answer)}$$

Thus, the current through each row is small, about two orders of magnitude smaller than the current through the water. This tends to spread the current through the eel's body, so that it need not stun or kill itself when it stuns or kills a fish.

28 Capacitance

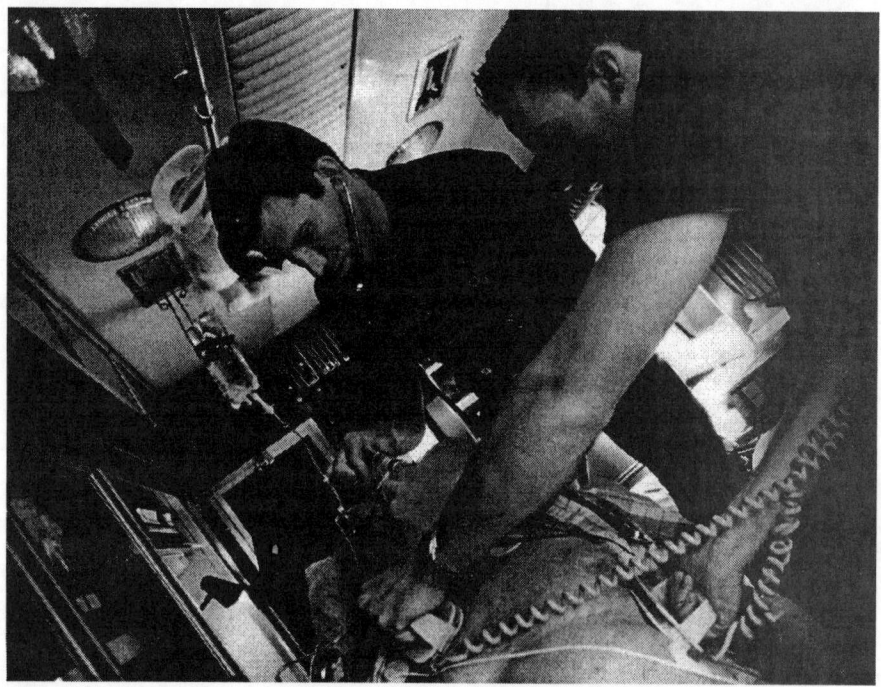

During ventricular fibrillation, a common type of heart
attack, the chambers of the heart fail to pump blood
because their muscle fibers randomly contract and relax.
To save a victim of ventricular fibrillation, the heart
muscle must be shocked to reestablish its normal rhythm.
For that, 20 A of current must be sent through the chest
cavity to transfer 200 J of electrical energy in about 2.0
ms. This requires about 100 kW of electric power. Such a
requirement may easily be met in a hospital, but not by,
say, the electrical system of an ambulance arriving to
help the victim.

**What, then, can provide the power needed for
defibrillation at remote locations?**

The answer is in this chapter.

28-1 The Uses of Capacitors

In Chapter 26 we discussed connecting an ordinary battery to a pair of metal plates as shown in Fig. 26-2. Such a pair of metal plates is one example of the basic component of a device known as a capacitor. A capacitor can be constructed using any two conductors separated by an insulator. If we connect each conductor making up a capacitor to one of the terminals of a source of potential difference such as a battery, one conductor acquires a net positive charge while the other conductor acquires a net negative charge of the same magnitude. The conductors can be any shape. Fig. 28-1 shows some other possible capacitor geometries. No matter what shape or size a capacitor's conductors are, we usually call the conductors *plates*.

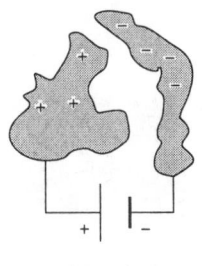 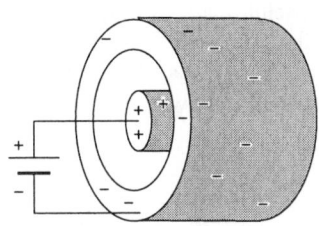 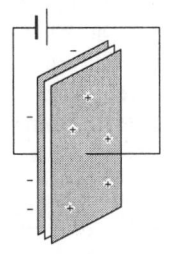

Amorphous capacitor (blobs) with air as an insulator

Cylindrical capacitor with air as an insulator

Parallel plate capacitor with paper and air as an insulator

Fig. 28-1: Three capacitors of different sizes and shapes have been connected to a battery. They each consist of a pair of conductors separated by an insulator. In each case the battery removes electrons from one of the two conductors, leaving it with excess positive charge and forces the same number of electrons to the opposite conductor.

Why construct and study capacitors? There are at least two reasons. First, capacitors are used in all sorts of circuits ranging from microcomputers to high power transmission devices. Second, capacitors are extremely useful as devices for storing electrical energy.

Capacitors in Electrical Circuits

Since a capacitor consists of conductors separated by an insulator, no current can flow through it. So at first glance, it doesn't seem to make sense to use a capacitor as a circuit element. Surprisingly, capacitors have very interesting and useful properties in circuits with changing currents through their other components. For example, variable capacitors are vital elements that enable us to tune radio and television receivers. They are found in most household electrical devices. Capacitors are used to control the frequency of the flashing lights used for warning signals at construction sites. The coaxial cables used to carry high frequency microwave and radio signals are cylindrical capacitors. Microscopic capacitors are used in communications and computers to shape the timing and strength of time varying signal transmissions. Fig. 28-2 shows some of the many sizes and shapes of a few of the capacitors commonly found in electric circuits.

Fig. 28-2: An assortment of capacitors commonly found in electrical circuits. The structures of these devices are hidden inside cases.

Capacitors as Energy Storage Devices

Just as you can store potential energy by pulling a bowstring, stretching a spring, compressing a gas, or lifting a book, you can also store electrical energy in the electric field found inside a "charged" **capacitor** as shown in Figs. 28-3 and 28-4. For example, energy storage in microscopic capacitors enable them to function as memory devices in modern digital computers and in the charge-coupled devices (CCDs) used in video cameras. Energy stored in capacitors can also be used to keep computer circuits running smoothly during brief power outages. A much larger capacitor lies at the heart of a battery-powered photoflash unit. This capacitor accumulates electrical energy relatively slowly during the time between flashes, building up an electric field as it does so. The electric field across the capacitor plates stores energy that can be released rapidly to

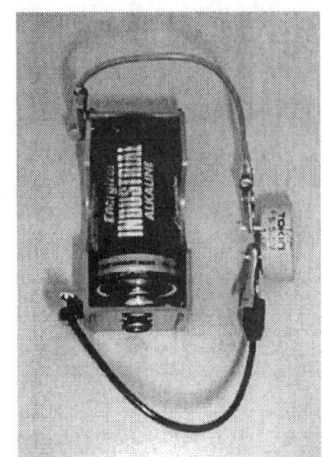

Fig. 28-3: When a battery is connected across the terminals of a capacitor, it can store electrical energy.

create an intensive flash of light. (It is important to note because capacitors are storehouses for electrical energy, some electrical devices can give you a nasty shock if you open them and accidentally touch both terminals of a capacitor—even when the device is turned off.)

28-2 Capacitance

Fig. 28-5 shows a conventional arrangement of a pair of metal plates we discussed previously. A device consisting of two parallel conducting plates of area A separated by a distance d is called a *parallel-plate capacitor*. The circuit symbol we use to represent a capacitor ($\dashv\vdash$) is based on the structure of a parallel-plate capacitor but is used for capacitors of all shapes. For the purpose of defining capacitance in a simple manner, we will consider an ideal parallel-plate capacitor with a perfect insulator between its plates. This perfect insulator allows absolutely no current to pass between capacitor plates. In fact, it is also a vacuum, so there is no matter between the capacitor plates such as air, glass or plastic. We further assume we will charge our capacitor with an ideal battery. Recall an ideal battery has no internal resistance so its emf and the potential difference across its terminals are always the same. In Section 28-6 and those following we will relax some of these idealized restrictions.

The Excess Charges on Plates are Equal and Opposite

When capacitor plates of any shape are connected to a battery, or some other voltage source, electrons flow from the negative terminal of the battery through the connecting wire and onto one plate of the capacitor. Meanwhile the positive terminal of the battery attracts electrons from the other plate, leaving an excess of positive metal atoms with missing electrons. Since we cannot find an electric field outside of a capacitor while it is connected to a battery, the overall capacitor always seems to be electrically neutral. We must conclude at any given time one plate has net or excess charge of $+q$ while the other has a net charge of $-q$. The battery appears to be separating charge by transferring electrons from one capacitor plate to the other, but this is not exactly what happens. The chemical reactions taking place in the battery are complex, so the electrons pulled off one plate are not necessarily the same ones being pushed onto the other plate. However, the battery deposits an electron on the negative plate for every one it pulls off the positive plate. In describing the behavior of capacitors, it is useful to think of the charge separation process as a transfer of charges from one capacitor plate to another.

Why Does Charge Transfer between Capacitor Plates Stop?

Observations show the battery eventually stops transferring charge because as electrons build up on the negative plate, they oppose the battery's action and start repelling the flow of additional electrons. Similarly, it becomes harder and harder for the battery to pull electrons off the positive plate as the atoms carrying positive net charge pull back on them. When enough charge has accumulated on the plates, the force exerted on an electron by the battery and the oppositely-directed forces exerted on it by the other charges on a plate cancel each other. No more electrons can flow from one plate to the other. We can use a high quality voltmeter to measure the potential difference across a capacitor just disconnected from a battery. This measurement shows that *charge separation stops when the potential difference across a capacitor has the same magnitude as the potential difference across the battery.*

Factors Affecting Charge Separation Capacity

By convention we refer to the *charge of a capacitor* as being $|q|$, the absolute value of the net charge on each plate. Although we refer to a capacitor with charges $+q$ and $-q$ on its plates as "charged," a capacitor is electrically neutral so we are actually describing its

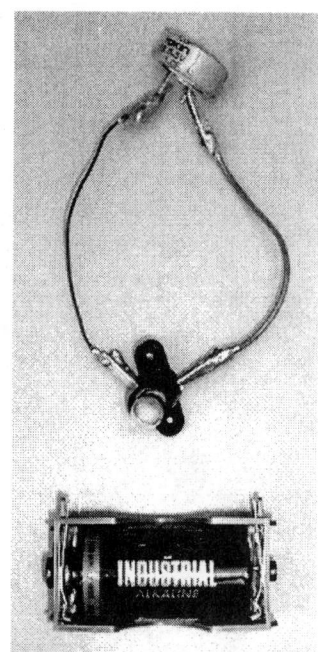

Fig. 28-4: When a "charged" capacitor is disconnected from its battery and wired in series with a bulb, the energy stored in it can light the bulb for a short period of time.

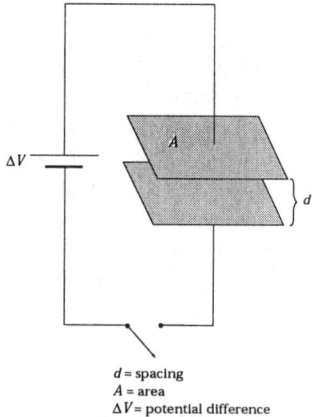

d = spacing
A = area
ΔV = potential difference

Fig. 28-5: A parallel-plate capacitor with identical plates of area A and spacing d is connected to a battery with potential difference ΔV. The plates have equal and opposite charges of magnitude $|q|$ on their facing surfaces.

charge separation created by a voltage source. What factors might affect the capacity for charge separation in a parallel-plate capacitor? We can use our knowledge of electrostatics to explore the effects of several factors. In particular, we will explore how we expect charge to depend on the potential difference across the battery terminals and on geometric factors such as the area of the plates and their spacing (Fig. 28-6):

1. *Potential Difference, ΔV :* For a given capacitor of any shape, we would expect the charge separation to be larger when the potential difference the battery places across the capacitor plates is larger. How much larger? Consider a group of n charges distributed on the plates of a capacitor. Since the plates are conductors, each one is an equipotential surface. According to Eq. 25-24 we can find the electric potential at a given point on a plate relative to infinity. We just need to know the locations of the group of n charges distributed on the capacitor plates. The potential is given by

$$V = \frac{1}{4\pi\varepsilon_0}\sum_{i=1}^{n}\frac{q_i}{r_i}$$ (Eq. 25-24)

where r_i represents the radial distance between the point where the potential is being calculated and the location of the ith charge. By examining this equation we can see if the potential is to be doubled, there needs to be twice as much charge at each location on the capacitor plates. We expect the amount of the charge separation on a capacitor to be proportional to the potential difference across its plates. We predict

$$|q| \propto |\Delta V|.$$

As you will see in the next subsection, the constant of proportionality between the charge magnitude on each plate and the potential difference across the plates for a given capacitor is known as its *capacitance*. We will deal more formally with the definition of capacitance and its units in the next section.

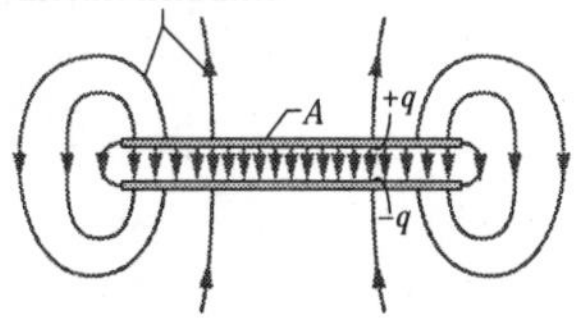

2. *Influence of Plate Area, A:* Consider a parallel-plate capacitor. For a given potential difference and plate spacing, d, how do we expect the charge separation capacity to depend on the area of the plates? When the plates have a large area, the electrons the battery is trying to push on the negative plate have more room to spread out. Likewise the un-neutralized atoms left behind when electrons are pulled off the positive plate can be distributed further apart. We expect as the area of the plates increases, so will the number of charges that can be transferred from one plate to the other.

A simple experiment can be done to show the charge separation capacity is in fact directly proportional to area. In this experiment, two sheets of aluminum foil are placed opposite to each other and separated by the insulating pages of a book. A multimeter like that described in Section 26-4 can be used to measure capacitance. Measurements are taken for different areas of foil. The results are shown in Fig. 28-7. We will derive this relationship theoretically in the next section.

Electric field lines

Fig. 28-6: As the field lines show, the electric field due to the charged plates is uniform in the central region between the plates. The field is not uniform at the edges of the plates, as indicated by the "fringing" of the field lines there.

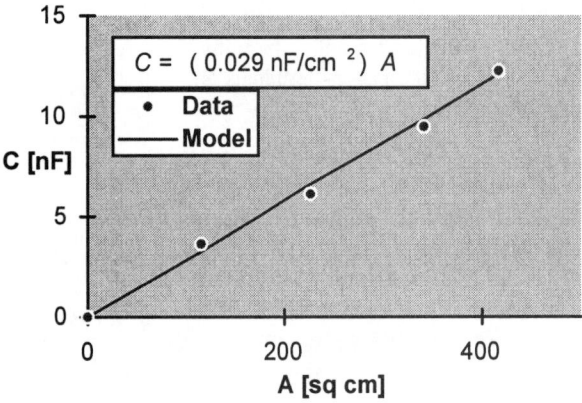

Fig. 28-7: Two rectangular pieces of aluminum foil are wedged between the insulating pages of a book. The capacitance of the system is measured as a function of the area of the plates. The result shows capacitance increasing in direct proportion to the area of the conducting aluminum plates.

3. *Influence of Plate Spacing, d*: Once again we consider a parallel-plate capacitor. For a given potential difference and plate area, A, how do we expect the charge separation capacity to depend on the spacing between the plates? When the plates have a small spacing, the excess positive charges on one plate are quite close to the excess negative charges on the other plate. Since opposite charges attract each other, these charges pull on each other across the insulating gap even though they cannot cross the gap. This attraction helps to counterbalance the repulsion between the like charges on each plate. As the spacing between plates becomes smaller, we expect the overall capacity for the charge separation caused by the battery action to become larger.

A simple experiment can be done to show the charge separation capacity does in fact increase as the spacing between plates decreases. In this experiment, two sheets of aluminum foil are placed opposite to each other and separated by the insulating pages of a book. A multimeter is used to measure capacitance as different numbers of pages are inserted between the foil plates. The results are shown in Fig. 28-8. This graph shows that the capacitance of the foil plate system is inversely proportional to the spacing, d, between the plates. We will derive this relationship theoretically in the next section.

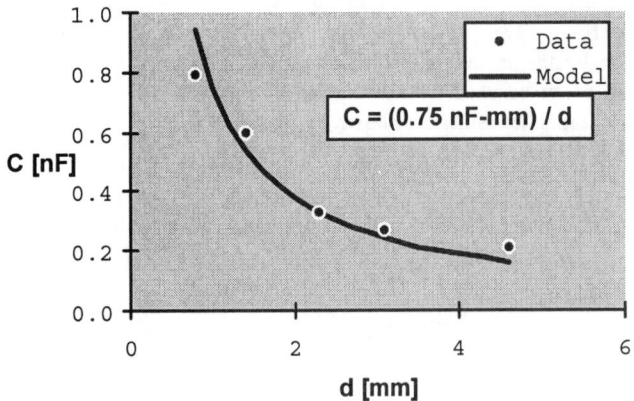

Fig. 28-8: Two rectangular pieces of aluminum foil are wedged between the insulating pages of a book. The capacitance of the system is measured as a function of the spacing between the plates. The result shows capacitance is inversely proportional to the spacing.

Defining Capacitance

As we just discussed in the last subsection, the magnitude of the excess charge on each plate of a capacitor, $|q|$, and the magnitude of the potential difference, $|\Delta V|$, across it should be proportional to each other, so

$$|q| = C|\Delta V|. \qquad (28\text{-}1)$$

The proportionality constant C is defined as the **capacitance**. The capacitance is a measure of how much charge must be put on the plates to produce a certain potential difference between them: the greater the capacitance, the larger the charge separation created by a given potential difference.

For a parallel-plate capacitor, experimental results have shown us its capacitance depends directly on the plate areas and inversely on the spacing between plates. We will see in Sections 28-6 and 28-7 that capacitance will also depend on the nature of the insulating material inserted between the plates. Capacitors having different shapes will not have the same simple relationships between plate area and spacing. In the next section, we will use the definition of electric potential and Gauss' Law to identify the theoretical geometric factors for several different types of capacitors including parallel-plate, cylindrical, and spherical capacitors.

Capacitance Units

The SI unit of capacitance following from this expression is the coulomb per volt. This unit occurs so often that it is given a special name, the *farad* (F):

$$1 \text{ farad} = 1 \text{ F} = 1 \text{ coulomb per volt} = 1 \text{ C/V}. \qquad (28\text{-}2)$$

As you will see, the farad is a very large unit. Fractions of the farad, such as the microfarad $(1\,\mu F = 10^{-6}\,F)$ and the picofarad $(1\,pF = 10^{-12}\,F)$, are more convenient units in practice. A summary of units and their common notations is shown in Table 28-1.

Table 28-1: Units of Capacitance

microfarad: 10^{-6} F = 1 μF = 1 UF
picofarad: 10^{-12} F = 1 pF = 1 $\mu\mu$F = 1 UUF
nanofarad: 10^{-9} F = 1 nF = 1000 $\mu\mu$F = 1000 UUF

READING EXERCISE 28-1: Does the capacitance C of a capacitor increase, decrease, or remain the same (a) when the charge magnitude $|q|$ on its plates is doubled and (b) when the potential difference ΔV_c across it is tripled?

28-3 Calculating the Capacitance

Our task here is to calculate the capacitance of a capacitor once we know its geometry. Because we will consider a number of different geometries, it seems wise to develop a general plan to simplify the work. In brief, our plan is as follows:

1. Assume a charge q on the plates;

2. Calculate the electric field $\vec{E}$ between the plates in terms of this charge, using Gauss' law;

3. Knowing $\vec{E}$, calculate the potential difference ΔV between the plates from

$$V_f - V_i = -\int_i^f \vec{E} \cdot d\vec{s} \qquad (\text{Eq. 25-17});$$

4. Calculate C from $|q| = C|\Delta V|$ (Eq. 28-1).

Before we start, we can simplify the calculation of both the electric field and the potential difference by making certain assumptions. We discuss each in turn.

Calculating the Electric Field

To relate the electric field $\vec{E}$ between the plates of a capacitor to the charge magnitude $|q|$ on either plate, we shall use Gauss' law:

$$\varepsilon_0 \oint \vec{E} \cdot d\vec{A} = q. \qquad (28\text{-}3)$$

Here q is the net charge enclosed by a Gaussian surface, and $\oint \vec{E} \cdot d\vec{A}$ is the net electric flux through that surface. In all cases we shall consider, the Gaussian surface will be such whenever electric flux passes through it, $\vec{E}$ will have a uniform magnitude E and the vectors $\vec{E}$ and $d\vec{A}$ will be parallel. This equation will then reduce to

$$|q| = \varepsilon_0 |\vec{E}| A \qquad \text{(special case of Eq. 28-3)}, \qquad (28\text{-}4)$$

in which A is the area of the part of the Gaussian surface through which flux passes. For convenience, we shall always draw the Gaussian surface in such a way it completely encloses the charge on the positive plate; see Fig. 28-9 for an example.

Calculating the Potential Difference

In the notation of Chapter 25 (Eq. 25-17), the potential difference between the plates of a capacitor is related to the field $\vec{E}$ by

$$V_f - V_i = -\int_i^f \vec{E} \cdot d\vec{s}, \qquad (28\text{-}5)$$

in which the integral is to be evaluated along any path starting on one plate and ends on the other. We shall always choose a path following an electric field line, from the negative plate to the positive plate. For this path, the vectors $\vec{E}$ and $d\vec{s}$ will have opposite directions, so the dot product $\vec{E} \cdot d\vec{s}$ will be equal to $-|\vec{E}||d\vec{s}|$. The right side of this equation will then be positive. Letting ΔV represent the difference, $V_f - V_i$, we can then recast the relationship as

$$\Delta V = -\int_-^+ |\vec{E}||d\vec{s}| \qquad \text{(special case of Eq. 28-5)}, \qquad (28\text{-}6)$$

in which the "–" and "+" remind us our path of integration starts on the negative plate and ends on the positive plate.

We are now ready to apply $|q| = \varepsilon_0 |\vec{E}| A$ (Eq. 28-4) and $\Delta V = \int_-^+ |\vec{E}||d\vec{s}|$ (Eq. 28-6) to some particular cases.

A Parallel-Plate Capacitor

We assume, as Fig. 28-9 suggests, the plates of our parallel-plate capacitor are so large and so close together we can neglect the fringing of the electric field at the edges of the plates, taking $\vec{E}$ to be constant throughout the region between the plates. This configuration was used in old time radios. As we will see in Chapter 33, the frequency of an oscillating circuit depends on the capacitance. In old radios the dial was connected to a set of nested metal plates. When the dial was turned, some of the plates rotated while others stayed fixed. By turning the dial, the overlap of the plates changed, changing the capacitance and thereby the frequency of the signal selected.

We draw a Gaussian surface enclosing just the excess charge q on the positive plate, as in Fig. 28-9. Recall from above

$$|q| = \varepsilon_0 |\vec{E}| A, \qquad (28\text{-}7)$$

where A is the area of each of the plates.

Equation 28-6 yields

$$\Delta V = \int_-^+ |\vec{E}||d\vec{s}| = |\vec{E}| \int_0^d ds = |\vec{E}| d. \qquad (28\text{-}8)$$

Here, $|\vec{E}|$ can be placed outside the integral because it is a constant; the second integral then is simply the plate separation d.

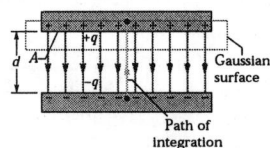

Fig. 28-9: A charged parallel-plate capacitor. A Gaussian surface encloses the charge on the positive plate. The integration of Eq. 28-6 is taken along a path extending directly from the negative plate to the positive plate.

Combining these two expressions with the relation $|q| = C|\Delta V|$ (Eq. 28-1), we find

$$C = \frac{\varepsilon_0 A}{d} \qquad \text{(parallel-plate capacitor).} \qquad (28\text{-}9)$$

This theoretical relationship matches the results of the experiments we presented in the last section! The capacitance does indeed depend only on geometrical factors—namely, the plate area A and the plate separation d. Note C increases as we increase the plate area A or decrease the separation d.

As an aside, we point out this expression suggests one of our reasons for writing the electrostatic constant in Coulomb's law in the form $1/4\pi\varepsilon_0$. If we had not done so, the expression for the capacitance of a parallel-plate capacitor above—which is used more often in engineering practice than Coulomb's law—would have been less simple in form. We note further it permits us to express the permittivity constant ε_0 in a unit more appropriate for use in problems involving capacitors; namely,

$$\varepsilon_0 = 9.0 \times 10^{-12} \ \text{F/m} = 9.0 \ \text{pF/m}. \qquad (28\text{-}10)$$

We have previously expressed this constant as

$$\varepsilon_0 = 9.0 \times 10^{-12} \ \text{C}^2/\text{N} \cdot \text{m}^2. \qquad (28\text{-}11)$$

A Cylindrical Capacitor

Fig. 28-11 shows, in cross section, a cylindrical capacitor of length L formed by two coaxial cylinders of radii a and b. We assume $L \gg b$ so we can neglect the fringing of the electric field occuring at the ends of the cylinders. Each plate contains a charge of magnitude $|q|$. This configuration is important because coaxial cables are used in the communications industry for the long distance transmission of electrical signals (Fig 28-10).

The electric field inside the cylinder is highly symmetrical, so we can use Gauss's Law to determine its values. As a Gaussian surface, we choose a cylinder of length L and radius r, closed by end caps and placed as is shown in Fig. 28-11. Then

$$|q| = \varepsilon_0 |\vec{E}| A = \varepsilon_0 |\vec{E}| (2\pi r L),$$

in which $2\pi r L$ is the area of the curved part of the Gaussian surface. There is no flux through the end caps. Solving for $|\vec{E}|$ yields

$$|\vec{E}| = \frac{|q|}{2\pi\varepsilon_0 L r}. \qquad (28\text{-}12)$$

Substitution of this result into our general expression for potential difference yields

$$\Delta V = \int_-^+ \vec{E} \cdot d\vec{s} = -\frac{q}{2\pi\varepsilon_0 L} \int_b^a \frac{dr}{r} = \frac{q}{2\pi\varepsilon_0 L} \ln\left(\frac{b}{a}\right), \qquad (28\text{-}13)$$

where here $ds = -dr$ (we integrated radially inward). From the relation $C = |q/\Delta V|$, we then have

$$C = 2\pi\varepsilon_0 \frac{L}{\ln(b/a)} \qquad \text{(cylindrical capacitor).} \qquad (28\text{-}14)$$

We see the capacitance of a cylindrical capacitor, like that of a parallel-plate capacitor, depends only on geometrical factors, in this case L, b, and a.

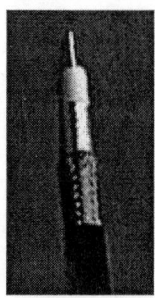

(a) cable

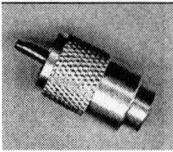

(b) connector

Fig. 28-10: Coaxial cables and connectors are cylindrical capacitors used for long distance transmission of television and radio signals. The cable consists of a central conducting wire surrounded by a layer of insulation and then a cylindrical conductor. All three elements are centered on the same axis.

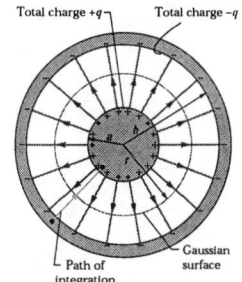

Fig. 28-11: A cross section of a long cylindrical capacitor, showing a cylindrical Gaussian surface of radius r (that encloses the positive plate) and the radial path of integration along which Eq. 28-6 is to be applied. This figure also serves to illustrate a spherical capacitor in a cross section through its center.

A Spherical Capacitor

Fig. 28-11 can also serve as a central cross section of a capacitor consisting of two concentric spherical shells, of radii a and b. As a Gaussian surface we draw a sphere of radius r concentric with the two shells; then

$$|q| = \varepsilon_0 |\vec{E}| A = \varepsilon_0 |\vec{E}| (4\pi r^2),$$

in which $4\pi r^2$ is the area of the spherical Gaussian surface. We solve this equation for $|\vec{E}|$, obtaining

$$|\vec{E}| = \frac{1}{4\pi\varepsilon_0} \frac{|q|}{r^2}, \tag{28-15}$$

which we recognize as the expression for the electric field due to a uniform spherical charge distribution from Chapter 24.

If we substitute this expression into Eq. 28-6, we find

$$\Delta V = \int_{-}^{+} \vec{E} \cdot d\vec{s} = -\frac{|q|}{4\pi\varepsilon_0} \int_b^a \frac{dr}{r^2} = \frac{|q|}{4\pi\varepsilon_0}\left(\frac{1}{a} - \frac{1}{b}\right) = \frac{|q|}{4\pi\varepsilon_0}\frac{b-a}{ab}, \tag{28-16}$$

where again we have substituted $-dr$ for ds. If we now substitute this into $|q| = C|\Delta V|$ (Eq. 28-1) and solve for C, we find

$$C = 4\pi\varepsilon_0 \frac{ab}{b-a} \qquad \text{(spherical capacitor).} \tag{28-17}$$

An Isolated Sphere

We can assign a capacitance to a *single* isolated spherical conductor of radius R by assuming that the "missing plate" is a conducting sphere of infinite radius. After all, the field lines leaving the surface of a positively charged isolated conductor must end somewhere; the walls of the room in which the conductor is housed can serve effectively as our sphere of infinite radius.

To find the capacitance of the isolated conductor, we first rewrite the expression for a spherical capacitor above as

$$C = 4\pi\varepsilon_0 \frac{a}{1 - a/b}.$$

If we then let $b \to \infty$ and substitute R for a, we find

$$C = 4\pi\varepsilon_0 R \qquad \text{(isolated sphere).} \tag{28-18}$$

Note this formula and the others we have derived for capacitance (Eqs. 28-9, 28-14, and 28-17) involve the constant ε_0 multiplied by a quantity having the dimensions of a length.

READING EXERCISE 28-2: Consider capacitors charged by and then removed from the same battery. Does the charge on the capacitor plates increase, decrease, or remain the same in each of the following situations? (a) The plate separation of a parallel-plate capacitor is increased. (b) The radius of the inner cylinder of a cylindrical capacitor is increased. (c) The radius of the outer spherical shell of a spherical capacitor is increased.

READING EXERCISE 28-3: Consider capacitors charged by identical batteries. If the capacitors stay connected to the batteries, does the magnitude of the excess charge on the capacitor plates increase, decrease, or remain the same in each of the following situations? (a) The plate separation of a parallel-plate capacitor is increased. (b) The radius of the inner cylinder of a cylindrical capacitor is increased. (c) The radius of the outer spherical shell of a spherical capacitor is increased.

28-4 Capacitors in Parallel and in Series

When there is a combination of capacitors in a circuit, we can sometimes replace that combination with an **equivalent capacitor**—that is, a single capacitor having the same behavior as the actual combination of capacitors. With such a replacement, we can simplify circuits. In addition, circuits often have what is termed *stray capacitance* due to the presence of conductors and insulators in other types of circuit elements. Knowing how the effective capacitances of such elements might combine with each other and other capacitors in the vicinity is vital to the design of high performance circuits. In this section we discuss the behavior of two basic types of capacitor combinations—parallel and series.

Capacitors in Parallel

Fig. 28-12*a* shows an electric circuit in which three capacitors are connected *in parallel* with battery B. This description has little to do with where the capacitor plates appear in the diagram. Rather, "in parallel" means one plate of each capacitor is wired directly to one plate of the other capacitors. The opposite plates of the capacitors are also wired to each other. When the parallel combination is connected to a battery, the battery's potential difference ΔV_b is applied across all three capacitors as shown in Fig. 28-12*a*.

We can anticipate how the parallel combination will behave by considering the special case in which all three capacitors are parallel-plate capacitors with the same spacing. What happens in this case is the effective area of the plates of the combined network of capacitors is equal to the sum of the three areas. Using Eq. 28-9 we see

$$C_{eq} = \frac{\varepsilon_0 A}{d} = \frac{\varepsilon_0 (A_1 + A_2 + A_3)}{d} = \frac{\varepsilon_0 A_1}{d} + \frac{\varepsilon_0 A_2}{d} + \frac{\varepsilon_0 A_3}{d} = C_1 + C_2 + C_3.$$

Even if the three capacitors are of different types with each having a different geometry, we expect the effective area of the combination will be increased. The proof of the pudding is in the experiment. It turns out a multimeter set to measure capacitance can be used to verify

$$C_{eq} = C_1 + C_2 + C_3,$$

for parallel combinations of three capacitors of all sorts of different types. Since the potential difference across a parallel combination of capacitors connected to a voltage source is the same, we can use the expression $|q| = C|\Delta V|$ (Eq. 28-1) to show if $C_{eq} = C_1 + C_2 + C_3$, then

$$\frac{|q_{eq}|}{|\Delta V|} = \frac{|q_1|}{|\Delta V|} + \frac{|q_2|}{|\Delta V|} + \frac{|q_3|}{|\Delta V|} \text{ so that } |q_{eq}| = |q_1| + |q_2| + |q_3|$$

In general,

▶When a potential difference ΔV is applied across several capacitors connected in parallel, that potential difference ΔV is applied across each capacitor. The total magnitude of the charge $|q|$ found on each plate of the equivalent capacitor is equal to the sum of the charge magnitude on each of the capacitors.

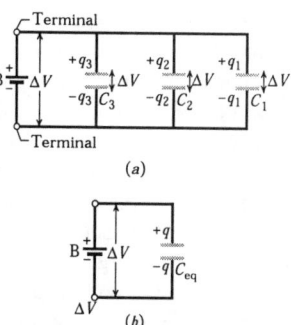

Fig. 28-12: (*a*) Three capacitors connected in parallel to battery B. The battery maintains potential difference ΔV across its terminals and thus across each capacitor. (*b*) The equivalent capacitor, with capacitance C_{eq}, replaces the parallel combination.

When we analyze a circuit of capacitors in parallel, we can simplify it with this mental replacement:

➤Capacitors connected in parallel can be replaced with an equivalent capacitor that has the same total charge $|q|$ and the same potential difference ΔV as the actual capacitors.

We can easily extend our method for finding the equivalent capacitance for three capacitors to any number of capacitors. For n capacitors wired in parallel,

$$C_{eq} = \sum_{j=1}^{n} C_j \qquad\qquad \text{(n capacitors in parallel).} \qquad (28\text{-}19)$$

In order to find the equivalent capacitance of a parallel combination, we simply add the individual capacitances.

Capacitors in Series

Fig. 28-13a shows three capacitors connected *in series* to battery B. This description has little to do with where the capacitors are located on the drawing. Rather, "in series" means the capacitors are wired serially, one after the other, so a battery can set up a potential difference ΔV across the two ends of the series as shown in Fig. 28-13a.

Let's consider what goes on with the charges on the capacitor plates of arbitrary geometries by following a *chain reaction* of events, in which the charging of each capacitor causes the charging of the next capacitor. We start with capacitor 3 and work upward to capacitor 1. When the battery is first connected to the series of capacitors, it produces charge $-q$ on the bottom plate of capacitor 3. That charge then repels negative charge from the top plate of capacitor 3 (leaving it with charge $+q$). The repelled negative charge moves to the bottom plate of capacitor 2 (giving it charge $-q$). That charge on the bottom plate of capacitor 2 then repels negative charge from the top plate of capacitor 2 (leaving it with charge $+q$) to the bottom plate of capacitor 1 (giving it charge $-q$). Finally the charge on the bottom plate of capacitor 1 helps move negative charge from the top plate of capacitor 1 to the battery, leaving that top plate with charge $+q$. We see then the potential differences existing across the capacitors in the series produce identical charges of magnitude $|q|$ on their plates.

Since the magnitudes of the charges on each pair of plates in a series connection are the same, we can use Eq. 28-1, $|q| = C|\Delta V|$, to summarize our reasoning in equation form:

$$|q_1| = |q_2| = |q_3| = |q| \quad \text{and so} \quad |\Delta V_1| = \frac{|q|}{C_1}, \; |\Delta V_2| = \frac{|q|}{C_2}, \; \text{and} \; |\Delta V_3| = \frac{|q|}{C_3}.$$

The total potential difference ΔV due to the battery is the sum of these three potential differences. Thus,

$$|\Delta V| = |\Delta V_1| + |\Delta V_2| + |\Delta V_3| \quad \text{so that} \quad \frac{|q|}{C_{eq}} = \frac{|q|}{C_1} + \frac{|q|}{C_2} + \frac{|q|}{C_3}.$$

The equivalent capacitance is then

$$C_{eq} = \frac{|q|}{|\Delta V|} \quad \text{and also} \quad \frac{1}{C_{eq}} = \frac{1}{C_1} + \frac{1}{C_2} + \frac{1}{C_3}.$$

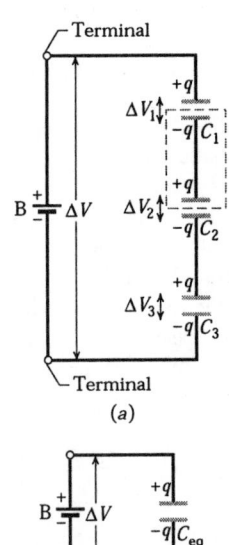

Fig. 28-13: (*a*) Three capacitors connected in series to battery B. The battery maintains potential difference ΔV between the top and bottom plates of the series combination. (*b*) The equivalent capacitor, with capacitance C_{eq}, replaces the series combination.

➤When a potential difference of magnitude $|\Delta V|$ is applied across several capacitors connected in series, each of the capacitors has the same magnitude of charge $|q|$ on its two plates. The sum of the potential differences across the entire network of capacitors is equal to the magnitude of the applied potential difference $|\Delta V|$.

Here is an important point about capacitors in series: when charge is shifted from one capacitor to another in a series of capacitors, it can move along only one route, such as from capacitor 3 to capacitor 2 in Fig. 28-13a. If there are additional routes, the capacitors are not in series. Hence, when we analyze a circuit of capacitors in series, we can simplify it with this mental replacement:

➤Capacitors connected in series can be replaced with an equivalent capacitor having the same magnitude of charge $|q|$ and the same magnitude of total potential difference $|\Delta V|$ as the actual capacitors.

We can easily extend our method of determining the equivalent capacitance of a set of capacitors wired in series from three capacitors to n capacitors by using the expression

$$\frac{1}{C_{eq}} = \sum_{j=1}^{n} \frac{1}{C_j} \qquad (n \text{ capacitors in series}). \qquad (28\text{-}20)$$

Using this expression, you can show the equivalent of a series of capacitances is always less than the least capacitance in the series.

Table 28-2 summarizes the equivalence relations for resistors and capacitors in series and in parallel. It also presents the information about potential differences and charges on the combinations we determined by thinking about the physics of how the charges move and distribute themselves in these different geometrical configurations.

TABLE 28-2: Series and Parallel Resistors and Capacitors

Series	Parallel	Series	Parallel
Resistors		Capacitors	
$R_{eq} = \sum_{j=1}^{n} R_j$	$\frac{1}{R_{eq}} = \sum_{j=1}^{n} \frac{1}{R_j}$	$\frac{1}{C_{eq}} = \sum_{j=1}^{n} \frac{1}{C_j}$	$C_{eq} = \sum_{j=1}^{n} C_j$
Eq. 27-4	Eq. 27-12	Eq. 28-20	Eq. 28-19
Same current through all resistors	Same potential difference across all resistors	Same charge on all capacitors	Same potential difference across all capacitors

READING EXERCISE 28-4: A battery with a potential difference ΔV stores charge of magnitude $|q|$ on each of two identical capacitors. What is the potential difference across, and the charge on either capacitor if the capacitors are wired (a) in parallel and (b) in series?

Touchstone Examples 28-4-1 and 28-4-2, at the end of this chapter, illustrate how to use what you learned in this section.

TE

28-5 Energy Stored in an Electric Field

Work must be done by an external agent to charge a capacitor. Starting with an uncharged capacitor, for example, imagine—using "magic tweezers"—you remove electrons from one plate and transfer them one at a time to the other plate. The electric field building up in the space between the plates has a direction that tends to oppose further transfer. As charge accumulates on the capacitor plates, you have to do increasingly larger amounts of work to transfer additional electrons. In practice, this work is done not by "magic tweezers" but by a battery, at the expense of its store of chemical energy.

We visualize the work required to charge a capacitor as being stored in the form of **electric potential energy** U in the electric field between the plates. You can recover this energy at will, by discharging the capacitor in a circuit, just as you can recover the potential energy stored in a stretched bow by releasing the bowstring to transfer the energy to the kinetic energy of an arrow. It's like carrying rocks up a hill against gravity.

Energy is stored in the height and can be recovered by letting the rocks fall down again. In a capacitor, we can recover the stored energy by connecting wires to the ends.

Suppose at a given instant, a charge $|q'|$ has been transferred from one plate of a capacitor to the other. The potential difference $|\Delta V'|$ between the plates at that instant will be $|q'|/C$. If an extra increment of charge $|dq'|$ is then transferred, the increment of work required will be, (from Chapter 25),

$$|dW| = |\Delta V'||dq'| = \frac{|q'|}{C}|dq'|.$$

The work required to bring the total capacitor charge separation up to a final value $|q|$ is

$$W = \int dW = \frac{1}{C}\int_0^q q'\,dq' = \frac{q^2}{2C}.$$

This work is stored as potential energy U in the capacitor, so

$$U = \frac{q^2}{2C} \qquad \text{(potential energy).} \qquad (28\text{-}21)$$

From $|q| = C|\Delta V|$, we can also write this as

$$U = \tfrac{1}{2}C(\Delta V)^2 \quad \text{(potential energy).} \qquad (28\text{-}22)$$

These relations hold no matter what the geometry of the capacitor is.

To gain some physical insight into energy storage, consider two parallel-plate capacitors identical except capacitor 1 has twice the plate separation of capacitor 2. Then capacitor 1 has twice the volume between its plates and also, from Eq. 28-9, half the capacitance of capacitor 2. Equation 28-4 tells us if both capacitors have the same charge q, the electric fields between their plates are identical. Equation 28-21 tells us capacitor 1 has twice the stored potential energy of capacitor 2. Of two otherwise identical capacitors with the same charge and same electric field, the one with twice the volume between its plates has twice the stored potential energy. Arguments like this tend to verify our earlier assumption:

▶The potential energy of a charged capacitor may be viewed as being stored in the electric field between its plates.

The Medical Defibrillator

The ability of a capacitor to store potential energy is the basis of *defibrillator* devices, which are used by emergency medical teams to stop the fibrillation of heart attack victims. In the portable version, a battery charges a capacitor to a high potential difference, storing a large amount of energy in less than a minute. The battery maintains only a modest potential difference; an electronic circuit repeatedly uses that potential difference to greatly increase the potential difference of the capacitor. The power, or rate of energy transfer, during this process is also modest.

Conducting leads ("paddles") are placed on the victim's chest. When a control switch is closed, the capacitor sends a portion of its stored energy from paddle to paddle through the victim. As an example, when a $70\ \mu F$ capacitor in a defibrillator is charged to 5000 V, Eq. 28-22 gives the energy stored in the capacitor as

$$U = \tfrac{1}{2}C(\Delta V)^2 = \tfrac{1}{2}(70 \times 10^{-6}\ \text{F})(5000\ \text{V})^2 = 875\ \text{J}.$$

About 200 J of this energy is sent through the victim during a pulse of about 2.0 ms. The power of the pulse is

$$P = \frac{U}{t} = \frac{200\ \text{J}}{2.0 \times 10^{-3}\ \text{s}} = 100\ \text{kW},$$

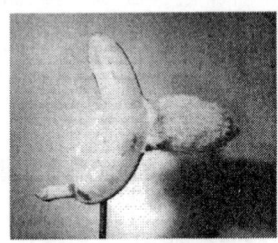

Fig. 28-14: To photograph a bullet blowing apart a banana, Harold Edgerton, the inventor of the stroboscope, used a capacitor to dump electrical energy into one of his stroboscopic lamps, which then brightly illuminated the banana for only $0.3\ \mu s$.

which is much greater than the power of the battery itself. This same technique of slowly charging a capacitor with a battery and then discharging the capacitor at a much higher power is commonly used in flash photography and stroboscopic photography (Fig. 28-14).

Energy Density

In a parallel-plate capacitor, neglecting fringing, the electric field has the same value at all points between the plates. The **energy density** u—that is, the potential energy per unit volume between the plates—should also be uniform. We can find u by dividing the total potential energy by the volume Ad of the space between the plates. Using Eq. 28-22, we obtain

$$u = \frac{U}{Ad} = \frac{C(\Delta V)^2}{2Ad}.$$

With Eq. 28-9 $(C = \varepsilon_0 A/d)$, this result becomes

$$u = \tfrac{1}{2}\varepsilon_0\left(\frac{\Delta V}{d}\right)^2.$$

However, from Eq. 25-36, $\Delta V/d$ equals the electric field magnitude $|\vec{E}|$, so

$$u = \tfrac{1}{2}\varepsilon_0 E^2 \qquad \text{(energy density).} \qquad (28\text{-}23)$$

Although we derived this result for the special case of a parallel-plate capacitor, it holds generally, whatever may be the source of the electric field. If an electric field $\vec{E}$ exists at any point in space, we can think of that point as a site of electric potential energy whose amount per unit volume is given by Eq. 28-23.

Touchstone Example 28-5-1, at the end of this chapter, illustrates how to use what you learned in this section.

TE

28-6 Capacitor with a Dielectric

If you fill the space between the plates of a capacitor with a *dielectric*, which is usually an insulating material such as mineral oil or plastic, what happens to the capacitance? Michael Faraday—to whom the whole concept of capacitance is largely due and for whom the SI unit of capacitance is named—first looked into this matter in 1837. Using simple equipment much like that shown in Fig. 28-15, he found that the capacitance *increased* by a numerical factor κ, which he called the dielectric constant of the insulating material. Table 28-3 shows some dielectric materials and their dielectric constants. The dielectric constant of a vacuum is unity by definition. Because air is mostly empty space, its measured dielectric constant is only slightly greater than unity.

Fig. 28-15: The simple electrostatic apparatus used by Faraday. An assembled apparatus (second from left) forms a spherical capacitor consisting of a central brass ball and a concentric brass shell. Faraday placed dielectric materials in the space between the ball and the shell.

TABLE 28-3: Some Properties of Dielectrics[a]

Material	Dielectric Constant κ	Dielectric Strength (kV/mm)
Air (1 atm)	1.00054	3
Polystyrene	2.6	24
Paper	3.5	16
Transformer oil	4.5	
Pyrex	4.7	14
Ruby mica	5.4	
Porcelain	6.5	
Silicon	12	
Germanium	16	
Ethanol	25	
Water (20°C)[b]	80.4	
Water (25°C)[b]	78.5	
Titania ceramic	130	
Strontium titanate	310	8

For a vacuum, κ = unity.

[a]Measured at room temperature, except for the water.
[b]Note that water is not an insulating material. It is listed because it has dielectric properties.

Another effect of the introduction of a dielectric is to limit the potential difference that can be applied between the plates to a certain value ΔV_{max}, called the *breakdown potential*. If this value is substantially exceeded, the dielectric material will break down and form a conducting path between the plates. That is, when the capacitor is filled with a dielectric, the charge separation you can maintain with a given potential difference increases. Every dielectric material has a characteristic *dielectric strength*, which is the maximum value of the electric field that it can tolerate without breakdown. A few such values are listed in Table 28-3.

As we discussed in connection with Eq. 28-18, the capacitance of any capacitor can be written in the form

$$C = \varepsilon_0 L, \qquad (28\text{-}24)$$

in which L has the dimensions of a length. For example, $L = A/d$ for a parallel-plate capacitor. Faraday's discovery was, with a dielectric *completely* filling the space between the plates, Eq. 28-24 becomes

$$C = \kappa \varepsilon_0 L = \kappa C_{air}, \qquad (28\text{-}25)$$

where C_{air} is the value of the capacitance with only air between the plates.

Fig. 28-16 provides some insight into Faraday's experiments. In Fig. 28-16a the battery ensures the potential difference ΔV between the plates will remain constant. When a dielectric slab is inserted between the plates, the charge of magnitude $|q|$ on the plates increases by a factor of κ; the additional charge is delivered to the capacitor plates by the battery. In Fig. 28-16b there is no battery and therefore the charge magnitude $|q|$ must remain constant when the dielectric slab is inserted; then the potential difference ΔV between the plates decreases by a factor of κ. Both these observations are consistent (through the relation $|q| = C|\Delta V|$) with the increase in capacitance caused by the dielectric.

Comparison of Eqs. 28-24 and 28-25 suggests that the effect of a dielectric can be summed up in more general terms:

➤In a region completely filled by a dielectric material of dielectric constant κ, all electrostatic equations containing the permittivity constant ε_0 are to be modified by replacing ε_0 with $\kappa\varepsilon_0$.

A point charge inside a dielectric produces an electric field that, by Coulomb's law, has the magnitude

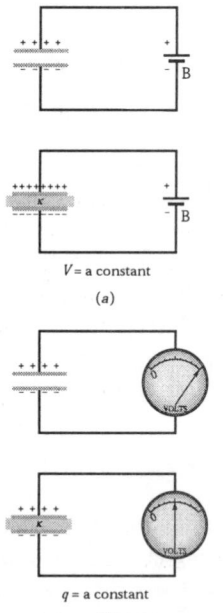

Fig. 28-16: (*a*) If the potential difference between the plates of a capacitor is maintained, as by battery B, the effect of a dielectric is to increase the charge on the plates. (*b*) If the charge on the capacitor plates is maintained, as in this case, the effect of a dielectric is to reduce the potential difference between the plates. The scale shown is that of a *potentiometer*, a device used to measure potential difference (here, between the plates). A capacitor cannot discharge through a potentiometer.

$$|\vec{E}| = \frac{1}{4\pi\kappa\varepsilon_0}\frac{|q|}{r^2}. \qquad (28\text{-}26)$$

Also, the expression for the electric field just outside an isolated conductor immersed in a dielectric (see Eq. 24-18) becomes

$$|\vec{E}| = \frac{|\sigma|}{\kappa\varepsilon_0}. \qquad (28\text{-}27)$$

Both these equations show *for a fixed distribution of charges, the effect of a dielectric is to weaken the electric field that would otherwise be present.*

Touchstone Example 28-6-1, at the end of this chapter, illustrates how to use what you learned in this section.

TE

28-7 Dielectrics: An Atomic View

What happens, in atomic and molecular terms, when we put a dielectric in an electric field? There are two possibilities, depending on the nature of the molecules:

1. *Polar dielectrics.* The molecules of some dielectrics, like water, have permanent electric dipole moments. In such materials (called *polar dielectrics*), the electric dipoles tend to line up with an external electric field as in Fig. 28-17. Because the molecules are continuously jostling each other as a result of their random thermal motion, this alignment is not complete, but it becomes more complete as the magnitude of the applied field is increased (or as the temperature, and thus the jostling, is decreased). The alignment of the electric dipoles produces an electric field directed opposite the applied field and smaller in magnitude.

2. *Nonpolar dielectrics.* Regardless of whether they have permanent electric dipole moments, molecules acquire dipole moments by induction when placed in an external electric field. In Section 25-9 (see Fig. 25-13), we saw this occurs because the external field tends to "stretch" the molecules, slightly separating the centers of negative and positive charge.

Fig. 28-18a shows a nonpolar dielectric slab with no external electric field applied. In Fig. 28-18b, an electric field E_0 is applied via a capacitor, whose plates are charged as shown. The result is a slight separation of the centers of the positive and negative charge distributions within the slab, producing positive charge on one face of the slab (due to the positive ends of dipoles there) and negative charge on the opposite face (due to the negative ends of dipoles there). The slab as a whole remains electrically neutral and—within the slab—there is no excess charge in any volume element.

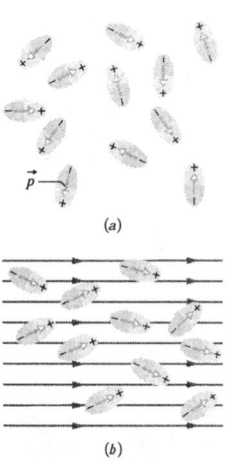

Fig. 28-17: (a) Molecules with a permanent electric dipole moment, showing their random orientation in the absence of an external electric field. (b) An electric field is applied, producing partial alignment of the dipoles. Thermal agitation prevents complete alignment.

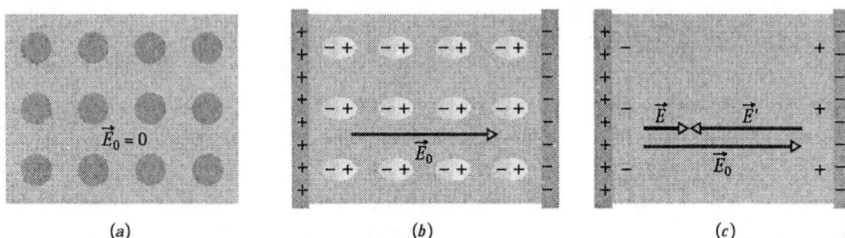

Fig. 28-18: (a) A nonpolar dielectric slab. The circles represent the electrically neutral atoms within the slab. (b) An electric field is applied via charged capacitor plates; the field slightly stretches the atoms, separating the centers of positive and negative charge. (c) The separation produces surface charges on the slab faces. These charges set up a field E', which opposes the applied field E_0. The resultant field E inside the dielectric (the vector sum of E_0 and E') has the same direction as E_0 but less magnitude.

Fig. 28-18c shows the induced surface charges on the faces produce an electric field $\vec{E}'$, in the direction opposite the applied electric field E_0. The resultant field $\vec{E}$ inside the dielectric (the vector sum of fields E_0 and $\vec{E}'$) has the direction of E_0 but is smaller in magnitude.

Both the field $\vec{E}'$ produced by the surface charges in Fig. 28-18c and the electric field produced by the permanent electric dipoles in Fig. 28-17 act in the same way—they oppose the applied field $\vec{E}$. (Inside the material, the E field fluctuates wildly, depending on whether you are close to one side of a molecule or another. The effects we are looking at are the average effects of the molecules.) Thus, the effect of both polar and nonpolar dielectrics is to weaken any applied field within them, as between the plates of a capacitor.

We can now see why the dielectric porcelain slab in Touchstone Example 28-6-1 is pulled into the capacitor: As it enters the space between the plates, the surface charge appearing on each slab face has the sign opposite the charge on the nearby capacitor plate. Thus, slab and plates attract each other.

28-8 Dielectrics and Gauss' Law

In our discussion of Gauss' law in Chapter 24, we assumed the charges existed in a vacuum. Here we shall see how to modify and generalize that law if dielectric materials, such as those listed in Table 28-3, are present. Fig. 28-19 shows a parallel-plate capacitor of plate area A, both with and without a dielectric. We assume the charge magnitude $|q|$ on the plates is the same in both situations. Note the field between the plates induces charges on the faces of the dielectric by one of the methods of Section 28-7.

For the situation of Fig. 28-19a, without a dielectric, we can find the electric field E_0 between the plates as we did in Fig. 28-9: We enclose the charge $+q$ on the top plate with a Gaussian surface and then apply Gauss' law. Letting E_0 represent the magnitude of the field, we find

$$\varepsilon_0 \oint \vec{E} \cdot d\vec{A} = \varepsilon_0 E_0 A = q, \tag{28-28}$$

or

$$\left|\vec{E}_0\right| = \frac{|q|}{\varepsilon_0 A}. \tag{28-29}$$

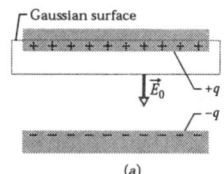

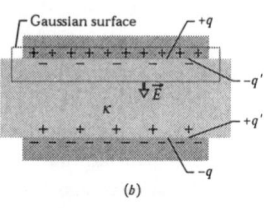

Fig. 28-19: A parallel-plate capacitor (a) without and (b) with a dielectric slab inserted. The charge q on the plates is assumed to be the same in both cases.

In Fig. 28-19b, with the dielectric in place, we can find the electric field between the plates (and within the dielectric) by using the same Gaussian surface. However, now the surface encloses two types of charge: it still encloses charge $+q$ on the top plate, but it now also encloses the induced charge $-q'$ on the top face of the dielectric. The charge on the conducting plate is said to be *free charge* because it can move if we change the electric potential of the plate; the induced charge on the surface of the dielectric is bound charge. It's stuck to the molecules of a non-conductor. It can only be displaced from its original position by microscopic amounts and cannot move from the surface.

The magnitude of the net charge enclosed by the Gaussian surface in Fig. 28-19b is $|q - q'|$, so Gauss' law now gives

$$\varepsilon_0 \oint \vec{E} \cdot d\vec{A} = \varepsilon_0 |\vec{E}| A = |q - q'|, \tag{28-30}$$

or

$$\left|\vec{E}\right| = \frac{|q - q'|}{\varepsilon_0 A}. \tag{28-31}$$

The effect of the dielectric is to weaken the original field $|\vec{E}_0|$ by a factor of κ, so we may write

$$|\vec{E}| = \frac{|\vec{E}_0|}{\kappa} = \frac{|q|}{\kappa \varepsilon_0 A}.$$ (28-32)

Comparison of Eqs. 28-31 and 28-32 shows

$$|q - q'| = \frac{|q|}{\kappa}.$$ (28-33)

Equation 28-33 shows correctly the magnitude $|q'|$ of the induced surface charge is less than the free charge q and is zero if no dielectric is present (then, $\kappa = 1$ in Eq. 28-33).

By substituting for $|q - q'|$ from Eq. 28-33 in Eq. 28-30, we can write Gauss' law in the form

$$\varepsilon_0 \oint \kappa \vec{E} \cdot d\vec{A} = |q| \qquad \text{(Gauss' law with dielectric).} \qquad (28\text{-}34)$$

This important equation, although derived for a parallel-plate capacitor, is true generally and is the most general form in which Gauss' law can be written. Note the following:

1. The flux integral now involves $\kappa \vec{E}$, not just $\vec{E}$. (The vector $\varepsilon_0 \kappa E$ is sometimes called the electric displacement $\vec{D}$, so Eq. 28-34 can be written in the form $\left| \oint \vec{D} \cdot d\vec{A} \right| = |q|$).

2. The charge $|q|$ enclosed by the Gaussian surface is now taken to be the free charge only. The induced surface charge is deliberately ignored on the right side of Eq. 28-34, having been taken fully into account by introducing the dielectric constant κ on the left side.

3. Equation 28-34 differs from Eq. 24-6, our original statement of Gauss' law, only in that ε_0 in the latter equation has been replaced by $\kappa \varepsilon_0$. We keep κ inside the integral of Eq. 28-34 to allow for cases in which κ is not constant over the entire Gaussian surface.

Gauss's law still holds when charged molecules are present, but it's hard to use, since we don't know where those molecular charges are. We only know their average affect, which is summarized by the measured constant κ. Here, we will see how to create a form of Gauss's law including the affect of the molecules automatically and allows us to work only with the charges we control directly—the "free" charges.

Touchstone Example 28-8-1, at the end of this chapter, illustrates how to use what you learned in this section.

TE

28-9 *RC* Circuits

In preceding sections we dealt only with circuits in which the currents did not vary with time. Here we begin a discussion of time-varying currents.

Charging a Capacitor

The capacitor of capacitance C in Fig. 28-20 is initially uncharged. To charge it, we close switch S on point a. This completes an *RC series circuit* consisting of the capacitor, an ideal battery of emf ε, and a resistance R. Since an ideal battery has no internal resistance, its emf is the same as the potential difference across the battery, ΔV_B.

From Section 28-2, we already know as soon as the circuit is complete, charge begins to flow (current exists) between a capacitor plate and a battery terminal on each side of the capacitor. This current increases the magnitude of the charge on the plates, $|q|$ and the magnitude of the potential difference $|\Delta V_C| = |q| / C$ across the capacitor. When that potential difference equals the potential difference across the battery (which here is equal to the potential difference of the battery, $|\Delta V_B|$), the current is zero. From

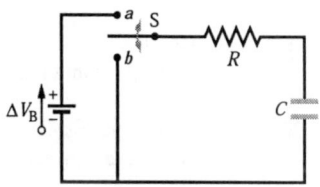

Fig. 28-20: When switch S is closed on a, the capacitor is *charged* through the resistor. When the switch is afterward closed on b, the capacitor *discharges* through the resistor.

Equation 28-1 $(|q| = C|\Delta V_C|)$, the *equilibrium* (final) *charge* on the then fully charged capacitor is equal to $C|\Delta V_B|$.

Here we want to examine the charging process. In particular we want to know how the magnitude of the excess charge $|q(t)|$ on the capacitor plates, the potential difference $\Delta V_C(t)$ across the capacitor, and the current $i(t)$ in the circuit vary with time during the charging process. We begin by applying the loop rule to the circuit, traversing it clockwise from the negative terminal of the battery. We find

$$|\Delta V_B| - iR - \frac{|q|}{C} = 0. \tag{28-35}$$

The last term on the left side represents the potential difference across the capacitor. The term is negative because the capacitor's top plate, which is connected to the battery's positive terminal, is at a higher potential than the lower plate. Thus, there is a drop in potential as we move down through the capacitor.

We cannot immediately solve Eq. 28-35 because it contains two variables, i and q. However, those variables are not independent but are related by

$$i = \frac{dq}{dt}. \tag{28-36}$$

Substituting this for i and rearranging, we find

$$R\frac{dq}{dt} + \frac{|q|}{C} = |\Delta V_B|. \qquad \text{(charging equation). } (28\text{-}37)$$

This differential equation describes the time variation of the magnitude of the excess charge of magnitude $|q|$ on the capacitor plates in Fig. 28-20. To solve it, we need to find the function $q(t)$ satisfying this equation and also satisfies the condition the capacitor be initially uncharged: $q = 0$ at $t = 0$.

We shall show next the solution to Eq. 28-37 is

$$|q| = C|\Delta V_B|(1 - e^{-t/RC}) \qquad \text{(charging a capacitor).} \tag{28-38}$$

(Here e is the exponential base, 2.718..., and not the elementary charge.) Note this expression does indeed satisfy our required initial condition, because at $t = 0$ the term $e^{-t/RC}$ is unity; so the equation gives $q = 0$. Note also as t goes to ∞ (that is, a long time later), the term $e^{-t/RC}$ goes to zero; so the equation gives the proper value for the full (equilibrium) charge on the capacitor—namely, $|q| = C|\Delta V_B|$. A plot of $q(t)$ for the charging process is given in Fig. 28-21a.

The derivative of $q(t)$ is the current $i(t)$ charging the capacitor:

$$i = \frac{d|q|}{dt} = \left(\frac{|\Delta V_B|}{R}\right)e^{-t/RC} \quad \text{(charging a capacitor).} \tag{28-39}$$

A plot of $i(t)$ for the charging process is given in Fig. 28-21b. Note the current has the initial value $|\Delta V_B|/R$ and it decreases to zero as the capacitor becomes fully charged.

▶ A capacitor being charged initially acts like ordinary connecting wire relative to the charging current. A long time later, it acts like a broken wire.

By combining $|q| = C|\Delta V|$ (Eq. 28-1) and $|q| = C|\Delta V_B|(1 - e^{-t/RC})$ (Eq. 28-38), we find the potential difference $\Delta V_C(t)$ across the capacitor during the charging process is

$$|\Delta V_C| = \frac{|q|}{C} = |\Delta V_B|(1 - e^{-t/RC}) \qquad \text{(charging a capacitor).} \tag{28-40}$$

This tells us $\Delta V_C = 0$ at $t = 0$ and $|\Delta V_C| = |\Delta V_B|$ when the capacitor is fully charged at $t \to \infty$.

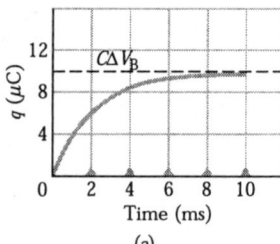

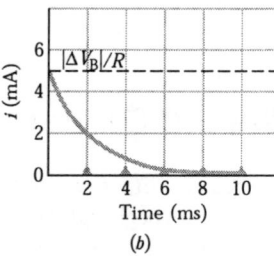

Fig. 28-21: (a) A plot of Eq. 28-38, which shows the buildup of charge on the capacitor of Fig. 28-20. (b) A plot of Eq. 28-39, which shows the decline of the charging current in the circuit of Fig. 28-20. The curves are plotted for $R = 2000\,\Omega$, $C = 1\,\mu F$, and $|\Delta V_B| = 10$ V. The small triangles represent successive intervals of one time constant τ.

The Time Constant

The product RC appearing in the equations above has the dimensions of time (both because the argument of an exponential must be dimensionless and because, in fact, $1.0\,\Omega \times 1.0\,\text{F} = 1.0\,\text{s}$). RC is called the capacitive time constant of the circuit and is represented with the symbol τ :

$$\tau = RC \qquad \text{(time constant).} \qquad (28\text{-}41)$$

From the expression for the magnitude of the excess charge as a function of time in a charging capacitor $|q| = C|\Delta V_B|(1 - e^{-t/RC})$ (Eq. 28-38), we can now see at time $t = \tau (= RC)$, the magnitude of the excess charge on the initially uncharged capacitor of Fig. 28-20 has increased from zero to

$$|q| = C|\Delta V_B|(1 - e^{-1}) = 0.63\,C|\Delta V_B|. \qquad (28\text{-}42)$$

In words, during the first time constant τ the magnitude of the excess charge has increased from zero to 63% of its final value $C|\Delta V_B|$. In Fig. 28-21, the small triangles along the time axes mark successive intervals of one time constant during the charging of the capacitor. The charging times for RC circuits are often stated in terms of τ; the greater τ is, the greater the charging time.

Discharging a Capacitor

Assume now the capacitor of Fig. 28-20 is fully charged to a potential ΔV_0 equal to the potential difference, $|\Delta V_B|$, of the battery. At a new time $t = 0$, switch S is thrown from a to b so the capacitor can *discharge* through resistance R. How do the magnitude of the excess charge $q(t)$ on the capacitor and the current $i(t)$ through the discharge loop of capacitor and resistance now vary with time?

The differential equation describing $q(t)$ in this case is similar to the one we worked with for the case of charging Eq. 28-37 except now, there is no battery in the discharge loop and so $\Delta V_B = 0$. Thus,

$$R\frac{d|q|}{dt} + \frac{|q|}{C} = 0 \qquad \text{(discharging equation).} \qquad (28\text{-}43)$$

The solution to this differential equation is

$$|q| = |q_0| e^{-t/RC} \qquad \text{(discharging a capacitor),} \qquad (28\text{-}44)$$

where $|q_0|(= C|\Delta V_0|)$ is the initial charge on the capacitor. You can verify by substitution that Eq. 28-44 is indeed a solution of Eq. 28-43.

Equation 28-44 tells us that q decreases exponentially with time, at a rate set by the capacitive time constant $\tau = RC$. At time $t = \tau$, the capacitor's charge has been reduced to $|q_0| e^{-1}$, or about 37% of the initial value. Note a greater τ means a greater discharge time.

Differentiating Eq. 28-44 gives us the current $i(t)$:

$$i = \frac{d|q|}{dt} = -\left(\frac{|q_0|}{RC}\right) e^{-t/RC} \qquad \text{(discharging a capacitor).} \qquad (28\text{-}45)$$

This tells us the current also decreases exponentially with time, at a rate set by τ. The initial current i_0 is equal to $|q_0|/RC$. Note you can find i_0 by simply applying the loop rule to the circuit at $t = 0$ just when the capacitor's initial potential ΔV_0 is connected across the resistance R, so the current must be

$$i_0 = \frac{\Delta V_0}{R} = \frac{(|q_0|/C)}{R} = \frac{|q_0|}{RC}.$$

28-20

The minus sign in the discharging capacitor expression can be ignored; it merely means the capacitor's charge q is decreasing.

READING EXERCISE 28-5: The table gives four sets of values for the circuit elements in Fig. 28-20. Rank the sets according to (a) the initial current (as the switch is closed on a) and (b) the time required for the current to decrease to half its initial value, greatest first.

	1	2	3	4
$|\Delta V_b|$ (V)	12	12	10	10
R ( )	2	3	10	5
C (μF)	3	2	0.5	2

Touchstone Example 28-9-1, at the end of this chapter, illustrates how to use what you learned in this section.

TE

Touchstone Example 28-4-1

(a) **Find the equivalent capacitance for the combination of capacitances shown in Fig. TE 28-1a, across which potential difference ΔV is applied. Assume**

$$C_1 = 12.0 \ \mu F, \quad C_2 = 5.30 \ \mu F, \quad \text{and} \quad C_3 = 4.50 \ \mu F.$$

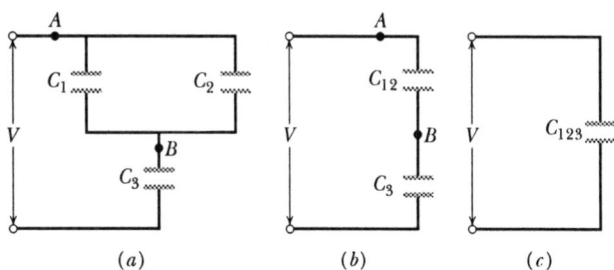

(a)   (b)   (c)

Fig. TE 28-1 (a) Three capacitors. (b) C_1 and C_2, a parallel combination, are replaced by C_{12}. (c) C_{12} and C_3, a series combination, are replaced by the equivalent capacitance C_{123}.

SOLUTION: The **Key Idea** here is that any capacitors connected in series can be replaced with their equivalent capacitor, and any capacitors connected in parallel can be replaced with their equivalent capacitor. Therefore, we should first check whether any of the capacitors in Fig. TE 28-1a are in parallel or series.

Capacitors 1 and 3 are connected one after the other, but are they in series? No. The potential ΔV that is applied to the capacitors produces charge on the bottom plate of capacitor 3. That charge causes charge to shift from the top plate of capacitor 3. However, note that the shifting charge can move to the bottom plates of both capacitor 1 and capacitor 2. Because there is more than one route for the shifting charge, capacitor 3 is *not* in series with capacitor 1 (or capacitor 2).

Are capacitor 1 and capacitor 2 in parallel? Yes. Their top plates are directly wired together and their bottom plates are directly wired together, and electric potential is applied between the top-plate pair and the bottom-plate pair. Thus, capacitor 1 and capacitor 2 are in parallel, and Eq. 28-19 tells us that their equivalent capacitance C_{12} is

$$C_{12} = C_1 + C_2 = 12.0 \ \mu F + 5.30 \ \mu F = 17.3 \ \mu F.$$

In Fig. TE 28-1b, we have replaced capacitors 1 and 2 with their equivalent capacitor, call it capacitor 12 (say "one two"). (The connections at points A and B are exactly the same in Figs. TE 28-1a and b.)

Is capacitor 12 in series with capacitor 3? Again applying the test for series capacitances, we see that the charge that shifts from the top plate of capacitor 3 must entirely go to the bottom plate of capacitor 12. Thus, capacitor 12 and capacitor 3 are in series, and we can replace them with their equivalent C_{123}, as shown in Fig. TE 28-1c. From Eq. 28-20, we have

$$\frac{1}{C_{123}} = \frac{1}{C_{12}} + \frac{1}{C_3} = \frac{1}{17.3 \ \mu F} + \frac{1}{4.50 \ \mu F} = 0.280 \ \mu F^{-1},$$

from which

$$C_{123} = \frac{1}{0.280 \ \mu F^{-1}} = 3.57 \ \mu F. \qquad \text{(Answer)}$$

(b) **The potential difference that is applied to the input terminals in Fig. TE 28-1a is $\Delta V = 12.5$ V. What is the charge on C_1?**

SOLUTION: One **Key Idea** here is that, to get the charge q_1 on capacitor 1, we now have to work backward to that capacitor, starting with the equivalent capacitor 123. Since the given potential difference ΔV (= 12.5 V) is applied across the actual combination of three capacitors in Fig. TE 28-1a, it is also applied across capacitor 123 in Fig. TE 28-1c. Thus, Eq. 28-1 ($|q| = C|\Delta V|$) gives us

$$|q_{123}| = C_{123}|\Delta V| = (3.57 \ \mu F)(12.5 \ V) = 44.6 \ \mu C.$$

A second **Key Idea** is that the series capacitors 12 and 3 in Fig. TE 28-1*b* have the same charge as their equivalent capacitor 123. Thus, capacitor 12 has charge $q_{12} = q_{123} = 44.6 \ \mu C$. From Eq. 28-1, the potential difference across capacitor 12 must be

$$|\Delta V_{12}| = \frac{|q_{12}|}{C_{12}} = \frac{44.6 \ \mu C}{17.3 \ \mu F} = 2.58 \ V.$$

A third **Key Idea** is that the parallel capacitors 1 and 2 both have the same potential difference as their equivalent capacitor 12. Thus, capacitor 1 has the potential difference $\Delta V_1 = \Delta V_{12} = 2.58$ V. Thus, from Eq. 28-1, the charge on capacitor 1 must be

$$|q_1| = C_1|\Delta V_1| = (12.0 \ \mu F)(2.58 \ V)$$
$$= 31.0 \ \mu C. \qquad \text{(Answer)}$$

Touchstone Example 28-4-2

Capacitor 1, with $C_1 = 3.55 \ \mu F$, is charged to a potential difference $\Delta V_0 = 6.30$ V, using a 6.30 V battery. The battery is then removed and the capacitor is connected as in Fig. TE 28-2 to an uncharged capacitor 2, with $C_2 = 8.95 \ \mu F$. When switch S is closed, charge flows between the capacitors until they have the same potential difference ΔV. Find ΔV.

SOLUTION: The situation here differs from the previous example because an applied electric potential is *not* maintained across a combination of capacitors by a battery or some other source. Here, just after switch S is closed, the only applied electric potential is that of capacitor 1 on capacitor 2, and that potential is decreasing. Thus, although the capacitors in Fig. TE 28-2 are connected end to end, in this situation they are not *in series;* and although they are drawn parallel, in this situation they are not *in parallel.*

To find the final electric potential (when the system comes to equilibrium and charge stops flowing), we use this **Key Idea**: After the switch is closed, the original charge q_0 on capacitor 1 is redistributed (shared) between capacitor 1 and capacitor 2. When equilibrium is reached, we can relate the original charge q_0 with the final charges q_1 and q_2 by writing

$$q_0 = q_1 + q_2.$$

Applying the relation $|q| = C|\Delta V|$ to each term of this equation yields

$$C_1|\Delta V_0| = C_1|\Delta V| + C_2|\Delta V|,$$

from which

$$\Delta V = \Delta V_0 \frac{C_1}{C_1 + C_2} = \frac{(6.30 \ V)(3.55 \ \mu F)}{3.55 \ \mu F + 8.95 \ \mu F}$$
$$= 1.79 \ V. \qquad \text{(Answer)}$$

When the capacitors reach this value of electric potential difference, the charge flow stops.

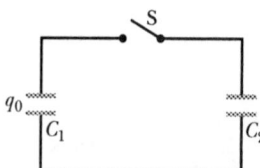

Fig. TE 28-2 A potential difference ΔV_0 is applied to capacitor 1 and the charging battery is removed. Switch S is then closed so that the charge on capacitor 1 is shared with capacitor 2.

Touchstone Example 28-5-1

An isolated conducting sphere whose radius R is 6.85 cm has a charge $q = 1.25$ nC.

(a) How much potential energy is stored in the electric field of this charged conductor?

SOLUTION: The **Key Idea** here is that the energy U stored in a capacitor depends on the charge q on the capacitor and the capacitance C of the capacitor, according to Eq. 28-21. Substituting from Eq. 28-18 for C, Eq. 28-21 gives us

$$U = \frac{q^2}{2C} = \frac{q^2}{8\pi\varepsilon_0 R}$$

$$= \frac{(1.25 \times 10^{-9}\,\text{C})^2}{(8\pi)(8.85 \times 10^{-12}\,\text{F/m})(0.0685\,\text{m})}$$

$$= 1.03 \times 10^{-7}\,\text{J} = 103\,\text{nJ}. \qquad \text{(Answer)}$$

(b) What is the energy density at the surface of the sphere?

SOLUTION: The **Key Idea** here is that the density u of the energy stored in an electric field depends on the magnitude E of the field, according to Eq. 28-23 ($u = \frac{1}{2}\varepsilon_0 E^2$), so we must first find E at the surface of the sphere. This is given by Eq. 24-8:

$$E = \frac{1}{4\pi\varepsilon_0}\frac{q}{R^2}.$$

The energy density is then

$$u = \frac{1}{2}\varepsilon_0 E^2 = \frac{q^2}{32\pi^2\varepsilon_0 R^4}$$

$$= \frac{(1.25 \times 10^{-9}\,\text{C})^2}{(32\pi^2)(8.85 \times 10^{-12}\,\text{C}^2/\text{N}\cdot\text{m}^2)(0.0685\,\text{m})^4}$$

$$= 2.54 \times 10^{-5}\,\text{J/m}^3 = 25.4\,\mu\text{J/m}^3. \qquad \text{(Answer)}$$

Touchstone Example 28-6-1

A parallel-plate capacitor whose capacitance C is 13.5 pF is charged by a battery to a potential difference $\Delta V = 12.5$ V between its plates. The charging battery is now disconnected and a porcelain slab ($\kappa = 6.50$) is slipped between the plates. What is the potential energy of the capacitor–slab device, both before and after the slab is put into place?

SOLUTION: The **Key Idea** here is that we can relate the potential energy U of the capacitor to the capacitance C and either the potential ΔV (with Eq. 28-22) or the charge q (with Eq. 28-21):

$$U_i = \frac{1}{2}C(\Delta V)^2 = \frac{q^2}{2C}.$$

Because we are given the initial potential ΔV (= 12.5 V), we use Eq. 28-22 to find the initial stored energy:

$$U_i = \frac{1}{2}C\Delta V^2 = \frac{1}{2}(13.5 \times 10^{-12}\,\text{F})(12.5\,\text{V})^2$$

$$= 1.055 \times 10^{-9}\,\text{J} = 1055\,\text{pJ} \approx 1100\,\text{pJ}. \qquad \text{(Answer)}$$

To find the final potential energy U_f (after the slab is introduced), we need another **Key Idea**: Because the battery has been disconnected, the charge on the capacitor cannot change when the dielectric is inserted. However, the potential *does* change. Thus, we must now use

Eq. 28-21 (based on q) to write the final potential energy U_f, but now that the slab is within the capacitor, the capacitance is κC. We then have

$$U_f = \frac{q^2}{2\kappa C} = \frac{U_i}{\kappa} = \frac{1055 \text{ pJ}}{6.50} = 162 \text{ pJ} \approx 160 \text{ pJ}. \qquad \text{(Answer)}$$

When the slab is introduced, the potential energy decreases by a factor of κ.

The "missing" energy, in principle, would be apparent to the person who introduced the slab. The capacitor would exert a tiny tug on the slab and would do work on it, in amount

$$W = U_i - U_f = (1055 - 162) \text{ pJ} = 893 \text{ pJ}.$$

If the slab were allowed to slide between the plates with no restraint and if there were no friction, the slab would oscillate back and forth between the plates with a (constant) mechanical energy of 893 pJ, and this system energy would transfer back and forth between kinetic energy of the moving slab and potential energy stored in the electric field.

Touchstone Example 28-8-1

Figure TE 28-3 shows a parallel-plate capacitor of plate area A and plate separation d. A potential difference ΔV_0 is applied between the plates. The battery is then disconnected, and a dielectric slab of thickness b and dielectric constant k is placed between the plates as shown. Assume

$$A = 115 \text{ cm}^2, \qquad d = 1.24 \text{ cm}, \qquad \Delta V_0 = 85.5 \text{ V},$$
$$b = 0.780 \text{ cm}, \qquad \kappa = 2.61.$$

(a) What is the capacitance C_0 before the dielectric slab is inserted?

SOLUTION: From Eq. 28-9 we have

$$C_0 = \frac{\varepsilon_0 A}{d} = \frac{(8.85 \times 10^{-12} \text{ F/m})(115 \times 10^{-4} \text{ m}^2)}{1.24 \times 10^{-2} \text{ m}}$$
$$= 8.21 \times 10^{-12} \text{ F} = 8.21 \text{ pF}. \qquad \text{(Answer)}$$

(b) What free charge appears on the plates?

SOLUTION: From Eq. 28-1,

$$|q| = C_0|\Delta V_0| = (8.21 \times 10^{-12} \text{ F})(85.5 \text{ V})$$
$$= 7.02 \times 10^{-10} \text{ C} = 702 \text{ pC}. \qquad \text{Answer)}$$

Because the charging battery was disconnected before the slab was introduced, the free charge remains unchanged as the slab is put into place.

(c) What is the electric field E_0 in the gaps between the plates and the dielectric slab?

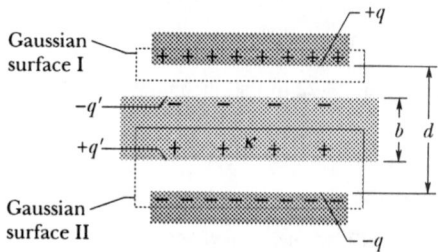

Gaussian surface I

$-q'$
$+q'$

Gaussian surface II

$+q$

b d

$-q$

Fig. TE 28-3 A parallel-plate capacitor containing a dielectric slab that only partially fills the space between the plates.

SOLUTION: A **Key Idea** here is to apply Gauss' law, in the form of Eq. 28-34, to Gaussian surface I in Fig. TE 28-3 — that surface passes through the gap, and so it encloses *only* the free charge on the upper capacitor plate. Because the area vector $d\vec{A}$ and the field vector $\vec{E}_0$ are both directed downward, the dot product in Eq. 28-34 becomes

$$\vec{E}_0 \cdot d\vec{A} = |\vec{E}_0| dA \cos 0° = E_0 \, dA.$$

Equation 28-34 then becomes

$$\varepsilon_0 \kappa |\vec{E}_0| \oint dA = q.$$

The integration now simply gives the surface area A of the plate. Thus, we obtain

$$\varepsilon_0 \kappa |\vec{E}_0| A = |\vec{q}|,$$

or

$$|\vec{E}_0| = \frac{|q|}{\varepsilon_0 \kappa A}.$$

One more **Key Idea** is needed before we evaluate E_0; that is, we must put $\kappa = 1$ here because Gaussian surface I does not pass through the dielectric. Thus, we have

$$|\vec{E}_0| = \frac{|q|}{\varepsilon_0 \kappa A} = \frac{7.02 \times 10^{-10}\,\text{C}}{(8.85 \times 10^{-12}\,\text{F/m})(1)(115 \times 10^{-4}\,\text{m}^2)}$$

$$= 6900\,\text{V/m} = 6.90\,\text{kV/m}. \qquad \text{(Answer)}$$

Note that the value of E_0 does not change when the slab is introduced because the amount of charge enclosed by Gaussian surface I in Fig. TE 28-3 does not change.

(d) What is the electric field E_1 in the dielectric slab?

SOLUTION: The **Key Idea** here is to apply Eq. 28-34 to Gaussian surface II in Fig. TE 28-3. That surface encloses free charge $-q$ and induced charge $+q'$, but we ignore the latter when we use Eq. 28-34. We find

$$\varepsilon_0 \oint \kappa \vec{E}_1 \cdot d\vec{A} = -\varepsilon_0 \kappa |\vec{E}_1| A = -q. \qquad \text{(TE28-1)}$$

(The first minus sign in this equation comes from the dot product $\vec{E}_1 \cdot d\vec{A}$, because now the field vector $\vec{E}_1$ is directed downward and the area vector $d\vec{A}$ is directed upward.) Equation TE28-1 gives us

$$|\vec{E}_1| = \frac{|q|}{\varepsilon_0 \kappa A} = \frac{|\vec{E}_0|}{\kappa} = \frac{6.90\,\text{kV/m}}{2.61} = 2.64\,\text{kV/m}. \qquad \text{(Answer)}$$

(e) What is the potential difference ΔV between the plates after the slab has been introduced?

SOLUTION: The **Key Idea** here is to find ΔV by integrating along a straight-line path extending directly from the bottom plate to the top plate. Within the dielectric, the path length is b and the electric field is E_1. Within the two gaps above and below the dielectric, the total path length is $d - b$ and the electric field is E_0. Equation 28-6 then yields

$$\Delta V = \int_-^+ |\vec{E}| |d\vec{s}| = |\vec{E}_0|(d - b) + |\vec{E}_1| b$$

$$= (6900\,\text{V/m})(0.0124\,\text{m} - 0.00780\,\text{m})$$

$$+ (2640\,\text{V/m})(0.00780\,\text{m})$$

$$= 52.3\,\text{V}. \qquad \text{(Answer)}$$

This is less than the original potential difference of 85.5 V.

(f) What is the capacitance with the slab in place?

SOLUTION: The **Key Idea** now is that the capacitance C is related to the free charge q and the potential difference ΔV via Eq. 28-1, just as when a dielectric is not in place. Taking q from (b) and ΔV from (e), we have

$$C = \frac{|q|}{|\Delta V|} = \frac{7.02 \times 10^{-10}\,C}{52.3\,V}$$
$$= 1.34 \times 10^{-11}\,F = 13.4\,pF. \qquad \text{(Answer)}$$

This is greater than the original capacitance of 8.21 pF.

Touchstone Example 28-9-1

A capacitor of capacitance C is discharging through a resistor of resistance R.

(a) In terms of the time constant $\tau = RC$, when will the charge on the capacitor be half its initial value?

SOLUTION: The **Key Idea** here is that the charge on the capacitor varies according to Eq. 28-44,

$$|q| = |q_0|e^{-t/RC},$$

in which $|q_0|$ is the initial charge. We are asked to find the time t at which $|q| = \frac{1}{2}|q_0|$, or at which

$$\tfrac{1}{2}|q_0| = |q_0|e^{-t/RC}. \qquad \text{(TE28-2)}$$

After canceling q_0, we realize that the time t we seek is "buried" inside an exponential function. To expose the symbol t in Eq. TE28-2, we take the natural logarithms of both sides of the equation. (The natural logarithm is the inverse function of the exponential function.) We find

$$\ln \tfrac{1}{2} = \ln(e^{-t/RC}) = -\frac{t}{RC},$$

or
$$t = (-\ln \tfrac{1}{2})RC = 0.69RC = 0.69\tau. \qquad \text{(Answer)}$$

(b) When will the energy stored in the capacitor be half its initial value?

SOLUTION: There are two **Key Ideas** here. First, the energy U stored in a capacitor is related to the charge q on the capacitor according to Eq. 28-21 ($U = Q^2/2C$). Second, that charge is decreasing according to Eq. 28-44. Combining these two ideas gives us

$$U = \frac{q^2}{2C} = \frac{q_0^2}{2C} e^{-2t/RC} = U_0 e^{-2t/RC},$$

in which U_0 is the initial stored energy. We are asked to find the time at which $U = \frac{1}{2}U_0$, or at which

$$\tfrac{1}{2}U_0 = U_0 e^{-2t/RC}.$$

Canceling U_0 and taking the natural logarithms of both sides, we obtain

$$\ln \tfrac{1}{2} = -\frac{2t}{RC},$$

or
$$t = -RC\frac{\ln\tfrac{1}{2}}{2} = 0.35RC = 0.35\tau. \qquad \text{(Answer)}$$

It takes longer (0.69τ versus 0.35τ) for the *charge* to fall to half its initial value than for the *stored energy* to fall to half its initial value. Doesn't this result surprise you?

29 Magnetic Fields

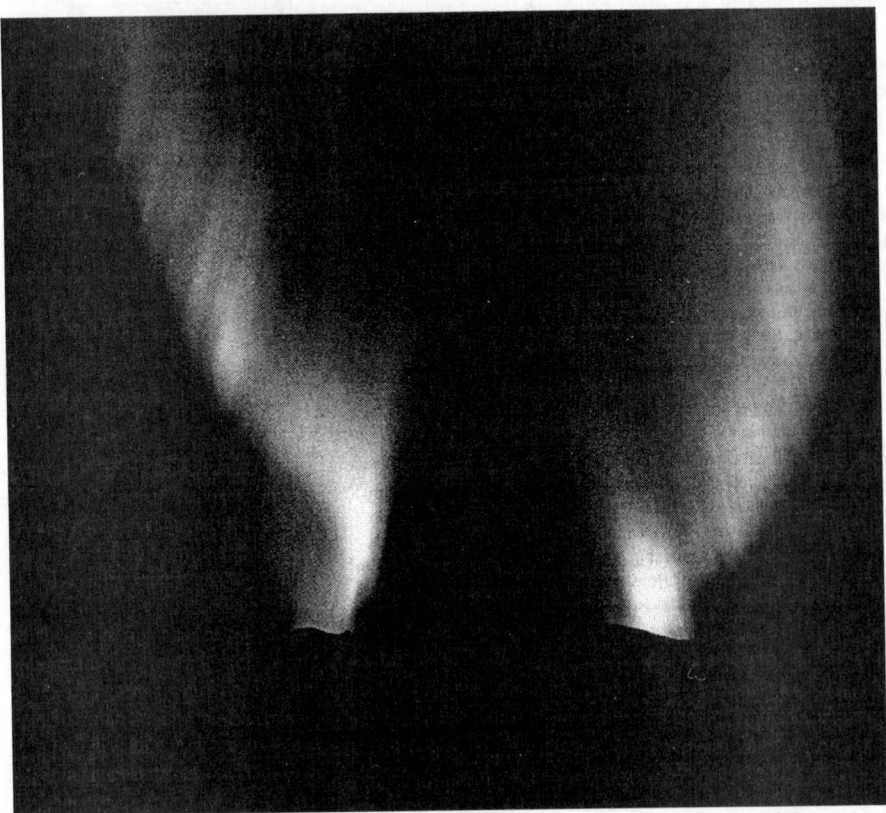

If you are outside on a dark night in the middle to high latitudes, you might be able to see an aurora, a ghostly "curtain" of light that hangs down from the sky. This curtain is not only local; it may be several hundred kilometers high and several thousand kilometers long, stretching around Earth in an arc. However, it is less than 1 km thick.

What produces this huge display, and what makes it so thin?

The answer is in this chapter.

29-1 The Magnetic Field

We have discussed how a charged plastic rod produces a vector field—the electric field $\vec{E}$—at all points in the space around it. Similarly, a magnet produces a vector field—the **magnetic field** $\vec{B}$—at all points in the space around it. You get a hint of that magnetic field whenever you attach a note to a refrigerator door with a small magnet, or accidentally erase a computer disk by bringing it near a magnet. The magnet acts on the door or disk *by means of* its magnetic field.

Fig. 29-1: Using an electromagnet to collect and transport scrap metal at a steel mill.

In a familiar type of magnet, a wire coil is wound around an iron core and a current is sent through the coil; the strength of the magnetic field is determined by the size of the current. In industry, such **electromagnets** are used for sorting scrap metal (Fig. 29-1) among many other things. You are probably more familiar with **permanent magnets**—magnets, like the refrigerator-door type, that do not need current to have a magnetic field.

In Chapter 23 we saw that an *electric charge* sets up an electric field that can then affect other electric charges. Here, we might reasonably expect that a *magnetic charge* sets up a magnetic field that can then affect other magnetic charges. Although such magnetic charges, called *magnetic monopoles*, are predicted by certain theories, their existence has not been confirmed.

How then are magnetic fields set up? There are two ways. (1) Moving electrically charged particles, such as a current in a wire, create magnetic fields. (2) Elementary particles such as electrons have an *intrinsic* magnetic field around them; that is, this field is a basic characteristic of the particles, just as are their mass and electric charge (or lack of charge). As we shall discuss in Chapter 32, the magnetic fields of the electrons in certain materials add together to give a net magnetic field around the material. This is true for the material in permanent magnets (which is good, because they can then hold notes to a refrigerator door). In other materials, the magnetic fields of all the electrons cancel out, giving no net magnetic field surrounding the material. This is true for the material in your body (which is also good, because otherwise you might be slammed up against a refrigerator door every time you passed one).

Experimentally we find that when a charged particle (either alone or as part of a current) moves through a magnetic field, a force due to the field can act on the particle. In this chapter we focus on the relation between the magnetic field and this force.

29-2 The Definition of $\vec{B}$

We determined the electric field $\vec{E}$ at a point by putting a test particle of charge q at rest at that point and measuring the electric force $\vec{F}_E$ acting on the particle. We then defined $\vec{E}$ as

$$\vec{E} = \frac{\vec{F}_E}{q}. \tag{29-1}$$

If a magnetic monopole were available, we could define $\vec{B}$ in a similar way. Because such particles have not been found, we must define $\vec{B}$ in another way, in terms of the magnetic force $\vec{F}_B$ exerted on a moving electrically charged test particle.

In principle, we do this by firing a charged particle through the point at which $\vec{B}$ is to be defined, using various directions and speeds for the particle and determining the force $\vec{F}_B$ that acts on the particle at that point. After many such trials we would find that when the particle's velocity $\vec{v}$ is along a particular axis through the point, force $\vec{F}_B$ is zero. For all other directions of $\vec{v}$, the magnitude of $\vec{F}_B$ is always proportional to $v \sin \phi$, where ϕ is the angle between the zero-force axis and the direction of $\vec{v}$. Furthermore, the direction of $\vec{F}_B$ is always perpendicular to the direction of $\vec{v}$. (These results suggest that a cross product is involved.)

We can then define a magnetic field $\vec{B}$ to be a vector quantity that is directed along the zero-force axis. We can next measure the magnitude of $\vec{F}_B$ when $\vec{v}$ is directed perpendicular to that axis and then define the magnitude of $\vec{B}$ in terms of that force magnitude:

$$|\vec{B}| = \frac{|\vec{F}_B|}{|q\vec{v}|},$$

where q is the charge of the particle.

We can summarize all these results with the following vector equation:

$$\vec{F}_B = q\vec{v} \times \vec{B}; \qquad (29\text{-}2)$$

that is, the force $\vec{F}_B$ on the particle is equal to the charge q times the cross product of its velocity $\vec{v}$ and the magnetic field $\vec{B}$. Using Eq. 12-9 to evaluate the cross product, we can write the magnitude of $\vec{F}_B$ as

$$|\vec{F}_B| = |q\vec{v}||\vec{B}|\sin\phi, \qquad (29\text{-}3)$$

where ϕ is the angle between the directions of velocity $\vec{v}$ and magnetic field $\vec{B}$.

Finding the Magnetic Force on a Particle

Equation 29-3 tells us that the magnitude of the force $\vec{F}_B$ acting on a particle in a magnetic field is proportional to the charge q and speed $|\vec{v}|$ of the particle. Thus, the force is equal to zero if the charge is zero or if the particle is stationary. Equation 29-3 also tells us that the magnitude of the force is zero if $\vec{v}$ and $\vec{B}$ are either parallel $(\phi = 0°)$ or antiparallel $(\phi = 180°)$, and the force is at its maximum when $\vec{v}$ and $\vec{B}$ are perpendicular to each other.

Equation 29-2 tells us all this plus the direction of $\vec{F}_B$. From Section 12-4, we know that the cross product $\vec{v} \times \vec{B}$ in Eq. 29-2 is a vector that is perpendicular to the two vectors $\vec{v}$ and $\vec{B}$. The right-hand rule (Fig. 29-2a) tells us that the thumb of the right hand points in the direction of $\vec{v} \times \vec{B}$ when the fingers sweep $\vec{v}$ into $\vec{B}$. If q is positive, then (by Eq. 29-2) the force $\vec{F}_B$ has the same sign as $\vec{v} \times \vec{B}$ and thus must be in the same direction; that is, for positive q, $\vec{F}_B$ is directed along the thumb as in Fig. 29-2b. If q is negative, then the force $\vec{F}_B$ and the cross product $\vec{v} \times \vec{B}$ have opposite signs and thus must be in opposite directions. For negative q, $\vec{F}_B$ is directed opposite the thumb as in Fig. 29-2c.

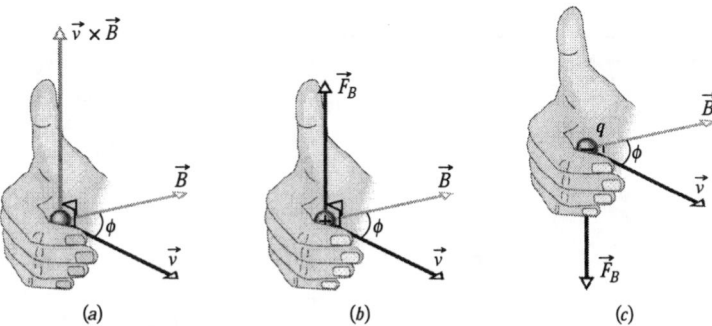

(a)　　　　　(b)　　　　　(c)

Fig. 29-2: (a) The right-hand rule (in which $\vec{v}$ is swept into $\vec{B}$ through the smaller angle ϕ between them) gives the direction of $\vec{v} \times \vec{B}$ as the direction of the thumb. (b) If q is positive, then the direction of $\vec{F}_B = q\vec{v} \times \vec{B}$ is in the direction of $\vec{v} \times \vec{B}$. (c) If q is negative, then the direction of $\vec{F}_B$ is opposite that of $\vec{v} \times \vec{B}$.

Regardless of the sign of the charge, however,

▶The force $\vec{F}_B$ acting on a charged particle moving with velocity $\vec{v}$ through a magnetic field $\vec{B}$ is *always* perpendicular to $\vec{v}$ and $\vec{B}$.

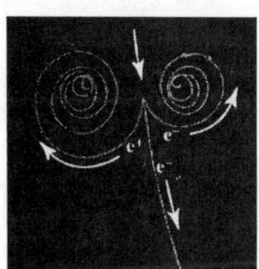

Fig. 29-3: The tracks of two electrons (e⁻) and a positron (e⁺) in a bubble chamber that is immersed in a uniform magnetic field that is directed out of the plane of the page.

Thus, $\vec{F}_B$ *never* has a component parallel to $\vec{v}$. This means that $\vec{F}_B$ cannot change the particle's speed v (and thus it cannot change the particle's kinetic energy). The force can change only the direction of $\vec{v}$ (and thus the direction of travel); only in this sense can $\vec{F}_B$ accelerate the particle.

To develop a feeling for Eq. 29-2, consider Fig. 29-3, which shows some tracks left by charged particles moving rapidly through a *bubble chamber* at the Lawrence Berkeley Laboratory. The chamber, which is filled with liquid hydrogen, is immersed in a strong uniform magnetic field that is directed out of the plane of the figure. An incoming gamma ray particle—which leaves no track because it is uncharged—transforms into an electron (spiral track marked e⁻) and a positron (track marked e⁺) while it knocks an electron out of a hydrogen atom (long track marked e⁻). Check with Eq. 29-2 and Fig. 29-2 that the three tracks made by these two negative particles and one positive particle curve in the proper directions.

The SI unit for $\vec{B}$ that follows from Eqs. 29-2 and 29-3 is the newton per coulomb-meter per second. For convenience, this is called the tesla (T):

$$1 \text{ tesla} = 1 \text{ T} = 1\frac{\text{newton}}{(\text{coulomb})(\text{meter/second})}.$$

Recalling that a coulomb per second is an ampere, we have

$$1 \text{ T} = 1\frac{\text{newton}}{(\text{coulomb/second})(\text{meter})} = 1\frac{\text{N}}{\text{A} \cdot \text{m}}. \tag{29-4}$$

An earlier (non-SI) unit for $\vec{B}$, still in common use, is the *gauss* (G), and

$$1 \text{ tesla} = 10^4 \text{ gauss.} \tag{29-5}$$

Table 29-1 lists the magnetic fields that occur in a few situations. Note that Earth's magnetic field near the planet's surface is about 10^{-4} T ($= 100\,\mu$T or 1 gauss).

TABLE 29-1: Some Approximate Magnetic Fields

At the surface of a neutron star	10^8 T
Near a big electromagnet	1.5 T
Near a small bar magnet	10^{-2} T
At Earth's surface	10^{-4} T
In interstellar space	10^{-10} T
Smallest value in a magnetically shielded room	10^{-14} T

READING EXERCISE 29-1: The figure shows three situations in which a charged particle with velocity $\vec{v}$ travels through a uniform magnetic field $\vec{B}$. In each situation, what is the direction of the magnetic force $\vec{F}_B$ on the particle?

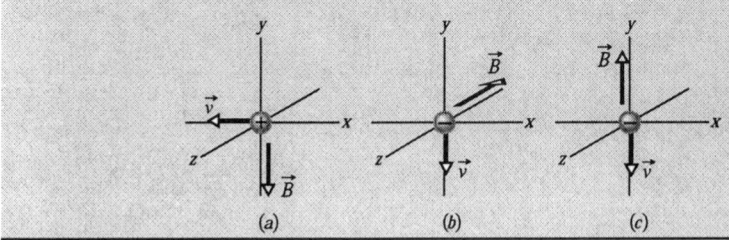

(a)   (b)   (c)

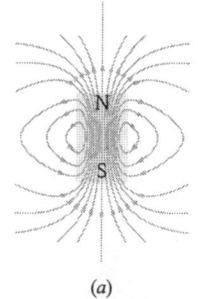

(a)

(b)

Fig. 29-4: (a) The magnetic field lines for a bar magnet. (b) A "cow magnet"—a bar magnet that is intended to be slipped down into the rumen of a cow to prevent accidentally ingested bits of scrap iron from reaching the cow's intestines. The iron filings at its ends reveal the magnetic field lines.

Magnetic Field Lines

We can represent magnetic fields with field lines, as we did for electric fields. Similar rules apply; that is, (1) the direction of the tangent to a magnetic field line at any point gives the direction of $\vec{B}$ at that point, and (2) the spacing of the lines represents the magnitude of $\vec{B}$—the magnetic field is stronger where the lines are closer together, and conversely.

Figure 29-4*a* shows how the magnetic field near a *bar magnet* (a permanent magnet in the shape of a bar) can be represented by magnetic field lines. The lines all pass through the magnet, and they all form closed loops (even those that are not shown closed in the figure). The external magnetic effects of a bar magnet are strongest near its ends, where the field lines are most closely spaced. Thus, the magnetic field of the bar magnet in Fig. 29-4*b* collects the iron filings mainly near the two ends of the magnet.

The (closed) field lines enter one end of a magnet and exit the other end. The end of a magnet from which the field lines emerge is called the *north pole* of the magnet; the other end, where field lines enter the magnet, is called the *south pole*. The magnets we use to fix notes on refrigerators are short bar magnets. Figure 29-5 shows two other common shapes for magnets: a *horseshoe magnet* and a magnet that has been bent around into the shape of a **C** so that the *pole faces* are facing each other. (The magnetic field between the pole faces can then be approximately uniform.) Regardless of the shape of the magnets, if we place two of them near each other we find:

▶Opposite magnetic poles attract each other, and like magnetic poles repel each other.

Earth has a magnetic field that is produced in its core by still unknown mechanisms. On Earth's surface, we can detect this magnetic field with a compass, which is essentially a slender bar magnet on a low-friction pivot. This bar magnet, or this needle, turns because its north-pole end is attracted toward the Arctic region of Earth. Thus, the *south pole* of Earth's magnetic field must be located toward the Arctic. Logically, we then should call the pole there a south pole. However, because we call that direction north, we are trapped into the statement that Earth has a *geomagnetic north pole* in that direction.

With more careful measurement we would find that in the northern hemisphere, the magnetic field lines of Earth generally point down into Earth and toward the Arctic. In the southern hemisphere, they generally point up out of Earth and away from the Antarctic—that is, away from Earth's *geomagnetic south pole*.

Touchstone Example 29-2-1, at the end of this chapter, illustrates how to use what you learned in this section.

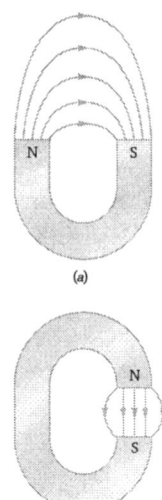

Fig. 29-5: (*a*) A horseshoe magnet and (*b*) a C-shaped magnet. (Only some of the external field lines are shown.)

TE

29-3 Crossed Fields: Discovery of the Electron

Both an electric field $\vec{E}$ and a magnetic field $\vec{B}$ can produce a force on a charged particle. When the two fields are perpendicular to each other, they are said to be *crossed fields*. Here we shall examine what happens to charged particles—namely, electrons—as they move through crossed fields. We use as our example the experiment that led to the discovery of the electron in 1897 by J. J. Thomson at Cambridge University.

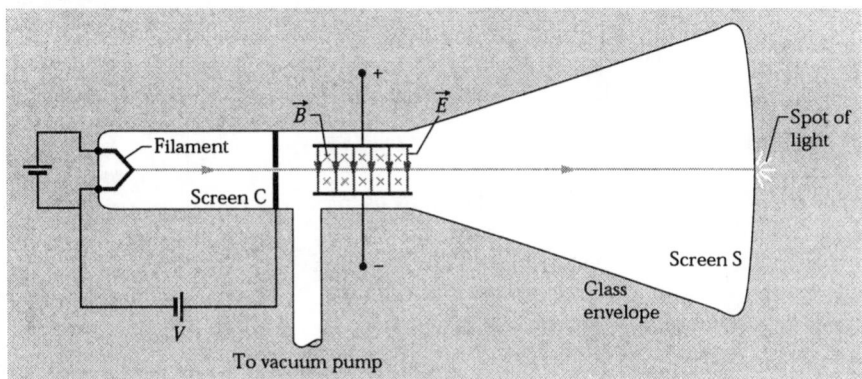

Fig. 29-6: A modern version of J. J. Thomson's apparatus for measuring the ratio of mass to charge for the electron. The electric field $\vec{E}$ is established by connecting a battery across the deflecting-plate terminals. The magnetic field $\vec{B}$ is set up by means of a current in a system of coils (not shown). The magnetic field shown is into the plane of the figure, as represented by the array of Xs (which resemble the feathered ends of arrows).

Figure 29-6 shows a modern, simplified version of Thomson's experimental apparatus—a *cathode ray tube* (which is like the picture tube in a standard television set). Charged particles (which we now know as electrons) are emitted by a hot filament at the rear of the evacuated tube and are accelerated by an applied potential difference V. After they pass through a slit in screen C, they form a narrow beam. They then pass through a region of crossed $\vec{E}$ and $\vec{B}$ fields, headed toward a fluorescent screen S, where they produce a spot of light (on a television screen the spot is part of the picture). The forces on the charged particles in the crossed-fields region can deflect them from the center of the screen. By controlling the magnitudes and directions of the fields, Thomson could thus control where the spot of light appeared on the screen. Recall that the force on a negatively charged particle due to an electric field is directed opposite the field. Thus, for the particular field arrangement of Fig. 29-6, electrons are forced up the page by the electric field $\vec{E}$ and down the page by the magnetic field $\vec{B}$; that is, the forces are *in opposition*. Thomson's procedure was equivalent to the following series of steps.

1. Set $\vec{E} = 0$ and $\vec{B} = 0$ and note the position of the spot on screen S due to the undeflected beam.

2. Turn on $\vec{E}$ and measure the resulting beam deflection.

3. Maintaining $\vec{E}$, now turn on $\vec{B}$ and adjust its value until the beam returns to the undeflected position. (With the forces in opposition, they can be made to cancel.)

We discussed the deflection of a charged particle moving through an electric field $\vec{E}$ between two plates (step 2 here) in Touchstone Example 23-9-1. We found that the magnitude of the deflection of the particle at the far end of the plates is

$$|y| = \frac{|q||\vec{E}|L^2}{2m|\vec{v}|^2},\tag{29-6}$$

where $|\vec{v}|$ is the particle's speed, m its mass, and q its charge, and L is the length of the plates. We can apply this same equation to the beam of electrons in Fig. 29-6; if need be, we can calculate the deflection by measuring the deflection of the beam on screen S and then working back to calculate the deflection y at the end of the plates. (Because the direction of the deflection is set by the sign of the particle's charge, Thomson was able to show that the particles that were lighting up his screen were negatively charged.)

When the two fields in Fig. 29-6 are adjusted so that the two deflecting forces cancel (step 3), we have from Eqs. 29-1 and 29-3

$$|q|E = |q\vec{v}||\vec{B}|\sin(90°) = |q||v||\vec{B}|$$

or

$$|\vec{v}| = \frac{|\vec{E}|}{|\vec{B}|}.\tag{29-7}$$

Thus, the crossed fields allow us to measure the speed of the charged particles passing through them. Substituting Eq. 29-7 for $|\vec{v}|$ in Eq. 29-6 and rearranging yield

$$\frac{m}{|q|} = \frac{B^2 L^2}{2|y||\vec{E}|},\tag{29-8}$$

in which all quantities on the right can be measured. Thus, the crossed fields allow us to measure the ratio $m/|q|$ of the particles moving through Thomson's apparatus.

Thomson claimed that these particles are found in all matter. He also claimed that they are lighter than the lightest known atom (hydrogen) by a factor of more than 1000. (The exact ratio proved later to be 1836.15.) His $m/|q|$ measurement, coupled with the boldness of his two claims, is considered to be the "discovery of the electron."

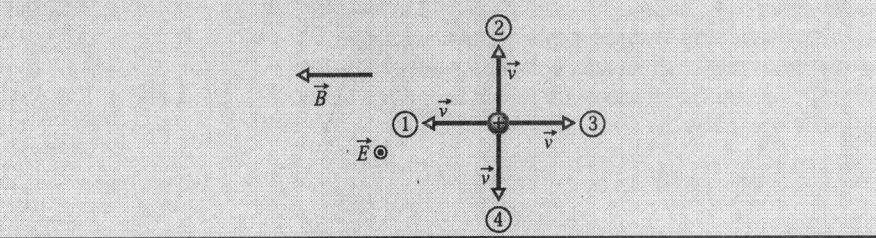

29-4 Crossed Fields: The Hall Effect

As we just discussed, a beam of electrons in a vacuum can be deflected by a magnetic field. Can the drifting conduction electrons in a copper wire also be deflected by a magnetic field? In 1879, Edwin H. Hall, then a 24-year-old graduate student at the Johns Hopkins University, showed that they can. This **Hall effect** allows us to find out whether the charge carriers in a conductor are positively or negatively charged. Beyond that, we can measure the number of such carriers per unit volume of the conductor.

Figure 29-7a shows a copper strip of width d, carrying a current i whose conventional direction is from the top of the figure to the bottom. The charge carriers are electrons and, as we know, they drift (with drift speed v_d) in the opposite direction, from bottom to top. At the instant shown in Fig. 29-7a, an external magnetic field $\vec{B}$, pointing into the plane of the figure, has just been turned on. From Eq. 29-2 we see that a magnetic deflecting force $\vec{F}_B$ will act on each drifting electron, pushing it toward the right edge of the strip.

As time goes on, electrons move to the right, mostly piling up on the right edge of the strip, leaving uncompensated positive charges in fixed positions at the left edge. The separation of positive and negative charges produces an electric field $\vec{E}$ within the strip, pointing from left to right in Fig. 29-7b. This field exerts an electric force $\vec{F}_E$ on each electron, tending to push it to the left.

An equilibrium quickly develops in which the electric force on each electron builds up until it just cancels the magnetic force. When this happens, as Fig. 29-7b shows, the force due to $\vec{B}$ and the force due to $\vec{E}$ are in balance. The drifting electrons then move along the strip toward the top of the page at velocity $\vec{v}_d$, with no further collection of electrons on the right edge of the strip and thus no further increase in the electric field $\vec{E}$.

A *Hall potential difference* ΔV is associated with the electric field across strip width d. From Eq. 25-36, the magnitude of that potential difference is

$$|\Delta V| = |\vec{E}|d. \tag{29-9}$$

By connecting a voltmeter across the width, we can measure the potential difference between the two edges of the strip. Moreover, the voltmeter can tell us which edge is at higher potential. For the situation of Fig. 29-7a, we would find that the left edge is at higher potential, which is consistent with our assumption that the charge carriers are negatively charged.

For a moment, let us make the opposite assumption, that the charge carriers in current i are positively charged (Fig. 29-7c). Convince yourself that as these charge carriers move from top to bottom in the strip, they are pushed to the right edge by $\vec{F}_B$ and thus that the *right* edge is at higher potential. Because that last statement is contradicted by our voltmeter reading, the charge carriers must be negatively charged.

Now for the quantitative part. When the electric and magnetic forces are in balance (Fig. 29-7b), Eqs. 29-1 and 29-3 give us

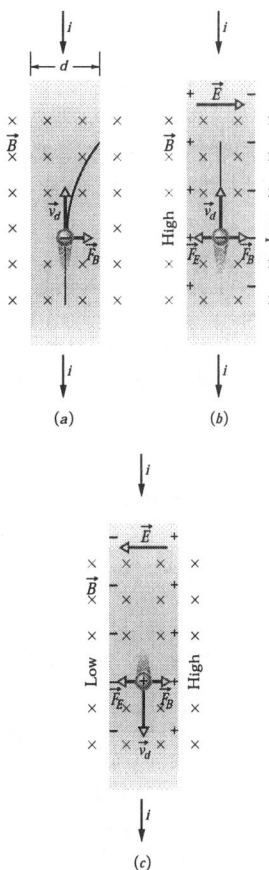

Fig. 29-7: A strip of copper carrying a current i is immersed in a magnetic field $\vec{B}$. (a) The situation immediately after the magnetic field is turned on. The curved path that will then be taken by an electron is shown. (b) The situation at equilibrium, which quickly follows. Note that negative charges pile up on the right side of the strip, leaving uncompensated positive charges on the left. Thus, the left side is at a higher potential than the right side. (c) For the same current direction, if the charge carriers were positively charged, *they* would pile up on the right side, and the right side would be at the higher potential.

$$e|\vec{E}| = e|v_d||\vec{B}|. \qquad (29\text{-}10)$$

where e is the magnitude of the charge on the electron. From Eq. 26-12, the drift speed v_d is

$$|\vec{v}_d| = \frac{|\vec{J}|}{ne} = \frac{|i|}{neA}, \qquad (29\text{-}11)$$

in which $|\vec{J}|(=|i|/A)$ is the current density in the strip, A is the cross-sectional area of the strip, and n is the *number density* of charge carriers (their number per unit volume).

In Eq. 29-10, substituting for $|\vec{E}|$ with Eq. 29-9 and substituting for $|\vec{v}_d|$ with Eq. 29-11, we obtain

$$n = \frac{|i\vec{B}|}{e\ell|\Delta V|}, \qquad (29\text{-}12)$$

in which $\ell = A/d$ is the thickness of the strip. With this equation we can find n from measurable quantities.

It is also possible to use the Hall effect to measure directly the drift speed v_d of the charge carriers, which you may recall is of the order of centimeters per hour. In this clever experiment, the metal strip is moved mechanically through the magnetic field in a direction opposite that of the drift velocity of the charge carriers. The speed of the moving strip is then adjusted until the Hall potential difference vanishes. At this condition, with no Hall effect, the velocity of the charge carriers *with respect to the laboratory frame* must be zero, so the velocity of the strip must be equal in magnitude but opposite in direction to the velocity of the negative charge carriers.

Touchstone Example 29-4-1, at the end of this chapter, illustrates how to use what you learned in this section.

TE

29-5 A Circulating Charged Particle

If a particle moves in a circle at constant speed, we can be sure that the net force acting on the particle is constant in magnitude and points toward the center of the circle, always perpendicular to the particle's velocity. Think of a stone tied to a string and whirled in a circle on a smooth horizontal surface, or of a satellite moving in a circular orbit around Earth. In the first case, the tension in the string provides the necessary force and centripetal acceleration. In the second case, Earth's gravitational attraction provides the force and acceleration.

Figure 29-8 shows another example: A beam of electrons is projected into a chamber by an *electron gun* G. The electrons enter in the plane of the page with speed v and move in a region of uniform magnetic field $\vec{B}$ directed out of the plane of the figure. As a result, a magnetic force $\vec{F}_B = q\vec{v} \times \vec{B}$ continually deflects the electrons, and because $\vec{v}$ and $\vec{B}$ are always perpendicular to each other, this deflection causes the electrons to follow a circular path. The path is visible in the photo because atoms of gas in the chamber emit light when some of the circulating electrons collide with them.

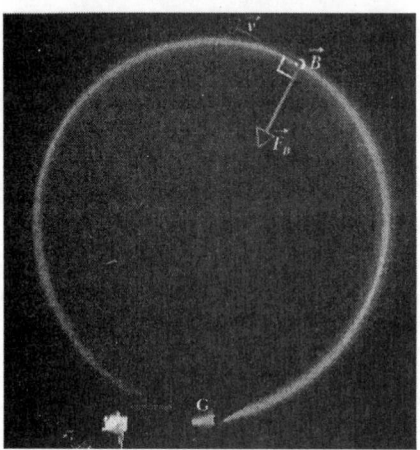

Fig. 29-8: Electrons circulating in a chamber containing gas at low pressure (their path is the glowing circle). A uniform magnetic field $\vec{B}$, pointing directly out of the plane of the page, fills the chamber. Note the radially directed magnetic force $\vec{F}_B$; for circular motion to occur, $\vec{F}_B$ must point toward the center of the circle. Use the right-hand rule for cross products to confirm that $\vec{F}_B = q\vec{v} \times \vec{B}$ gives $\vec{F}_B$ the proper direction. (Don't forget the sign of q.)

We would like to determine the parameters that characterize the circular motion of these electrons, or of any particle of charge magnitude q and mass m moving perpendicular to a uniform magnetic field $\vec{B}$ at speed v. From Eq. 29-3, the force acting on the particle has a magnitude of qvB. From Newton's Second Law ($\vec{F} = m\vec{a}$) applied to uniform circular motion (Eq. 5-19),

$$F = m\frac{v^2}{r}, \tag{29-13}$$

we have

$$qvB = \frac{mv^2}{r}. \tag{29-14}$$

Solving for r, we find the radius of the circular path as

$$r = \frac{mv}{qB} \qquad \text{(radius)}. \tag{29-15}$$

The period T (the time for one full revolution) is equal to the circumference divided by the speed:

$$T = \frac{2\pi r}{v} = \frac{2\pi}{v}\frac{mv}{qB} = \frac{2\pi m}{qB} \qquad \text{(period)}. \tag{29-16}$$

The frequency f (the number of revolutions per unit time) is

$$f = \frac{1}{T} = \frac{qB}{2\pi m} \qquad \text{(frequency)}. \tag{29-17}$$

The angular frequency ω of the motion is then

$$\omega = 2\pi f = \frac{qB}{m} \qquad \text{(angular frequency)}. \tag{29-18}$$

The quantities $T, f,$ and ω do not depend on the speed of the particle (provided that speed is much less than the speed of light). Fast particles move in large circles and slow ones in small circles, but all particles with the same charge-to-mass ratio q/m take the same time T (the period) to complete one round trip. Using Eq. 29-2, you can show that if you are looking in the direction of $\vec{B}$, the direction of rotation for a positive particle is always counterclockwise; that for a negative particle is always clockwise.

Helical Paths

If the velocity of a charged particle has a component parallel to the (uniform) magnetic field, the particle will move in a helical path about the direction of the field vector. Figure 29-9a, for example, shows the velocity vector $\vec{v}$ of such a particle resolved into two components, one parallel to $\vec{B}$ and one perpendicular to it:

$$v_\parallel = v \cos\phi \quad \text{and} \quad v_\perp = v \sin\phi. \tag{29-19}$$

The parallel component determines the *pitch p* of the helix—that is, the distance between adjacent turns (Fig. 29-9b). The perpendicular component determines the radius of the helix and is the quantity to be substituted for v in Eq. 29-15.

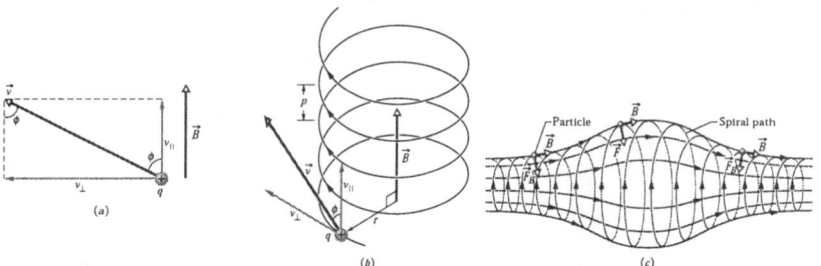

Figure 29-9c shows a charged particle spiraling in a nonuniform magnetic field. The more closely spaced field lines at the left and right sides indicate that the magnetic field is stronger there. When the field at an end is strong enough, the particle "reflects" from that end. If the particle reflects from both ends, it is said to be trapped in a *magnetic bottle*.

Electrons and protons are trapped in this way by the terrestrial magnetic field; the trapped particles form the *Van Allen radiation belts*, which loop well above Earth's atmosphere between Earth's north and south geomagnetic poles. These particles bounce back and forth, from one end of this magnetic bottle to the other, within a few seconds.

When a large solar flare shoots additional energetic electrons and protons into the radiation belts, an electric field is produced in the region where electrons normally reflect. This field eliminates the reflection and instead drives electrons down into the atmosphere, where they collide with atoms and molecules of air, causing that air to emit light. This light forms the aurora—a curtain of light that hangs down to an altitude of about 100 km. Green light is emitted by oxygen atoms, and pink light is emitted by nitrogen molecules, but often the light is so dim that we perceive only white light.

Aurora extend in arcs above Earth and can occur in a region called the *auroral oval* that is shown in Figs. 29-10 and 29-11 as seen from space. Although an aurora is long, it is less than 1 km thick (north to south) because the paths of the electrons producing it converge as the electrons spiral down the converging magnetic field lines (Fig. 29-10).

READING EXERCISE 29-3: The figure shows the circular paths of two particles that travel at the same speed in a uniform magnetic field $\vec{B}$, which is directed into the page. One particle is a proton; the other is an electron (which is less massive). (a) Which particle follows the smaller circle, and (b) does that particle travel clockwise or counterclockwise?

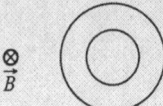

Touchstone Example 29-5-1, at the end of this chapter, illustrates how to use what you learned in this section.

Fig. 29-9: (a) A charged particle moves in a uniform magnetic field B, its velocity $\vec{v}$ making an angle ϕ with the field direction. (b) The particle follows a helical path, of radius r and pitch p. (c) A charged particle spiraling in a nonuniform magnetic field. (The particle can become trapped, spiraling back and forth between the strong field regions at either end.) Note that the magnetic force vectors at the left and right sides have a component pointing toward the center of the figure.

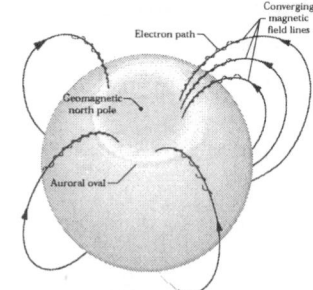

Fig. 29-10: The auroral oval surrounding Earth's geomagnetic north pole (in northwestern Greenland). Magnetic field lines converge toward that pole. Electrons moving toward Earth are "caught by" and spiral around these field lines, entering the terrestrial atmosphere at high latitudes and producing aurora within the oval.

29-6 Cyclotrons and Synchrotrons

What is the structure of matter on the smallest scale? This question has always intrigued physicists. One way of getting at the answer is to allow an energetic charged particle (a proton, for example) to slam into a solid target. Better yet, allow two such energetic protons to collide head-on. Then analyze the debris from many such collisions to learn the nature of the subatomic particles of matter. The Nobel Prizes in physics for 1976 and 1984 were awarded for just such studies.

How can we give a proton enough kinetic energy for such an experiment? The direct approach is to allow the proton to "fall" through a potential difference V, thereby increasing its kinetic energy by eV. As we want higher and higher energies, however, it becomes more and more difficult to establish the necessary potential difference.

A better way is to arrange for the proton to circulate in a magnetic field, and to give it a modest electrical "kick" once per revolution. For example, if a proton circulates 100 times in a magnetic field and receives an energy boost of 100 keV every time it completes an orbit, it will end up with a kinetic energy of (100)(100 keV) or 10 MeV. Two very useful accelerating devices are based on this principle.

Fig. 29-11: A false-color image of aurora inside the north auroral oval, recorded by the satellite Dynamic Explorer, using ultraviolet light emitted by oxygen atoms excited in the aurora. The sun-lit portion of Earth is the crescent at the left.

The Cyclotron

Figure 29-12 is a top view of the region of a *cyclotron* in which the particles (protons, say) circulate. The two hollow **D**-shaped objects (open on their straight edges) are made of sheet copper. These *dees*, as they are called, are part of an electrical oscillator that alternates the electric potential difference across the gap between the dees. The electrical signs of the dees are alternated so that the electric field in the gap alternates in direction, first toward one dee and then toward the other dee, back and forth. The dees are immersed in a magnetic field ($B = 1.5$ T) whose direction is out of the plane of the page and that is set up by a large electromagnet.

Suppose that a proton, injected by source S at the center of the cyclotron in Fig. 29-12, initially moves toward a negatively charged dee. It will accelerate toward this dee and enter it. Once inside, it is shielded from electric fields by the copper walls of the dee; that is, the electric field does not enter the dee. The magnetic field, however, is not screened by the (nonmagnetic) copper dee, so the proton moves in a circular path whose radius, which depends on its speed, is given by Eq. 29-15 $(r = mv/qB)$.

Let us assume that at the instant the proton emerges into the center gap from the first dee, the potential difference between the dees is reversed. Thus, the proton *again* faces a negatively charged dee and is *again* accelerated. This process continues, the circulating proton always being in step with the oscillations of the dee potential, until the proton has spiraled out to the edge of the dee system. There a deflector plate sends it out through a portal.

The key to the operation of the cyclotron is that the frequency f at which the proton circulates in the field (and that does not depend on its speed) must be equal to the fixed frequency f_{osc} of the electrical oscillator, or

$$f = f_{osc} \quad \text{(resonance condition).} \quad (29\text{-}20)$$

This *resonance condition* says that, if the energy of the circulating proton is to increase, energy must be fed to it at a frequency f_{osc} that is equal to the natural frequency f at which the proton circulates in the magnetic field.

Combining Eqs. 29-17 and 29-20 allows us to write the resonance condition as

$$\left| q\vec{B} \right| = 2\pi m f_{osc}. \quad (29\text{-}21)$$

For the proton, q and m are fixed. The oscillator (we assume) is designed to work at a single fixed frequency f_{osc}. We then "tune" the cyclotron by varying $\vec{B}$ until Eq. 29-21 is satisfied and then many protons circulate through the magnetic field, to emerge as a beam.

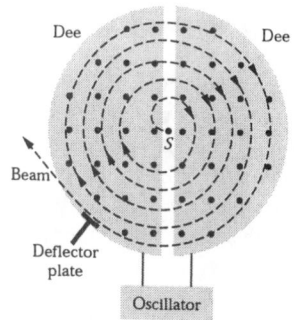

Fig. 29-12: The elements of a cyclotron, showing the particle source S and the dees. A uniform magnetic field is directed up from the plane of the page. Circulating protons spiral outward within the hollow dees, gaining energy every time they cross the gap between the dees.

The Proton Synchrotron

At proton energies above 50 MeV, the conventional cyclotron begins to fail because one of the assumptions of its design—that the frequency of revolution of a charged particle circulating in a magnetic field is independent of the particle's speed—is true only for speeds that are much less than the speed of light. At greater proton speeds (above about 10% of the speed of light), we must treat the problem relativistically. According to relativity theory, as the speed of a circulating proton approaches that of light, the proton's frequency of revolution decreases steadily. Thus, the protons get out of step with the cyclotron's oscillator—whose frequency remains fixed at f_{osc}—and eventually the energy of the circulating proton stops increasing.

There is another problem. For a 500 GeV proton in a magnetic field of magnitude 1.5 T, the path radius is 1.1 km. The corresponding magnet for a conventional cyclotron of the proper size would be impossibly expensive, the area of its pole faces being about $4 \times 10^6 \ m^2$.

The *proton synchrotron* is designed to meet these two difficulties. The magnetic field B and the oscillator frequency f_{osc}, instead of having fixed values as in the conventional cyclotron, are made to vary with time during the accelerating cycle. When this is done properly, (1) the frequency of the circulating protons remains in step with the oscillator at all times, and (2) the protons follow a circular—not a spiral—path. Thus, the magnet need extend only along that circular path, not over some $4 \times 10^6 \ m^2$. The circular path, however, still must be large if high energies are to be achieved. The proton synchrotron at the Fermi National Accelerator Laboratory (Fermilab) in Illinois has a circumference of 6.3 km and can produce protons with energies of about 1 TeV ($= 10^{12}$ eV).

29-7 Magnetic Force on a Current-Carrying Wire

We have already seen (in connection with the Hall effect) that a magnetic field exerts a sideways force on electrons moving in a wire. This force must then be transmitted to the wire itself, because the conduction electrons cannot escape sideways out of the wire.

In Fig. 29-13a, a vertical wire, carrying no current and fixed in place at both ends, extends through the gap between the vertical pole faces of a magnet. The magnetic field between the faces is directed outward from the page. In Fig. 29-13b, a current is sent upward through the wire; the wire deflects to the right. In Fig. 29-13c, we reverse the direction of the current and the wire deflects to the left.

Figure 29-14 shows what happens inside the wire of Fig. 29-13. We see one of the conduction electrons, drifting downward with an assumed drift speed v_d. Equation 29-3, in which we must put $\phi = 90°$, tells us that a force $\vec{F}_B$ of magnitude $|e\vec{v}_d||\vec{B}|$ must act on each such electron. From Eq. 29-2 we see that this force must be directed to the right. We expect then that the wire as a whole will experience a force to the right, in agreement with Fig. 29-13b.

If, in Fig. 29-14, we were to reverse *either* the direction of the magnetic field *or* the direction of the current, the force on the wire would reverse, being directed now to the left. Note too that it does not matter whether we consider negative charges drifting downward in the wire (the actual case) or positive charges drifting upward. The direction of the deflecting force on the wire is the same. We are safe then in dealing with a current of positive charge.

Consider a length L of the wire in Fig. 29-14. All the conduction electrons in this section of wire will drift past the plane that is parallel to xx' in Fig. 29-14 in a time $t = L/|\vec{v}_d|$. Thus, in that time the charge that will pass through the plane is given by

$$q = it = i\frac{\vec{L}}{|\vec{v}_d|}.$$

Substituting this into Eq. 29-3 yields

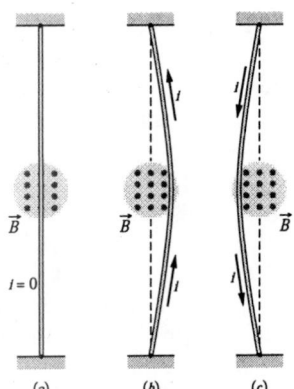

Fig. 29-13: A flexible wire passes between the pole faces of a magnet (only the farther pole face is shown). (*a*) Without current in the wire, the wire is straight. (*b*) With upward current, the wire is deflected rightward. (*c*) With downward current, the deflection is leftward. The connections for getting the current into the wire at one end and out of it at the other end are not shown.

$$\left|\vec{F}_B\right| = \left|q\vec{v}_d\right|\left|\vec{B}\right|\sin\phi = \frac{\left|i\vec{L}\right|\left|\vec{v}_d\right|\left|\vec{B}\right|}{\left|\vec{v}_d\right|}\sin 90°$$

or
$$\left|\vec{F}_B\right| = \left|i\vec{L}\right|\left|\vec{B}\right|. \qquad (29\text{-}22)$$

This equation gives the magnetic force that acts on a length L of straight wire carrying a current i and immersed in a magnetic field $\vec{B}$ that is perpendicular to the wire.

If the magnetic field is *not* perpendicular to the wire, as in Fig. 29-15, the magnetic force is given by a generalization of Eq. 29-22:

$$\vec{F}_B = i\vec{L} \times \vec{B} \qquad \text{(force on a current). (29-23)}$$

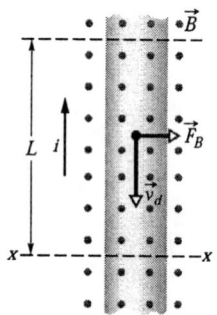

Here $\vec{L}$ is a *length vector* that has magnitude L and is directed along the wire segment in the direction of the (conventional) current. The magnitude of F_B is

$$\vec{F}_B = \left|i\vec{L}\right|\left|\vec{B}\right|\sin\phi, \qquad (29\text{-}24)$$

where ϕ is the angle between the directions of $\vec{L}$ and $\vec{B}$. The direction of $\vec{F}_B$ is that of the cross product $\vec{L} \times \vec{B}$, because we take current i to be a positive quantity. Equation 29-23 tells us that $\vec{F}_B$ is always perpendicular to the plane defined by $\vec{L}$ and $\vec{B}$, as indicated in Fig. 29-15.

Equation 29-23 is equivalent to Eq. 29-2 in that either can be taken as the defining equation for $\vec{B}$. In practice, we define $\vec{B}$ from Eq. 29-23. It is much easier to measure the magnetic force acting on a wire than that on a single moving charge.

If a wire is not straight or the field is not uniform, we can imagine it broken up into small straight segments and apply Eq. 29-23 to each segment. The force on the wire as a whole is then the vector sum of all the forces on the segments that make it up. In the differential limit, we can write

$$d\vec{F}_B = i\,d\vec{L} \times \vec{B}, \qquad (29\text{-}24)$$

and we can find the resultant force on any given arrangement of currents by integrating Eq. 29-24 over that arrangement.

In using Eq. 29-24, bear in mind that there is no such thing as an isolated current-carrying wire segment of length $d\vec{L}$. There must always be a way to introduce the current into the segment at one end and take it out at the other end.

Fig. 29-14: A close-up view of a section of the wire of Fig. 29-13b. The current direction is upward, which means that electrons drift downward. A magnetic field that emerges from the plane of the page causes the electrons and the wire to be deflected to the right.

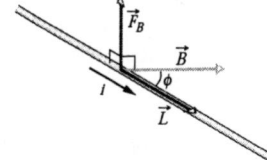

Fig. 29-15: A wire carrying current i makes an angle ϕ with magnetic field $\vec{B}$. The wire has length L in the field and length vector $\vec{L}$ (in the direction of the current). A magnetic force $\vec{F}_B = i\vec{L} \times \vec{B}$ acts on the wire.

READING EXERCISE 29-4: The figure shows a current i through a wire in a uniform magnetic field $\vec{B}$, as well as the magnetic force $\vec{F}_R$ acting on the wire. The field is oriented so that the force is maximum. In what direction is the field?

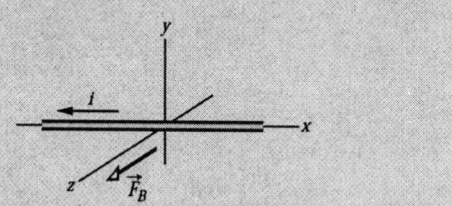

Touchstone Example 29-7-1, at the end of this chapter, illustrates how to use what you learned in this section.

TE

29-8 Torque on a Current Loop

Much of the world's work is done by electric motors. The forces behind this work are the magnetic forces that we studied in the preceding section—that is, the forces that a magnetic field exerts on a wire that carries a current.

Figure 29-16 shows a simple motor, consisting of a single current-carrying loop immersed in a magnetic field $\vec{B}$. The two magnetic forces $\vec{F}$ and $-\vec{F}$ produce a torque on the loop, tending to rotate it about its central axis. Although many essential details have been omitted, the figure does suggest how the action of a magnetic field on a current loop produces rotary motion. Let us analyze that action.

Figure 29-17a shows a rectangular loop of sides a and b, carrying a current i and immersed in a uniform magnetic field $\vec{B}$. We place it in the field so that its long sides, labeled 1 and 3, are perpendicular to the field direction (which is into the page), but its short sides, labeled 2 and 4, are not. Wires to lead the current into and out of the loop are needed but, for simplicity, they are not shown.

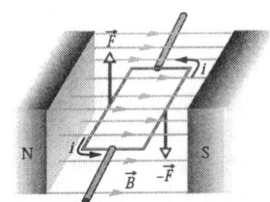

Fig. 29-16: The elements of an electric motor. A rectangular loop of wire, carrying a current and free to rotate about a fixed axis, is placed in a magnetic field. Magnetic forces on the wire produce a torque that rotates it. A commutator (not shown) reverses the direction of the current every half-revolution so that the torque always acts in the same direction.

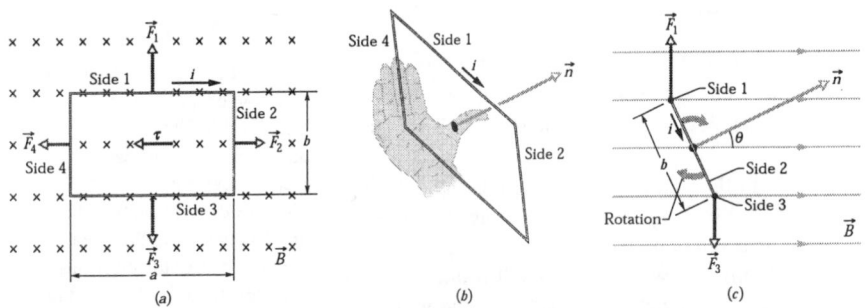

(a)   (b)   (c)

Fig. 29-17: A rectangular loop, of length a and width b and carrying a current i, is located in a uniform magnetic field. A torque τ acts to align the normal vector $\vec{n}$ with the direction of the field. (a) The loop as seen by looking in the direction of the magnetic field. (b) A perspective of the loop showing how the right-hand rule gives the direction of $\vec{n}$, which is perpendicular to the plane of the loop. (c) A side view of the loop, from side 2. The loop rotates as indicated.

To define the orientation of the loop in the magnetic field, we use a normal vector $\vec{n}$ that is perpendicular to the plane of the loop. Figure 29-17b shows a right-hand rule for finding the direction of $\vec{n}$. Point or curl the fingers of your right hand in the direction of the current at any point on the loop. Your extended thumb then points in the direction of the normal vector $\vec{n}$.

In Fig. 29-17c, the normal vector of the loop is shown at an arbitrary angle θ to the direction of the magnetic field $\vec{B}$. We wish to find the net force and net torque acting on the loop in this orientation.

The net force on the loop is the vector sum of the forces acting on its four sides. For side 2 the vector $\vec{L}$ in Eq. 29-23 points in the direction of the current and has magnitude b. The angle between $\vec{L}$ and $\vec{B}$ for side 2 (see Fig. 29-17c) is $90°-\theta$. Thus, the magnitude of the force acting on this side is

$$\left|\vec{F_2}\right| = \left|\vec{B}\right|\sin(90°-\theta) = \left|ib\right|\left|\vec{B}\right|\cos\theta. \qquad (29\text{-}25)$$

You can show that the force $\vec{F_4}$ acting on side 4 has the same magnitude as $\vec{F_2}$ but acts in the opposite direction. Thus, $\vec{F_2}$ and $\vec{F_4}$ cancel out exactly. Their net force is zero and, because their common line of action is through the center of the loop, their net torque is also zero.

The situation is different for sides 1 and 3. For them, $\vec{L}$ is perpendicular to $\vec{B}$, so the forces $\vec{F_1}$ and $\vec{F_3}$ have the common magnitude $\left|ia\vec{B}\right|$. Because these two forces have opposite directions, they do not tend to move the loop up or down. However, as Fig. 29-17c shows, these two forces do *not* share the same line of action so they *do* produce a net torque. The torque tends to rotate the loop so as to align its normal vector $\vec{n}$ with the direction of the magnetic field $\vec{B}$. That torque has moment arm of magnitude

$(b/2)\sin\theta$ about the central axis of the loop. The magnitude $|\vec{\tau}'|$ of the torque due to forces $\vec{F}_1$ and $\vec{F}_3$ is then (see Fig. 29-17c)

$$|\vec{\tau}'| = \left| iaB\frac{b}{2}\sin\theta \right| + \left| iaB\frac{b}{2}\sin\theta \right| = \left| iab\vec{B} \right| \sin\theta. \qquad (29\text{-}26)$$

Suppose we replace the single loop of current with a *coil* of N loops, or *turns*. Further, suppose that the turns are wound tightly enough that they can be approximated as all having the same dimensions and lying in a plane. Then the turns form a *flat coil* and a torque $\vec{\tau}'$ with the magnitude given in Eq. 29-26 acts on each of them. The total torque on the coil then has magnitude

$$|\vec{\tau}| = N|\vec{\tau}'| = N|iab||\vec{B}|\sin\phi = (N|i|A)|\vec{B}|\sin\phi, \qquad (29\text{-}27)$$

in which $A(= ab)$ is the area enclosed by the coil. The quantities in parentheses $(N|i|A)$ are grouped together because they are all properties of the coil: its number of turns, its area, and the current it carries. Equation 29-27 holds for all flat coils, no matter what their shape, provided the magnetic field is uniform.

Instead of focusing on the motion of the coil, it is simpler to keep track of the vector $\vec{n}$, which is normal to the plane of the coil. Equation 29-27 tells us that a current-carrying flat coil placed in a magnetic field will tend to rotate so that $\vec{n}$ has the same direction as the field.

In a motor, the current in the coil is reversed as $\vec{n}$ begins to line up with the field direction, so that a torque continues to rotate the coil. This automatic reversal of the current is accomplished with a commutator that electrically connects the rotating coil with the stationary contacts on the wires that supply the current from some source.

29-9 The Magnetic Dipole Moment

We can describe the current-carrying coil of the preceding section with a single vector $\vec{\mu}$, its magnetic dipole moment. We take the direction of $\vec{\mu}$ to be that of the normal vector $\vec{n}$ to the plane of the coil, as in Fig. 29-17c. We define the magnitude of $\vec{\mu}$ as

$$|\mu| = N|i|A \quad \text{(magnetic moment)}, \qquad (29\text{-}28)$$

in which N is the number of turns in the coil, i is the current through the coil, and A is the area enclosed by each turn of the coil. (Equation 29-28 tells us that the unit of $\vec{\mu}$ is the ampere-square meter.) Using $\vec{\mu}$, we can rewrite Eq. 29-27 for the magnitude of the torque on the coil due to a magnetic field as

$$|\vec{\tau}| = |\mu||\vec{B}|\sin\phi, \qquad (29\text{-}29)$$

in which ϕ is the angle between the vectors $\vec{\mu}$ and $\vec{B}$.

We can generalize this to the vector relation

$$\vec{\tau} = \vec{\mu} \times \vec{B}, \qquad (29\text{-}30)$$

which reminds us very much of the corresponding equation for the torque exerted by an *electric* field on an *electric* dipole—namely, Eq. 23-31:

$$\vec{\tau} = \vec{p} \times \vec{E}.$$

In each case the torque exerted by the external field—either magnetic or electric—is equal to the vector product of the corresponding dipole moment and the field vector.

A magnetic dipole in an external magnetic field has a **magnetic potential energy** that depends on the dipole's orientation in the field. For electric dipoles,

$$U(\theta) = -\vec{p} \cdot \vec{E}.$$

In strict analogy, we can write for the magnetic case

$$U(\theta) = -\vec{\mu} \cdot \vec{B}. \qquad (29\text{-}31)$$

A magnetic dipole has its lowest energy $(-\mu B \cos 0 = -\mu B)$ when its dipole moment $\vec{\mu}$ is lined up with the magnetic field (Fig. 29-18). It has its highest energy $(= -\mu B \cos 180° = +\mu B)$ when $\vec{\mu}$ is directed opposite the field.

When a magnetic dipole rotates from an initial orientation θ_i to another orientation θ_f, the work W done on the dipole by the magnetic field is

$$W = -\Delta U = -(U_f - U_i), \qquad (29\text{-}32)$$

where U_f and U_i are calculated with Eq. 29-31. If an applied torque (due to "an external agent") acts on the dipole during the change in its orientation, then work W_a is done on the dipole by the applied torque. *If the dipole is stationary* before and after the change in its orientation, then work W_a is the negative of the work done on the dipole by the field. Thus,

$$W_a = -W = U_f - U_i. \qquad (29\text{-}33)$$

So far, we have identified only a current-carrying coil as a magnetic dipole. However, a simple bar magnet is also a magnetic dipole, as is a rotating sphere of charge. Earth itself is (approximately) a magnetic dipole. Finally, most subatomic particles, including the electron, the proton, and the neutron, have magnetic dipole moments. As you will see in Chapter 32, all these quantities can be viewed as current loops. For comparison, some approximate magnetic dipole moments are shown in Table 29-2.

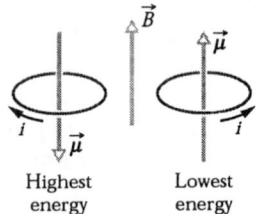

Fig. 29-18: The orientations of highest and lowest energy of a magnetic dipole in an external magnetic field $\vec{B}$. The direction of the current i gives the direction of the magnetic dipole moment $\vec{\mu}$ via the right-hand rule shown for $\vec{n}$ in Fig. 29-17b.

TABLE 29-2: Some Magnetic Dipole Moments

A small bar magnet	5 J/T
Earth	8.0×10^{22} J/T
A proton	1.4×10^{-26} J/T
An electron	9.3×10^{-24} J/T

READING EXERCISE 29-5: The figure shows four orientations, at angle θ, of a magnetic dipole moment $\vec{\mu}$ in a magnetic field. Rank the orientations according to (a) the magnitude of the torque on the dipole and (b) the potential energy of the dipole, greatest first.

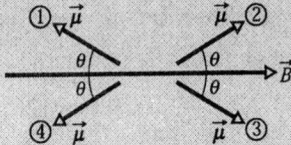

Touchstone Example 29-9-1, at the end of this chapter, illustrates how to use what you learned in this section.

TE

Touchstone Example 29-2-1

A uniform magnetic field $\vec{B}$, with magnitude 1.2 mT, is directed vertically upward throughout the volume of a laboratory chamber. A proton with kinetic energy 5.3 MeV enters the chamber, moving horizontally from south to north. What is the magnitude of the magnetic deflecting force acting on the proton as it enters the chamber? The proton mass is 1.67×10^{-27} kg. (Neglect Earth's magnetic field.)

SOLUTION: Because the proton is charged and moving through a magnetic field, a magnetic force $\vec{F}_B$ can act on it. The **Key Idea** here is that, because the initial direction of the proton's velocity is not along a magnetic field line, $\vec{F}_B$ is not simply zero. To find the magnitude of $\vec{F}_B$, we can use Eq. 29-3 provided we first find the proton's speed $|\vec{v}|$. We can find $|\vec{v}|$ from the given kinetic energy, since $K = \frac{1}{2}mv^2$. Solving for v, we obtain

$$|\vec{v}| = \sqrt{\frac{2K}{m}} = \sqrt{\frac{(2)(5.3 \text{ MeV})(1.60 \times 10^{-13} \text{ J/MeV})}{1.67 \times 10^{-27} \text{ kg}}}$$

$$= 3.2 \times 10^7 \text{ m/s}.$$

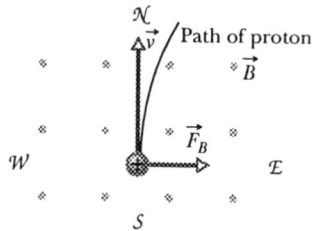

Fig. TE29-1 An overhead view of a proton moving from south to north with velocity $\vec{v}$ in a chamber. A magnetic field is directed vertically upward in the chamber, as represented by the array of dots (which resemble the tips of arrows). The proton is deflected toward the east.

Equation 29-3 then yields

$$|\vec{F}_B| = |q\vec{v}||\vec{B}|\sin\phi$$
$$= (1.60 \times 10^{-19} \text{ C})(3.2 \times 10^7 \text{ m/s})$$
$$\times (1.2 \times 10^{-3} \text{ T})|(\sin 90°)|$$
$$= 6.1 \times 10^{-15} \text{ N}. \qquad \text{(Answer)}$$

This may seem like a small force, but it acts on a particle of small mass, producing a large acceleration; namely,

$$|\vec{a}| = \frac{|\vec{F}_B|}{m} = \frac{6.1 \times 10^{-15} \text{ N}}{1.67 \times 10^{-27} \text{ kg}} = 3.7 \times 10^{12} \text{ m/s}^2.$$

To find the direction of $\vec{F}_B$, we use the **Key Idea** that $\vec{F}_B$ has the direction of the cross product $q\vec{v} \times \vec{B}$. Because the charge q is positive, $\vec{F}_B$ must have the same direction as $\vec{v} \times \vec{B}$, which can be determined with the right-hand rule for cross products (as in Fig. 29-2b). We know that $\vec{v}$ is directed horizontally from south to north and $\vec{B}$ is directed vertically up. The right-hand rule shows us that the deflecting force $\vec{F}_B$ must be directed horizontally from west to east, as Fig. TE29-1 shows. (The array of dots in the figure represents a magnetic field directed out of the plane of the figure. An array of **X**s would have represented a magnetic field directed into that plane.)

If the charge of the particle were negative, the magnetic deflecting force would be directed in the opposite direction—that is, horizontally from east to west. This is predicted automatically by Eq. 29-2 if we substitute a negative value for q.

Touchstone Example 29-4-1

Figure TE29-2 shows a solid metal cube, of edge length $d = 1.5$ cm, moving in the positive y direction at a constant velocity $\vec{v}$ of magnitude 4.0 m/s. The cube moves through a uniform magnetic field of magnitude 0.050 T directed toward positive z.

(a) Which cube face is at a lower electric potential and which is at a higher electric potential because of the motion through the field?

SOLUTION: One **Key Idea** here is that, because the cube is moving through a magnetic field $\vec{B}$, a magnetic force $\vec{F}_B$ acts on its charged particles, including its conduction electrons. A second **Key Idea** is how $\vec{F}_B$ causes an electric potential difference between certain faces of the cube. When the cube first begins to move through the magnetic field, its electrons do also. Because each electron has charge $q = -e$ and is moving through a magnetic field with velocity $\vec{v}$ the magnetic force $\vec{F}_B$ acting on it is given by Eq. 29-2. Because q is negative, the direction of $\vec{F}_B$ is opposite the cross product $\vec{v} \times \vec{B}$, which is in the positive direction of the x axis in Fig. TE29-2. Thus, $\vec{F}_B$ acts in the negative direction of the x axis, toward the left face of the cube (which is hidden from view in Fig. TE29-2).

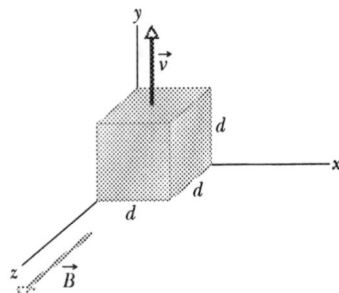

Fig. TE29-2 A solid metal cube of edge length d, at constant velocity $\vec{v}$ through a uniform magnetic field $\vec{B}$.

Most of the electrons are fixed in place in the molecules of the cube. However, because the cube is a metal, it contains conduction electrons that are free to move. Some of those conduction electrons are deflected by $\vec{F}_B$ to the left cube face, making that face negatively charged and leaving the right face positively charged. This charge separation produces an electric field $\vec{E}$ directed from the positively charged right face to the negatively charged left face. Thus, the left face is at a lower electric potential, and the right face is at a higher electric potential.

(b) What is the potential difference between the faces of higher and lower electric potential?

SOLUTION: The **Key Ideas** here are these:

1. The electric field $\vec{E}$ created by the charge separation produces an electric force $\vec{F}_E = q\vec{E}$ on each electron. Because q is negative, this force is directed opposite the field $\vec{E}$—that is, toward the right. Thus on each electron, $\vec{F}_E$ acts toward the right and $\vec{F}_B$ acts toward the left.

2. When the cube had just begun to move through the magnetic field and the charge separation had just begun, the magnitude of $\vec{E}$ began to increase from zero. Thus, the magnitude of $\vec{F}_E$ also began to increase from zero and was initially smaller than the magnitude $\vec{F}_B$. During this early stage, the net force on any electron was dominated by $\vec{F}_B$, which continuously moved additional electrons to the left cube face, increasing the charge separation.

3. However, as the charge separation increased, eventually magnitude $|\vec{F}_E|$ became equal to magnitude $|\vec{F}_B|$. The net force on any electron was then zero, and no additional electrons were moved to the left cube face. Thus, the magnitude of F_E could not increase further, and the electrons were then in equilibrium.

We seek the potential difference ΔV between the left and right cube faces after equilibrium was reached (which occurred quickly). We can obtain the magnitude of ΔV with Eq. 29-9 ($|\Delta V| = |\vec{E}|d$) provided we first find the magnitude $|\vec{E}|$ of the electric field at equilibrium. We can do so with the equation for the balance of forces ($|\vec{F}_E| = |\vec{F}_B|$).

For $|\vec{F}_E|$, we substitute $|q\vec{E}|$. For $|\vec{F}_B|$, we substitute $|qv||\vec{B}|\sin\phi$ from Eq. 29-3. From Fig. TE29-2, we see that the smallest angle ϕ between $\vec{v}$ and $\vec{B}$ is 90°; so $\sin\phi = 1$. We can now write ($|\vec{F}_E| = |\vec{F}_B|$) as

$$|q\vec{E}| = |q\vec{v}||\vec{B}|\sin 90° = |q\vec{v}||\vec{B}|.$$

This gives us $|\vec{E}| = |\vec{v}||\vec{B}|$, so Eq. 29-9 $|\Delta V| = |\vec{E}|d$ becomes

$$|\Delta V| = |V_{\text{left}} - V_{\text{right}}| = |\vec{v}||\vec{B}|d. \qquad \text{(TE29-1)}$$

Substituting known values gives us

$$|\Delta V| = (4.0 \text{ m/s})(0.050 \text{ T})(0.015 \text{ m})$$
$$= 0.0030 \text{ V} = 3.0 \text{ mV}.$$

Since the left face of the cube has excess negatives charges, the right face is at a higher potential than the left face by 3.0 mV. (Answer)

Touchstone Example 29-5-1

Figure TE29-3 shows the essentials of a *mass spectrometer*, which can be used to measure the mass of an ion; an ion of mass m (to be measured) and charge q is produced in source S. The initially stationary ion is accelerated by the electric field due to a potential difference ΔV. The ion leaves S and enters a separator chamber in which a uniform magnetic field $\vec{B}$ is perpendicular to the path of the ion. The magnetic field causes the ion to move in a semicircle, striking (and thus altering) a photographic plate at distance x from the entry slit. Suppose that in a certain trial $B = 80.000$ mT and $\Delta V = 1000.0$ V, and ions of charge $q = +1.6022 \times 10^{-19}$ C strike the plate at $x = 1.6254$ m. What is the mass m of the individual ions, in unified atomic mass units (1 u $= 1.6605 \times 10^{-27}$ kg)?

SOLUTION: One **Key Idea** here is that, because the (uniform) magnetic field causes the (charged) ion to follow a circular path, we can relate the ion's mass m to the path's radius r with Eq. 29-15 ($r = m|\vec{v}|/|q\vec{B}|$). From Fig. TE29-3 we see that $r = x/2$, and we are given the magnitude $|\vec{B}|$ of the magnetic field. However, we need to know the ion's speed $|\vec{v}|$ in the magnetic field, after it has been accelerated due to the potential difference ΔV.

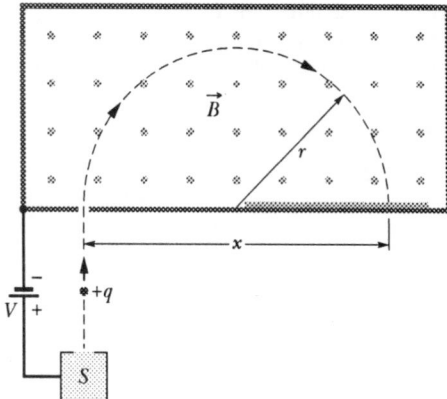

Fig. TE29-3 Essentials of a mass spectrometer. A positive ion, after being accelerated from its source S by potential difference ΔV, enters a chamber of uniform magnetic field $\vec{B}$. There it travels through a semicircle of radius r and strikes a photographic plate at a distance x from where it entered the chamber.

To relate $|\vec{v}|$ and ΔV, we use the **Key Idea** that mechanical energy ($E_{mec} = K + U$) is conserved during the acceleration. When the ion emerges from the source, its kinetic energy is approximately zero. At the end of the acceleration, its kinetic energy is $\frac{1}{2}mv^2$. Also, during the acceleration, the positive ion moves through a change in potential of $-\Delta V$. Thus, because the ion has positive charge q, its potential energy changes by $-q\Delta V$. If we now write the conservation of mechanical energy as

$$\Delta K + \Delta U = 0,$$

we get

$$\frac{1}{2}mv^2 - (q\Delta V) = 0$$

or

$$|\vec{v}| = \sqrt{\frac{2|q\Delta V|}{m}}. \tag{TE29-2}$$

Substituting this into Eq. 29-15 gives us

$$r = \frac{m|\vec{v}|}{|q\vec{B}|} = \frac{m}{|q\vec{B}|}\sqrt{\frac{2|q\Delta V|}{m}} = \frac{1}{|\vec{B}|}\sqrt{\frac{2m\Delta V}{q}}.$$

Thus,

$$|x| = 2r = \frac{2}{|\vec{B}|}\sqrt{\frac{2m|\Delta V|}{|q|}}.$$

Solving this for m and substituting the given data yield

$$m = \frac{B^2|q|x^2}{8|\Delta V|}$$

$$= \frac{(0.080000\ \text{T})^2(1.6022 \times 10^{-19}\ \text{C})(1.6254\ \text{m})^2}{8(1000.0\ \text{V})}$$

$$= 3.3863 \times 10^{-25}\ \text{kg} = 203.93\ \text{u}. \qquad \text{(Answer)}$$

Touchstone Example 29-7-1

A straight, horizontal length of copper wire has a current $i = 28$ A through it. If this current is directed out of the page as shown in Fig. TE29-4, what are the magnitude and direction of the minimum magnetic field $\vec{B}$ needed to suspend the wire—that is, to balance the gravitational force on it? The linear density (mass per unit length) of the wire is 46.6 g/m.

SOLUTION: One **Key Idea** is that, because the wire carries a current, a magnetic force $\vec{F}_B$ can act on the wire if we place it in a magnetic field $\vec{B}$. To balance the downward gravitational force $\vec{F}_g$ on the wire, we want $\vec{F}_B$ to be directed upward (Fig. TE29-4).

A second **Key Idea** is that the direction of $\vec{F}_B$ is related to the directions of $\vec{B}$ and the wire's length vector $\vec{L}$ by Eq. 29-23. Because $\vec{L}$ is directed horizontally (and the current is taken to be positive), Eq. 29-23 and the right-hand rule for cross products tell us that $\vec{B}$ must be horizontal and rightward (in Fig. TE29-4) to give the required upward $\vec{F}_B$.

The magnitude of $\vec{F}_B$ is given by Eq. 29-24 ($|\vec{F}_B| = |i\vec{L}||\vec{B}|\sin\phi$). Because we want $\vec{F}_B$ to balance $\vec{F}_g$ we want

$$iLB \sin\phi = mg, \tag{TE29-3}$$

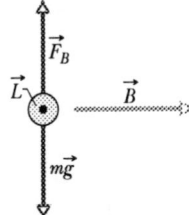

Fig. TE29-4 A current-carrying wire (shown in cross section) can be made to "float" in a magnetic field. The current in the wire emerges from the plane of the page, and the magnetic field is directed to the right.

where mg is the magnitude of $\vec{F}_g$ and m is the mass of the wire. We also want the minimal field magnitude B for $\vec{F}_B$ to balance $\vec{F}_g$. Thus, we need to maximize $\sin \phi$ in Eq. TE29-3. To do so, we set $\phi = 90°$, thereby arranging for $\vec{B}$ to be perpendicular to the wire. We then have $\sin \phi = 1$, so Eq. TE29-3 yields a magnetic field magnitude of

$$|\vec{B}| = \frac{mg}{|i\vec{L}| \sin \phi} = \frac{(m/|\vec{L}|)g}{|i|}. \tag{TE29-4}$$

We write the result this way because we know m/L, the linear density of the wire. Substituting known data then gives us a magnitude of

$$|\vec{B}| = \frac{(46.6 \times 10^{-3}\ \text{kg/m})(9.8\ \text{m/s}^2)}{28\ \text{A}}$$

$$= 1.6 \times 10^{-2}\ \text{T.} \qquad\qquad \text{(Answer)}$$

This is about 160 times the strength of Earth's magnetic field. As stated in the second paragraph of this solution, the right-hand rule tells us that $|\vec{B}|$ must point to the right.

Touchstone Example 29-9-1

Figure TE29-5 shows a circular coil with 250 turns, an area A of $2.52 \times 10^{-4}\ \text{m}^2$, and a current of 100 μA. The coil is at rest in a uniform magnetic field of magnitude $|\vec{B}| = 0.85\ \text{T}$, with its magnetic dipole moment $\vec{\mu}$ initially aligned with $\vec{B}$.

(a) In Fig. TE29-5, what is the direction of the current in the coil?

SOLUTION: The **Key Idea** here is to apply the following right-hand rule to the coil: Imagine cupping the coil with your right hand so that your right thumb is outstretched in the direction of $\vec{\mu}$. The direction in which your fingers curl around the coil is the direction of the current in the coil. This, in the wires on the rear side of the coil—those we see in Fig. TE29-5—the current is from top to bottom.

(b) How much work would the torque applied by an external agent have to do on the coil to rotate it 90° from its initial orientation, so that $\vec{\mu}$ is perpendicular to $\vec{B}$ and the coil is again at rest?

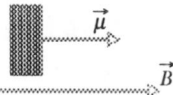

Fig. TE29-5 A side view of a circular coil carrying a current and oriented so that its magnetic dipole moment $\vec{\mu}$ is aligned with magnetic field $\vec{B}$.

SOLUTION: The **Key Idea** here that the work W_a done by the applied torque would be equal to the change in the coil's potential energy due to its change in orientation. From Eq. 29-33 ($W_a = U_f - U_i$), we find

$$W_a = U(90°) - U(0°)$$
$$= -|\vec{\mu}||\vec{B}| \cos 90° - (-|\vec{\mu}||\vec{B}| \cos 0°) = 0 + |\vec{\mu}||\vec{B}|$$
$$= |\vec{\mu}||\vec{B}|.$$

Substituting for $|\vec{\mu}|$ from Eq. 29-28 ($\mu = N|i|A$), we find that

$$W_a = (N|i|A|\vec{B}|)$$
$$= (250)(100 \times 10^{-6}\ \text{A})(2.52 \times 10^{-4}\ \text{m}^2)(0.85\ \text{T})$$
$$= 5.356 \times 10^{-6}\ \text{J} \approx 5.4\ \mu\text{J.} \qquad\qquad \text{(Answer)}$$

30 Magnetic Fields Due to Currents

This is the way we presently launch materials into space. However, when we begin mining the Moon and the asteroids, where we will not have a source of fuel for such conventional rockets, we shall need a more effective way. Electromagnetic launchers may be the answer. A small prototype, the *electromagnetic rail gun*, can presently accelerate a projectile from rest to a speed of 10 km/s (36 000 km/h) within 1 ms.

How can such rapid acceleration possibly be accomplished?

The answer is in this chapter

30-1 Calculating the Magnetic Field Due to a Current

As we discussed in Section 29-1, one way to produce a magnetic field is with moving charges—that is, with a current. Our goal in this chapter is to calculate the magnetic field that is produced by a given distribution of currents. We shall use the same basic procedure we used in Chapter 23 to calculate the electric field produced by a given distribution of charged particles.

Let us quickly review that basic procedure. We first mentally divide the charge distribution into charge elements dq, as is done for a charge distribution of arbitrary shape in Fig. 30-1a. We then calculate the field $d\vec{E}$ produced at some point P by a typical charge element. Because the electric fields contributed by different elements can be superimposed, we calculate the net field $\vec{E}$ at P by summing, via integration, the contributions $d\vec{E}$ from all the elements.

Recall that we express the magnitude of $d\vec{E}$ as

$$|d\vec{E}| = \frac{1}{4\pi\varepsilon_0}\frac{|dq|}{r^2},$$ (30-1)

in which r is the distance between the charge element dq and point P. For a positively charged element, the direction of $d\vec{E}$ is that of $\vec{r}$, where $\vec{r}$ is the vector that extends from the charge element dq to the point P. Using $\vec{r}$, we can rewrite Eq. 30-1 in vector form as

$$d\vec{E} = \frac{1}{4\pi\varepsilon_0}\frac{dq}{r^3}\vec{r},$$ (30-2)

which indicates that the direction of the vector $d\vec{E}$ produced by a positively charged element is the direction of the vector $\vec{r}$. Note that Eq. 30-2 is an inverse-square law ($d\vec{E}$ depends on inverse r^2) in spite of the exponent 3 in the denominator. That exponent is in the equation only because we added a factor of magnitude r in the numerator.

Now let us use the same basic procedure to calculate the magnetic field due to a current. Figure 30-1b shows a wire of arbitrary shape carrying a current i. We want to find the magnetic field $\vec{B}$ at a nearby point P. We first mentally divide the wire into differential elements ds and then define for each element a length vector $d\vec{s}$ that has length ds and whose direction is the direction of the current in ds. We can then define a differential *current-length element* to be $i\,d\vec{s}$; we wish to calculate the field $d\vec{B}$ produced at P by a typical current-length element. From experiment we find that magnetic fields, like electric fields, can be superimposed to find a net field. Thus, we can calculate the net field $\vec{B}$ at P by summing, via integration, the contributions $d\vec{B}$ from all the current-length elements. However, this summation is more challenging than the process associated with electric fields because of a complexity; whereas a charge element dq producing an electric field is a scalar, a current-length element $i\,d\vec{s}$ producing a magnetic field is the product of a scalar and a vector.

The magnitude of the field $d\vec{B}$ produced at point P by a current-length element $i\,d\vec{s}$ turns out to be

$$|d\vec{B}| = \frac{\mu_0}{4\pi}\frac{|i\,d\vec{s}|\sin\theta}{r^2},$$ (30-3)

where θ is the angle between the directions of $d\vec{s}$ and $\vec{r}$, the vector that extends from ds to P. Symbol μ_0 is a constant, called the *permeability constant*, whose value is defined to be exactly

$$\mu_0 = 4\pi\times10^{-7} \text{ T}\cdot\text{m/A} \approx 1.26\times10^{-6}\text{ T}\cdot\text{m/A}.$$ (30-4)

The direction of $d\vec{B}$, shown as being into the page in Fig. 30-1b, is that of the cross product $d\vec{s}\times\vec{r}$. We can therefore write Eq. 30-3 in vector form as

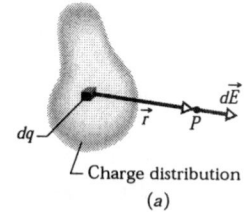

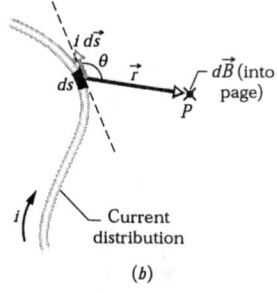

Fig. 30-1: (a) A charge element dq produces a differential electric field dE at point P. (b) A current-length element $i\,d\vec{s}$ produces a differential magnetic field dB at point P. The green × (the tail of an arrow) at the dot for point P indicates that dB is directed into the page there.

$$dB = \frac{\mu_0}{4\pi} \frac{i\,d\vec{s} \times \vec{r}}{r^3} \qquad \text{(Biot-Savart law).} \qquad (30\text{-}5)$$

This vector equation and its scalar form, Eq. 30-3, are known as the **law of Biot and Savart** (rhymes with "Leo and bazaar"). The law, which is experimentally deduced, is an inverse-square law (the exponent in the denominator of Eq. 30-5 is 3 only because of the factor $\vec{r}$ in the numerator). We shall use this law to calculate the net magnetic field $\vec{B}$ produced at a point by various distributions of current.

Magnetic Field Due to a Current in a Long Straight Wire

Shortly we shall use the law of Biot and Savart to prove that the magnitude of the magnetic field at a perpendicular distance R from a long (infinite) straight wire carrying a current i is given by

$$\left|\vec{B}\right| = \frac{\mu_0 |i|}{2\pi R} \qquad \text{(long straight wire).} \qquad (30\text{-}6)$$

The field magnitude $\left|\vec{B}\right|$ in Eq. 30-6 depends only on the current and the perpendicular distance R of the point from the wire. We shall show in our derivation that the field lines of $\vec{B}$ form concentric circles around the wire, as Fig. 30-2 shows and as the iron filings in Fig. 30-3 suggest. The increase in the spacing of the lines in Fig. 30-2 with increasing distance from the wire represents the $1/R$ decrease in the magnitude of $\vec{B}$ predicted by Eq. 30-6. The lengths of the two vectors $\vec{B}$ in the figure also show the $1/R$ decrease.

Here is a simple right-hand rule for finding the direction of the magnetic field set up by a current-length element, such as a section of a long wire:

▶*Right-hand rule:* Grasp the element in your right hand with your extended thumb pointing in the direction of the current. Your fingers will then naturally curl around in the direction of the magnetic field lines due to that element.

The result of applying this right-hand rule to the current in the straight wire of Fig. 30-2 is shown in a side view in Fig. 30-4*a*. To determine the direction of the magnetic field $\vec{B}$ set up at any particular point by this current, mentally wrap your right hand around the wire with your thumb in the direction of the current. Let your fingertips pass through the point; their direction is then the direction of the magnetic field at that point. In the view of Fig. 30-2, $\vec{B}$ at any point is *tangent to a magnetic field line;* in the view of Fig. 30-4, it is *perpendicular to a dashed radial line connecting the point and the current.*

Proof of Equation 30-6

Figure 30-5, which is just like Fig. 30-1*b* except that now the wire is straight and of infinite length, illustrates the task at hand; we seek the field $\vec{B}$ at point P, a perpendicular distance R from the wire. The magnitude of the differential magnetic field produced at P by the current-length element $\left|i\,d\vec{s}\right|$ located a distance r from P is given by Eq. 30-3:

$$\left|d\vec{B}\right| = \frac{\mu_0}{4\pi} \frac{\left|i\,d\vec{s}\right| \sin\theta}{r^2}.$$

The direction of $d\vec{B}$ in Fig. 30-5 is that of the vector $d\vec{s} \times \vec{r}$ —namely, directly into the page.

Note that $d\vec{B}$ at point P has this same direction for all the current-length elements into which the wire can be divided. Thus, we can find the magnitude of the magnetic field produced at P by the current-length elements in the upper half of the infinitely long wire by integrating dB in Eq. 30-3 from 0 to ∞.

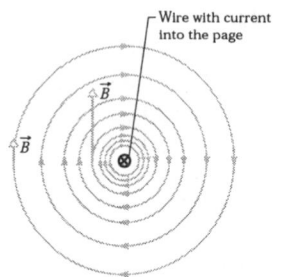

Fig. 30-2: The magnetic field lines produced by a current in a long straight wire form concentric circles around the wire. Here the current is into the page, as indicated by the $\times$.

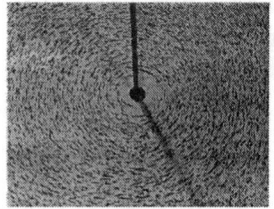

Fig. 30-3: Iron filings that have been sprinkled onto cardboard collect in concentric circles when current is sent through the central wire. The alignment, which is along magnetic field lines, is caused by the magnetic field produced by the current.

Now consider a current-length element in the lower half of the wire, one that is as far below P as $d\vec{s}$ is above P. By Eq. 30-5, the magnetic field produced at P by this current-length element has the same magnitude and direction as that from $i\,d\vec{s}$ in Fig. 30-5. Further, the magnetic field produced by the lower half of the wire is exactly the same as that produced by the upper half. To find the magnitude of the total magnetic field $\vec{B}$ at P, we need only multiply the result of our integration by 2. We get

$$|\vec{B}| = 2\int_0^\infty |d\vec{B}| = \frac{\mu_0 |i|}{2\pi} \int_0^\infty \frac{\sin\theta |d\vec{s}|}{r^2}. \tag{30-7}$$

The variables θ, s, and r in this equation are not independent but (see Fig. 30-5) are related by

$$r = \sqrt{s^2 + R^2}$$

and

$$\sin\theta = \sin(\pi - \theta) = \frac{R}{\sqrt{s^2 + R^2}}.$$

With these substitutions and integral 19 in Appendix E, Eq. 30-7 describing the magnitude of the electric field becomes

$$|\vec{B}| = \frac{\mu_0 |i|}{2\pi} \int_0^\infty \frac{R|d\vec{s}|}{(s^2 + R^2)^{3/2}}$$

$$= \frac{\mu_0 i}{2\pi R} \left[\frac{s}{(s^2 + R^2)^{1/2}} \right]_0^\infty = \frac{\mu_0 |i|}{2\pi R}, \tag{30-8}$$

which is the relation we set out to prove. Note that the magnetic field at P due to either the lower half or the upper half of the infinite wire in Fig. 30-5 is half this value; that is,

$$|\vec{B}| = \frac{\mu_0 |i|}{4\pi R} \qquad \text{(semi-infinite straight wire).} \tag{30-9}$$

Magnetic Field Due to a Current in a Circular Arc of Wire

To find the magnetic field produced at a point by a current in a curved wire, we would again use Eq. 30-3 to write the magnitude of the field produced by a single current-length element, and we would again integrate to find the net field produced by all the current-length elements. That integration can be difficult, depending on the shape of the wire; it is fairly straightforward, however, when the wire is a circular arc and the point is the center of curvature.

Figure 30-6a shows such an arc-shaped wire with central angle ϕ, radius R, and center C, carrying current i. At C, each current-length element $i\,d\vec{s}$ of the wire produces a magnetic field element of magnitude dB given by Eq. 30-3. Moreover, as Fig. 30-6b shows, no matter where the element is located on the wire, the angle θ between the vectors $d\vec{s}$ and $\vec{r}$ is 90°; also, $r = R$. Thus, by substituting R for r and 90° for θ, we obtain from Eq. 30-3,

$$d\vec{B} = \frac{\mu_0}{4\pi} \frac{|i\,d\vec{s}|\sin 90°}{R^2} = \frac{\mu_0}{4\pi} \frac{|i\,d\vec{s}|}{R^2}. \tag{30-10}$$

The field at C due to each current-length element in the circular arc has this same magnitude.

An application of the right-hand rule anywhere along the wire (as in Fig. 30-6c) will show that all the differential fields $d\vec{B}$ have the same direction at C—directly out of the page. Thus, the total field at C is simply the sum (via integration) of all the fields $d\vec{B}$.

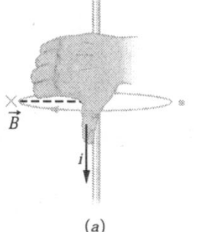

(a)

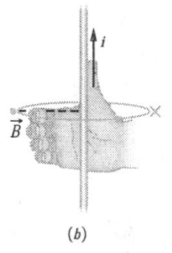

(b)

Fig. 30-4: A right-hand rule gives the direction of the magnetic field due to a current in a wire. (a) The situation of Fig. 30-2, seen from the side. The magnetic field $\vec{B}$ at any point to the left of the wire is perpendicular to the dashed radial line and directed into the page, in the direction of the fingertips, as indicated by the $\times$. (b) If the current is reversed, $\vec{B}$ at any point to the left is still perpendicular to the dashed radial line but now is directed out of the page, as indicated by the dot.

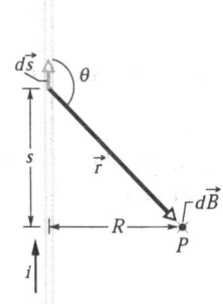

Fig. 30-5: Calculating the magnetic field produced by a current i in a long straight wire. The field dB at P associated with the current-length element $i\,d\vec{s}$ is directed into the page, as shown.

We use the identity $ds = R\,d\phi$ to change the variable of integration from ds to $d\phi$ and obtain, from Eq. 30-10 a magnitude of

$$\vec{B} = \int d\vec{B} = \int_0^\phi \frac{\mu_0}{4\pi} \frac{|i|R\,d\phi}{R^2} = \frac{\mu_0|i|}{4\pi R} \int_0^\phi d\phi.$$

Integrating, we find that

$$|\vec{B}| = \frac{\mu_0|i|\phi}{4\pi R} \qquad \text{(at center of circular arc).} \qquad (30\text{-}11)$$

Note that this equation gives us the magnetic field *only* at the center of curvature of a circular arc of current. When you insert data into the equation, you must be careful to express ϕ in radians rather than degrees. For example, to find the magnitude of the magnetic field at the center of a full circle of current, you would substitute 2π rad for ϕ in Eq. 30-11, finding

$$|\vec{B}| = \frac{\mu_0|i|(2\pi)}{4\pi R} = \frac{\mu_0|i|}{2R} \qquad \text{(at center of full circle).} \qquad (30\text{-}12)$$

Touchstone Examples 30-1-1 and 30-1-2, at the end of this chapter, illustrate how to use what you learned in this section.

TE

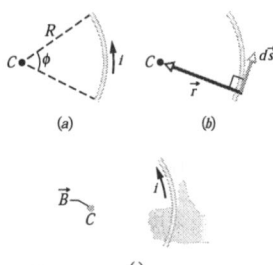

30-2 Force Between Two Parallel Currents

Two long parallel wires carrying currents exert forces on each other. Figure 30-7 shows two such wires, separated by a distance d and carrying currents i_a and i_b. Let us analyze the forces on these wires due to each other.

We seek first the force on wire b in Fig. 30-7 due to the current in wire a. That current produces a magnetic field $\vec{B}_a$ and it is this magnetic field that actually causes the force we seek. To find the force, then, we need the magnitude and direction of the field $\vec{B}_a$ *at the site of wire b*. The magnitude of $\vec{B}_a$ at every point of wire b is, from Eq. 30-6,

$$|\vec{B}_a| = \frac{\mu_0|i_a|}{2\pi d}. \qquad (30\text{-}13)$$

A (curled-straight) right-hand rule tells us that the direction of $\vec{B}_a$ at wire b is down, as Fig. 30-7 shows.

Now that we have the field, we can find the force that wire a produces on wire b. Equation 29-23 tells us that the force $\vec{F}_{b\rightarrow a}$ on a length L of wire b due to the external magnetic field $\vec{B}_a$ is

$$\vec{F}_{b\rightarrow a} = i_b\vec{L}\times\vec{B}_a \qquad (30\text{-}14)$$

where $\vec{L}$ is the length vector of the wire. In Fig. 30-7, vectors $\vec{L}$ and $\vec{B}_a$ are perpendicular, so with Eq. 30-13, we can write

$$|\vec{F}_{b\rightarrow a}| = |i_b\vec{L}||\vec{B}_a||\sin 90°| = \frac{\mu_0|\vec{L}i_a i_b|}{2\pi d}. \qquad (30\text{-}15)$$

The direction of $\vec{F}_{b\rightarrow a}$ is the direction of the cross product $\vec{L}\times\vec{B}_a$. Applying the right-hand rule for cross products to $\vec{L}$ and $\vec{B}_a$ in Fig. 30-7, we find that $\vec{F}_{b\rightarrow a}$ points directly toward wire a, as shown.

The general procedure for finding the force on a current-carrying wire is this:

Fig. 30-6: (*a*) A wire in the shape of a circular arc with center *C* carries current *i*. (*b*) For any element of wire along the arc, the angle between the directions of $d\vec{s}$ and $\vec{r}$ is 90°. (*c*) Determining the direction of the magnetic field at the center *C* due to the current in the wire; the field is out of the page, in the direction of the fingertips, as indicated by the colored dot at *C*.

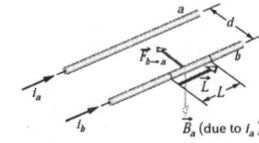

Fig. 30-7: Two parallel wires carrying currents in the same direction attract each other. $\vec{B}_a$ is the magnetic field at wire *b* produced by the current in wire *a*. $\vec{F}_{b\rightarrow a}$ is the resulting force acting on wire *b* because it carries current in field $\vec{B}_a$.

▶To find the force on a current-carrying wire due to a second current-carrying wire, first find the field due to the second wire at the site of the first wire. Then find the force on the first wire due to that field.

We could now use this procedure to compute the force on wire *a* due to the current in wire *b*. We would find that the force is directly toward wire *b*; hence, the two wires with parallel currents attract each other. Similarly, if the two currents were antiparallel, we could show that the two wires repel each other. Thus,

▶Parallel currents attract, and antiparallel currents repel.

The force acting between currents in parallel wires is the basis for the definition of the ampere, which is one of the seven SI base units. The definition, adopted in 1946, is this:

▶The ampere is that constant current which, if maintained in two straight, parallel conductors of infinite length, of negligible circular cross section, and placed 1 m apart in vacuum, would produce on each of these conductors a force of magnitude 2×10^{-7} newton per meter of length.

Rail Gun

A rail gun is a device in which a magnetic force can accelerate a projectile to a high speed in a short time. The basics of a rail gun are shown in Fig. 30-8*a*. A large current is sent out along one of two parallel conducting rails, across a conducting "fuse" (such as a narrow piece of copper) between the rails, and then back to the current source along the second rail. The projectile to be fired lies on the far side of the fuse and fits loosely between the rails. Immediately after the current begins, the fuse element melts and vaporizes, creating a conducting gas between the rails where the fuse had been.

The curled-straight right-hand rule of Fig. 30-4 reveals that the currents in the rails of Fig. 30-8*a* produce magnetic fields that are directed downward between the rails. The net magnetic field $\vec{B}$ exerts a force $\vec{F}$ on the gas due to the current *i* through the gas (Fig. 30-8*b*). With Eq. 30-14 and the right-hand rule for cross products, we find that $\vec{F}$ points outward along the rails. As the gas is forced outward along the rails, it pushes the projectile, accelerating it by as much as $5 \times 10^6 g$, and then launches it with a speed of 10 km/s, all within 1 ms.

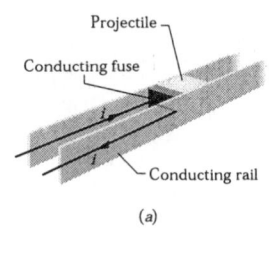
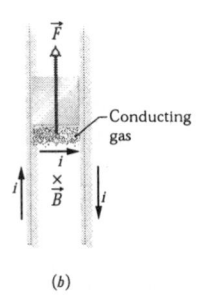

(*a*)

(*b*)

Fig. 30-8: (*a*) A rail gun, as a current *i* is set up in it. The current rapidly causes the conducting fuse to vaporize. (*b*) The current produces a magnetic field $\vec{B}$ between the rails, and the field causes a force $\vec{F}$ to act on the conducting gas, which is part of the current path. The gas propels the projectile along the rails, launching it.

READING EXERCISE 30-1: The figure shows three long, straight, parallel, equally spaced wires with identical currents either into or out of the page. Rank the wires according to the magnitude of the force on each due to the currents in the other two wires, greatest first.

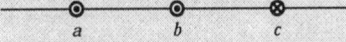

30-3 Ampere's Law

We can find the net electric field due to *any* distribution of charges with the inverse-square law for the differential field $d\vec{E}$ (Eq. 30-2), but if the distribution is complicated, we may have to use a computer. Recall, however, that if the distribution has planar, cylindrical, or spherical symmetry, we can apply Gauss' law to find the net electric field with considerably less effort.

Similarly, we can find the net magnetic field due to any distribution of currents with the inverse-square law for the differential field $d\vec{B}$ (Eq. 30-5), but again we may have to use a computer for a complicated distribution. However, if the distribution has some symmetry, we may be able to apply **Ampere's law** to find the magnetic field with considerably less effort. This law, which can be derived from the Biot-Savart law, has traditionally been credited to André Marie Ampère (1775-1836), for whom the SI unit of current is named. However, the law actually was advanced by English physicist James Clerk Maxwell.

Ampere's law is

$$\oint \vec{B} \cdot d\vec{s} = \mu_0 i_{enc} \qquad \text{(Ampere's law).} \qquad (30\text{-}16)$$

The circle on the integral sign means that the scalar (or dot) product $\vec{B} \cdot d\vec{s}$ is to be integrated around a *closed* loop, called an *Amperian loop*. The current i_{enc} on the right is the *net* current encircled by that loop.

To see the meaning of the scalar product $\vec{B} \cdot d\vec{s}$ and its integral, let us first apply Ampere's law to the general situation of Fig. 30-9. The figure shows cross sections of three long straight wires that carry currents i_1, i_2, and i_3 either directly into or directly out of the page. An arbitrary Amperian loop lying in the plane of the page encircles two of the currents but not the third. The counterclockwise direction marked on the loop indicates the arbitrarily chosen direction of integration for Eq. 30-16.

To apply Ampere's law, we mentally divide the loop into differential vector elements that are everywhere directed along the tangent to the loop in the direction of integration. Assume that at the location of the element $d\vec{s}$ shown in Fig. 30-9, the net magnetic field due to the three currents is $\vec{B}$. Because the wires are perpendicular to the page, we know that the magnetic field at $d\vec{s}$ due to each current is in the plane of Fig. 30-9; thus, their net magnetic field $\vec{B}$ at $d\vec{s}$ must also be in that plane. However, we do not know the orientation of $\vec{B}$ within the plane. In Fig. 30-9, $\vec{B}$ is arbitrarily drawn at an angle θ to the direction of $d\vec{s}$.

The scalar product $\vec{B} \cdot d\vec{s}$ on the left side of Eq. 30-16 is then equal to $B\cos\theta\, ds$. Thus, Ampere's law can be written as

$$\oint \vec{B} \cdot d\vec{s} = |\vec{B}|\cos\theta |d\vec{s}| = \mu_0 i_{enc}. \qquad (30\text{-}17)$$

We can now interpret the scalar product $\vec{B} \cdot d\vec{s}$ as being the product of a length ds of the Amperian loop and the field component $|\vec{B}|\cos\theta$ that is tangent to the loop. Then we can interpret the integration as being the summation of all such products around the entire loop.

When we can actually perform this integration, we do not need to know the direction of $\vec{B}$ before integrating. Instead, we arbitrarily assume $\vec{B}$ to be generally in the direction of integration (as in Fig. 30-9). Then we use the following curled-straight right-hand rule to assign a plus sign or a minus sign to each of the currents that make up the net encircled current i_{enc}:

> ▶**Right-hand rule for Ampere's Law:** Curl your right hand around the Amperian loop, with the fingers pointing in the direction of integration. A current through the loop in the general direction of your outstretched thumb is assigned a plus sign, and a current generally in the opposite direction is assigned a minus sign.

Finally, we solve Eq. 30-17 for the magnitude of $\vec{B}$. Once we have chosen a coordinate system to describe the system, we can use the Ampere's law right-hand rule to decide whether $\vec{B}$ is positive or negative.

In Fig. 30-10 we apply the curled-straight rule for Ampere's law to the situation of Fig. 30-9. With the indicated counterclockwise direction of integration, the net current encircled by the loop is

$$i_{enc} = i_1 - i_2.$$

(Current i_3 is not encircled by the loop.) We can then rewrite Eq. 30-17 as

$$\oint |\vec{B}|\cos\theta |d\vec{s}| = \mu_0 |(i_2 - i_1)|. \qquad (30\text{-}18)$$

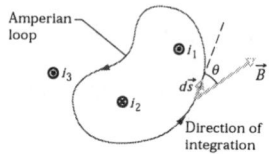

Fig. 30-9: Ampere's law applied to an arbitrary Amperian loop that encircles two long straight wires but excludes a third wire. Note the directions of the currents.

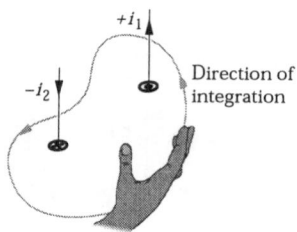

Fig. 30-10: A right-hand rule for Ampere's law, to determine the signs for currents encircled by an Amperian loop. The situation is that of Fig. 30-9.

You might wonder why, since current i_3 contributes to the magnetic-field magnitude B on the left side of Eq. 30-18, it is not needed on the right side. The answer is that the contributions of current i_3 to the magnetic field cancel out because the integration in Eq. 30-18 is made around the full loop. In contrast, the contributions of an encircled current to the magnetic field do not cancel out.

We cannot solve Eq. 30-18 for the magnitude $|\vec{B}|$ of the magnetic field, because for the situation of Fig. 30-9 we do not have enough information to simplify and solve the integral. However, we do know the outcome of the integration; it must be equal to the value of $\mu_0 |(i_1 - i_2)|$, which is set by the net current passing through the loop.

We shall now apply Ampere's law to two situations in which symmetry does allow us to simplify and solve the integral, hence to find the magnetic field.

The Magnetic Field Outside a Long Straight Wire with Current

Figure 30-11 shows a long straight wire that carries current i directly out of the page. Equation 30-6 tells us that the magnetic field $\vec{B}$ produced by the current has the same magnitude at all points that are the same distance r from the wire; that is, the field $\vec{B}$ has cylindrical symmetry about the wire. We can take advantage of that symmetry to simplify the integral in Ampere's law (Eqs. 30-16 and 30-17) if we encircle the wire with a concentric circular Amperian loop of radius r, as in Fig. 30-11. The magnetic field $\vec{B}$ then has the same magnitude $|\vec{B}|$ at every point on the loop. We shall integrate counterclockwise, so that $d\vec{s}$ has the direction shown in Fig. 30-11.

We can further simplify the quantity $|\vec{B}| \cos\theta$ in Eq. 30-17 by noting that $\vec{B}$ is tangent to the loop at every point along the loop, as is $d\vec{s}$. Thus, $\vec{B}$ and $d\vec{s}$ are either parallel or antiparallel at each point of the loop, and we shall arbitrarily assume the former. Then at every point the angle θ between $d\vec{s}$ and $\vec{B}$ is 0°, so $\cos\theta = \cos 0° = 1$. The integral in Eq. 30-17 then becomes

$$\oint \vec{B} \cdot d\vec{s} = \oint |\vec{B}| \cos\theta \, |d\vec{s}| = |\vec{B}| \oint |d\vec{s}| = |\vec{B}|(2\pi r).$$

Note that $\oint |d\vec{s}|$ above is the summation of all the line segment lengths $|d\vec{s}|$ around the circular loop; that is, it simply gives the circumference $2\pi r$ of the loop.

Our right-hand rule gives us a plus sign for the current of Fig. 30-11. The right side of Ampere's law becomes $+\mu_0 i$ and we then have

$$|\vec{B}|(2\pi r) = \mu_0 i$$

or

$$|\vec{B}| = \frac{\mu_0 i}{2\pi r} \qquad \text{(when } i \text{ is positive)} \quad (30\text{-}19)$$

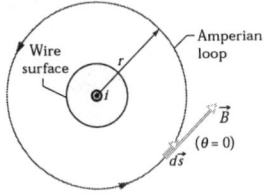

Fig. 30-11: Using Ampere's law to find the magnetic field produced by a current i in a long straight wire. The Amperian loop is a concentric circle that lies outside the wire.

With a slight change in notation, this is Eq. 30-6, which we derived earlier—with considerably more effort—using the law of Biot and Savart. In addition, because we assumed i is positive, we know that the correct direction of $\vec{B}$ must be the one shown in Fig. 30-11.

The Magnetic Field Inside a Long Straight Wire with Current

Figure 30-12 shows the cross section of a long straight wire of radius R that carries a uniformly distributed current i directly out of the page. Because the current is uniformly distributed over a cross section of the wire, the magnetic field $\vec{B}$ that it produces must be cylindrically symmetrical. Thus, to find the magnetic field at points inside the wire, we can again use an Amperian loop of radius r, as shown in Fig. 30-12, where now $r < R$. Symmetry again suggests that $\vec{B}$ is tangent to the loop, as shown, so the left side of Ampere's law again yields

$$\oint \vec{B} \cdot d\vec{s} = |\vec{B}| \left| \oint d\vec{s} \right| = |\vec{B}|(2\pi r). \qquad (30\text{-}20)$$

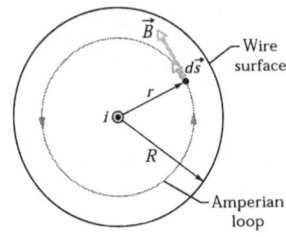

Fig. 30-12: Using Ampere's law to find the magnetic field that a current i produces inside a long straight wire of circular cross section. The current is uniformly distributed over the cross section of the wire and emerges from the page. An Amperian loop is drawn inside the wire.

To find the right side of Ampere's law, we note that because the current is uniformly distributed, the current i_{enc} encircled by the loop is proportional to the area encircled by the loop; that is,

$$i_{enc} = i \frac{\pi r^2}{\pi R^2}. \qquad (30\text{-}21)$$

Our right-hand rule tells us that i_{enc} gets a plus sign. Then Ampere's law gives us

$$|\vec{B}|(2\pi r) = \mu_0 i \frac{\pi r^2}{\pi R^2}$$

or $\qquad\qquad\qquad\qquad |\vec{B}| = \left(\frac{\mu_0 i}{2\pi R^2}\right) r. \qquad (30\text{-}22)$

Thus, inside the wire, the magnitude $|\vec{B}|$ of the magnetic field is proportional to r; that magnitude is zero at the center and a maximum at the surface, where $r = R$. Note that Eqs. 30-19 and 30-22 give the same value for B at $r = R$; that is, the expressions for the magnetic field outside the wire and inside the wire yield the same result at the surface of the wire.

READING EXERCISE 30-2: The figure shows three equal currents i (two parallel and one antiparallel) and four Amperian loops. Rank the loops according to the magnitude of $\oint \vec{B} \cdot d\vec{s}$ along each, greatest first.

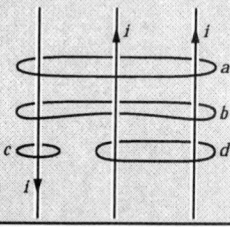

Touchstone Example 30-3-1, at the end of this chapter, illustrates how to use what you learned in this section.

TE

30-4 Solenoids and Toroids

Magnetic Field of a Solenoid

We now turn our attention to another situation in which Ampere's law proves useful. It concerns the magnetic field produced by the current in a long, tightly wound helical coil of wire. Such a coil is called a **solenoid** (Fig. 30-13). We assume that the length of the solenoid is much greater than the diameter.

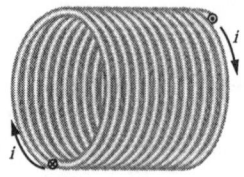

Fig. 30-13: A solenoid carrying current i.

Figure 30-14 shows a section through a portion of a "stretched-out" solenoid. The solenoid's magnetic field is the vector sum of the fields produced by the individual turns (loops) that make up the solenoid. For points very close to each turn, the wire behaves magnetically almost like a long straight wire, and the lines of $\vec{B}$ there are almost concentric circles. Figure 30-14 suggests that the field tends to cancel between adjacent turns. It also suggests that, at points inside the solenoid and reasonably far from the wire, $\vec{B}$ is approximately parallel to the (central) solenoid axis. In the limiting case of an *ideal solenoid*, which is infinitely long and consists of tightly packed (*close-packed*) turns of square wire, the field inside the coil is uniform and parallel to the solenoid axis.

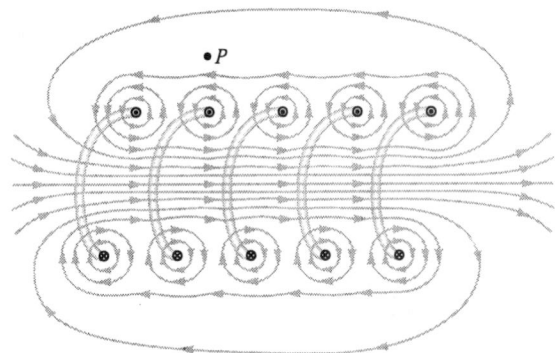

Fig. 30-14: A vertical cross section through the central axis of a "stretched-out" solenoid. The back portions of five turns are shown, as are the magnetic field lines due to a current through the solenoid. Each turn produces circular magnetic field lines near it. Near the solenoid's axis, the field lines combine into a net magnetic field that is directed along the axis. The closely spaced field lines there indicate a strong magnetic field. Outside the solenoid the field lines are widely spaced; the field there is very weak.

At points above the solenoid, such as P in Fig. 30-14, the field set up by the upper parts of the solenoid turns (marked $\odot$) is directed to the left (as drawn near P) and tends to cancel the field set up by the lower parts of the turns (marked $\otimes$), which is directed to the right (not drawn). In the limiting case of an ideal solenoid, the magnetic field outside the solenoid is zero. Taking the external field to be zero is an excellent assumption for a real solenoid if its length is much greater than its diameter and if we consider external points such as point P that are not at either end of the solenoid. The direction of the magnetic field along the solenoid axis is given by a curled-straight right-hand rule: Grasp the solenoid with your right hand so that your fingers follow the direction of the current in the windings; your extended right thumb then points in the direction of the axial magnetic field.

Figure 30-15 shows the lines of $\vec{B}$ for a real solenoid. The spacing of the lines of $\vec{B}$ in the central region shows that the field inside the coil is fairly strong and uniform over the cross section of the coil. The external field, however, is relatively weak.

Let us now apply Ampere's law,

$$\oint \vec{B} \cdot d\vec{s} = \mu_0 i_{enc}, \qquad (30\text{-}23)$$

to the ideal solenoid of Fig. 30-16, where $\vec{B}$ is uniform within the solenoid and zero outside it, using the rectangular Amperian loop $abcda$. We write $\oint \vec{B} \cdot d\vec{s}$ as the sum of four integrals, one for each loop segment:

$$\oint \vec{B} \cdot d\vec{s} = \int_a^b \vec{B} \cdot d\vec{s} + \int_b^c \vec{B} \cdot d\vec{s}$$
$$+ \int_c^d \vec{B} \cdot d\vec{s} + \int_d^a \vec{B} \cdot d\vec{s}. \qquad (30\text{-}24)$$

The first integral on the right of Eq. 30-24 is $|\vec{B}|h$, where $|\vec{B}|$ is the magnitude of the uniform field $\vec{B}$ inside the solenoid and h is the (arbitrary) length of the segment from a to b. The second and fourth integrals are zero because for every element ds of these segments, $\vec{B}$ either is perpendicular to ds or is zero, and thus $\vec{B} \cdot d\vec{s}$ is zero. The third integral, which is taken along a segment that lies outside the solenoid, is zero because $B = 0$ at all external points. Thus, $\oint \vec{B} \cdot d\vec{s}$ for the entire rectangular loop has the value $|\vec{B}|h$.

The net current i_{enc} encircled by the rectangular Amperian loop in Fig. 30-16 is not the same as the current i in the solenoid windings because the windings pass more than once through this loop. Let n be the number of turns per unit length of the solenoid; then the loop encloses nh turns and

$$i_{enc} = i(nh)$$

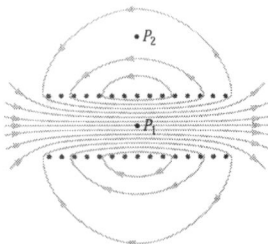

Fig. 30-15: Magnetic field lines for a real solenoid of finite length. The field is strong and uniform at interior points such as P_1 but relatively weak at external points such as P_2.

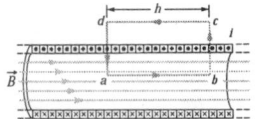

Fig. 30-16: Application of Ampere's law to a section of a long ideal solenoid carrying a current i. The Amperian loop is the rectangle $abcd$.

Ampere's law then gives us

$$|\vec{B}|h = \mu_0|i|nh.$$

or $\qquad\qquad |\vec{B}| = \mu_0|i|n \qquad\qquad$ (ideal solenoid). $\qquad\qquad$ (30-25)

Although we derived Eq. 30-25 for an infinitely long ideal solenoid, it holds quite well for actual solenoids if we apply it only at interior points, well away from the solenoid ends. Equation 30-25 is consistent with the experimental fact that the magnetic field magnitude $|\vec{B}|$ within a solenoid does not depend on the diameter or the length of the solenoid and that $|\vec{B}|$ is uniform over the solenoidal cross section. A solenoid thus provides a practical way to set up a known uniform magnetic field for experimentation, just as a parallel-plate capacitor provides a practical way to set up a known uniform electric field.

Magnetic Field of a Toroid

Figure 30-17a shows a **toroid**, which we may describe as a solenoid bent into the shape of a hollow doughnut. What magnetic field $\vec{B}$ is set up at its interior points (within the hollow of the doughnut)? We can find out from Ampere's law and the symmetry.

From the symmetry, we see that the lines of $\vec{B}$ form concentric circles inside the toroid, directed as shown in Fig. 30-17b. Let us choose a concentric circle of radius r as an Amperian loop and traverse it in the clockwise direction. Ampere's law (Eq. 30-16) yields

$$|\vec{B}|(2\pi r) = \mu_0|i|N,$$

where i is the current in the toroid windings (and is positive for those windings enclosed by the Amperian loop) and N is the total number of turns. This gives

$$\boxed{|\vec{B}| = \frac{\mu_0|i|N}{2\pi}\frac{1}{r}} \qquad\qquad \text{(toroid).} \qquad\qquad (30\text{-}26)$$

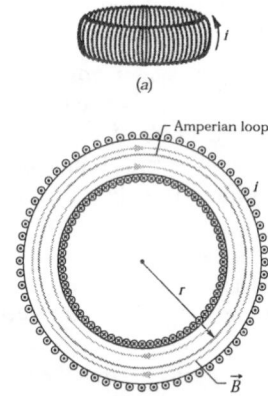

(a)

(b)

Fig. 30-17: (a) A toroid carrying a current i. (b) A horizontal cross section of the toroid. The interior magnetic field (inside the doughnut-shaped tube) can be found by applying Ampere's law with the Amperian loop shown.

In contrast to the situation for a solenoid, $\vec{B}$ is not constant over the cross section of a toroid. It is easy to show, with Ampere's law, that $\vec{B} = 0$ for points outside an ideal toroid (as if the toroid were made from an ideal solenoid).

The direction of the magnetic field within a toroid follows from our curled-straight right-hand rule: Grasp the toroid with the fingers of your right hand curled in the direction of the current in the windings; your extended right thumb points in the direction of the magnetic field.

30-5 A Current-Carrying Coil as a Magnetic Dipole

So far we have examined the magnetic fields produced by current in a long straight wire, a solenoid, and a toroid. We turn our attention here to the field produced by a coil carrying a current. You saw in Section 29-9 that such a coil behaves as a magnetic dipole in that, if we place it in an external magnetic field $\vec{B}$, a torque $\vec{\tau}$ given by

$$\vec{\tau} = \vec{\mu} \times \vec{B} \qquad\qquad (30\text{-}27)$$

acts on it. Here $\vec{\mu}$ is the magnetic dipole moment of the coil and has the magnitude NiA, where N is the number of turns (or loops), i is the current in each turn, and A is the area enclosed by each turn.

Recall that the direction of $\vec{\mu}$ is given by a curled-straight right-hand rule: Grasp the coil so that the fingers of your right hand curl around it in the direction of the current; your extended thumb then points in the direction of the dipole moment $\vec{\mu}$.

Magnetic Field of a Coil

We turn now to the other aspect of a current-carrying coil as a magnetic dipole. What magnetic field does *it* produce at a point in the surrounding space? The problem does not have enough symmetry to make Ampere's law useful, so we must turn to the law of Biot and Savart. For simplicity, we first consider only a coil with a single circular loop and only points on its central axis, which we take to be a z axis. We shall show that the magnitude of the magnetic field at such points only has a z-component vector that is given by

$$B(z)\hat{k} = \frac{\mu_0 i R^2}{2(R^2 + z^2)^{3/2}}\hat{k}, \qquad (30\text{-}28)$$

in which R is the radius of the circular loop and z is the distance of the point in question from the center of the loop. Furthermore, the direction of the magnetic field $\vec{B}$ is the same as the direction of the magnetic dipole moment $\vec{\mu}$ of the loop.

For axial points far from the loop, we have $z \gg R$ in Eq. 30-28. With that approximation, the equation for the z-component of $\vec{B}$, which is a function of z only, reduces to

$$B(z) \approx \frac{\mu_0 i R^2}{2z^3}.$$

Recalling that πR^2 is the area A of the loop and extending our result to include a coil of N turns, we can write this equation as

$$B(z) = \frac{\mu_0}{2\pi}\frac{NiA}{z^3}.$$

Further, since $\vec{B}$ and $\vec{\mu}$ have the same direction, we can write the equation in vector form, substituting from the identity $\mu = NiA$:

$$\vec{B}(z) = \frac{\mu_0}{2\pi}\frac{\vec{\mu}}{z^3} \qquad \text{(current-carrying coil).} \qquad (30\text{-}29)$$

Thus, we have two ways in which we can regard a current-carrying coil as a magnetic dipole: (1) it experiences a torque when we place it in an external magnetic field; (2) it generates its own intrinsic magnetic field, given, for distant points along its axis, by Eq. 30-29. Figure 30-18 shows the magnetic field of a current loop; one side of the loop acts as a north pole (in the direction of $\vec{\mu}$) and the other side as a south pole, as suggested by the lightly drawn magnet in the figure.

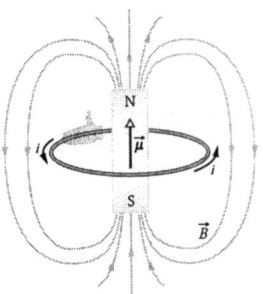

Fig. 30-18: A current loop produces a magnetic field like that of a bar magnet and thus has associated north and south poles. The magnetic dipole moment $\vec{\mu}$ of the loop, given by a curled-straight right-hand rule, points from the south pole to the north pole, in the direction of the field $\vec{B}$ within the loop.

READING EXERCISE 30-3: The figure here shows four arrangements of circular loops of radius r or $2r$, centered on vertical axes (perpendicular to the loops) and carrying identical currents in the directions indicated. Rank the arrangements according to the magnitude of the net magnetic field at the dot, midway between the loops on the central axis, greatest first.

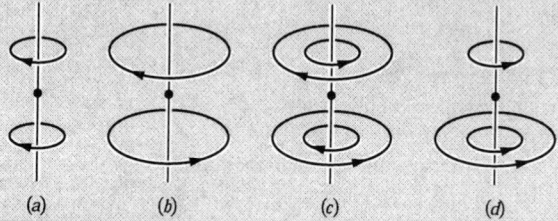

(a)   (b)   (c)   (d)

Proof of Equation 30-28

Figure 30-19 shows the back half of a circular loop of radius R carrying a current i. Consider a point P on the axis of the loop, a distance z from its plane. Let us apply the law of Biot and Savart to a differential element ds of the loop, located at the left side of the loop. The length vector $d\vec{s}$ for this element points perpendicularly out of the page. The angle θ between $d\vec{s}$ and $\vec{r}$ in Fig. 30-19 is 90°; the plane formed by these two vectors is perpendicular to the plane of the figure and contains both $\vec{r}$ and $d\vec{s}$. From the law of Biot and Savart (and the right-hand rule), the differential field $d\vec{B}$ produced at point P by the current in this element is perpendicular to this plane and thus lies in the plane of the figure, perpendicular to $\vec{r}$, as indicated in Fig. 30-19.

Let us resolve $d\vec{B}$ into two components: $d\vec{B}_{\parallel}$ along the axis of the loop and $d\vec{B}_{\perp}$ perpendicular to this axis. From the symmetry, the vector sum of all the perpendicular components $d\vec{B}_{\perp}$ due to all the loop elements ds is zero. This leaves only the axial components $d\vec{B}_{\perp}$ and we have the axial component given by

$$B_{\parallel} = \int dB_{\parallel}$$

For the element $d\vec{s}$ in Fig. 30-19, the law of Biot and Savart (Eq. 30-3) tells us that the axial magnetic field component at distance r is

$$dB_{\parallel} = \frac{\mu_0}{4\pi}\frac{i\,|d\vec{s}|\sin 90°}{r^2}.$$

We also have

$$dB_{\parallel} = |d\vec{B}|\cos\alpha.$$

Combining these two relations, we obtain

$$dB_{\parallel} = \frac{\mu_0 i \cos\alpha\,|d\vec{s}|}{4\pi r^2}. \tag{30-30}$$

Figure 30-19 shows that r and α are not independent but are related to each other. Let us express each in terms of the variable z, the distance between point P and the center of the loop. The relations are

$$r = \sqrt{R^2 + z^2} \tag{30-31}$$

and

$$\cos\alpha = \frac{R}{r} = \frac{R}{\sqrt{R^2 + z^2}}. \tag{30-32}$$

Substituting Eqs. 30-31 and 30-32 into Eq. 30-30, we find

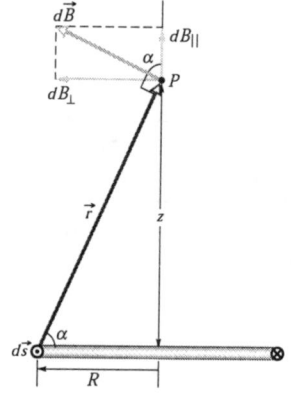

Fig. 30-19: A current loop of radius R. The plane of the loop is perpendicular to the page and only the back half of the loop is shown. We use the law of Biot and Savart to find the magnetic field at point P on the central axis of the loop.

$$dB_\parallel = \frac{\mu_0 iR}{4\pi(R^2 + z^2)^{3/2}} |d\vec{s}|.$$

Note that i, R, and z have the same values for all elements ds around the loop, so when we integrate this equation, we find that the axial field component is given as

$$B = \int dB_\parallel$$
$$= \frac{\mu_0 iR}{4\pi(R^2 + z^2)^{3/2}} \int |d\vec{s}|$$

or, since $\int |d\vec{s}|$ is simply the circumference $2\pi R$ of the loop, the axial or z-component of the magnetic field is

$$\vec{B}(z)\hat{k} = \frac{\mu_0 iR^2}{2(R^2 + z^2)^{3/2}} \hat{k},$$

This is Eq. 30-28, the relation we sought to prove.

Touchstone Example 30-1-1

The wire in Fig. TE30-1a carries a current i and consists of a circular arc of radius R and central angle $\pi/2$ rad, and two straight sections whose extensions intersect the center C of the arc. What magnetic field $\vec{B}$ does the current produce at C?

SOLUTION: One **Key Idea** here is that we can find the magnetic field $\vec{B}$ at point C by applying the Biot–Savart law of Eq. 30-5 to the wire. A second **Key Idea** is that the application of Eq. 30-5 can be simplified by evaluating $\vec{B}$ separately for the three distinguishable sections of the wire—namely, (1) the straight section at the left, (2) the straight section at the right, and (3) the circular arc.

Straight sections. For any current-length element in section 1, the angle θ between $d\vec{s}$ and $\vec{r}$ is zero (Fig. 30-7b), so Eq. 30-3 gives us

$$|d\vec{B_1}| = \frac{\mu_0}{4\pi} \frac{|i\,d\vec{s}|\sin\theta}{r^2} = \frac{\mu_0}{4\pi} \frac{|i\,d\vec{s}|\sin 0}{r^2} = 0.$$

Thus, the current along the entire length of wire in straight section 1 contributes no magnetic field at C:

$$\vec{B_1} = 0.$$

The same situation prevails in straight section 2, where the angle θ between $d\vec{s}$ and $\vec{r}$ for any current-length element is 180°. Thus,

$$\vec{B_2} = 0.$$

Circular arc. The **Key Idea** here is that application of the Biot–Savart law to evaluate the magnetic field at the center of a circular arc leads to Eq. 30-11 ($B = \mu_0 i\phi/4\pi R$). Here the central angle ϕ of the arc is $\pi/2$ rad. Thus from Eq. 30-11, the magnitude of the magnetic field $\vec{B_3}$ at the arc's center C is

$$|\vec{B_3}| = \frac{\mu_0 |i|(\pi/2)}{4\pi R} = \frac{\mu_0 |i|}{8R}.$$

To find the direction of $\vec{B_3}$, we apply the right-hand rule displayed in Fig. 30-4. Mentally grasp the circular arc with your right hand as in Fig. TE30-1c, with your thumb in the direction of the current. The direction in which your fingers curl around the wire indicates the direction of

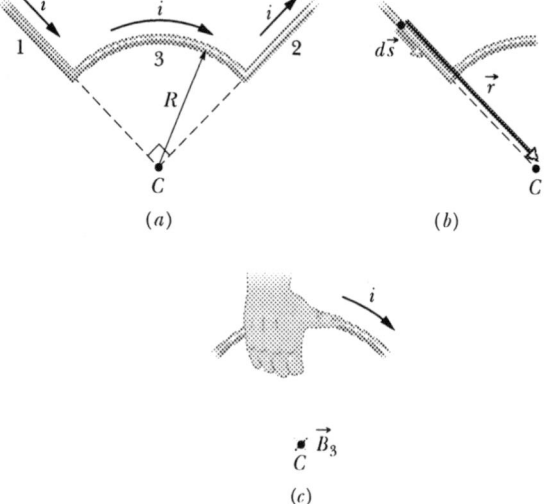

(a)   (b)

(c)

Fig. TE30-1 (a) A wire consists of two straight sections (1 and 2) and a circular arc (3), and carries current i. (b) For a current-length element in section 1, the angle between $d\vec{s}$ and $\vec{r}$ is zero. (c) Determining the direction of magnetic field $\vec{B_3}$ at C due to the current in the circular arc; the field is into the page there.

the magnetic field lines around the wire. In the region of point C (inside the circular arc), your fingertips point *into the plane* of the page. Thus, $\vec{B}_3$ is directed into that plane.

Net field. Generally, when we must combine two or more magnetic fields to find the net magnetic field, we must combine the fields as vectors and not simply add their magnitudes. Here, however, only the circular arc produces a magnetic field at point C. Thus, we can write the magnitude of the net field $\vec{B}$ as

$$|\vec{B}| = |\vec{B}_1 + \vec{B}_2 + \vec{B}_3| = 0 + 0 + \left| \frac{\mu_0 i}{8R} \right| = \frac{\mu_0 |i|}{8R}. \qquad \text{(Answer)}$$

The direction of $\vec{B}$ is the direction of $\vec{B}_3$—namely, into the plane of Fig. TE30-1.

Touchstone Example 30-1-2

Figure TE30-2a shows two long parallel wires carrying currents i_1 and i_2 in opposite directions. What are the magnitude and direction of the net magnetic field at point P? Assume the following values: $i_1 = 15$ A, $i_2 = 32$ A, and $d = 5.3$ cm.

SOLUTION: One **Key Idea** here is that the net magnetic field $\vec{B}$ at point P is the vector sum of the magnetic fields due to the currents in the two wires. A second **Key Idea** is that we can find the magnetic field due to any current by applying the Biot–Savart law to the current. For points near the current in a long straight wire, that law leads to Eq. 30-6.

In Fig. TE30-2a, point P is distance R from both currents i_1 and i_2. Thus, Eq. 30-6 tells us that at point P those currents produce magnetic fields $\vec{B}_1$ and $\vec{B}_2$ with magnitudes

$$|\vec{B}_1| = \frac{\mu_0 |i_1|}{2\pi R} \quad \text{and} \quad |\vec{B}_2| = \frac{\mu_0 |i_2|}{2\pi R}.$$

In the right triangle of Fig. TE30-2a, note that the base angles (between sides R and d) are both 45°. Thus, we may write $\cos 45° = R/d$ and replace R with $d \cos 45°$. Then the field magnitudes B_1 and B_2 become

$$|\vec{B}_1| = \frac{\mu_0 |i_1|}{2\pi d \cos 45°} \quad \text{and} \quad |\vec{B}_2| = \frac{\mu_0 |i_2|}{2\pi d \cos 45°}.$$

We want to combine $\vec{B}_1$ and $\vec{B}_2$ to find their vector sum, which is the net field $\vec{B}$ at P. To find the directions of $\vec{B}_1$ and $\vec{B}_2$, we apply the right-hand rule of Fig. 30-4 to each current in Fig. TE30-2a. For wire 1, with current out of the page, we mentally grasp the wire with the right hand, with the thumb pointing out of the page. Then the curled fingers indicate that the field lines run counterclockwise. In particular, in the region of point P, they are directed upward to

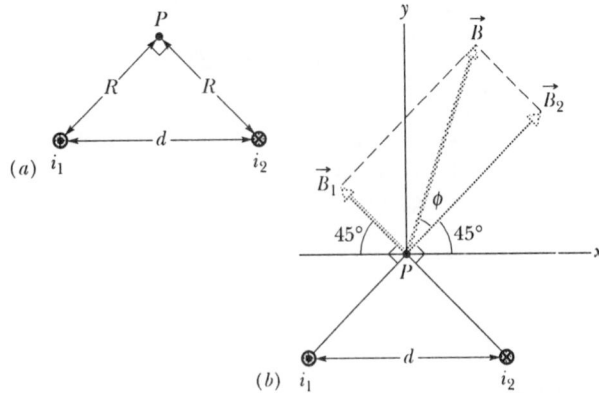

Fig. TE30-2 (a) Two wires carry currents i_1 and i_2 in opposite directions (out of and into the page). Note the right angle at P. (b) The separate fields $\vec{B}_1$ and $\vec{B}_2$ are combined vectorially to yield the net field $\vec{B}$.

the left. Recall that the magnetic field at a point near a long, straight current-carrying wire must be directed perpendicular to a radial line between the point and the current. Thus, $\vec{B}_1$ must be directed upward to the left as drawn in Fig. TE30-2b. (Note carefully the perpendicular symbol between vector $\vec{B}_1$ and the line connecting point P and wire 1.)

Repeating this analysis for the current in wire 2, we find that $\vec{B}_2$ is directed upward to the right as drawn in Fig. TE30-2b. (Note the perpendicular symbol between vector $\vec{B}_2$ and the line connecting point P and wire 2.)

We can now vectorially add $\vec{B}_1$ and $\vec{B}_2$ to find the net magnetic field $\vec{B}$ at point P, either by using a vector-capable calculator or by resolving the vectors into components and then combining the components of $\vec{B}$. However, in Fig. TE30-2b, there is a third method: Because $\vec{B}_1$ and $\vec{B}_2$ are perpendicular to each other, they form the legs of a right triangle, with $\vec{B}$ as the hypotenuse. The Pythagorean theorem then gives us

$$|\vec{B}| = \sqrt{B_1^2 + B_2^2} = \frac{\mu_0}{2\pi d(\cos 45°)}\sqrt{i_1^2 + i_2^2}$$

$$= \frac{(4\pi \times 10^{-7}\,\text{T}\cdot\text{m/A})\sqrt{(15\,\text{A})^2 + (32\,\text{A})^2}}{(2\pi)(5.3 \times 10^{-2}\,\text{m})(\cos 45°)}$$

$$|\vec{B}| = 1.89 \times 10^{-4}\,\text{T} \approx 190\,\mu\text{T}. \qquad \text{(Answer)}$$

The angle ϕ between the directions of $\vec{B}$ and $\vec{B}_2$ in Fig. TE30-2b follows from

$$\phi = \tan^{-1}\frac{B_1}{B_2},$$

which, with B_1 and B_2 as given above, yields

$$\phi = \tan^{-1}\frac{i_1}{i_2} = \tan^{-1}\frac{15\,\text{A}}{32\,\text{A}} = 25°.$$

The angle between the direction of $\vec{B}$ and the x axis shown in Fig. TE30-2b is then

$$\phi + 45° = 25° + 45° = 70°. \qquad \text{(Answer)}$$

Touchstone Example 30-3-1

Figure TE30-3a shows the cross section of a long conducting cylinder with inner radius $a = 2.0$ cm and outer radius $b = 4.0$ cm. The cylinder carries a current out of the page, and the magnitude of the current density in the cross section is given by $|\vec{J}| = cr^2$, with $c = 3.0 \times 10^6$ A/m^4 and r in meters. What is the magnetic field $|\vec{B}|$ at a point that is 3.0 cm from the central axis of the cylinder?

SOLUTION: The point at which we want to evaluate $\vec{B}$ is inside the material of the conducting cylinder, between its inner and outer radii. We note that the current distribution has cylindrical symmetry (it is the same all around the cross section for any given radius). Thus, the **Key Idea** here is that the symmetry allows us to use Ampere's law to find $\vec{B}$ at the point. We first draw the Amperian loop shown in Fig. TE30-3b. The loop is concentric with the cylinder and has radius $r = 3.0$ cm, because we want to evaluate $\vec{B}$ at that distance from the cylinder's central axis.

Next, we must compute the current i_{enc} that is encircled by the Amperian loop. However, a second **Key Idea** is that we *cannot* set up a proportionality as in Eq. 30-21, because here the current is not uniformly distributed. Instead, we must integrate the current density from the cylinder's inner radius a to the loop radius r:

$$i_{\text{enc}} = \int \vec{J} \cdot d\vec{A} = \int_a^r cr^2\,(2\pi r\,dr)$$

$$= 2\pi c \int_a^r r^3\,dr = 2\pi c \left[\frac{r^4}{4}\right]_a^r$$

$$= \frac{\pi c(r^4 - a^4)}{2}.$$

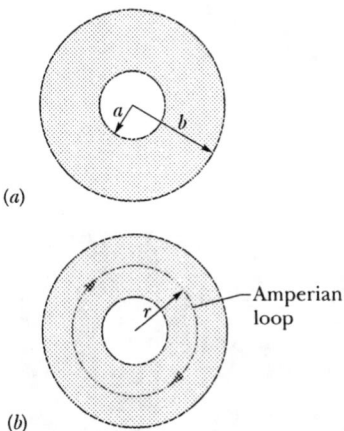

(a)

(b)

Fig. TE30-3 (a) Cross section of a conducting cylinder of inner radius a and outer radius b. (b) An Amperian loop of radius r is added to compute the magnetic field at points that are a distance r from the central axis.

The direction of integration indicated in Fig. TE30-3b is (arbitrarily) clockwise. Applying the right-hand rule for Ampere's law to that loop, we find that we should take i_{enc} as negative because the current is directed out of the page but our thumb is directed into the page.

We next evaluate the left side of Ampere's law exactly as we did in Fig. 30-12, and we again obtain Eq. 30-20. Then Ampere's law,

$$\oint \vec{B} \cdot d\vec{s} = \mu_0 i_{enc},$$

gives us

$$|\vec{B}|(2\pi r) = -\frac{\mu_0 \pi c}{2}(r^4 - a^4).$$

Solving for $|\vec{B}|$ and substituting known data yield

$$
\begin{aligned}
|\vec{B}| &= -\frac{\mu_0 c}{4r}(r^4 - a^4) \\
&= -\frac{(4\pi \times 10^{-7}\,\text{T}\cdot\text{m/A})(3.0 \times 10^6\,\text{A/m}^4)}{4(0.030\,\text{m})} \\
&\quad \times [(0.030\,\text{m})^4 - (0.020\,\text{m})^4] \\
&= -2.0 \times 10^{-5}\,\text{T}.
\end{aligned}
$$

Thus, the magnetic field $\vec{B}$ at a point 3.0 cm from the central axis has the magnitude

$$|\vec{B}| = 2.0 \times 10^{-5}\,\text{T} \qquad\qquad \text{(Answer)}$$

and forms magnetic field lines that are directed opposite our direction of integration, hence counterclockwise in Fig. TE30-3b.

31 Induction and Inductance

Soon after rock began in the mid-1950s, guitarists switched from acoustic guitars to electric guitars—but it was Jimi Hendrix who first understood the electric guitar as an electronic instrument. He exploded on the scene in the 1960s, ripping his pick along the strings, positioning himself and his guitar in front of a speaker to sustain feedback, and then laying down chords on top of the feedback. He shoved rock forward from the melodies of Buddy Holly into the psychedelia of the late 1960s and into the early heavy metal of Led Zeppelin and the raw energy of Joy Division in the 1970s, and his ideas continue to influence rock today.

What is it about an electric guitar that distinguishes it from an acoustic guitar and enabled Hendrix to make so much broader use of this electronic instrument?

The answer is in this chapter.

31-1 Two Symmetric Situations

In Section 29-8, we saw that if we put a closed conducting loop in a magnetic field and then send current through the loop, forces due to the magnetic field create a torque to turn the loop:

$$\text{current loop} + \text{ magnetic field} \Rightarrow \text{ torque.} \qquad (31\text{-}1)$$

Suppose that, instead, with the current off, we turn the loop by hand. Will the opposite of Eq. 31-1 occur? That is, will a current now appear in the loop:

$$\text{torque} + \text{magnetic field} \Rightarrow \text{current?} \qquad (31\text{-}2)$$

The answer is yes—a current does appear. The situations of Eqs. 31-1 and 31-2 are symmetric. The physical law on which Eq. 31-2 depends is called *Faraday's law of induction*. Whereas Eq. 31-1 is the basis for the electric motor, Eq. 31-2 and Faraday's law are the basis for the electric generator. This chapter is concerned with that law and the process it describes.

32-2 Two Experiments

Let us examine two simple experiments to prepare for our discussion of Faraday's law of induction.

First Experiment. Figure 31-1 shows a conducting loop connected to a sensitive current meter. Since there is no battery or other source of emf included, there is no current in the circuit. However, if we move a bar magnet toward the loop, a current suddenly appears in the circuit. The current disappears when the magnet stops. If we then move the magnet away, a current again suddenly appears, but now in the opposite direction. If we experimented for a while, we would discover the following:

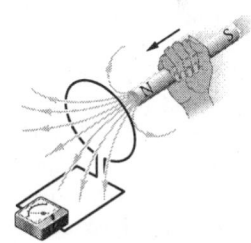

Fig. 31-1: A current meter registers a current in the wire loop when the magnet is moving with respect to the loop.

1. A current appears only if there is relative motion between the loop and the magnet (one must move relative to the other); the current disappears when the relative motion between them ceases.

2. Faster motion produces a greater current.

3. If moving the magnet's north pole toward the loop causes, say, clockwise current, then moving the north pole away causes counterclockwise current. Moving the south pole toward or away from the loop also causes currents, but in the reversed directions.

The current produced in the loop is called an **induced current;** the work done per unit charge to produce that current (to move the conduction electrons that constitute the current) is called an **induced emf;** and the process of producing the current and emf is called induction.

Second Experiment. For this experiment we use the apparatus of Fig. 31-2, with the two conducting loops close to each other but not touching. If we close switch S, to turn on a current in the right-hand loop, the meter suddenly and briefly registers a current—an induced current—in the left-hand loop. If we then open the switch, another sudden and brief induced current appears in the left-hand loop, but in the opposite direction. We get an induced current (and thus an induced emf) only when the current in the right-hand loop is changing (either turning on or turning off) and not when it is constant (even if it is large).

The induced emf and induced current in these experiments are apparently caused when something changes—but what is that "something"? Faraday knew.

Fig. 31-2: The current meter registers a current in the left-hand wire loop just as switch S is closed (to turn on the current in the right-hand wire loop) or opened (to turn off the current in the right-hand loop). No motion of the coils is involved.

31-3 Faraday's Law of Induction

Faraday realized that an emf and a current can be induced in a loop, as in our two experiments, by changing the *amount of magnetic field* passing through the loop. He further realized that the "amount of magnetic field" can be visualized in terms of the magnetic field lines passing through the loop. **Faraday's law of induction**, stated in terms of our experiments, is this:

➤An emf is induced in the left-hand loop in Figs. 31-1 and 31-2 when the number of magnetic field lines that pass through the loop is changing.

The actual number of field lines passing through the loop does not matter; the values of the induced emf and induced current are determined by the rate at which that number changes.

In our first experiment (Fig. 31-1), the magnetic field lines spread out from the north pole of the magnet. Thus, as we move the north pole closer to the loop, the number of field lines passing through the loop increases. That increase apparently causes conduction electrons in the loop to move (the induced current) and provides energy (the induced emf) for their motion. When the magnet stops moving, the number of field lines through the loop no longer changes and the induced current and induced emf disappear.

In our second experiment (Fig. 31-2), when the switch is open (no current), there are no field lines. However, when we turn on the current in the right-hand loop, the increasing current builds up a magnetic field around that loop and at the left-hand loop. While the field builds, the number of magnetic field lines through the left-hand loop increases. As in the first experiment, the increase in field lines through that loop apparently induces a current and an emf there. When the current in the right-hand loop reaches a final, steady value, the number of field lines through the left-hand loop no longer changes, and the induced current and induced emf disappear.

Faraday's law does not explain why a current and an emf are induced in either experiment; it is just a statement that helps us visualize the induction.

A Quantitative Treatment

To put Faraday's law to work, we need a way to calculate the amount of magnetic field that passes through a loop. In Chapter 24, in a similar situation, we needed to calculate the amount of an electric field that passes through a surface. There we defined an electric flux $\Phi_E = \int \vec{E} \cdot d\vec{A}$. Here we define a *magnetic flux*: Suppose a loop enclosing an area A is placed in a magnetic field $\vec{B}$. Then the magnetic flux through the loop is

$$\Phi_B = \int \vec{B} \cdot d\vec{A} \qquad \text{(magnetic flux through area } A\text{).} \qquad (31\text{-}3)$$

As in Chapter 24, $d\vec{A}$ is a vector of magnitude $|d\vec{A}|$ that is perpendicular to the plane of a differential area $d\vec{A}$.

As a special case of Eq. 31-3, suppose that the loop lies in a plane and that the magnetic field is perpendicular to the plane of the loop. Then we can write the dot product in Eq. 31-3 as $|\vec{B}||d\vec{A}|\cos 0° = |\vec{B}||d\vec{A}|$. If the magnetic field is also uniform, then B can be brought out in front of the integral sign. The remaining $\int |d\vec{A}|$ then gives just the area A of the loop. Thus, Eq. 31-3 reduces to

$$\Phi_B = |\vec{B}|A \qquad (\vec{B} \perp \text{ to plane of area } A, \ \vec{B} \text{ uniform).} \qquad (31\text{-}4)$$

From Eqs. 31-3 and 31-4, we see that the SI unit for magnetic flux is the tesla-square meter, which is called the weber (abbreviated Wb):

$$1 \text{ weber } = 1 \text{ Wb} = 1 \text{ T} \cdot \text{m}^2. \qquad (31\text{-}5)$$

With the notion of magnetic flux, we can state Faraday's law in a more quantitative and useful way:

➤The magnitude of the emf ε induced in a conducting loop is equal to the rate at which the magnetic flux Φ_B through that loop changes with time.

As you will see in the next section, the induced emf ε tends to oppose the flux change, so Faraday's law is formally written as

$$\varepsilon = -\frac{d\Phi_B}{dt} \qquad \text{(Faraday's law)}, \qquad (31\text{-}6)$$

with the minus sign indicating that opposition. We often neglect the minus sign in Eq. 31-6, seeking only the magnitude of the induced emf.

If we change the magnetic flux through a coil of N turns, an induced emf appears in every turn and the total emf induced in the coil is the sum of these individual induced emfs. If the coil is tightly wound (closely packed), so that the same magnetic flux Φ_B passes through all the turns, the total emf induced in the coil is

$$\varepsilon = -N\frac{d\Phi_B}{dt} \qquad \text{(coil of N turns)}. \qquad (31\text{-}7)$$

Here are the general means by which we can change the magnetic flux through a coil:

1. Change the magnitude $|\vec{B}|$ of the magnetic field within the coil.

2. Change the area of the coil, or the portion of that area that happens to lie within the magnetic field (for example, by expanding the coil or sliding it out of the field).

3. Change the angle between the direction of the magnetic field $\vec{B}$ and the area of the coil (for example, by rotating the coil so that $\vec{B}$ is first perpendicular to the plane of the coil and then is along that plane).

READING EXERCISE 31-1: The graph gives the magnitude $\vec{B}(t)$ of a uniform magnetic field that exists throughout a conducting loop, perpendicular to the plane of the loop. Rank the five regions of the graph according to the magnitude of the emf induced in the loop, greatest first.

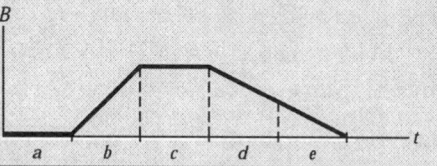

Touchstone Example 31-3-1, at the end of this chapter, illustrates how to use what you learned in this section.

TE

31-4 Lenz's Law

Soon after Faraday proposed his law of induction, Heinrich Friedrich Lenz devised a rule—now known as Lenz's law—for determining the direction of an induced current in a loop:

➤An induced current has a direction such that the magnetic field due to the current opposes the change in the magnetic flux that induces the current.

Furthermore, the direction of an induced emf is that of the induced current. To get a feel for Lenz's law, let us apply it in two different but equivalent ways to Fig. 31-3, where the north pole of a magnet is being moved toward a conducting loop.

1. **Opposition to Pole Movement.** The approach of the magnet's north pole in Fig. 31-3 increases the magnetic flux in the loop and thereby induces a current in the loop. From Fig. 30-18, we know that the loop then acts as a magnetic dipole with a south pole and a north pole, and that its magnetic dipole moment $\vec{\mu}$ is directed from south to north. To oppose the magnetic flux increase being caused by the approaching magnet, the loop's north pole (and thus $\vec{\mu}$) must face *toward* the approaching north pole so as to repel it (Fig. 31-3). Then the curled-straight right-hand rule for $\vec{\mu}$ (Fig. 30-18) tells us that the current induced in the loop must be counterclockwise in Fig. 31-3.

If we next pull the magnet away from the loop, a current will again be induced in the loop. Now, however, the loop will have a south pole facing the retreating north pole of the magnet, so as to oppose the retreat. Thus, the induced current will be clockwise.

2. **Opposition to Flux Change.** In Fig. 31-3, with the magnet initially distant, no magnetic flux passes through the loop. As the north pole of the magnet then nears the loop with its magnetic field $\vec{B}$ directed *toward the left*, the flux through the loop increases. To oppose this increase in flux, the induced current i must set up its own field $\vec{B}_i$ *directed toward the right* inside the loop, as shown in Fig. 31-4a; then the rightward flux of field $\vec{B}_i$ opposes the increasing leftward flux of field $\vec{B}$. The curled-straight right-hand rule of Fig. 30-4 then tells us that i must be counterclockwise in Fig. 31-4a.

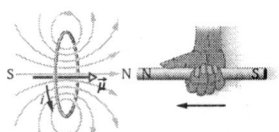

Fig. 31-3: Lenz's law at work. As the magnet is moved toward the loop, a current is induced in the loop. The current produces its own magnetic field, with magnetic dipole moment $\vec{\mu}$ oriented so as to oppose the motion of the magnet. Thus, the induced current must be counterclockwise as shown.

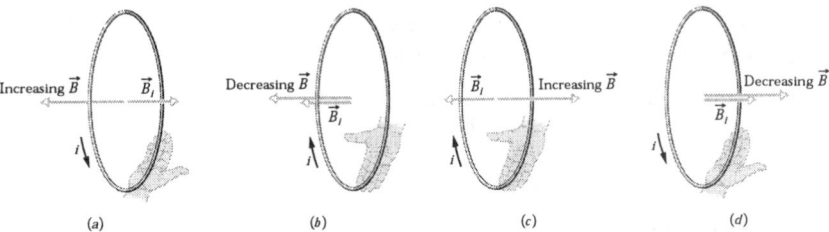

(a)   (b)   (c)   (d)

Fig. 31-4: The current i induced in a loop has the direction such that the current's magnetic field $\vec{B}_i$ opposes the change in the magnetic field $\vec{B}$ inducing i. The field $\vec{B}_i$ is always directed opposite an increasing field $\vec{B}$ (a, c) and in the same direction as a decreasing field $\vec{B}$ (b, d). The curled-straight right-hand rule gives the direction of the induced current based on the direction of the induced field.

Note carefully that the flux of $\vec{B}_i$ always opposes the *change* in the flux of $\vec{B}$, but that does not always mean that $\vec{B}_i$ points opposite $\vec{B}$. For example, if we next pull the magnet away from the loop, the flux Φ_B from the magnet is still directed to the left through the loop, but it is now decreasing. The flux of $\vec{B}_i$ must now be to the left inside the loop, to oppose the *decrease* in Φ_B, as shown in Fig. 31-4b. Thus, $\vec{B}_i$ and $\vec{B}$ are now in the same direction.

Figures 31-4c and d show the situations in which the south pole of the magnet approaches and retreats from the loop, respectively.

Electric Guitars

Figure 31-5 shows a Fender Stratocaster©, the type of electric guitar that was used by Jimi Hendrix and by many other musicians. Whereas an acoustic guitar depends for its sound on the acoustic resonance produced in the hollow body of the instrument by the oscillations of the strings, an electric guitar is a solid instrument, so there is no body resonance. Instead, the oscillations of the metal strings are sensed by electric "pickups" that send signals to an amplifier and a set of speakers.

The basic construction of a pickup is shown in Fig. 31-6. Wire connecting the instrument to the amplifier is coiled around a small magnet. The magnetic field of the

Fig. 31-5: A Fender Stratocaster© has three groups of six electric pickups each (within the wide part of the body). A toggle switch (at the bottom of the guitar) allows the musician to determine which group of pickups sends signals to an amplifier and thus to a speaker system.

magnet produces a north and south pole in the section of the metal string just above the magnet. That section of string then has its own magnetic field. When the string is plucked and thus made to oscillate, its motion relative to the coil changes the flux of its magnetic field through the coil, inducing a current in the coil. As the string oscillates toward and away from the coil, the induced current changes direction at the same frequency as the string's oscillations, thus relaying the frequency of oscillation to the amplifier and speaker.

On a Stratocaster©, there are three groups of pickups, placed at the near end of the strings (on the wide part of the body). The group closest to the near end better detects the high-frequency oscillations of the strings; the group farthest from the near end better detects the low-frequency oscillations. By throwing a toggle switch on the guitar, the musician can select which group or which pair of groups will send signals to the amplifier and speakers.

To gain further control over his music, Hendrix sometimes rewrapped the wire in the pickup coils of his guitar to change the number of turns. In this way, he altered the amount of emf induced in the coils and thus their relative sensitivity to string oscillations. Even without this additional measure, you can see that the electric guitar offers far more control over the sound that is produced than can be obtained with an acoustic guitar.

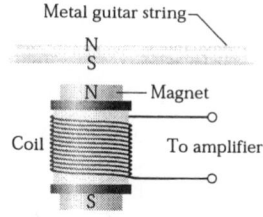

Fig. 31-6: A side view of an electric guitar pickup. When the metal string (which acts like a magnet) is made to oscillate, it causes a variation in magnetic flux that induces a current in the coil.

READING EXERCISE 31-2: The figure shows three situations in which identical circular conducting loops are in uniform magnetic fields that are either increasing (Inc) or decreasing (Dec) in magnitude at identical rates. In each, the dashed line coincides with a diameter. Rank the situations according to the magnitude of the current induced in the loops, greatest first.

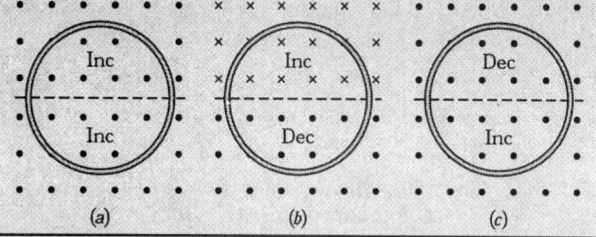

Touchstone Example 31-4-1, at the end of this chapter, illustrates how to use what you learned in this section.

TE

31-5 Induction and Energy Transfers

By Lenz's law, whether you move the magnet toward or away from the loop in Fig. 31-1, a magnetic force resists the motion, requiring your applied force to do positive work. At the same time, thermal energy is produced in the material of the loop because of the material's electrical resistance to the current that is induced. The energy you transfer to the closed *loop + magnet* system via your applied force ends up in this thermal energy. (For now, we neglect energy that is radiated away from the loop as electromagnetic waves during the induction.) The faster you move the magnet, the more rapidly your applied force does work, and the greater the rate at which your energy is transferred to thermal energy in the loop.

Regardless of how current is induced in a loop, energy is always transferred to thermal energy during the process because of the electrical resistance of the loop (unless the loop is superconducting). For example, in Fig. 31-2, when switch S is closed and a current is briefly induced in the left-hand loop, energy is transferred from the battery to thermal energy in that loop.

Figure 31-7 shows another situation involving induced current. A rectangular loop of wire of width L has one end in a uniform external magnetic field that is directed perpendicularly into the plane of the loop. This field may be produced, for example, by a large electromagnet. The dashed lines in Fig. 31-7 show the assumed limits of the

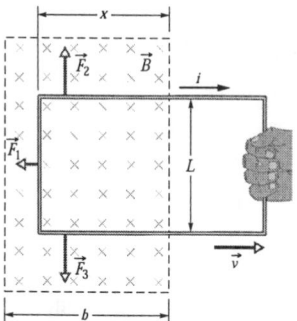

Fig. 31-7: You pull a closed conducting loop out of a magnetic field at constant velocity $\vec{v}$. While the loop is moving, a clockwise current i is induced in the loop, and the loop segments still within the magnetic field experience forces $\vec{F}_1$, $\vec{F}_2$, and $\vec{F}_3$.

magnetic field; the fringing of the field at its edges is neglected. You are asked to pull this loop to the right at a constant velocity $\vec{v}$.

The situation of Fig. 31-7 does not differ in any essential way from that of Fig. 31-1. In each case a magnetic field and a conducting loop are in relative motion; in each case the flux of the field through the loop is changing with time. It is true that in Fig. 31-1 the flux is changing because $\vec{B}$ is changing and in Fig. 31-7 the flux is changing because the area of the loop still in the magnetic field is changing, but that difference is not important. The important difference between the two arrangements is that the arrangement of Fig. 31-7 makes calculations easier. Let us now calculate the rate at which you do mechanical work as you pull steadily on the loop in Fig. 31-7.

As you will see, if you are to pull the loop at a constant velocity $\vec{v}$, you must apply a constant force $\vec{F}$ to the loop because a magnetic force of equal magnitude but opposite direction acts on the loop to oppose you. From Eq. 9-36, the rate at which you do work is then

$$P = |\vec{F}||\vec{v}|, \tag{31-8}$$

where $|\vec{F}|$ is the magnitude of your force and $|\vec{v}|$ is the speed of the loop. We wish to find an expression for the power, P, in terms of the magnitude $|\vec{B}|$ of the magnetic field and the characteristics of the loop—namely, its resistance R to current and its dimension L.

As you move the loop to the right in Fig. 31-7, the portion of its area within the magnetic field decreases. Thus, the flux through the loop also decreases and, according to Lenz's law, a current is produced in the loop. It is the presence of this current that causes the force that opposes your pull.

To find the current, we first apply Faraday's law. When x is the length of the loop still in the magnetic field, the area of the loop still in the field is Lx. Then from Eq. 31-4, the magnitude of the flux through the loop is

$$\Phi_B = |\vec{B}|A = |\vec{B}|L|x|. \tag{31-9}$$

As x decreases, the flux decreases. Faraday's law tells us that with this flux decrease, an emf is induced in the loop. Dropping the minus sign in Eq. 31-6 and using Eq. 31-9, we can write the magnitude of this emf as

$$|\varepsilon| = \left|\frac{d\Phi_B}{dt}\right| = \frac{d}{dt}|\vec{B}|L|x| = |\vec{B}|L\left|\frac{dx}{dt}\right| = |\vec{B}||\vec{v}|L, \tag{31-10}$$

in which we have replaced $|dx/dt|$ with $|\vec{v}|$, the speed at which the loop moves.

Figure 31-8 shows the loop as a circuit; induced emf ε is represented on the left, and the collective resistance R of the loop is represented on the right. The direction of the induced current i is obtained as in Fig. 31-4b; ε must have the same direction.

To find the magnitude of the induced current, we cannot apply the loop rule for potential differences in a circuit because, as you will see in Section 31-6, we cannot define a potential difference for an induced emf. However, we can apply the equation $i = \varepsilon/R$, as we did in Touchstone Example 31-4-1. With Eq. 31-10, this becomes

$$|i| = \frac{|\vec{B}||\vec{v}|L}{R}. \tag{31-11}$$

Because three segments of the loop in Fig. 31-7 carry this current through the magnetic field, sideways deflecting forces act on those segments. From Eq. 29-23 we know that such a deflecting force is, in general notation,

$$\vec{F}_d = i\vec{L} \times \vec{B}. \tag{31-12}$$

In Fig. 31-7, the deflecting forces acting on the three segments of the loop are marked $\vec{F}_1$, $\vec{F}_2$, and $\vec{F}_3$. Note, however, that from the symmetry, $\vec{F}_2$ and $\vec{F}_3$ are equal in

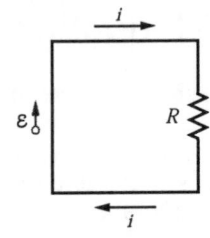

Fig. 31-8: A circuit diagram for the loop of Fig. 31-7 while the loop is moving.

magnitude and cancel. This leaves only $\vec{F}_1$, which is directed opposite your force $\vec{F}$ on the loop and thus is the force that opposes you. Therefore, $\vec{F} = -\vec{F}_1$.

Using Eq. 31-12 to obtain the magnitude of $\vec{F}_1$ and noting that the angle between $\vec{B}$ and the length vector $\vec{L}$ for the left segment is 90°, we write

$$|F| = |\vec{F}_1| = |i||\vec{B}|L \sin 90° = |i||B|L. \tag{31-13}$$

Substituting Eq. 31-11 for i in Eq. 31-13 then gives us

$$|\vec{F}| = \frac{B^2 |\vec{v}| L^2}{R}. \tag{31-14}$$

Since $|\vec{B}|$, L, and R are constants, the speed $|\vec{v}|$ at which you move the loop is constant if the magnitude $|\vec{F}|$ of the force you apply to the loop is also constant.

By substituting Eq. 31-14 into Eq. 31-8, we find the rate at which you do work on the loop as you pull it from the magnetic field:

$$P = |\vec{F}||\vec{v}| = \frac{B^2 v^2 L^2}{R} \qquad \text{(rate of doing work).} \tag{31-15}$$

To complete our analysis, let us find the rate at which thermal energy appears in the loop as you pull it along at constant speed. We calculate it from Eq. 26-21,

$$P = i^2 R. \tag{31-16}$$

Substituting for i from Eq. 31-11, we find

$$P = \left(\frac{|\vec{B}||\vec{v}|L}{R} \right)^2 R = \frac{B^2 v^2 L^2}{R} \qquad \text{(thermal energy rate),} \tag{31-17}$$

which is exactly equal to the rate at which you are doing work on the loop (Eq. 31-15). Thus, the work that you do in pulling the loop through the magnetic field appears as thermal energy in the loop, manifesting itself as a small increase in the temperature of the loop.

Eddy Currents

Suppose we replace the conducting loop of Fig. 31-7 with a solid conducting plate. If we then move the plate out of the magnetic field as we did the loop (Fig. 31-9a), the relative motion of the field and the conductor again induces a current in the conductor. Thus, we again encounter an opposing force and must do work because of the induced current. With the plate, however, the conduction electrons making up the induced current do not follow one path as they do with the loop. Instead, the electrons swirl about within the plate as if they were caught in an eddy (or whirlpool) of water. Such a current is called an *eddy current* and can be represented as in Fig. 31-9a *as if* it followed a single path.

To cook food on an induction stove, oscillating current is sent through a conducting coil that lies just below the cooking surface. The magnetic field produced by that current oscillates and induces an oscillating current in the conducting cooking pan. Because the pan has some resistance to that current, the electrical energy of the current is continuously transformed to thermal energy, resulting in a temperature increase of the pan and the food in it. The cooking surface itself might never get hot.

As with the conducting loop of Fig. 31-7, the current induced in the plate results in mechanical energy being dissipated as thermal energy. The dissipation is more apparent in the arrangement of Fig. 31-9b; a conducting plate, free to rotate about a pivot, is allowed to swing down through a magnetic field like a pendulum. Each time the plate enters and leaves the field, a portion of its mechanical energy is transferred to its thermal energy. After several swings, no mechanical energy remains and the warmed-up plate just hangs from its pivot.

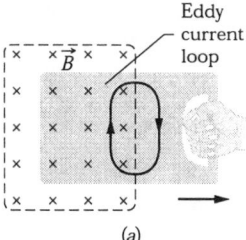

(a)

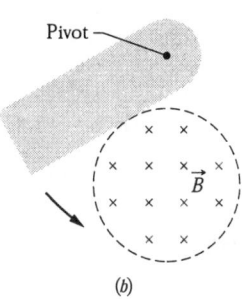

(b)

Fig. 31-9: (a) As you pull a solid conducting plate out of a magnetic field, *eddy currents* are induced in the plate. A typical loop of eddy current is shown; it has the same clockwise sense of circulation as the current in the conducting loop of Fig. 31-7. (b) A conducting plate is allowed to swing like a pendulum about a pivot and into a region of magnetic field. As it enters and leaves the field, eddy currents are induced in the plate.

READING EXERCISE 31-3: The figure shows four wire loops, with edge lengths of either L or $2L$. All four loops will move through a region of uniform magnetic field $\vec{B}$ (directed out of the page) at the same constant velocity. Rank the four loops according to the maximum magnitude of the emf induced as they move through the field, greatest first.

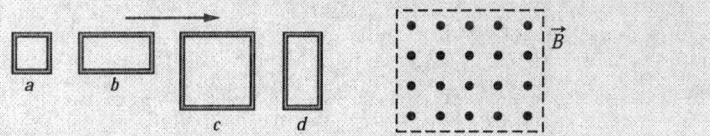

31-6 Induced Electric Fields

Let us place a copper ring of radius r in a uniform external magnetic field, as in Fig. 31-10a. The field—neglecting fringing—fills a cylindrical volume of radius R. Suppose that we increase the strength of this field at a steady rate, perhaps by increasing—in an appropriate way—the current in the windings of the electromagnet that produces the field. The magnetic flux through the ring will then change at a steady rate and—by Faraday's law—an induced emf and thus an induced current will appear in the ring. From Lenz's law we can deduce that the direction of the induced current is counterclockwise in Fig. 31-10a.

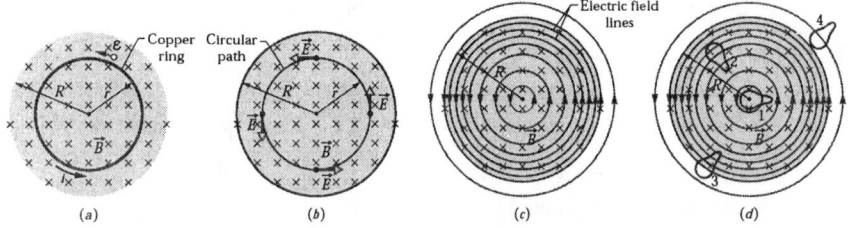

Fig. 31-10: (a) If the magnetic field increases at a steady rate, a constant induced current appears, as shown, in the copper ring of radius r. (b) An induced electric field exists even when the ring is removed; the electric field is shown at four points. (c) The complete picture of the induced electric field, displayed as field lines. (d) Four similar closed paths that enclose identical areas. Equal emfs are induced around paths 1 and 2, which lie entirely within the region of changing magnetic field. A smaller emf is induced around path 3, which only partially lies in that region. No emf is induced around path 4, which lies entirely outside the magnetic field.

If there is a current in the copper ring, an electric field must be present along the ring; an electric field is needed to do the work of moving the conduction electrons. Moreover, the electric field must have been produced by the changing magnetic flux. This **induced electric field** $\vec{E}$ is just as real as an electric field produced by static charges; either field will exert a force $q_0\vec{E}$ on a particle of charge q_0.

By this line of reasoning, we are led to a useful and informative restatement of Faraday's law of induction:

➤A changing magnetic field produces an electric field.

The striking feature of this statement is that the electric field is induced even if there is no copper ring.

To fix these ideas, consider Fig. 31-10b, which is just like Fig. 31-10a except the copper ring has been replaced by a hypothetical circular path of radius r. We assume, as previously, that the magnetic field $\vec{B}$ is increasing in magnitude at a constant rate dB/dt. The electric field induced at various points around the circular path must—from the symmetry—be tangent to the circle, as Fig. 31-10b shows.*

Hence, the circular path is an electric field line. There is nothing special about the circle of radius r, so the electric field lines produced by the changing magnetic field must be a set of concentric circles, as in Fig. 31-10c.

As long as the magnetic field is increasing with time, the electric field represented by the circular field lines in Fig. 31-10c will be present. If the magnetic field remains constant with time, there will be no induced electric field and thus no electric field lines.

---

*Arguments of symmetry would also permit the lines of $\vec{E}$ around the circular path to be radial, rather than tangential. However, such radial lines would imply that there are free charges, distributed symmetrically about the axis of symmetry, on which the electric field lines could begin or end; there are no such charges.

If the magnetic field is decreasing with time (at a constant rate), the electric field lines will still be concentric circles as in Fig. 31-10c, but they will now have the opposite direction. All this is what we have in mind when we say: "A changing magnetic field produces an electric field."

A Reformulation of Faraday's Law

Consider a particle of charge q_0 moving around the circular path of Fig. 31-10b. The work W done on it in one revolution by the induced electric field is εq_0, where ε is the induced emf—that is, the work done per unit charge in moving the test charge around the path. From another point of view, the work is

$$\int \vec{F} \cdot d\vec{s} = |q_o \vec{E}|(2\pi r), \tag{31-18}$$

where $|q_o \vec{E}|$ is the magnitude of the force acting on the test charge and $2\pi r$ is the distance over which that force acts. Setting these two expressions for W equal to each other and canceling q_0, we find that

$$\varepsilon = 2\pi r |\vec{E}|. \tag{31-19}$$

More generally, we can rewrite Eq. 31-18 to give the work done on a particle of charge q_0 moving along any closed path:

$$W = \oint \vec{F} \cdot d\vec{s} = q_0 \oint \vec{E} \cdot d\vec{s}. \tag{31-20}$$

(The circle indicates that the integral is to be taken around the closed path.) Substituting εq_0 for W, we find that

$$\varepsilon = \oint \vec{E} \cdot d\vec{s}. \tag{31-21}$$

This integral reduces at once to Eq. 31-19 if we evaluate it for the special case of Fig. 31-10b.

With Eq. 31-21, we can expand the meaning of induced emf. Previously, induced emf has meant the work per unit charge done in maintaining current due to a changing magnetic flux, or it has meant the work done per unit charge on a charged particle that moves around a closed path in a changing magnetic flux. However, with Fig. 31-10b and Eq. 31-21, an induced emf can exist without the need of a current or particle: An induced emf is the sum—via integration—of quantities $\vec{E} \cdot d\vec{s}$ around a closed path, where $\vec{E}$ is the electric field induced by a changing magnetic flux and $d\vec{s}$ is a differential length vector along the closed path.

If we combine Eq. 31-21 with Faraday's law in Eq. 31-6 ($\varepsilon = -d\Phi_B/dt$), we can rewrite Faraday's law as

$$\oint \vec{E} \cdot d\vec{s} = -\frac{d\Phi_B}{dt} \qquad \text{(Faraday's law)}. \tag{31-22}$$

This equation says simply that a changing magnetic field induces an electric field. The changing magnetic field appears on the right side of this equation, the electric field on the left.

Faraday's law in the form of Eq. 31-22 can be applied to *any* closed path that can be drawn in a changing magnetic field. Figure 31-10d, for example, shows four such paths, all having the same shape and area but located in different positions in the changing field. For paths 1 and 2, the induced emfs $\varepsilon (= \oint \vec{E} \cdot d\vec{s})$ are equal because these paths lie entirely in the magnetic field and thus have the same value of $d\Phi_B/dt$. This is true even though the electric field vectors at points along these paths are different, as indicated by the patterns of electric field lines in the figure. For path 3 the induced emf is smaller because the enclosed flux Φ_B (hence $d\Phi_B/dt$) is smaller, and for path 4 the induced emf is zero, even though the electric field is not zero at any point on the path.

A New Look at Electric Potential

Induced electric fields are produced not by static charges but by a changing magnetic flux. Although electric fields produced in either way exert forces on charged particles, there is an important difference between them. The simplest evidence of this difference is that the field lines of induced electric fields form closed loops, as in Fig. 31-10c. Field lines produced by static charges never do so but must start on positive charges and end on negative charges.

In a more formal sense, we can state the difference between electric fields produced by induction and those produced by static charges in these words:

▶Electric potential has meaning only for electric fields that are produced by static charges; it has no meaning for electric fields that are produced by induction.

You can understand this statement qualitatively by considering what happens to a charged particle that makes a single journey around the circular path in Fig. 31-10b. It starts at a certain point and, on its return to that same point, has experienced an emf ε of, let us say, 5 V; that is, work of 5 J/C has been done on the particle, and thus the particle should then be at a point that is 5 V greater in potential. However, that is impossible because the particle is back at the same point, which cannot have two different values of potential. We must conclude that potential has no meaning for electric fields that are set up by changing magnetic fields.

We can take a more formal look by recalling Eq. 25-17, which defines the potential difference between two points i and f in an electric field $\vec{E}$:

$$V_f - V_i = -\int_i^f \vec{E} \cdot d\vec{s}. \tag{31-23}$$

In Chapter 25 we had not yet encountered Faraday's law of induction, so the electric fields involved in the derivation of Eq. 25-17 were those due to static charges. If i and f in Eq. 31-23 are the same point, the path connecting them is a closed loop, V_i and V_f are identical, and Eq. 31-23 reduces to

$$\oint \vec{E} \cdot d\vec{s} = 0. \tag{31-24}$$

However, when a changing magnetic flux is present, this integral is *not* zero but is $-d\Phi_B/dt$, as Eq. 31-22 asserts. Thus, assigning electric potential to an induced electric field leads us to a contraction. We must conclude that electric potential has no meaning for electric fields associated with induction.

READING EXERCISE 31-4: The figure shows five lettered regions in which a uniform magnetic field extends either directly out of the page (as in region a) or into the page. The field is increasing in magnitude at the same steady rate in all five regions; the regions are identical in area. Also shown are four numbered paths along which $\oint \vec{E} \cdot d\vec{s}$ has the magnitudes given below in terms of a quantity mag. Determine whether the magnetic fields in regions b through e are directed into or out of the page.

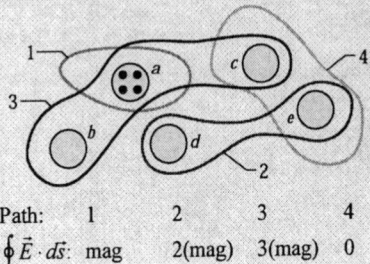

Path:	1	2	3	4
$\oint \vec{E} \cdot d\vec{s}$:	mag	2(mag)	3(mag)	0

Touchstone Example 31-6-1, at the end of this chapter, illustrates how to use what you learned in this section.

TE

31-7 Inductors and Inductance

We found in Chapter 28 that a capacitor can be used to produce a desired electric field. We considered the parallel-plate arrangement as a basic type of capacitor. Similarly, an **inductor** (symbol ⅏) can be used to produce a desired magnetic field. We shall consider a long solenoid (more specifically, a short length near the middle of a long solenoid) as our basic type of inductor.

If we establish a current i in the windings (or turns) of an inductor (a solenoid), the current produces a magnetic flux Φ_B through the central region of the inductor. The **inductance** of the inductor is then

$$L = \frac{N\Phi_B}{i} \qquad \text{(inductance defined)}, \qquad (31\text{-}25)$$

in which N is the number of turns. The windings of the inductor are said to be *linked* by the shared flux, and the product $N\Phi_B$ is called the *magnetic flux linkage*. The inductance L is thus a measure of the flux linkage produced by the inductor per unit of current.

Because the SI unit of magnetic flux is the tesla-square meter, the SI unit of inductance is the tesla-square meter per ampere $(\text{T} \cdot \text{m}^2/\text{A})$. We call this the **henry** (H), after American physicist Joseph Henry, the codiscoverer of the law of induction and a contemporary of Faraday. Thus,

$$1 \text{ henry} = 1 \text{ H} = 1 \text{T} \cdot \text{m}^2/\text{A}. \qquad (31\text{-}26)$$

Through the rest of this chapter we assume that all inductors, no matter what their geometric arrangement, have no magnetic materials such as iron in their vicinity. Such materials would distort the magnetic field of an inductor.

Inductance of a Solenoid

Consider a long solenoid of cross-sectional area A. What is the inductance per unit length near its middle?

To use the defining equation for inductance (Eq. 31-25), we must calculate the flux linkage set up by a given current in the solenoid windings. Consider a length l near the middle of this solenoid. The flux linkage for this section of the solenoid is

$$N\Phi_B = (nl)(|\vec{B}|A)$$

in which n is the number of turns per unit length of the solenoid and $|\vec{B}|$ is the magnitude of the magnetic field within the solenoid.

The magnitude $|\vec{B}|$ is given by Eq. 30-25,

$$|\vec{B}| = \mu_0 in,$$

so from Eq. 31-25,

$$L = \frac{N\Phi_B}{i} = \frac{(nl)(|\vec{B}|A)}{i} = \frac{(nl)(\mu_0|i|n)(A)}{i} \qquad (31\text{-}27)$$
$$= \mu_0 n^2 l A.$$

Thus, the inductance per unit length for a long solenoid near its center is

$$\frac{L}{l} = \mu_0 n^2 A \qquad \text{(solenoid).} \qquad (31\text{-}28)$$

The crude inductors with which Michael Faraday discovered the law of induction. In those days amenities such as insulated wire were not commercially available. It is said that Faraday insulated his wires by wrapping them with strips cut from one of his wife's petticoats.

Inductance—like capacitance—depends only on the geometry of the device. The dependence on the square of the number of turns per unit length is to be expected. If you, say, triple n, you not only triple the number of turns (N) but you also triple the flux $\left(\Phi_B = |\vec{B}|A = \mu_0 |i| n A\right)$ through each turn, multiplying the flux linkage $N\Phi_B$ and thus the inductance L by a factor of 9.

If the solenoid is very much longer than its radius, then Eq. 31-27 gives its inductance to a good approximation. This approximation neglects the spreading of the magnetic field lines near the ends of the solenoid, just as the parallel-plate capacitor formula $\left(C = \varepsilon_0 A/d\right)$ neglects the fringing of the electric field lines near the edges of the capacitor plates.

From Eq. 31-27, and recalling that n is a number per unit length, we can see that an inductance can be written as a product of the permeability constant μ_0 and a quantity with the dimensions of a length. This means that μ_0 can be expressed in the unit henry per meter:

$$\mu_0 = 4\pi \times 10^{-7} \text{ T} \cdot \text{m/A}$$
$$= 4\pi \times 10^{-7} \text{ H/m}. \tag{31-29}$$

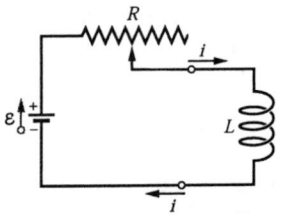

31-8 Self-Induction

If two coils—which we can now call inductors—are near each other, a current i in one coil produces a magnetic flux Φ_B through the second coil. We have seen that if we change this flux by changing the current, an induced emf appears in the second coil according to Faraday's law. An induced emf appears in the first coil as well.

➤ An induced emf ε_L appears in any coil in which the current is changing.

Fig. 31-11: If the current in a coil is changed by varying the contact position on a variable resistor, a self-induced emf ε_L will appear in the coil while the current is changing.

This process (see Fig. 31-11) is called **self-induction**, and the emf that appears is called a **self-induced emf**. It obeys Faraday's law of induction just as other induced emfs do.

For any inductor, Eq. 31-25 tells us that

$$N\Phi_B = Li. \tag{31-30}$$

Faraday's law tell us that

$$\varepsilon_L = -\frac{d(N\Phi_B)}{dt}. \tag{31-31}$$

By combining Eqs. 31-30 and 31-31 we can write:

$$\varepsilon_L = -L\frac{di}{dt} \qquad \text{(self-induced emf).} \tag{31-32}$$

Thus, in any inductor (such as a coil, a solenoid, or a toroid) a self-induced emf appears whenever the current changes with time. The magnitude of the current has no influence on the magnitude of the induced emf; only the rate of change of the current counts.

You can find the *direction* of a self-induced emf from Lenz's law. The minus sign in Eq. 31-32 indicates that—as the law states—the self-induced emf ε_L has the orientation such that it opposes the change in current i. We can drop the minus sign when we want only the magnitude of ε_L.

Suppose that, as in Fig. 31-12a, you set up a current i in a coil and arrange to have it increase with time at a rate di/dt. In the language of Lenz's law, this increase in the current is the "change" that the self-induction must oppose. For such opposition to occur, a self-induced emf must appear in the coil, pointing—as the figure shows—so as to oppose the increase in the current. If you cause the current to decrease with time, as in Fig. 31-12b, the self-induced emf must point in a direction that tends to oppose the decrease in the current, as the figure shows.

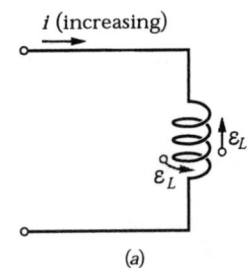

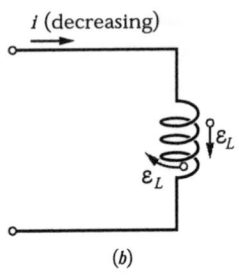

Fig. 31-12: (a) The current i is increasing and the self-induced emf ε_L appears along the coil in a direction such that it opposes the increase. The arrow representing ε_L can be drawn along a turn of the coil or alongside the coil. Both are shown. (b) The current i is decreasing and the self-induced emf appears in a direction such that it opposes the decrease.

In Section 31-6 we saw that we cannot define an electric potential for an electric field (and thus for an emf) that is induced by a changing magnetic flux. This means that when a self-induced emf is produced in the inductor of Fig. 31-11, we cannot define an electric potential within the inductor itself, where the flux is changing. However, potentials can still be defined at points of the circuit that are not within the inductor— points where the electric fields are due to charge distributions and their associated electric potentials.

Moreover, we can define a self-induced potential difference ΔV_L *across an inductor* (between its terminals, which we assume to be outside the region of changing flux). If the inductor is an ideal inductor (its wire has negligible resistance), the magnitude of ΔV_L is equal to the magnitude of the self-induced emf ε_L.

If, instead, the wire in the inductor has resistance r, we mentally separate the inductor into a resistance r (which we take to be outside the region of changing flux) and an ideal inductor of self-induced emf ε_L. As with a real battery of emf ε and internal resistance r, the potential difference across the terminals of a real inductor then differs from the emf. Unless otherwise indicated, we assume here that inductors are ideal.

READING EXERCISE 31-5: The figure shows an emf ε_L induced in a coil. Which of the following can describe the current through the coil: (a) constant and rightward, (b) constant and leftward, (c) increasing and rightward, (d) decreasing and rightward, (e) increasing and leftward, (f) decreasing and leftward?

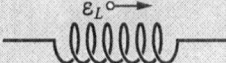

31-9 *RL* Circuits

In Section 28-9 we saw that if we suddenly introduce an emf ε into a single-loop circuit containing a resistor R and a capacitor C, the charge on the capacitor does not build up immediately to its final equilibrium value $C\varepsilon$ but approaches it in an exponential fashion:

$$q = C\varepsilon(1 - e^{-t/\tau_C}) \qquad (31\text{-}32)$$

The rate at which the charge builds up is determined by the capacitive time constant τ_C, defined in Eq. 28-41 as

$$\tau_C = RC. \qquad (31\text{-}33)$$

If we suddenly remove the emf from this same circuit, the charge does not immediately fall to zero but approaches zero in an exponential fashion:

$$q = q_0 e^{-t/\tau_C}. \qquad (31\text{-}34)$$

The time constant τ_C describes the fall of the charge as well as its rise.

An analogous slowing of the rise (or fall) of the current occurs if we introduce an emf ε into (or remove it from) a single-loop circuit containing a resistor R and an inductor L. When the switch S in Fig. 31-13 is closed on a, for example, the current in the resistor starts to rise. If the inductor were not present, the current would rise rapidly to a steady value ε/R. Because of the inductor, however, a self-induced emf ε_L appears in the circuit; from Lenz's law, this emf opposes the rise of the current, which means that it opposes the battery emf ε in polarity. Thus the current in the resistor responds to the difference between two emfs, a constant one ε due to the battery and a variable one $\varepsilon_L(=-L\,di/dt)$ due to self-induction. As long as ε_L is present, the current in the resistor will be less than ε/R.

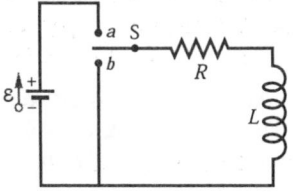

Fig. 31-13: An *RL* circuit. When switch S is closed on a, the current rises and approaches a limiting value of ε/R.

As time goes on, the rate at which the current increases becomes less rapid and the magnitude of the self-induced emf, which is proportional to di/dt, becomes smaller. Thus, the current in the circuit approaches ε/R asymptotically.

We can generalize these results as follows:

▶Initially, an inductor acts to oppose changes in the current through it. A long time later, it acts like ordinary connecting wire.

Now let us analyze the situation quantitatively. With the switch S in Fig. 31-13 thrown to a, the circuit is equivalent to that of Fig. 31-14. Let us apply the loop rule, starting at point x in this figure and moving clockwise around the loop along with current i.

1. *Resistor.* Because we move through the resistor in the direction of current i, the electric potential decreases by iR. Thus, as we move from point x to point y, we encounter a potential change of $-iR$.

2. *Inductor.* Because current i is changing, there is a self-induced emf ε_L in the inductor. The magnitude of ε_L is given by Eq. 31-32 as $L\,di/dt$. The direction of ε_L is upward in Fig. 31-14 because current i is downward through the inductor and increasing. Thus, as we move from point y to point z, opposite the direction of ε_L, we encounter a potential change of $-L\,di/dt$.

3. *Battery.* As we move from point z back to starting point x, we encounter a potential change of $+\varepsilon$ due to the battery's emf.

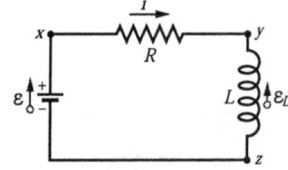

Fig. 31-14: The circuit of Fig. 31-13 with the switch closed on a. We apply the loop rule for circuits clockwise, starting at x.

Thus, the loop rule gives us

$$-iR - L\frac{di}{dt} + \varepsilon = 0$$

or

$$L\frac{di}{dt} + Ri = \varepsilon \qquad \text{(RL circuit).} \qquad (31\text{-}36)$$

Equation 31-36 is a differential equation involving the variable i and its first derivative di/dt. To solve it, we seek the function $i(t)$ such that when $i(t)$ and its first derivative are substituted in Eq. 31-36, the equation is satisfied and the initial condition $i(0) = 0$ is satisfied.

Equation 31-36 and its initial condition are of exactly the form of Eq. 28-37 for an *RC* circuit, with i replacing q, L replacing R, and R replacing $1/C$. The solution of Eq. 31-36 must then be of exactly the form of Eq. 28-38 with the same replacements. That solution is

$$i = \frac{\varepsilon}{R}(1 - e^{-Rt/L}), \qquad (31\text{-}37)$$

which we can rewrite as

$$i = \frac{\varepsilon}{R}(1 - e^{-t/\tau_L}) \qquad \text{(rise of current).} \qquad (31\text{-}38)$$

Here τ_L, the inductive time constant, is given by

$$\tau_L = \frac{L}{R} \qquad \text{(time constant).} \qquad (31\text{-}39)$$

Let's examine Eq. 31-38 for when the switch is closed (at time $t = 0$) and for a time long after the switch is closed $(t \to \infty)$. If we substitute $t = 0$ into Eq. 31-38, the exponential becomes $e^{-0} = 1$. Thus, Eq. 31-38 tells us that the current is initially $i = 0$, as we expected. Next, if we let t go to ∞, then the exponential goes to $e^{-\infty} = 0$. Thus, Eq. 31-38 tells us that the current goes to its equilibrium value of ε/R.

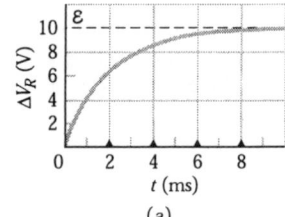

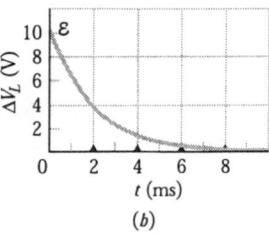

Fig. 31-15: The variation with time of (a) ΔV_R the potential difference across the resistor in the circuit of Fig. 31-14, and (b) ΔV_L, the potential difference across the inductor in that circuit. The small triangles represent successive intervals of one inductive time constant $\tau_L = L/R$. The figure is plotted for $r = 2000\,\Omega$, $L = 4.0$ H, and $\varepsilon = 10$ V.

We can also examine the potential differences in the circuit. Figure 31-15 shows how the potential differences $\Delta V_R (= iR)$ across the resistor and $\Delta V_L (= L\,di/dt)$ across the inductor vary with time for particular values of ε, L, and R. Compare this figure carefully with the corresponding figure for an RC circuit (Fig. 28-21).

To show that the quantity $\tau_L (= L/R)$ has the dimension of time, we convert from Henries per ohm as follows:

$$1\frac{H}{\Omega} = 1\frac{H}{\Omega}\left(\frac{1\,V\cdot s}{1\,H\cdot A}\right)\left(\frac{1\Omega\cdot A}{1\,V}\right) = 1\,s.$$

The first quantity in parentheses is a conversion factor based on Eq. 31-32, and the second one is a conversion factor based on the relation $\Delta V = iR$.

The physical significance of the time constant follows from Eq. 31-37. If we put $t = \tau_L = L/R$ in this equation, it reduces to

$$i = \frac{\varepsilon}{R}(1 - e^{-1}) = 0.63\frac{\varepsilon}{R}. \tag{31-45}$$

Thus, the time constant τ_L is the time it takes the current in the circuit to reach about 63% of its final equilibrium value ε/R. Since the potential difference ΔV_R across the resistor is proportional to the current i, a graph of the increasing current versus time has the same shape as that of ΔV_R in Fig. 31-15a.

If the switch S in Fig. 31-13 is closed on a long enough for the equilibrium current ε/R to be established, and then is thrown to b, the effect will be to remove the battery from the circuit. (The connection to b must actually be made an instant before the connection to a is broken. A switch that does this is called a *make-before-break* switch.)

With the battery gone, the current through the resistor will decrease. However, it cannot drop immediately to zero but must decay to zero over time. The differential equation that governs the decay can be found by putting $\varepsilon = 0$ in Eq. 31-35:

$$L\frac{di}{dt} + iR = 0. \tag{31-41}$$

By analogy with Eqs. 28-37 and 28-38, the solution of this differential equation that satisfies the initial condition $i(0) = i_0 = \varepsilon/R$ is

$$i = \frac{\varepsilon}{R}e^{-t/\tau_L} = i_0 e^{-t/\tau_L} \qquad \text{(decay of current).} \tag{31-42}$$

We see that both current rise (Eq. 31-38) and current decay (Eq. 31-42) in an RL circuit are governed by the same inductive time constant, τ_L.

We have used i_0 in Eq. 31-42 to represent the current at time $t = 0$. In our case that happened to be ε/R, but it could be any other initial value.

Touchstone Example 31-9-1, at the end of this chapter, illustrates how to use what you learned in this section.

TE

31-10 Energy Stored in a Magnetic Field

When we pull two particles with opposite signs of charge away from each other, we say that the resulting electric potential energy is stored in the electric field of the particles. We get it back from the field by letting the particles move closer together again. In the same way we can consider energy to be stored in a magnetic field.

To derive a quantitative expression for that stored energy, consider again Fig. 31-14, which shows a source of emf ε connected to a resistor R and an inductor L. Equation 31-36, restated here for convenience,

$$\varepsilon = L\frac{di}{dt} + iR \qquad\qquad (31\text{-}43)$$

is the differential equation that describes the growth of current in this circuit. We stress that this equation follows immediately from the loop rule and that the loop rule in turn is an expression of the principle of conservation of energy for single-loop circuits. If we multiply each side of Eq. 31-43 by i, we obtain

$$\varepsilon i = Li\frac{di}{dt} + i^2 R, \qquad\qquad (31\text{-}44)$$

which has the following physical interpretation in terms of work and energy:

1. If a charge dq passes through the battery of emf ε in Fig. 31-14 in time dt, the battery does work on it in the amount $\varepsilon\,dq$. The rate at which the battery does work is $(\varepsilon\,dq)/dt$, or εi. Thus, the left side of Eq. 31-44 represents the rate at which the emf device delivers energy to the rest of the circuit.

2. The rightmost term in Eq. 31-44 represents the rate at which energy appears as thermal energy in the resistor.

3. Energy that is delivered to the circuit but does not appear as thermal energy must, by the conservation-of-energy hypothesis, be stored in the magnetic field of the inductor. Since Eq. 31-44 represents conservation of energy for *RL* circuits, the middle term must represent the rate dU_B/dt at which energy is stored in the magnetic field.

Thus

$$\frac{dU_B}{dt} = Li\frac{di}{dt}. \qquad\qquad (31\text{-}45)$$

We can write this as

$$dU_B = Li\,di.$$

Integrating yields

$$\int_0^{U_B} dU_B = \int_0^i Li\,di$$

or $\qquad\qquad\qquad U_B = \tfrac{1}{2}Li^2 \qquad\qquad$ (magnetic energy), $\qquad (31\text{-}46)$

which represents the total energy stored by an inductor L carrying a current i. Note the similarity in form between this expression and the expression for the energy stored by a capacitor with capacitance C and charge q; namely,

$$U_E = \frac{q^2}{2C}. \qquad\qquad (31\text{-}47)$$

(The variable i^2 corresponds to q^2, and the constant L corresponds to $1/C$.)

31-11 Energy Density of a Magnetic Field

Consider a length l near the middle of a long solenoid of cross-sectional area A carrying current i; the volume associated with this length is Al. The energy U_B stored by the length l of the solenoid must lie entirely within this volume because the magnetic field outside such a solenoid is approximately zero. Moreover, the stored energy must be uniformly distributed within the solenoid because the magnetic field is (approximately) uniform everywhere inside.

Thus, the energy stored per unit volume of the field is

$$u_B = \frac{U_B}{Al}$$

or, since

$$U_B = \tfrac{1}{2}Li^2,$$

we have

$$u_B = \frac{Li^2}{2Al} = \frac{L}{l}\frac{i^2}{2A}.$$

Here L is the inductance of length l of the solenoid.

Substituting for L/l from Eq. 31-28, we find

$$u_B = \tfrac{1}{2}\mu_0 n^2 i^2, \tag{31-48}$$

where n is the number of turns per unit length. From Eq. 30-25 $\left(\left|\vec{B}\right| = \mu_0 |i|n\right)$ we can write this *energy density* as

$$u_B = \frac{B^2}{2\mu_0} \qquad \text{(magnetic energy density).} \tag{31-49}$$

This equation gives the density of stored energy at any point where the magnetic field is $\vec{B}$. Even though we derived it by considering a special case, the solenoid, Eq. 31-49 holds for all magnetic fields, no matter how they are generated. Equation 31-49 is comparable to Eq. 28-23; namely,

$$u_E = \tfrac{1}{2}\varepsilon_0 E^2, \tag{31-50}$$

which gives the energy density (in a vacuum) at any point in an electric field. Note that both u_B and u_E are proportional to the square of the appropriate field magnitude, $\left|\vec{B}\right|$ or $\left|\vec{E}\right|$.

READING EXERCISE 31-6: The table lists the number of turns per unit length, current, and cross-sectional area for three solenoids. Rank the solenoids according to the magnetic energy density within them, greatest first.

Solenoid	Turns per Unit Length	Current	Area
a	$2n_1$	i_1	$2A_1$
b	n_1	$2i_1$	A_1
c	n_1	i_1	$6A_1$

31-12 Mutual Induction

In this section we return to the case of two interacting coils, which we first discussed in Section 31-2, and we treat it in a somewhat more formal manner. We saw earlier that if two coils are close together as in Fig. 31-2, a steady current i in one coil will set up a magnetic flux Φ through the other coil (*linking* the other coil). If we change i with time, an emf ε given by Faraday's law appears in the second coil; we called this process *induction*. We could better have called it **mutual induction**, to suggest the mutual interaction of the two coils and to distinguish it from *self-induction*, in which only one coil is involved.

Let us look a little more quantitatively at mutual induction. Figure 31-16*a* shows two circular close-packed coils near each other and sharing a common central axis. There is a steady current i_1 in coil 1, produced by the battery in the external circuit. This current creates a magnetic field represented by the lines of $\vec{B}_1$ in the figure. Coil 2 is connected to a sensitive meter but contains no battery; a magnetic flux Φ_{21} (the flux through coil 2 associated with the current in coil 1) links the N_2 turns of coil 2.

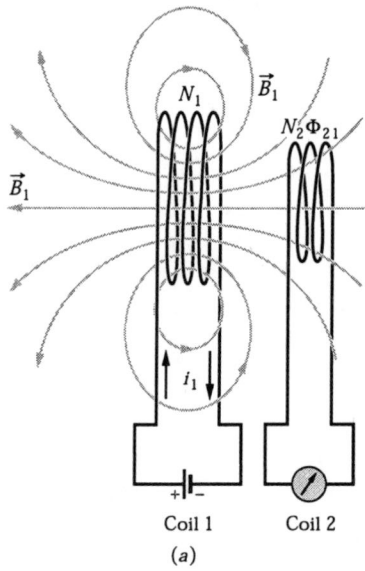 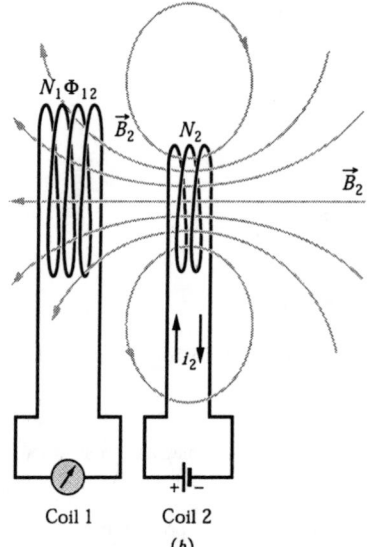

Fig. 31-26: Mutual induction. (*a*) If the current in coil 1 changes, an emf will be induced in coil 2. (*b*) If the current in coil 2 changes, an emf will be induced in coil 1.

We define the mutual inductance M_{21} of coil 2 with respect to coil 1 as

$$M_{21} = \frac{N_2 \Phi_{21}}{i_1}, \tag{31-51}$$

which has the same form as Eq. 31-25 $\left(L = N\Phi/i\right)$, the definition of (self) inductance. We can recast Eq. 31-51 as

$$M_{21}i_1 = N_2\Phi_{21}.$$

If, by external means, we cause i_1 to vary with time, we have

$$M_{21}\frac{di_1}{dt} = N_2 \frac{d\Phi_{21}}{dt}.$$

The right side of this equation is, according to Faraday's law, just the magnitude of the emf ε_2 appearing in coil 2 due to the changing current in coil 1. Thus, with a minus sign to indicate direction,

$$\varepsilon_2 = -M_{21}\frac{di_1}{dt}, \tag{31-52}$$

which you should compare with Eq. 31-32 for self-induction $\left(\varepsilon = -L\,di/dt\right)$.

Let us now interchange the roles of coils 1 and 2, as in Fig. 31-16*b*; that is, we set up a current i_2 in coil 2 by means of a battery, and this produces a magnetic flux Φ_{12} that links coil 1. If we change i_2 with time, we have, by the argument given above,

$$\varepsilon_1 = -M_{12}\frac{di_2}{dt}. \tag{31-53}$$

Thus, we see that the emf induced in either coil is proportional to the rate of change of current in the other coil. The proportionality constants M_{21} and M_{12} seem to be different. We assert, without proof, that they are in fact the same so that no subscripts are needed. (This conclusion is true but is in no way obvious.) Thus, we have

$$M_{21} = M_{12} = M, \qquad\qquad (31\text{-}54)$$

and we can rewrite Eqs. 31-52 and 31-53 as

$$\varepsilon_2 = -M\frac{di_1}{dt} \qquad\qquad (31\text{-}55)$$

and

$$\varepsilon_1 = -M\frac{di_2}{dt}. \qquad\qquad (31\text{-}56)$$

The induction is indeed mutual. The SI unit for M (as for L) is the henry.

Touchstone Example 31-12-1, at the end of this chapter, illustrates how to use what you learned in this section.

TE

Touchstone Example 31-3-1

The long solenoid S shown (in cross section) in Fig. TE31-1 has **220 turns/cm and carries a current $i = 1.5$ A; its diameter D is 3.2 cm. At its center we place a 130-turn closely packed coil C of diameter $d = 2.1$ cm. The current in the solenoid is reduced to zero at a steady rate in 25 ms. What is the magnitude of the emf that is induced in coil C while the current in the solenoid is changing?**

SOLUTION: The **Key Ideas** here are these:

1. Because coil C is located in the interior of the solenoid, it lies within the magnetic field produced by current i in the solenoid; thus, there is a magnetic flux Φ_B through coil C.

2. Because current i decreases, flux Φ_B also decreases.

3. As Φ_B decreases, emf ε is induced in coil C, according to Faraday's law.

Because coil C consists of more than one turn, we apply Faraday's law in the form of Eq. 31-7 ($\varepsilon = -N \, d\Phi_B/dt$), where the number of turns N is 130 and $d\Phi_B/dt$ is the rate at which the flux in each turn changes.

   Because the current in the solenoid decreases at a steady rate, flux Φ_B also decreases at a steady rate and we can write $d\Phi_B/dt$ as $\Delta\Phi_B/\Delta t$. Then, to evaluate $\Delta\Phi_B$, we need the final and initial flux. The final flux $\Phi_{B,f}$ is zero because the final current in the solenoid is zero. To find the initial flux $\Phi_{B,i}$, we need two more **Key Ideas:**

4. The flux through each turn of coil C depends on the area A and orientation of that turn in the solenoid's magnetic field $\vec{B}$. Because $\vec{B}$ is uniform and directed perpendicular to area A, the flux is given by Eq. 31-4 ($\Phi_B = |\vec{B}|A$).

5. The magnitude $|\vec{B}|$ of the magnetic field in the interior of a solenoid depends on the solenoid's current i and its number n of turns per unit length, according to Eq. 30-25 ($|\vec{B}| = (\mu_0| i |n)$).

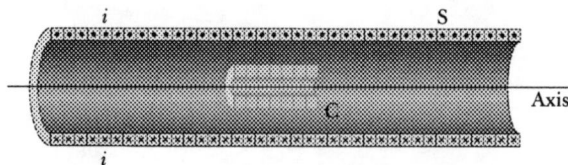

Fig. TE 31-1 A coil C is located inside a solenoid S, which carries current i.

For the situation of Fig. TE31-1, A is $\frac{1}{4}\pi d^2$ ($= 3.46 \times 10^{-4}$ m^2) and n is 220 turns/cm, or 22 000 turns/cm. Substituting Eq. 30-25 into Eq. 31-4 then leads to

$$\Phi_{B,i} = |\vec{B}|A = (\mu_0| i |n)A$$
$$= (4\pi \times 10^{-7} \, \text{T·m/A})(1.5 \, \text{A})(22\,000 \, \text{turns/m})$$
$$\times (3.46 \times 10^{-4} \, \text{m}^2)$$
$$= 1.44 \times 10^{-5} \, \text{Wb}.$$

Now we can write

$$\frac{d\Phi_B}{dt} = \frac{\Delta\Phi_B}{\Delta t} = \frac{\Phi_{B,f} - \Phi_{B,i}}{\Delta t}$$
$$= \frac{(0 - 1.44 \times 10^{-5} \, \text{Wb})}{25 \times 10^{-3} \, \text{s}}$$
$$= -5.76 \times 10^{-4} \, \text{Wb/s} = -5.76 \times 10^{-4} \, \text{V}.$$

We are interested only in magnitudes, so we ignore the minus signs here and in Eq. 31-7, writing

$$|\varepsilon| = \left| N \frac{d\Phi_B}{dt} \right| = (130 \, \text{turns})(5.76 \times 10^{-4} \, \text{V})$$
$$= 7.5 \times 10^{-2} \, \text{V} = 75 \, \text{mV}. \qquad \text{(Answer)}$$

Touchstone Example 31-4-1

Figure TE31-2 shows a conducting loop consisting of a half-circle of radius $r = 0.20$ m and three straight sections. The half-circle lies in a uniform magnetic field $|\vec{B}|$ that is directed out of the page; the field magnitude is given by $|\vec{B}| = (4.0 \text{ T/s}^2)t^2 + (2.0 \text{ T/s})t + 3.0$ T. An ideal battery with emf $\varepsilon_{bat} = 2.0$ V is connected to the loop. The resistance of the loop is $2.0 \, \Omega$.

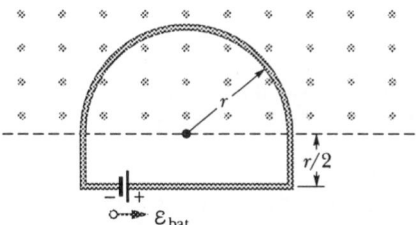

Fig. TE 31-2 A battery is connected to a conducting loop consisting of a half-circle of radius r that lies in a uniform magnetic field. The field is directed out of the page; its magnitude is changing.

(a) What are the magnitude and direction of the emf ε_{ind} induced around the loop by field $\vec{B}$ at $t = 10$ s?

SOLUTION: One **Key Idea** here is that, according to Faraday's law, the magnitude of ε_{ind} is equal to the rate $d\Phi_B/dt$ at which the magnetic flux through the loop changes. A second **Key Idea** is that the flux through the loop depends on the loop's area A and its orientation in the magnetic field $\vec{B}$. Because $\vec{B}$ is uniform and is perpendicular to the plane of the loop, the flux is given by Eq. 31-4 ($\Phi_B = |\vec{B}|A$). Using this equation and realizing that only the field magnitude B changes in time (not the area A), we rewrite Faraday's law, Eq. 31-6, as

$$\varepsilon_{ind} = \frac{d\Phi_B}{dt} = \frac{d(BA)}{dt} = A \left| \frac{d\vec{B}}{dt} \right|.$$

A third **Key Idea** is that, because the flux penetrates the loop only within the half-circle, the area A in this equation is $\frac{1}{2}\pi r^2$. Substituting this and the given expression for B yields

$$\varepsilon_{ind} = A \left| \frac{d\vec{B}}{dt} \right| = \frac{\pi r^2}{2} \frac{d}{dt}(4.0 \text{ T/s}^2)t^2 + (2.0 \text{ T/s})t + 3.0 \text{ T}$$

$$= \frac{\pi r^2}{2}(8.0t + 2.0).$$

At $t = 10$ s, then,

$$\varepsilon_{ind} = \frac{\pi(0.20 \text{ m})^2}{2}[8.0(10) + 2.0]$$

$$= 5.152 \text{ V} \approx 5.2 \text{ V}. \qquad \text{(Answer)}$$

To find the direction of ε_{ind}, we first note that in Fig. TE31-2 the flux through the loop is out of the page and increasing. Then the **Key Idea** here is that the induced field B_i (due to the induced current) must oppose that increase, and thus be *into* the page. Using the curled–straight right-hand rule (Fig. TE30-1c), we find that the induced current must be clockwise around the loop. The induced emf ε_{ind} must then also be clockwise.

(b) What is the current in the loop at $t = 10$ s?

SOLUTION: The **Key Idea** here is that two emfs tend to move charges around the loop. The induced emf ε_{ind} tends to drive a current clockwise around the loop; the battery's emf ε_{bat} tends to drive a current counterclockwise. Because ε_{ind} is greater than ε_{bat}, the net emf ε_{net} is clockwise, and thus so is the current. To find the current at $t = 10$ s, we use $i = \varepsilon/R$:

$$i = \frac{\varepsilon_{\text{net}}}{R} = \frac{\varepsilon_{\text{ind}} - \varepsilon_{\text{bat}}}{R}$$

$$= \frac{5.152 \text{ V} - 2.0 \text{ V}}{2.0 \, \Omega} = 1.58 \text{ A} \approx 1.6 \text{ A.} \qquad \text{(Answer)}$$

Touchstone Example 31-6-1

In Fig. TE31-3*b*, take $R = 8.5$ cm and $dB/dt = 0.13$ T/s.

(a) Find an expression for the magnitude E of the induced electric field at points within the magnetic field, at radius r from the center of the magnetic field. Evaluate the expression for $r = 5.2$ cm.

SOLUTION: The **Key Idea** here is that an electric field is induced by the changing magnetic field, according to Faraday's law. To calculate the field magnitude E, we apply Faraday's law in the form of Eq. 31-22. We use a circular path of integration with radius $r \leq R$ because we want E for points within the magnetic field. We assume from the symmetry that $\vec{E}$ in Fig. 31-10*b* is tangent to the circular path at all points. The path vector $d\vec{s}$ is also always tangent to the circular path, so the dot product $\vec{E} \cdot d\vec{s}$ in Eq. 31-22 must have the magnitude $E \, ds$ at all points on the path. We can also assume from the symmetry that E has the same value at all points along the circular path. Then the left side of Eq. 31-22 becomes

$$\oint \vec{E} \cdot d\vec{s} = \oint E \, ds = E \oint ds = |\vec{E}|(2\pi r). \qquad \text{(TE31-1)}$$

(The integral $\oint ds$ is the circumference $2\pi r$ of the circular path.)

Next, we need to evaluate the right side of Eq. 31-22. Because $\vec{B}$ is uniform over the area A encircled by the path of integration and is directed perpendicular to that area, the magnetic flux is given by Eq. 31-4:

$$\Phi_B = |\vec{B}|A = |\vec{B}|(\pi r^2). \qquad \text{(TE31-2)}$$

Substituting this and Eq. TE31-1 into Eq. 31-22 and dropping the minus sign, we find that the magnitude of the electric field is

$$|\vec{E}(2\pi r)| = (\pi r^2)\left|\frac{d\vec{B}}{dt}\right|$$

or

$$|\vec{E}| = \frac{r}{2}\left|\frac{d\vec{B}}{dt}\right|. \qquad \text{(Answer)} \quad \text{(TE31-3)}$$

Equation TE31-3 gives the magnitude of the electric field at any point for which $r \leq R$ (that is, within the magnetic field). Substituting given values yields, for the magnitude of $\vec{E}$ at $r = 5.2$ cm,

$$|\vec{E}| = \frac{(5.2 \times 10^{-2} \text{ m})}{2}(0.13 \text{ T/s})$$

$$= 0.0034 \text{ V/m} = 3.4 \text{ mV/m.} \qquad \text{(Answer)}$$

(b) Find an expression for the magnitude E of the induced electric field at points that are outside the magnetic field, at radius r. Evaluate the expression for $r = 12.5$ cm.

SOLUTION: The **Key Idea** of part (a) applies here also, except that we use a circular path of integration with radius $r \geq R$, because we want to evaluate E for points outside the magnetic field. Proceeding as in (a), we again obtain Eq. TE31-1. However, we do not then obtain Eq. TE31-2, because the new path of integration is now outside the magnetic field, and we need this **Key Idea**: The magnetic flux encircled by the new path is only that in the area πR^2 of the magnetic field region. Therefore,

$$\Phi_B = |\vec{B}|A = |\vec{B}|(\pi R^2). \qquad \text{(TE31-4)}$$

Substituting this and Eq. TE31-1 into Eq. 31-22 (without the minus sign) and solving for the magnitude of $|\vec{E}|$ yield

$$|\vec{E}| = \frac{R^2}{2r} \frac{dB}{dt}. \qquad \text{(Answer)} \quad \text{(TE31-5)}$$

Since $|\vec{E}|$ is not zero here, we know that an electric field is induced even at points that are outside the changing magnetic field, an important result that (as you shall see in Section 33-3) makes transformers possible. With the given data, Eq. TE31-5 yields the magnitude of $|\vec{E}|$ at $r = 12.5$ cm:

$$E = \frac{(8.5 \times 10^{-2} \text{ m})^2}{(2)(12.5 \times 10^{-2} \text{ m})} (0.13 \text{ T/s})$$

$$= 3.8 \times 10^{-3} \text{ V/m} = 3.8 \text{ mV/m}. \qquad \text{(Answer)}$$

Equations TE31-3 and TE31-5 give the same result, as they must, for $r = R$. Figure 31-3 shows a plot of $|\vec{E}(r)|$ based on these two equations.

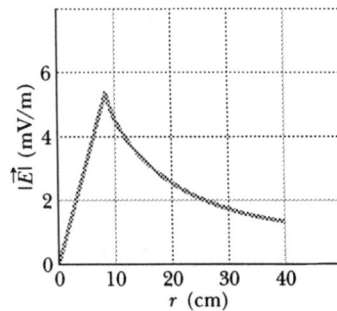

Fig. TE 31-3 A plot of the induced electric field $E(r)$ for the conditions of TE 31-6-1.

Touchstone Example 31-9-1

Figure TE31-4a shows a circuit that contains three identical resistors with resistance $R = 9.0\ \Omega$, two identical inductors with inductance $L = 2.0$ mH, and an ideal battery with emf $\varepsilon = 18$ V.

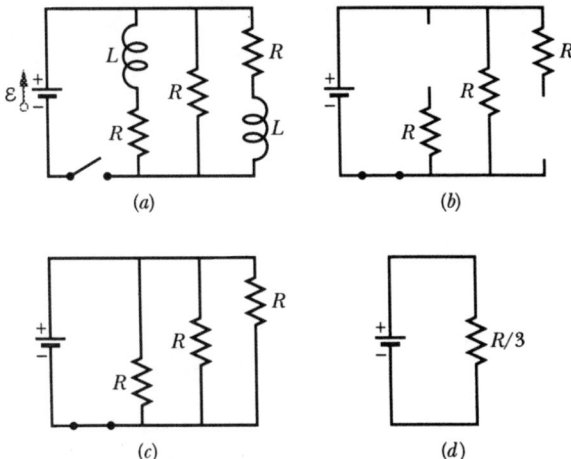

Fig. TE 31-4 (*a*) A multiloop *RL* circuit with an open switch. (*b*) The equivalent circuit just after the switch has been closed. (*c*) The equivalent circuit a long time later. (*d*) The single-loop circuit that is equivalent to circuit (*c*).

(a) What is the current i through the battery just after the switch is closed?

SOLUTION: The **Key Idea** here is that just after the switch is closed, the inductor acts to oppose a change in the current through it. Because the current through each inductor is zero before the switch is closed, it will also be zero just afterward. Thus, immediately after the switch is closed, the inductors act as broken wires, as indicated in Fig. TE31-4b. We then have a single-loop circuit for which the loop rule gives us

$$\varepsilon = iR = 0.$$

Substituting given data, we find that

$$i = \frac{\varepsilon}{R} = \frac{18 \text{ V}}{9.0 \, \Omega} = 2.0 \text{ A}.$$ (Answer)

(b) What is the current i through the battery long after the switch has been closed?

SOLUTION: The **Key Idea** here is that long after the switch has been closed, the currents in the circuit have reached their equilibrium values, and the inductors act as simple connecting wires, as indicated in Fig. TE31-4c. We then have a circuit with three identical resistors in parallel; from Eq. 29-11, their equivalent resistance is $R_{eq} = R/3 = (9.0 \, \Omega)/3 = 3.0 \, \Omega$. The equivalent circuit shown in Fig. TE31-4d then yields the loop equation $\varepsilon - iR_{eq} = 0$, or

$$i = \frac{\varepsilon}{R_{eq}} = \frac{18 \text{ V}}{3.0 \, \Omega} = 6.0 \text{ A}.$$ (Answer)

Touchstone Example 31-12-1

Figure TE31-5 shows two circular close-packed coils, the smaller (radius R_2, with N_2 turns) being coaxial with the larger (radius R_1, with N_1 turns) and in the same plane.

(a) Derive an expression for the magnitude of the mutual inductance M for this arrangement of these two coils, assuming that $R_1 \gg R_2$.

SOLUTION: The **Key Idea** here is that the mutual inductance M for these coils is the ratio of the flux linkage ($N\Phi$) through one coil to the current i in the other coil, which produces that flux linkage. Thus, we need to assume that currents exist in the coils; then we need to calculate the flux linkage in one of the coils.

The magnetic field through the larger coil due to the smaller coil is nonuniform in both magnitude and direction, so the flux through the larger coil due to the smaller coil is nonuniform and difficult to calculate. However, the smaller coil is small enough for us to assume that the

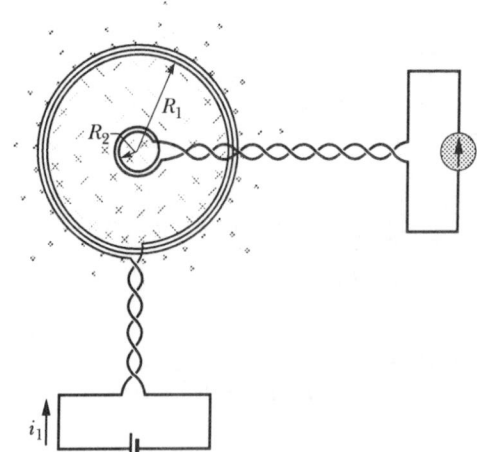

Fig. TE 31-5 A small coil is located at the center of a large coil. The mutual inductance of the coils can be determined by sending current i_1 through the large coil.

magnetic field through it due to the larger coil is approximately uniform. Thus, the flux through it due to the larger coil is also approximately uniform. Hence, to find M we shall assume a current i_1 in the larger coil and calculate the flux linkage $N_2\Phi_{21}$ in the smaller coil:

$$|M| = \left|\frac{N_2\Phi_{21}}{i_1}\right|. \qquad \text{(TE31-6)}$$

A second **Key Idea** is that the flux Φ_{21} through each turn of the smaller coil is, from Eq. 31-4,

$$\Phi_{21} = |\vec{B}_1|A_2,$$

where B_1 is the magnitude of the magnetic field at points within the small coil due to the larger coil, and $A_2 \; (= \pi R_2^2)$ is the area enclosed by the turn. Thus, the flux linkage in the smaller coil (with its N_2 turns) is

$$N_2\Phi_{21} = N_2|\vec{B}_1|A_2. \qquad \text{(TE31-7)}$$

A third **Key Idea** is that to find B_1 at points within the smaller coil, we can use Eq. 30-28, with z set to 0 because the smaller coil is in the plane of the larger coil. That equation tells us that each turn of the larger coil produces a magnetic field of magnitude $\mu_0 i_1 / 2R_1$ at points within the smaller coil. Thus, the larger coil (with its N_1 turns) produces a total magnetic field of magnitude

$$|\vec{B}_1| = N_1\frac{\mu_0|i_1|}{2R_1} \qquad \text{(TE31-8)}$$

at points within the smaller coil.

Substituting Eq. TE31-8 for B_1 and πR_2^2 for A_2 in Eq. TE31-7 yields

$$N_2\Phi_{21} = \frac{\pi\mu_0 N_1 N_2 R_2^2 |i_1|}{2R_1}.$$

Substituting this result into Eq. TE31-6, we find

$$|M| = \left|\frac{N_2\Phi_{21}}{i_1}\right| = \frac{\pi\mu_0 N_1 N_2 R_2^2}{2R_1}. \qquad \text{(Answer)} \quad \text{(TE31-9)}$$

(b) **What is the value of M for $N_1 = N_2 = 1200$ turns, $R_2 = 1.1$ cm, and $R_1 = 15$ cm?**

SOLUTION: Equation TE31-9 yields

$$|M| = \frac{(\pi)(4\pi \times 10^{-7}\ \text{H/m})(1200)(1200)(0.011\ \text{m})^2}{(2)(0.15\ \text{m})}$$

$$= 2.29 \times 10^{-3}\ \text{H} \approx 2.3\ \text{mH}. \qquad \text{(Answer)}$$

Consider the situation if we reverse the roles of the two coils—that is, if we produce a current i_2 in the smaller coil and try to calculate M from Eq. 31-50 in the form

$$|M| = \left|\frac{N_1\Phi_{12}}{i_2}\right|.$$

The calculation of Φ_{12} (the nonuniform flux of the smaller coil's magnetic field encompassed by the larger coil) is not simple. If we were to do the calculation numerically using a computer, we would find M to be 2.3 mH, as above! This emphasizes that Eq. 31-53 ($M_{21} = M_{12} = M$) is not obvious.

32 Magnetism of Matter: Maxwell's Equations

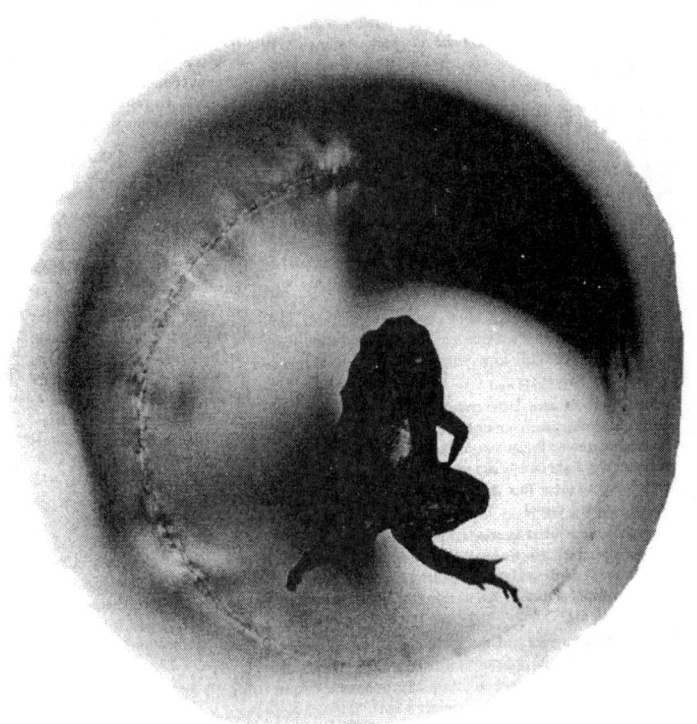

This is an overhead view of a frog that is being levitated in a magnetic field produced by current in a vertical solenoid below the frog. The solenoid's upward magnetic force on the frog balances the downward gravitational force on the frog. (The frog is not in discomfort; the sensation is like floating in water, which frogs like very much.) However, a frog is not magnetic (it would not, for example, stick to a refrigerator door).

How, then, can there be a magnetic force on the frog?

The answer is in this chapter.

32-1 Magnets

The first known magnets were *lodestones*, which are stones that have been magnetized (made magnetic) naturally. When the ancient Greeks and ancient Chinese discovered these rare stones, they were amused by the stones' ability to attract metal over a short distance, as if by magic. Only much later did they learn to use lodestones (and artificially magnetized pieces of iron) in compasses to determine direction.

Today, magnets and magnetic materials are ubiquitous. We find them in VCRs, audio cassettes, ATM and credit cards, audio headsets, and even in the inks for paper money. In fact, some breakfast cereals that are "iron fortified" contain small bits of magnetic materials (you can collect them from a slurry of cereal and water with a magnet). More important, the modern electronics industry as we know it (including the music and information sectors) would not exist without magnetic materials.

The magnetic properties of materials can be traced back to their atoms and electrons. We begin here, however, with the bar magnet in Fig. 32-1. As you have seen, iron filings sprinkled around such a magnet tend to align with the magnetic field of the magnet, and their pattern reveals the magnetic field lines. The clustering of the lines at the ends of the magnet suggests that one end is a *source* of the lines (the field diverges from it) and the other end is a *sink* of the lines (the field converges toward it). By convention, we call the source the *north pole* of the magnet and the opposite end the *south pole*, and we say that the magnet, with its two poles, is an example of a **magnetic dipole**.

Suppose we break apart a bar magnet the way we break a piece of chalk (Fig. 32-2). We should, it seems, be able to isolate a single pole, or *monopole*. Surprisingly, however, we cannot—not even if we break the magnet down to its individual atoms and then to its electrons and nuclei. Each fragment of the magnet has a north pole and a south pole. Thus, we must conclude the following:

▶The simplest magnetic structure that can exist is a magnetic dipole. Magnetic monopoles do not exist (as far as we know).

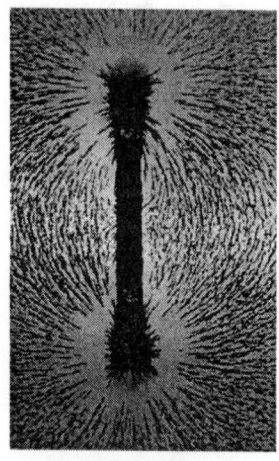

Fig. 32-1: A bar magnet is a magnetic dipole. The iron filings suggest the magnetic field lines. (The background is illuminated with colored light.)

32-2 Gauss' Law for Magnetic Fields

Gauss' law for magnetic fields is a formal way of saying that magnetic monopoles do not exist. The law asserts that the net magnetic flux Φ_B through any closed Gaussian surface is zero:

$$\Phi_B = \oint \vec{B} \cdot d\vec{A} = 0 \qquad \text{(Gauss' law for magnetic fields).} \qquad (32\text{-}1)$$

Contrast this with Gauss' law for electric fields,

$$\Phi_E = \oint \vec{E} \cdot d\vec{A} = \frac{q_{enc}}{\varepsilon_0} \qquad \text{(Gauss' law for electric fields).}$$

In both equations, the integral is taken over a *closed* Gaussian surface. Gauss' law for electric fields says that this integral (the net electric flux through the surface) is proportional to the net electric charge q_{enc} enclosed by the surface. Gauss' law for magnetic fields says that there can be no net magnetic flux through the surface because there can be no net "magnetic charge" (individual magnetic poles) enclosed by the surface. The simplest magnetic structure that can exist and thus be enclosed by a Gaussian surface is a dipole, which consists of both a source and a sink for the field lines. Thus, there must always be as much magnetic flux into the surface as out of it, and the net magnetic flux must always be zero.

Gauss' law for magnetic fields holds for more complicated structures than a magnetic dipole, and it holds even if the Gaussian surface does not enclose the entire

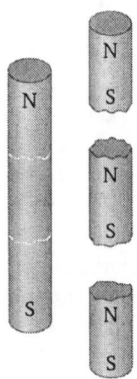

Fig. 32-2: If you break a magnet, each fragment becomes a separate magnet, with its own north and south poles.

structure. Gaussian surface II near the bar magnet of Fig. 32-3 encloses no poles, and we can easily conclude that the net magnetic flux through it is zero. Gaussian surface I is more difficult. It may seem to enclose only the north pole of the magnet because it encloses the label N and not the label S. However, a south pole must be associated with the lower boundary of the surface, because magnetic field lines enter the surface there. (The enclosed section is like one piece of the broken bar magnet in Fig. 32-2.) Thus, Gaussian surface I encloses a magnetic dipole and the net flux through the surface is zero.

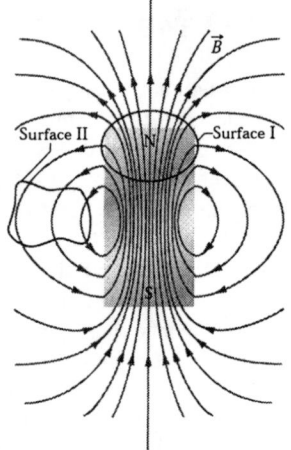

Fig. 32-3: The field lines for the magnetic field $\vec{B}$ of a short bar magnet. The red curves represent cross sections of closed, three-dimensional Gaussian surfaces.

READING EXERCISE 32-1: The figure shows four closed surfaces with flat top and bottom faces and curved sides. The table gives the areas A of the faces and the magnitudes $|\vec{B}|$ of the uniform and perpendicular magnetic fields through those faces; the units of A and $|\vec{B}|$ are arbitrary but consistent. Rank the surfaces according to the magnitudes of the magnetic flux through their curved sides, greatest first.

Surface	A_{top}	$|\vec{B}_{top}|$, direction	A_{bot}	$|\vec{B}_{bot}|$, direction
a	2	6, outward	4	3, inward
b	2	1, inward	4	2, inward
c	2	6, inward	2	8, outward
d	2	3, outward	3	2, outward

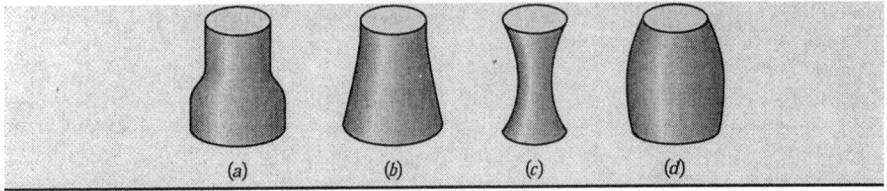

(a)   (b)   (c)   (d)

32-3 The Magnetism of Earth

Earth is a huge magnet; for points near Earth's surface, its magnetic field can be approximated as the field of a huge bar magnet—a magnetic dipole—that straddles the center of the planet. Figure 32-4 is an idealized symmetric depiction of the dipole field, without the distortion caused by passing charged particles from the Sun.

Because Earth's magnetic field is that of a magnetic dipole, a magnetic dipole moment $\vec{\mu}$ is associated with the field. For the idealized field of Fig. 32-4, the magnitude of $\vec{\mu}$ is 8.0×10^{22} J/T and the direction of $\vec{\mu}$ makes an angle of 11.5° with the rotation axis (RR) of Earth. The *dipole axis* (MM in Fig. 32-4) lies along $\vec{\mu}$ and intersects Earth's surface at the *geomagnetic north pole* in northwest Greenland and the *geomagnetic south pole* in Antarctica. The lines of the magnetic field $\vec{B}$ generally emerge in the southern hemisphere and reenter Earth in the northern hemisphere. Thus, the magnetic pole that is in Earth's northern hemisphere and known as a "north magnetic pole" *is really the south pole of Earth's magnetic dipole.*

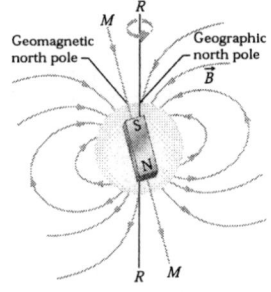

Fig. 32-4: Earth's magnetic field represented as a dipole field. The dipole axis MM makes an angle of 11.5° with Earth's rotational axis RR. The south pole of the dipole is in Earth's northern hemisphere.

The direction of the magnetic field at any location on Earth's surface is commonly specified in terms of two angles. The **field declination** is the angle (left or right) between geographic north (which is toward 90° latitude) and the horizontal component of the field. The **field inclination** is the angle (up or down) between a horizontal plane and the field's direction.

Magnetometers measure these angles and determine the field with much precision. However, you can do reasonably well with just a *compass* and a *dip meter*. A compass is simply a needle-shaped magnet that is mounted so it can rotate freely about a vertical axis. When it is held in a horizontal plane, the north-pole end of the needle points, generally, toward the geomagnetic north pole (really a south magnetic pole, remember). The angle between the needle and geographic north is the field declination. A dip meter is a similar magnet that can rotate freely about a horizontal axis. When its vertical plane of rotation is aligned with the direction of the compass, the angle between the meter's needle and the horizontal is the field inclination.

At any point on Earth's surface, the measured magnetic field may differ appreciably, in both magnitude and direction, from the idealized dipole field of Fig. 32-4. In fact, the point where the field is actually perpendicular to Earth's surface and inward is not located at the geomagnetic north pole in Greenland as we would expect; instead, this so-called *dip north pole* is located in the Queen Elizabeth Islands in northern Canada, far from Greenland.

In addition, the field observed at any location on the surface of Earth varies with time, by measurable amounts over a period of a few years and by substantial amounts over, say, 100 years. For example, between 1580 and 1820 the direction indicated by compass needles in London changed by 35°.

In spite of these local variations, the average dipole field changes only slowly over such relatively short time periods. Variations over longer periods can be studied by measuring the weak magnetism of the ocean floor on either side of the Mid-Atlantic Ridge (Fig. 32-5). This floor has been formed by molten magma that oozed up through the ridge from Earth's interior, solidified, and was pulled away from the ridge (by the drift of tectonic plates) at the rate of a few centimeters per year. As the magma solidified, it became weakly magnetized with its magnetic field in the direction of Earth's magnetic field at the time of solidification. Study of this solidified magma across the ocean floor reveals that Earth's field has reversed its *polarity* (directions of the north pole and south pole) about every million years. The reason for the reversals is not known. In fact, the mechanism that produces Earth's magnetic field is only vaguely understood.

32-4 Magnetism and Electrons

Magnetic materials, from lodestones to videotapes, are magnetic because of the electrons within them. We have already seen one way in which electrons can generate a magnetic field: Send them through a wire as an electric current, and their motion produces a magnetic field around the wire. There are two more ways, each involving a magnetic dipole moment that produces a magnetic field in the surrounding space. However, their explanation requires quantum physics that is beyond the physics presented in this book, so here we shall only outline the results.

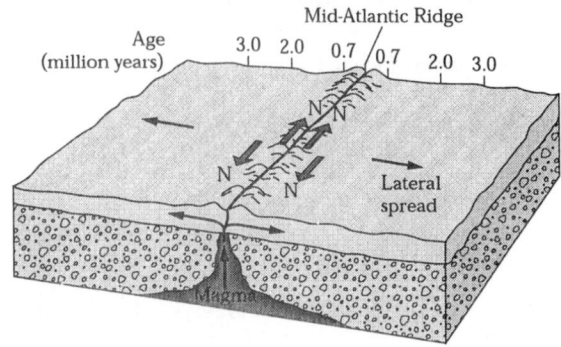

Fig. 32-5: A magnetic profile of the seafloor on either side of the Mid-Atlantic Ridge. The seafloor, extruded through the ridge and spreading out as part of the tectonic drift system, displays a record of the past magnetic history of Earth's core. The direction of the magnetic field produced by the core reverses about every million years.

Spin Magnetic Dipole Moment

An electron has an intrinsic angular momentum called its **spin angular momentum** (or just **spin**) $\vec{S}$; associated with this spin is an intrinsic **spin magnetic dipole moment** $\vec{\mu}_s$. (By *intrinsic*, we mean that $\vec{S}$ and $\vec{\mu}_s$ are basic characteristics of an electron, like its mass and electric charge.) $\vec{S}$ and $\vec{\mu}_s$ are related by

$$\vec{\mu}_s = -\frac{e}{m}\vec{S}, \qquad (32\text{-}2)$$

in which e is the elementary charge $(1.60\times10^{-19}\ \text{C})$ and m is the mass of an electron $(9.11\times10^{-31}\ \text{kg})$. The minus sign means that $\vec{\mu}_s$ and $\vec{S}$ are oppositely directed.

Spin $\vec{S}$ is quite different from the angular momenta of Chapter 12 in two respects:

1. $\vec{S}$ itself cannot be measured. However, its component along any axis can be measured.

2. A measured component of $\vec{S}$ is *quantized*, which is a general term that means it is restricted to certain values. A measured component of $\vec{S}$ can have only two values, which differ only in sign.

Let us assume that the component of spin $\vec{S}$ is measured along the z axis of a coordinate system. Then the measured component S_z can have only the two values given by

$$S_z = m_s \frac{h}{2\pi}, \qquad \text{for } m_s = \pm\tfrac{1}{2}, \qquad (32\text{-}3)$$

where m_s is called the *spin magnetic quantum number* and $h(= 6.63\times10^{-34} \text{ J}\cdot\text{s})$ is the Planck constant, the ubiquitous constant of quantum physics. The signs given in Eq. 32-3 have to do with the direction of S_z along the z axis. When S_z is parallel to the z axis, m_s is $+\tfrac{1}{2}$ and the electron is said to be spin up. When S_z is antiparallel to the z axis, m_s is $-\tfrac{1}{2}$ and the electron is said to be spin down.

The spin magnetic dipole moment $\vec{\mu}_s$ of an electron also cannot itself be measured; only a component can be measured, and that component too is quantized, with two possible values of the same magnitude but different signs. We can relate the component $\mu_{s,z}$ measured on the z axis to S_z by rewriting Eq. 32-2 in component form for the z axis as

$$\mu_{s,z} = -\frac{e}{m} S_z.$$

Substituting for S_z from Eq. 32-3 then gives us

$$\mu_{s,z} = \pm\frac{eh}{4\pi m}, \qquad (32\text{-}4)$$

where the plus and minus signs correspond to $\mu_{s,z}$ being parallel and antiparallel to the z axis, respectively.

The quantity on the right side of Eq. 32-4 is called the *Bohr magneton* μ_B:

$$\mu_{\text{B}} = \frac{eh}{4\pi m} = 9.27\times10^{-24} \text{ J/T} \qquad \text{(Bohr magneton)} \qquad (32\text{-}5)$$

Spin magnetic dipole moments of electrons and other elementary particles can be expressed in terms of μ_B. For an electron, the magnitude of the measured z component of $\vec{\mu}_s$ is

$$\mu_{s,z} = 1\,\mu_{\text{B}}. \qquad (32\text{-}6)$$

(The quantum physics of the electron, called *quantum electrodynamics*, or QED, reveals that $\mu_{s,z}$ is actually slightly greater than $1\,\mu_B$, but we shall neglect that fact.)

When an electron is placed in an external magnetic field $\vec{B}_{\text{ext}}$, a potential energy U can be associated with the orientation of the electron's spin magnetic dipole moment $\vec{\mu}_s$ just as a potential energy can be associated with the orientation of the magnetic dipole moment $\vec{\mu}_s$ of a current loop placed in $\vec{B}_{\text{ext}}$. From Eq. 29-31, the potential energy for the electron is

$$U = -\vec{\mu}_s \cdot \vec{B}_{\text{ext}} = -\mu_{s,z} B_{\text{ext}}, \qquad (32\text{-}7)$$

where the z axis is taken to be in the direction of $\vec{B}_{\text{ext}}$.

If we imagine an electron to be a microscopic sphere (which it is not), we can represent the spin $\vec{S}$, the spin magnetic dipole moment $\vec{\mu}_s$, and the associated magnetic dipole field as in Fig. 32-6. Although we use the word "spin" here, electrons do not spin

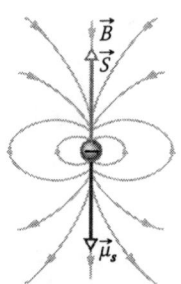

Fig. 32-6: The spin $\vec{S}$, spin magnetic dipole moment $\vec{\mu}_s$, and magnetic dipole field $\vec{B}$ of an electron represented as a microscopic sphere.

like tops. How, then, can something have angular momentum without actually rotating? Again, we would need quantum physics to provide the answer.

Protons and neutrons also have an intrinsic angular momentum called spin and an associated intrinsic spin magnetic dipole moment. For a proton those two vectors have the same direction, and for a neutron they have opposite directions. We shall not examine the contributions of these dipole moments to the magnetic fields of atoms because they are about a thousand times smaller than that due to an electron.

READING EXERCISE 32-2: The figure shows the spin orientations of two particles in an external magnetic field $\vec{B}_{ext}$. (a) If the particles are electrons, which spin orientation is at lower potential energy? (b) If, instead, the particles are protons, which spin orientation is at lower potential energy?

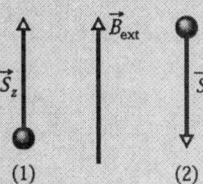

Orbital Magnetic Dipole Moment

When it is in an atom, an electron has an additional angular momentum called its **orbital angular momentum** $\vec{L}_{orb}$. Associated with $\vec{L}_{orb}$ is an **orbital magnetic dipole moment** $\vec{\mu}_{orb}$; the two are related by

$$\vec{\mu}_{orb} = -\frac{e}{2m}\vec{L}_{orb}. \tag{32-8}$$

The minus sign means that $\vec{\mu}_{orb}$ and $\vec{L}_{orb}$ have opposite directions.

Orbital angular momentum $\vec{L}_{orb}$ cannot be measured; only components along an axis can be measured, and those components are quantized. The components along, say, a z axis can have only the values given by

$$L_{orb,z} = m_l \frac{h}{2\pi}, \text{ for } m_l = 0, \pm1, \pm2, \dots, \pm(\text{limit}), \tag{32-9}$$

in which m_l is called the *orbital magnetic quantum number* and "limit" refers to some largest allowed integer value for m_l. The signs in Eq. 32-9 have to do with the direction of $L_{orb,z}$ along the z axis.

The orbital magnetic dipole moment $\vec{\mu}_{orb}$ of an electron also cannot itself be measured; only its component along an axis can be measured, and that component is again quantized. By writing Eq. 32-8 for a component along the same z axis as above and then substituting for $L_{orb,z}$ from Eq. 32-9, we can write the z component $\vec{\mu}_{orb,z}$ of the orbital magnetic dipole moment as

$$\mu_{orb,z} = -m_l \frac{eh}{4\pi m} \tag{32-10}$$

and, in terms of the Bohr magneton, as

$$\mu_{orb,z} = -m_l \mu_B. \tag{32-11}$$

When an atom is placed in an external magnetic field $\vec{B}_{ext}$, a potential energy U can be associated with the orientation of the orbital magnetic dipole moment of each electron in the atom. Its value is

$$U = -\vec{\mu}_{orb} \cdot \vec{B}_{ext} = -\mu_{orb,z} B_{ext}, \tag{32-12}$$

where the z axis is taken in the direction of $\vec{B}_{ext}$.

Although we have used the words "orbit" and "orbital" here, electrons do not orbit the nucleus of an atom like planets orbiting the Sun. How can an electron have an orbital angular momentum without orbiting in the common meaning of the term? Once again, this can be explained only with quantum physics.

Loop Model for Electron Orbits

We can obtain Eq. 32-8 with the nonquantum derivation that follows, in which we assume that an electron moves along a circular path with a radius that is much larger than an atomic radius (hence the name "loop model"). However, the derivation does not apply to an electron within an atom (for which we need quantum physics).

We imagine an electron moving at constant speed $|\vec{v}|$ in a circular path of radius r, counterclockwise as shown in Fig. 32-7. The motion of the negative charge of the electron is equivalent to a conventional current i (of positive charge) that is clockwise, as also shown in Fig. 32-7. The magnitude of the orbital magnetic dipole moment of such a *current loop* is obtained from Eq. 29-28 with $N=1$:

$$\mu_{orb} = iA, \tag{32-13}$$

where A is the area enclosed by the loop. The direction of this magnetic dipole moment is, from the right-hand rule of Fig. 30-18, downward in Fig. 32-7.

To evaluate Eq. 32-13, we need the current i. Current is, generally, the rate at which charge passes some point in a circuit. Here, the charge of magnitude e takes a time $T = 2\pi r/v$ to circle from any point back through that point, so

$$i = \frac{\text{charge}}{\text{time}} = \frac{e}{2\pi r/v}. \tag{32-14}$$

Substituting this and the area $A = \pi r^2$ of the loop into Eq. 32-13 gives us

$$\mu_{orb} = \frac{e}{2\pi r/v}\pi r^2 = \frac{evr}{2}. \tag{32-15}$$

To find the orbital angular momentum $\vec{L}_{orb}$ of the electron, we use Eq. 12-21, $\vec{\ell} = m(\vec{r}\times\vec{v})$. Because $\vec{r}$ and $\vec{v}$ are perpendicular, $\vec{L}_{orb}$ has the magnitude

$$\vec{L}_{orb} = mr|\vec{v}|\sin 90° = mr|\vec{v}|. \tag{32-16}$$

$\vec{L}_{orb}$ is directed upward in Fig. 32-7 (see Fig. 12-8). Combining Eqs. 32-15 and 32-16, generalizing to a vector formulation, and indicating the opposite directions of the vectors with a minus sign yield

$$\vec{\mu}_{orb} = -\frac{e}{2m}\vec{L}_{orb},$$

which is Eq. 32-8. Thus, by "classical" (nonquantum) analysis we have obtained the same result, in both magnitude and direction, given by quantum physics. You might wonder, since this derivation gives the correct result for an electron within an atom, why the derivation is invalid for that situation. The answer is that this line of reasoning yields other results that are contradicted by experiments.

Loop Model in a Nonuniform Field

We continue to consider an electron orbit as a current loop, as we did in Fig. 32-7. Now, however, we draw the loop in a nonuniform magnetic field $\vec{B}_{ext}$ as shown in Fig. 32-8a. (This field could be the diverging field near the north pole of the magnet in Fig. 32-3.) We make this change to prepare for the next several sections, in which we shall discuss

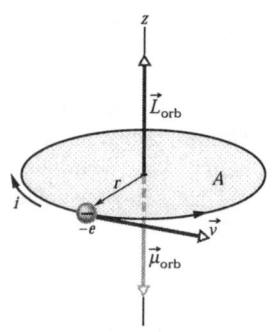

Fig. 32-7: An electron moving at constant speed $|\vec{v}|$ in a circular path of radius r that encloses an area A. The electron has an orbital angular momentum $\vec{L}_{orb}$ and an associated orbital magnetic dipole moment $\vec{\mu}_{orb}$. A clockwise current i (of positive charge) is equivalent to the counterclockwise circulation of the negatively charged electron.

the forces that act on magnetic materials when the materials are placed in a nonuniform magnetic field. We shall discuss these forces by assuming that the electron orbits in the materials are tiny current loops like that in Fig. 32-8a.

Here we assume that the magnetic field vectors all around the electron's circular path have the same magnitude and form the same angle with the vertical, as shown in Figs. 32-8b and d. We also assume that all the electrons in an atom move either counterclockwise (Fig. 32-8b) or clockwise (Fig. 32-8d). The associated conventional current i around the current loop and the orbital magnetic dipole moment $\vec{\mu}_{orb}$ produced by i are shown for each of these directions of motion.

Figures 32-8c and e show diametrically opposite views of an element $d\vec{L}$ of the loop with the same direction as i, as seen from the plane of the orbit. Also shown are the field $\vec{B}_{ext}$ and the resulting magnetic force $d\vec{F}$ on $d\vec{L}$. Recall that a current along an element $d\vec{L}$ in a magnetic field $\vec{B}_{ext}$ experiences a magnetic force $d\vec{F}$ as given by Eq. 29-24:

$$d\vec{F} = i\, d\vec{L} \times \vec{B}_{ext}.$$
(32-17)

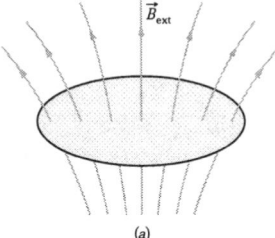
(a)

On the left side of Fig. 32-8c, Eq. 32-17 tells us that the force $d\vec{F}$ is directed upward and rightward. On the right side, the force $d\vec{F}$ is just as large and is directed upward and leftward. Because their angles are the same, the horizontal components of these two forces cancel and the vertical components add. The same is true at any other two symmetric points on the loop. Thus, the net force on the current loop of Fig. 32-8b must be upward. The same reasoning leads to a downward net force on the loop in Fig. 32-8d. We shall use these two results shortly when we examine the behavior of magnetic materials in nonuniform magnetic fields.

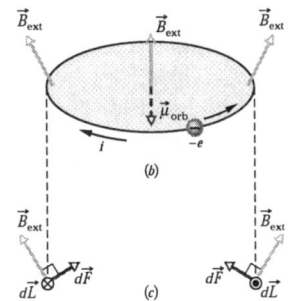
(b)

(c)

32-5 Magnetic Materials

Each electron in an atom has an orbital magnetic dipole moment and a spin magnetic dipole moment that combine vectorially. The resultant of these two vector quantities combines vectorially with similar resultants for all other electrons in the atom, and the resultant for each atom combines with those for all the other atoms in a sample of a material. If the combination of all these magnetic dipole moments produces a magnetic field, then the material is magnetic. There are three general types of magnetism: diamagnetism, paramagnetism, and ferromagnetism.

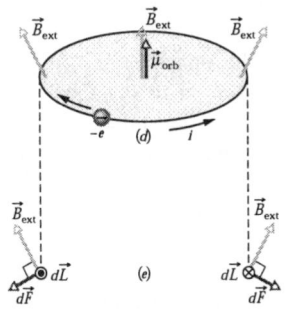
(d)

(e)

1. *Diamagnetism* is exhibited by all common materials but is so feeble that it is masked if the material also exhibits magnetism of either of the other two types. In diamagnetism, weak magnetic dipole moments are produced in the atoms of the material when the material is placed in an external magnetic field $\vec{B}_{ext}$; the combination of all those induced dipole moments gives the material as a whole only a feeble net magnetic field. The dipole moments and thus their net field disappear when $\vec{B}_{ext}$ is removed. The term *diamagnetic material* usually refers to materials that exhibit only diamagnetism.

2. *Paramagnetism* is exhibited by materials containing transition elements, rare earth elements, and actinide elements (see Appendix G). Each atom of such a material has a permanent resultant magnetic dipole moment, but the moments are randomly oriented in the material and the material as a whole lacks a net magnetic field. However, an external magnetic field $\vec{B}_{ext}$ can partially align the atomic magnetic dipole moments to give the material a net magnetic field. The alignment and thus its field disappear when $\vec{B}_{ext}$ is removed. The term *paramagnetic material* usually refers to materials that exhibit primarily paramagnetism.

3. *Ferromagnetism* is a property of iron, nickel, and certain other elements (and of compounds and alloys of these elements). Some of the electrons in these materials have their resultant magnetic dipole moments aligned, which produces regions with strong magnetic dipole moments. An external field $\vec{B}_{ext}$ can then align the magnetic moments of such regions, producing a strong magnetic field for a sample of the material; the field partially persists when $\vec{B}_{ext}$ is removed. We usually use the term

Fig. 32-8: (a) A loop model for an electron orbiting in an atom while in a nonuniform magnetic field B_{ext}. (b) Charge $-e$ moves counterclockwise; the associated conventional current i is clockwise. (c) The magnetic forces $d\vec{F}$ on the left and right sides of the loop, as seen from the plane of the loop. The net force on the loop is upward. (d) Charge $-e$ now moves clockwise. (e) The net force on the loop is now downward.

ferromagnetic material, and even the common term magnetic material, to refer to materials that exhibit primarily ferromagnetism.

The next three sections explore these three types of magnetism.

32-6 Diamagnetism

We cannot yet discuss the quantum physical explanation of diamagnetism, but we can provide a classical explanation with the loop model of Figs. 32-7 and 32-8. To begin, we assume that in an atom of a diamagnetic material each electron can orbit only clockwise as in Fig. 32-8d or counterclockwise as in Fig. 32-8b. To account for the lack of magnetism in the absence of an external magnetic field $\vec{B}_{ext}$, we assume the atom lacks a net magnetic dipole moment. This implies that before $\vec{B}_{ext}$ is applied, as many electrons orbit one way as orbit the other, with the result that the net upward magnetic dipole moment of the atom equals the net downward magnetic dipole moment.

Now let's turn on the nonuniform field $\vec{B}_{ext}$ of Fig. 32-8a, in which $\vec{B}_{ext}$ is directed upward but is diverging (the magnetic field lines are diverging). We could do this by increasing the current through an electromagnet or by moving the north pole of a bar magnet closer to, and below, the orbits. As the magnitude of $\vec{B}_{ext}$ increases from zero to its final maximum, steady-state value, a clockwise electric field is induced around each electron's orbital loop according to Faraday's law and Lenz's law. Let us see how this induced electric field affects the orbiting electrons in Figs. 32-8b and d.

In Fig. 32-8b, the counterclockwise electron is accelerated by the clockwise electric field. Thus, as the magnetic field $\vec{B}_{ext}$ increases to its maximum value, the electron speed increases to a maximum value. This means that the associated conventional current i and the downward magnetic dipole moment $\vec{\mu}$ due to i also increase.

In Fig. 32-8d, the clockwise electron is decelerated by the clockwise electric field. Thus, here, the electron speed, the associated current i, and the upward magnetic dipole moment $\vec{\mu}$ due to i all decrease. By turning on field $\vec{B}_{ext}$, we have given the atom a net magnetic dipole moment that is upward. This would also be so if the magnetic field were uniform.

The nonuniformity of field $\vec{B}_{ext}$ also affects the atom. Because the current i in Fig. 32-8b increases, the upward magnetic forces $d\vec{F}$ in Fig. 32-8c also increase, as does the net upward force on the current loop. Because current i in Fig. 32-8d decreases, the downward magnetic forces $d\vec{F}$ in Fig. 32-8e also decrease, as does the net downward force on the current loop. Thus, by turning on the *nonuniform* field $\vec{B}_{ext}$ we have produced a net force on the atom; moreover, that force is directed *away* from the region of greater magnetic field.

We have argued with fictitious electron orbits (current loops), but we have ended up with exactly what happens to a diamagnetic material: If we apply the magnetic field of Fig. 32-8, the material develops a downward magnetic dipole moment and experiences an upward force. When the field is removed, both the dipole moment and the force disappear. The external field need not be positioned as shown; similar arguments can be made for other orientations of $\vec{B}_{ext}$. In general,

▶A diamagnetic material placed in an external magnetic field $\vec{B}_{ext}$ develops a magnetic dipole moment directed opposite $\vec{B}_{ext}$. If the field is nonuniform, the diamagnetic material is repelled from a region of greater magnetic field toward a region of lesser field.

The frog in the photograph opening this chapter is diamagnetic (as is any other animal). When the frog was placed in the diverging magnetic field near the top end of a vertical current-carrying solenoid, every atom in the frog was repelled upward, away from the region of stronger magnetic field at that end of the solenoid. The frog moved upward into weaker and weaker magnetic field until the upward magnetic force balanced the gravitational force on it, and there it hung in midair. If we built a solenoid that was large enough, we could similarly levitate a person in midair owing to the person's diamagnetism.

READING EXERCISE 32-3: The figure shows two diamagnetic spheres located near the south pole of a bar magnet. Are (a) the magnetic forces on the spheres and (b) the magnetic dipole moments of the spheres directed toward or away from the bar magnet? (c) Is the magnetic force on sphere 1 greater than, less than, or equal to that on sphere 2?

● ● [S ▬▬▬▬ N]
1   2

32-7 Paramagnetism

In paramagnetic materials, the spin and orbital magnetic dipole moments of the electrons in each atom do not cancel but add vectorially to give the atom a net (and permanent) magnetic dipole moment $\vec{\mu}$. In the absence of an external magnetic field, these atomic dipole moments are randomly oriented, and the net magnetic dipole moment of the material is zero. However, if a sample of the material is placed in an external magnetic field $\vec{B}_{ext}$, the magnetic dipole moments tend to line up with the field, which gives the sample a net magnetic dipole moment. This alignment with the external field is the opposite of what we saw with diamagnetic materials.

Liquid oxygen is suspended between the two pole faces of a magnet because the liquid is paramagnetic and is magnetically attracted to the magnet.

➤A paramagnetic material placed in an external magnetic field $\vec{B}_{ext}$ develops a magnetic dipole moment in the direction of $\vec{B}_{ext}$. If the field is nonuniform, the paramagnetic material is attracted toward a region of greater magnetic field from a region of lesser field.

A paramagnetic sample with N atoms would have a magnetic dipole moment with a magnitude of $N\mu$ if the alignment of its atomic dipoles were complete. However, random collisions of atoms due to thermal agitation transfer energy among them, disrupting their alignment and thus reducing the sample's magnetic dipole moment.

The importance of thermal agitation may be measured by comparing two energies. One, from Eq. 20-24, is the mean translational kinetic energy $K\left(=\frac{3}{2}kT\right)$ of an atom at temperature T, where k is the Boltzmann constant $\left(1.38\times10^{-23} \text{ J/K}\right)$ and T is in kelvins (not Celsius degrees). The other, from Eq. 29-31, is the difference in energy $\Delta U_B\left(=2\mu B_{ext}\right)$ between parallel alignment and antiparallel alignment of the magnetic dipole moment of an atom and the external field. As we shall show below, $K >> \Delta U_B$, even for ordinary temperatures and field magnitudes. Thus, energy transfers during collisions among atoms can significantly disrupt the alignment of the atomic dipole moments, keeping the magnetic dipole moment of a sample much less than $N\mu$.

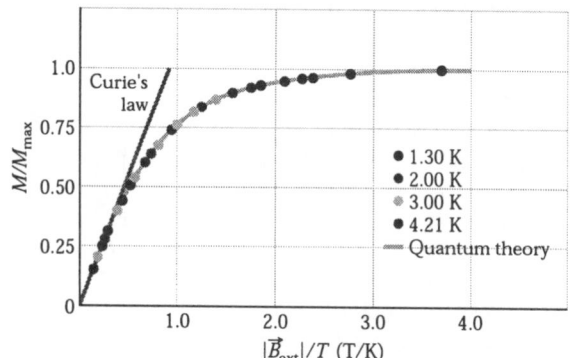

Fig. 32-9: A *magnetization curve* for potassium chromium sulfate, a paramagnetic salt. The ratio of magnetization M of the salt to the maximum possible magnetization M_{max} is plotted versus the ratio of the magnitude of the applied magnetic field $\left|\vec{B}_{ext}\right|$ to the temperature T. Curie's law fits the data at the left; quantum theory fits all the data. After W. E. Henry.

We can express the extent to which a given paramagnetic sample is magnetized by finding the ratio of its magnetic dipole moment to its volume V. This vector quantity, the magnetic dipole moment per unit volume, is called the **magnetization** $\vec{M}$ of the sample, and its magnitude is

$$\left|\vec{M}\right| = \frac{\text{measured magnetic moment}}{V}. \qquad (32\text{-}18)$$

The unit of $\vec{M}$ is the ampere-square meter per cubic meter, or ampere per meter (A/m). Complete alignment of the atomic dipole moments, called *saturation* of the sample, corresponds to the maximum value $M_{max} = N\mu/V$.

In 1895 Pierre Curie discovered experimentally that the magnitude of the magnetization of a paramagnetic sample is directly proportional to the magnitude of the external magnetic field $\vec{B}_{ext}$ and inversely proportional to the temperature T in kelvins; that is,

$$|\vec{M}| = C\frac{|\vec{B}_{ext}|}{T}. \qquad (32\text{-}19)$$

Equation 32-19 is known as *Curie's law*, and C is called the *Curie constant*. Curie's law is reasonable in that increasing $|\vec{B}_{ext}|$ tends to align the atomic dipole moments in a sample and thus to increase $|\vec{M}|$, whereas increasing T tends to disrupt the alignment via thermal agitation and thus to decrease $|\vec{M}|$. However, the law is actually an approximation that is valid only when the ratio $|\vec{B}_{ext}|/T$ is not too large.

Figure 32-9 shows the ratio M/M_{max} as a function of $|\vec{B}_{ext}|/T$ for a sample of the salt potassium chromium sulfate, in which chromium ions are the paramagnetic substance. The plot is called a *magnetization curve*. The straight line for Curie's law fits the experimental data at the left, for $|\vec{B}_{ext}|/T$ below about $0.5 \, T/K$. The curve that fits all the data points is based on quantum physics. The data on the right side, near saturation, are very difficult to obtain because they require very strong magnetic fields (about 100 000 times Earth's field), even at the very low temperatures noted in Fig. 32-9.

READING EXERCISE 32-4: The figure here shows two paramagnetic spheres located near the south pole of a bar magnet. Are (a) the magnetic forces on the spheres and (b) the magnetic dipole moments of the spheres directed toward or away from the bar magnet? (c) Is the magnetic force on sphere 1 greater than, less than, or equal to that on sphere 2?

32-8 Ferromagnetism

When we speak of magnetism in everyday conversation, we almost always have a mental picture of a bar magnet or a disk magnet (probably clinging to a refrigerator door). That is, we picture a ferromagnetic material having strong, permanent magnetism, and not a diamagnetic or paramagnetic material having weak, temporary magnetism.

Iron, cobalt, nickel, gadolinium, dysprosium, and alloys of these and other elements exhibit ferromagnetism because of a quantum physical effect called *exchange coupling*. In this process the spins of the electrons in one atom interact with those of neighboring atoms. The result is an alignment of the magnetic dipole moments of the atoms, in spite of the randomizing tendency of atomic collisions. This persistent alignment is what gives ferromagnetic materials their permanent magnetism.

If the temperature of a ferromagnetic material is raised above a certain critical value, called the *Curie temperature*, the exchange coupling ceases to be effective. Most such materials then become simply paramagnetic; that is, the dipoles still tend to align with an external field but much more weakly, and thermal agitation can now more easily disrupt the alignment. The Curie temperature for iron is 1043 K (= 770°C).

The magnetization of a ferromagnetic material such as iron can be studied with an arrangement called a *Rowland ring* (Fig. 32-10). The material is formed into a thin toroidal core of circular cross section. A primary coil P having n turns per unit length is wrapped around the core and carries current i_P. (The coil is essentially a long solenoid bent into a circle.) If the iron core were not present, the magnitude of the magnetic field inside the coil would be, from Eq. 30-25,

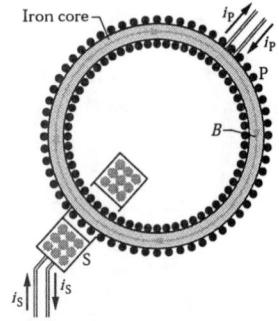

Fig. 32-10: A Rowland ring. Current i_P is sent through a primary coil P whose core is the ferromagnetic material to be studied (here iron) and that is magnetized by the current. (The turns of the coil are represented by dots.) The extent of magnetization of the core determines the total magnetic field $\vec{B}$ within coil P. Field $\vec{B}$ can be measured by means of a secondary coil S.

$$\left|\vec{B}_0\right| = \mu_0 |i_P| n. \qquad (32\text{-}20)$$

However, with the iron core present, the magnetic field $\vec{B}$ inside the coil is greater than $\vec{B}_0$, usually by a large amount. We can write the magnitude of this field as

$$\left|\vec{B}\right| = \left|\vec{B}_0 + \vec{B}_M\right| = \left|\vec{B}_0\right| - \left|\vec{B}_M\right| \qquad (32\text{-}21)$$

where $\left|\vec{B}_M\right|$ is the magnitude of the magnetic field contributed by the iron core. This contribution results from the alignment of the atomic dipole moments within the iron, due to exchange coupling and to the applied magnetic field $\vec{B}_0$, and is proportional to the magnetization $\vec{M}$ of the iron. That is, the contribution $\left|\vec{B}_M\right|$ is proportional to the magnetic dipole moment per unit volume of the iron. To determine $\left|\vec{B}_M\right|$ we use a secondary coil S to measure $\vec{B}$, compute $\vec{B}_0$ with Eq. 32-20, and subtract as suggested by Eq. 32-21.

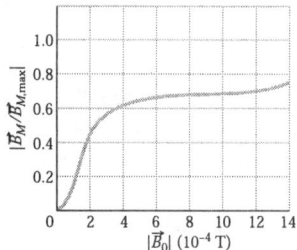

Figure 32-11 shows a magnetization curve for a ferromagnetic material in a Rowland ring: the ratio $\left|\vec{B}_M / \vec{B}_{M,\max}\right|$, where $\left|\vec{B}_{M,\max}\right|$ is the maximum possible value of $\vec{B}_M$, corresponding to saturation, is plotted versus $\left|\vec{B}_0\right|$. The curve is similar to Fig. 32-9, the magnetization curve for a paramagnetic substance, in that both curves are measures of the extent to which an applied magnetic field can align the atomic dipole moments of a material.

For the ferromagnetic core yielding Fig. 32-11, the alignment of the dipole moments is about 70% complete for $\left|\vec{B}_0\right| \approx 1 \times 10^{-3}$ T. If $\left|\vec{B}_0\right|$ were increased to 1 T, the alignment would be almost complete (but $\left|\vec{B}_0\right| = 1$ T, and thus almost complete saturation, is quite difficult to obtain).

Fig. 32-11: A magnetization curve for a ferromagnetic core material in the Rowland ring of Fig. 32-10. On the vertical axis, 1.0 corresponds to complete alignment (saturation) of the atomic dipoles within the material.

Magnetic Domains

Exchange coupling produces strong alignment of adjacent atomic dipoles in a ferromagnetic material at a temperature below the Curie temperature. Why, then, isn't the material naturally at saturation even when there is no applied magnetic field $\vec{B}_0$? That is, why isn't every piece of iron, such as an iron nail, a naturally strong magnet?

To understand this, consider a specimen of a ferromagnetic material such as iron that is in the form of a single crystal; that is, the arrangement of the atoms that make it up—its crystal lattice—extends with unbroken regularity throughout the volume of the specimen. Such a crystal will, in its normal state, be made up of a number of *magnetic domains*. These are regions of the crystal throughout which the alignment of the atomic dipoles is essentially perfect. The domains, however, are not all aligned. For the crystal as a whole, the domains are so oriented that they largely cancel each other as far as their external magnetic effects are concerned.

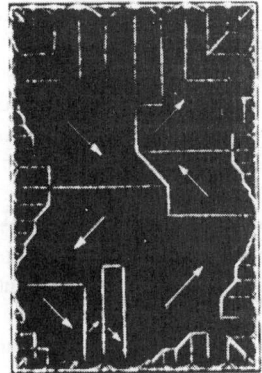

Figure 32-12 is a magnified photograph of such an assembly of domains in a single crystal of nickel. It was made by sprinkling a colloidal suspension of finely powdered iron oxide on the surface of the crystal. The domain boundaries, which are thin regions in which the alignment of the elementary dipoles changes from a certain orientation in one domain to a different orientation in the other, are the sites of intense, but highly localized and nonuniform, magnetic fields. The suspended colloidal particles are attracted to these boundaries and show up as the white lines (not all the domain boundaries are apparent in Fig. 32-12). Although the atomic dipoles in each domain are completely aligned as shown by the arrows, the crystal as a whole may have a very small resultant magnetic moment.

Actually, a piece of iron as we ordinarily find it is not a single crystal but an assembly of many tiny crystals, randomly arranged; we call it a polycrystalline solid. Each tiny crystal, however, has its array of variously oriented domains, just as in Fig. 32-12. If we magnetize such a specimen by placing it in an external magnetic field of gradually increasing strength, we produce two effects; together they produce a magnetization curve of the shape shown in Fig. 32-11. One effect is a growth in size of the domains that are oriented along the external field at the expense of those that are not.

Fig. 32-12: A photograph of domain patterns within a single crystal of nickel; white lines reveal the boundaries of the domains. The white arrows superimposed on the photograph show the orientations of the magnetic dipoles within the domains and thus the orientations of the net magnetic dipoles of the domains. The crystal as a whole is unmagnetized if the net magnetic field (the vector sum over all the domains) is zero.

The second effect is a shift of the orientation of the dipoles within a domain, as a unit, to become closer to the field direction.

Exchange coupling and domain shifting give us the following result:

> ▶A ferromagnetic material placed in an external magnetic field $\vec{B}_{ext}$ develops a strong magnetic dipole moment in the direction of $\vec{B}_{ext}$. If the field is nonuniform, the ferromagnetic material is attracted toward a region of greater magnetic field from a region of lesser field.

You can actually hear sound produced by shifting domains: Put an audio cassette player into its play mode without a cassette in place (or with a blank cassette) and turn the volume control to maximum. Then bring a strong magnet up to the play head (which is ferromagnetic). The magnetic field causes the domains in the play head to shift abruptly, which abruptly alters the magnetic field through a coil wrapped around the play head. The resulting suddenly induced currents in the coil are amplified and fed to the speaker, producing a fizzing sound.

Hysteresis

Magnetization curves for ferromagnetic materials are not retraced as we increase and then decrease the external magnetic field $\vec{B}_0$. Figure 32-13 is a plot of $|\vec{B}_M|$ versus $|\vec{B}_0|$ during the following operations with a Rowland ring: (1) Starting with the iron unmagnetized (point a), increase the current in the toroid until $|\vec{B}_0|(= \mu_0 in)$ has the value corresponding to point b; (2) reduce the current in the toroid winding (and thus $|\vec{B}_0|$) back to zero (point c); (3) reverse the toroid current and increase it in magnitude until $|\vec{B}_0|$ has the value corresponding to point d; (4) reduce the current to zero again (point e); (5) reverse the current once more until point b is reached again.

The lack of retraceability shown in Fig. 32-13 is called **hysteresis**, and the curve $bcdeb$ is called a *hysteresis loop*. Note that at points c and e the iron core is magnetized, even though there is no current in the toroid windings; this is the familiar phenomenon of permanent magnetism.

Hysteresis can be understood through the concept of magnetic domains. Evidently the motions of the domain boundaries and the reorientations of the domain directions are not totally reversible. When the applied magnetic field $\vec{B}_0$ is increased and then decreased back to its initial value, the domains do not return completely to their original configuration but retain some "memory" of their alignment after the initial increase. This memory of magnetic materials is essential for the magnetic storage of information, as on cassette tapes and computer disks.

This memory of the alignment of domains can also occur naturally. When lightning sends currents along multiple tortuous paths through the ground, the currents produce intense magnetic fields that can suddenly magnetize any ferromagnetic material in nearby rock. Because of hysteresis, such rock material retains some of that magnetization after the lightning strike (after the currents disappear). Pieces of the rock, later exposed, broken, and loosened by weathering, are then lodestones.

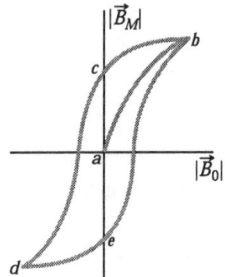

Fig. 32-13: A magnetization curve (ab) for a ferromagnetic specimen and an associated hysteresis loop ($bcdeb$).

Touchstone Example 32-8-1, at the end of this chapter, illustrates how to use what you learned in this section.

TE

32-9 Induced Magnetic Fields

In Chapter 31 you saw that a changing magnetic flux induces an electric field, and we ended up with Faraday's law of induction in the form

$$\oint \vec{E} \cdot d\vec{s} = -\frac{d\Phi_B}{dt} \qquad \text{(Faraday's law of induction).} \quad (32\text{-}22)$$

Here $\vec{E}$ is the electric field induced along a closed loop by the changing magnetic flux Φ_B encircled by that loop. Because symmetry is often so powerful in physics, we should be tempted to ask whether induction can occur in the opposite sense; that is, can a changing electric flux induce a magnetic field?

The answer is that it can; furthermore, the equation governing the induction of a magnetic field is almost symmetric with Eq. 32-22. We often call it Maxwell's law of induction after James Clerk Maxwell, and we write it as

$$\oint \vec{B} \cdot d\vec{s} = \mu_0 \varepsilon_0 \frac{d\Phi_E}{dt} \qquad \text{(Maxwell's law of induction)}. \qquad (32\text{-}23)$$

Here $\vec{B}$ is the magnetic field induced along a closed loop by the changing electric flux Φ_E in the region encircled by that loop.

As an example of this sort of induction, we consider the charging of a parallel-plate capacitor with circular plates, as shown in Fig. 32-14a. (Although we shall focus on this particular arrangement, a changing electric flux will always induce a magnetic field whenever it occurs.) We assume that the charge on the capacitor is being increased at a steady rate by a constant current i in the connecting wires. Then the magnitude of the electric field between the plates must also be increasing at a steady rate.

Figure 32-14b is a view of the right-hand plate of Fig. 32-14a from between the plates. The electric field is directed into the page. Let us consider a circular loop through point 1 in Figs. 32-14a and b, concentric with the capacitor plates and with a radius smaller than that of the plates. Because the electric field through the loop is changing, the electric flux through the loop must also be changing. According to Eq. 32-23, this changing electric flux induces a magnetic field around the loop.

Experiment proves that a magnetic field $\vec{B}$ *is* indeed induced around such a loop, directed as shown. This magnetic field has the same magnitude at every point around the loop and thus has circular symmetry about the central axis of the capacitor plates.

If we now consider a larger loop, say through point 2 outside the plates in Figs. 32-14a and b, we find that a magnetic field is induced around that loop as well. Thus, while the electric field is changing, magnetic fields are induced between the plates, both inside and outside the gap. When the electric field stops changing, these induced magnetic fields disappear.

Although Eq. 32-23 is similar to Eq. 32-22, the equations differ in two ways. First, Eq. 32-23 has the two extra symbols, μ_0 and ε_0, but they appear only because we employ SI units. Second, Eq. 32-23 lacks the minus sign of Eq. 32-22. That difference in sign means that the induced electric field $\vec{E}$ and the induced magnetic field $\vec{B}$ have opposite directions when they are produced in otherwise similar situations.

To see this opposition of directions, examine Fig. 32-15, in which an increasing magnetic field $\vec{B}$, directed into the page, induces an electric field $\vec{E}$. The induced field $\vec{E}$ is counterclockwise, whereas the induced magnetic field $\vec{B}$ in Fig. 32-14b is clockwise.

Ampere-Maxwell Law

Now recall that the left side of Eq. 32-23, the integral of the dot product $\vec{B} \cdot d\vec{s}$ around a closed loop, appears in another equation—namely, Ampere's law:

$$\oint \vec{B} \cdot d\vec{s} = \mu_0 i_{\text{enc}} \qquad \text{(Ampere's law)}, \qquad (32\text{-}24)$$

where i_{enc} is the current encircled by the closed loop. Thus, our two equations that specify the magnetic field $\vec{B}$ produced by means other than a magnetic material (that is, by a current and by a changing electric field) give the field in exactly the same form. We can combine the two equations into the single equation

$$\oint \vec{B} \cdot d\vec{s} = \mu_0 \varepsilon_0 \frac{d\Phi_E}{dt} + \mu_0 i_{\text{enc}} \qquad \text{(Ampere-Maxwell law)}. \qquad (32\text{-}25)$$

When there is a current but no change in electric flux (such as with a wire carrying a constant current), the first term on the right side of Eq. 32-25 is zero, and Eq. 32-25

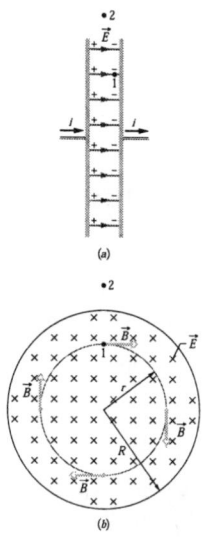

Fig. 32-14: (a) A circular parallel-plate capacitor, shown in side view, is being charged by a constant current i. (b) A view from within the capacitor, toward the plate at the right. The electric field $\vec{E}$ is uniform, is directed into the page (toward the plate), and grows in magnitude as the charge on the capacitor increases. The magnetic field $\vec{B}$ induced by this changing electric field is shown at four points on a circle with a radius r less than the plate radius R.

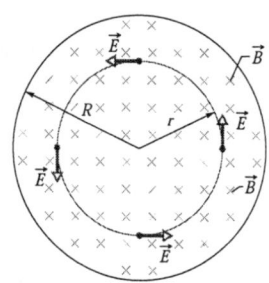

Fig. 32-15: A uniform magnetic field $\vec{B}$ in a circular region. The field, directed into the page, is increasing in magnitude. The electric field $\vec{E}$ induced by the changing magnetic field is shown at four points on a circle concentric with the circular region. Compare this situation with that of Fig. 32-14b.

reduces to Eq. 32-24, Ampere's law. When there is a change in electric flux but no current (such as inside or outside the gap of a charging capacitor), the second term on the right side of Eq. 32-25 is zero, and Eq. 32-25 reduces to Eq. 32-23, Maxwell's law of induction.

Touchstone Example 32-9-1, at the end of this chapter, illustrates how to use what you learned in this section.

TE

32-10 Displacement Current

If you compare the two terms on the right side of Eq. 32-25, you will see that the product $\varepsilon_0(d\Phi_E/dt)$ in the first term must have the dimension of a current. In fact, historically, that product has been treated as being a fictitious current called the **displacement current** i_d:

$$i_d = \varepsilon_0 \frac{d\Phi_E}{dt} \qquad \text{(displacement current).} \qquad (32\text{-}26)$$

"Displacement" is poorly chosen in that nothing is being displaced, but we are stuck with the word. Nevertheless, we can now rewrite Eq. 32-25 as

$$\oint \vec{B} \cdot d\vec{s} = \mu_0 i_{d,\text{enc}} + \mu_0 i_{\text{enc}} \qquad \text{(Ampere-Maxwell law),} \qquad (32\text{-}27)$$

in which $i_{d,\text{enc}}$ is the displacement current that is encircled by the integration loop.

Let us again focus on a charging capacitor with circular plates, as in Fig. 32-16a. The real current i that is charging the plates changes the electric field $\vec{E}$ between the plates. The fictitious displacement current i_d between the plates is associated with that changing field $\vec{E}$. Let us relate these two currents.

The charge q on the plates at any time is related to the magnitude $|\vec{E}|$ of the field between the plates at that time by Eq. 28-4:

$$|q| = \varepsilon_0 A |\vec{E}|, \qquad (32\text{-}28)$$

in which A is the plate area. To get the real current i, we differentiate Eq. 32-28 with respect to time, finding

$$\frac{dq}{dt} = i = \varepsilon_0 A \frac{dE}{dt}. \qquad (32\text{-}29)$$

To get the displacement current i_d, we can use Eq. 32-26. Assuming that the electric field $\vec{E}$ between the two plates is uniform (we neglect any fringing), we can replace the electric flux Φ_E in that equation with EA. Then Eq. 32-26 becomes

$$|i_d| = \varepsilon_0 \left| \frac{d\Phi_E}{dt} \right| = \varepsilon_0 \frac{d(|\vec{E}|A)}{dt} = \varepsilon_0 A \left| \frac{dE}{dt} \right|. \qquad (32\text{-}30)$$

Comparing Eqs. 32-29 and 32-30, we see that the real current i charging the capacitor and the fictitious displacement current i_d between the plates have the same magnitude:

$$i_d = i \qquad \text{(displacement current in a capacitor).} \qquad (32\text{-}31)$$

Thus, we can consider the fictitious displacement current i_d to be simply a continuation of the real current i from one plate, across the capacitor gap, to the other plate. Because the electric field is uniformly spread over the plates, the same is true of this fictitious displacement current i_d, as suggested by the spread of current arrows in Fig. 32-16a. Although no charge actually moves across the gap between the plates, the idea of the

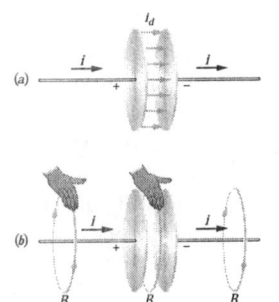

Fig. 32-16: (a) The displacement current i_d between the plates of a capacitor that is being charged by a current i. (b) The right-hand rule for finding the direction of the magnetic field around a wire with a real current (as at the left) also gives the direction of the magnetic field around a displacement current (as in the center).

fictitious current i_d can help us to quickly find the direction and magnitude of an induced magnetic field, as follows.

Finding the Induced Magnetic Field

In Chapter 30 we found the direction of the magnetic field produced by a real current i by using the right-hand rule of Fig. 30-4. We can apply the same rule to find the direction of an induced magnetic field produced by a fictitious displacement current i_d, as is shown in the center of Fig. 32-16b for a capacitor.

We can also use i_d to find the magnitude of the magnetic field induced by a charging capacitor with parallel circular plates of radius R. We simply consider the space between the plates to be an imaginary circular wire of radius R carrying the imaginary current i_d. Then, from Eq. 30-22, the magnitude of the magnetic field at a point inside the capacitor at radius r from the center is

$$|\vec{B}| = \left(\frac{\mu_0 |i_d|}{2\pi R^2}\right) r \qquad \text{(inside a circular capacitor).} \qquad (32\text{-}32)$$

Similarly, from Eq. 30-19, the magnitude of the magnetic field at a point outside the capacitor at radius r is

$$|\vec{B}| = \frac{\mu_0 |i_d|}{2\pi r} \qquad \text{(outside a circular capacitor).} \qquad (32\text{-}33)$$

Touchstone Example 32-10-1, at the end of this chapter, illustrates how to use what you learned in this section.

TE

32-11 Maxwell's Equations

Equation 32-25 is the last of the four fundamental equations of electromagnetism, called *Maxwell's equations* and displayed in Table 32-1. These four equations explain a diverse range of phenomena, from why a compass needle points north to why a car starts when you turn the ignition key. They are the basis for the functioning of such electromagnetic devices as electric motors, cyclotrons, television transmitters and receivers, telephones, fax machines, radar, and microwave ovens.

Maxwell's equations are the basis from which many of the equations you have seen since Chapter 22 can be derived. They are also the basis of many of the equations you will see in Chapters 34 through 37, which introduce you to optics, as well as such optical devices as telescopes and eyeglasses.

TABLE 32-1: Maxwell's Equations[a]

Name	Equation	
Gauss' law for electricity	$\oint \vec{E} \cdot d\vec{A} = q/\varepsilon_0$	Relates net electric flux to net enclosed electric charge
Gauss' law for magnetism	$\oint \vec{B} \cdot d\vec{A} = 0$	Relates net magnetic flux to net enclosed magnetic charge
Faraday's law	$\oint \vec{E} \cdot d\vec{s} = -\dfrac{d\Phi_B}{dt}$	Relates induced electric field to changing magnetic flux
Ampere-Maxwell law	$\oint \vec{B} \cdot d\vec{s} = \mu_0 \varepsilon_0 \dfrac{d\Phi_E}{dt} + \mu_0 i_{enc}$	Relates induced magnetic field to changing electric flux and to current

[a]Written on the assumption that no dielectric or magnetic materials are present.

Touchstone Example 32-8-1

A compass needle made of pure iron (with density 7900 kg/m³) has a length L of 3.0 cm, a width of 1.0 mm, and a thickness of 0.50 mm. The magnitude of the magnetic dipole moment of an iron atom is $\mu_{Fe} = 2.1 \times 10^{-23}$ J/T. If the magnetization of the needle is equivalent to the alignment of 10% of the atoms in the needle, what is the magnitude of the needle's magnetic dipole moment $\vec{\mu}$?

SOLUTION: One **Key Idea** here is that alignment of all N atoms in the needle would give a magnitude of $N|\vec{\mu}_{Fe}|$ for the needle's magnetic dipole moment $\vec{\mu}$. However, the needle has only 10% alignment (the random orientation of the rest does not give any net contribution to $\vec{\mu}$). Thus,

$$|\vec{\mu}| = 0.10 N |\vec{\mu}_{Fe}|. \qquad \text{(TE32-1)}$$

A second **Key Idea** is that we can find the number of atoms N in the needle from the needle's mass:

$$N = \frac{\text{needle's mass}}{\text{iron's atomic mass}}. \qquad \text{(TE32-2)}$$

Iron's atomic mass is not listed in Appendix F, but its molar mass M is. Thus, we write

$$\text{iron's atomic mass} = \frac{\text{iron's molar mass } M}{\text{Avogadro's number } N_A}. \qquad \text{(TE32-3)}$$

Equation TE32-2 then becomes

$$N = \frac{m N_A}{M}. \qquad \text{(TE32-4)}$$

The needle's mass m is the product of its density and its volume. The volume works out to be 1.5×10^{-8} m³, so we can write

$$
\begin{aligned}
\text{needle's mass } m &= (\text{needle's density})(\text{needle's volume}) \\
&= (7900 \text{ kg/m}^3)(1.5 \times 10^{-8} \text{ m}^3) \\
&= 1.185 \times 10^{-4} \text{ kg}.
\end{aligned}
$$

Substituting into Eq. TE32-4 with this value for m, and also 55.847 g/mole ($= 0.055\ 847$ kg/mole) for M and 6.02×10^{23} for N_A, we find

$$
\begin{aligned}
N &= \frac{(1.185 \times 10^{-4} \text{ kg})(6.02 \times 10^{23})}{0.055\ 847 \text{ kg/mole}} \\
&= 1.2774 \times 10^{21}.
\end{aligned}
$$

Substituting this and the value of μ_{Fe} into Eq. 32-22 then yields

$$
\begin{aligned}
\mu &= (0.10)(1.2774 \times 10^{21})(2.1 \times 10^{-23} \text{ J/T}) \\
&= 2.682 \times 10^{-3} \text{ J/T} \approx 2.7 \times 10^{-3} \text{ J/T}. \qquad \text{(Answer)}
\end{aligned}
$$

Touchstone Example 32-9-1

A parallel-plate capacitor with circular plates of radius R is being charged as in Fig. 32-14a.

(a) Derive an expression for the magnitude of the magnetic field at radii r for the case $r \leq R$.

SOLUTION: The **Key Idea** here is that a magnetic field can be set up by a current and by induction due to a changing electric flux; both effects are included in Eq. 32-25. There is no current between the capacitor plates of Fig. 32-14, but the electric flux there is changing. Thus, Eq. 32-25 reduces to

$$\oint \vec{B} \cdot d\vec{s} = \mu_0 \varepsilon_0 \frac{d\Phi_E}{dt}. \qquad \text{(TE32-5)}$$

We shall separately evaluate the left and right sides of this equation.

Left side of Eq. TE32-5: We choose a circular Amperian loop with a radius $r \leq R$ as shown in Fig. 32-14, because we want to evaluate the magnetic field for $r \leq R$—that is, inside the capacitor. The magnetic field $\vec{B}$ at all points along the loop is tangent to the loop, as is the path element $d\vec{s}$. Thus, $\vec{B}$ and $d\vec{s}$ are either parallel or antiparallel at each point of the loop. For simplicity, assume they are parallel (the choice does not alter our outcome here). Then

$$\oint \vec{B} \cdot d\vec{s} = \oint |\vec{B}| |d\vec{s}| \cos 0° = \oint B \, ds.$$

Due to the circular symmetry of the plates, we can also assume that $\vec{B}$ has the same magnitude at every point around the loop. Thus, B can be taken outside the integral on the right side of the above equation. The integral that remains is $\oint ds$, which simply gives the circumference $2\pi r$ of the loop. The left side of Eq. TE32-5 is then $(B)(2\pi r)$.

Right side of Eq. TE32-5: We assume that the electric field $\vec{E}$ is uniform between the capacitor plates and directed perpendicular to the plates. Then the electric flux Φ_E through the Amperian loop is EA, where A is the area encircled by the loop within the electric field. Thus, the right side of Eq. TE32-5 is $\mu_0 \varepsilon_0 \, d(EA)/dt$.

Substituting our results for the left and right sides into Eq. TE32-5, we get

$$|(\vec{B})|(2\pi r) = \mu_0 \varepsilon_0 \frac{d(|\vec{E}|A)}{dt}.$$

Because A is a constant, we write $d(EA)$ as $A \, dE$, so we have

$$|(\vec{B})|(2\pi r) = \mu_0 \varepsilon_0 A \left| \frac{d\vec{E}}{dt} \right|. \qquad \text{(TE32-6)}$$

We next use this **Key Idea**: The area A that is encircled by the Amperian loop within the electric field is the full area πr^2 of the loop, because the loop's radius r is less than (or equal to) the plate radius R. Substituting πr^2 for A in Eq. TE32-6 and solving the result for B give us, for $r \leq r$,

$$|\vec{B}| = \frac{\mu_0 \varepsilon_0 r}{2} \left| \frac{d\vec{E}}{dt} \right|. \qquad \text{(Answer)} \quad \text{(TE32-7)}$$

This equation tells us that, inside the capacitor, B increases linearly with increased radial distance r, from zero at the center of the plates to a maximum value at the plate edges (where $r = R$).

(b) Evaluate the field magnitude $|\vec{B}|$ for $r = R/5 = 11.0$ mm and $dE/dt = 1.50 \times 10^{12}$ V/m · s.

SOLUTION: From the answer to (a), we have

$$\begin{aligned}
|\vec{B}| &= \frac{1}{2} \mu_0 \varepsilon_0 r \left| \frac{d\vec{E}}{dt} \right| \\
&= \tfrac{1}{2}(4\pi \times 10^{-7} \text{ T} \cdot \text{m/A})(8.85 \times 10^{-12} \text{ C}^2/\text{N} \cdot \text{m}^2) \\
&\quad \times (11.0 \times 10^{-3} \text{ m})(1.50 \times 10^{12} \text{ V/m} \cdot \text{s}) \\
&= 9.18 \times 10^{-8} \text{ T}. \qquad \text{(Answer)}
\end{aligned}$$

(c) Derive an expression for the induced magnetic field for the case $r \geq R$.

SOLUTION: Our procedure is the same as in (a) except we now use an Amperian loop with a radius r that is greater than the plate radius R, to evaluate B outside the capacitor. Evaluating the left and right sides of Eq. TE32-5 again leads to Eq. TE32-6. However, we then need this subtle **Key Idea**: The electric field exists only between the plates, *not* outside the plates. Thus, the area A that is encircled by the Amperian loop in the electric field is *not* the full area πr^2 of the loop. Rather, A is only the plate area πR^2.

Substituting πR^2 for A in Eq. TE32-6 and solving the result for $|\vec{B}|$ give us, for $r \geq R$,

$$|\vec{B}| = \frac{\mu_0 \varepsilon_0 R^2}{2r} \left| \frac{d\vec{E}}{dt} \right|.$$ (Answer)   (TE32-8)

This equation tells us that, outside the capacitor, $|\vec{B}|$ decreases with increased radial distance r, from a maximum value at the plate edges (where $r = R$). By substituting $r = R$ into Eqs. TE32-7 and TE32-8, you can show that these equations are consistent; that is, they give the same maximum value of $|\vec{B}|$ at the plate radius.

The magnitude of the induced magnetic field calculated in (b) is so small that it can scarcely be measured with simple apparatus. This is in sharp contrast to the magnitudes of induced electric fields (Faraday's law), which can be measured easily. This experimental difference exists partly because induced emfs can easily be multiplied by using a coil of many turns. No technique of comparable simplicity exists for multiplying induced magnetic fields. In any case, the experiment suggested by this sample problem has been done, and the presence of the induced magnetic fields has been verified quantitatively.

Touchstone Example 32-10-1

The circular parallel-plate capacitor in Touchstone Example 32-9-1 is being charged with a current i.

(a) Between the plates, what is the magnitude of $\oint \vec{B} \cdot d\vec{s}$, in terms of μ_0 and i, at a radius $r = R/5$ from their center?

SOLUTION: The first **Key Idea** of TE32-9-1a holds here too. However, now we can replace the product $\varepsilon_0 \, d\Phi_E/dt$ in Eq. 32-25 with a fictitious displacement current i_d. Then integral $\oint \vec{B} \cdot d\vec{s}$ is given by Eq. 32-27, but because there is no real current i between the capacitor plates, the equation reduces to

$$\oint \vec{B} \cdot d\vec{s} = \mu_0 i_{d,\mathrm{enc}}.$$ (TE32-9)

Because we want to evaluate $\oint \vec{B} \cdot d\vec{s}$ at radius $r = R/5$ (within the capacitor), the integration loop encircles only a portion $i_{d,\mathrm{enc}}$ of the total displacement current i_d. A second **Key Idea** is to assume that i_d is uniformly spread over the full plate area. Then the portion of the displacement current encircled by the loop is proportional to the area encircled by the loop:

$$\frac{\left(\begin{array}{c} \text{encircled displacement} \\ \text{current } i_{d,\mathrm{enc}} \end{array} \right)}{\left(\begin{array}{c} \text{total displacement} \\ \text{current } i_d \end{array} \right)} = \frac{\text{encircled area } pr^2}{\text{full plate area } pR^2}.$$

This gives us a current magnitude of

$$|i_{d,\mathrm{enc}}| = \frac{\pi r^2}{\pi R^2}.$$

Substituting this into Eq. TE32-9, we obtain

$$\oint \vec{B} \cdot d\vec{s} = \mu_0 i_d \frac{\pi r^2}{\pi R^2}.$$ (TE32-10)

Now substituting $i_d = i$ (from Eq. 32-31) and $r = R/5$ into Eq. TE32-10 leads to

$$\oint \vec{B} \cdot d\vec{s} = \mu_0 i \frac{(R/5)^2}{R^2} = \frac{\mu_0 i}{25}.$$ (Answer)

(b) In terms of the maximum induced magnetic field, what is the magnitude of the magnetic field induced at $r = R/5$, inside the capacitor?

32-19

SOLUTION: The **Key Idea** here is that, because the capacitor has parallel circular plates, we can treat the space between the plates as an imaginary wire of radius R carrying the imaginary current i_d. Then we can use Eq. 32-32 to find the induced magnetic field magnitude $|\vec{B}|$ at any point inside the capacitor. At $r = R/5$, that equation yields

$$|\vec{B}| = \left(\frac{\mu_0 |i_d|}{2\pi R^2}\right) r = \frac{\mu_0 |i_d| (R/5)}{2\pi R^2} = \frac{\mu_0 |i_d|}{10\pi R}. \qquad \text{(TE32-11)}$$

The maximum field magnitude $|\vec{B}_{\max}|$ within the capacitor occurs at $r = R$. It is

$$|\vec{B}_{\max}| = \left(\frac{\mu_0 |i_d|}{2\pi R^2}\right) R = \frac{\mu_0 |i_d|}{2\pi R}. \qquad \text{(TE32-12)}$$

Dividing Eq. 32-44 by Eq. 32-45 and rearranging the result, we find

$$|\vec{B}| = \frac{|\vec{B}_{\max}|}{5}. \qquad \text{(Answer)}$$

We should be able to obtain this result with a little reasoning and less work. Equation 32-32 tells us that inside the capacitor, $|\vec{B}|$ increases linearly with r. Therefore, a point $\frac{1}{5}$ the distance out to the full radius R of the plates, where $|\vec{B}_{\max}|$ occurs, should have a field $|\vec{B}|$ that is $\frac{1}{5}|\vec{B}_{\max}|$.

33 Electromagnetic Oscillations and Alternating Current

When a high-voltage power transmission line requires repair, a utility company cannot just shut it down, perhaps blacking out an entire city. Repairs must be made while the lines are electrically "hot." The man outside the helicopter in this photograph has just replaced a spacer between 500 kV lines *by hand*, a procedure that requires considerable expertise.

How does he manage this repair without being electrocuted?

The answer is in this chapter.

33-1 New Physics—Old Mathematics

In this chapter you will see how the electric charge q varies with time in a circuit made up of an inductor L, a capacitor C, and a resistor R. From another point of view, we shall discuss how energy shuttles back and forth between the magnetic field of the inductor and the electric field of the capacitor, while it is being gradually dissipated as thermal energy in the resistor.[*]

We have discussed oscillations before, in another context. In Chapter 16 we saw how displacement x varies with time in a mechanical oscillating system made up of a block of mass m, a spring of spring constant k, and a viscous or frictional element such as oil; Fig. 16-17 shows such a system. You also saw how energy shuttles back and forth between the kinetic energy of the oscillating mass and the potential energy of the spring, being gradually dissipated as thermal energy.

The parallel between these two idealized systems is exact, and the controlling differential equations are identical. Thus, there is no new mathematics to be learned; we can simply change the symbols and give our full attention to the physics of the situation.

33-2 *LC* Oscillations, Qualitatively

Of the three circuit elements, resistance R, capacitance C, and inductance L, we have so far discussed the series combinations RC (in Section 28-9) and RL (in Section 31-9). In these two kinds of circuit we found that the charge, current, and potential difference grow and decay exponentially. The time scale of the growth or decay is given by a *time constant* τ, which is either capacitive or inductive.

We now examine the remaining two-element circuit combination LC. You will see that in this case the charge, current, and potential difference do not decay exponentially with time but vary sinusoidally (with period T and angular frequency ω). The resulting oscillations of the capacitor's electric field and the inductor's magnetic field are said to be **electromagnetic oscillations**. Such a circuit is said to oscillate.

Parts *a* through *h* of Fig. 33-1 show succeeding stages of the oscillations in a simple *LC* circuit. From Eq. 28-21, the energy stored in the electric field of the capacitor at any time is

$$U_E = \frac{q^2}{2C},\qquad\qquad (33\text{-}1)$$

where q is the charge on the capacitor at that time. From Eq. 31-46, the energy stored in the magnetic field of the inductor at any time is

$$U_B = \frac{Li^2}{2},\qquad\qquad (33\text{-}2)$$

where i is the current through the inductor at that time.

---

[*] The method of repairing high-voltage lines shown in the opening photograph is patented by Scott H. Yenzer and is licensed exclusively to Haverfield Corporation of Gettysburg, PA. As the lineman approaches a hot line, the electric field surrounding the line brings his body to nearly the potential of the line. To match the two potentials, he then extends a conducting "wand" to the line. To avoid being electrocuted, he must be isolated from anything electrically connected to the ground. To ensure that his body is always at a single potential—that of the line he is working on—he wears a conducting suit, hood, and gloves, all of which are electrically connected to the line via the wand.

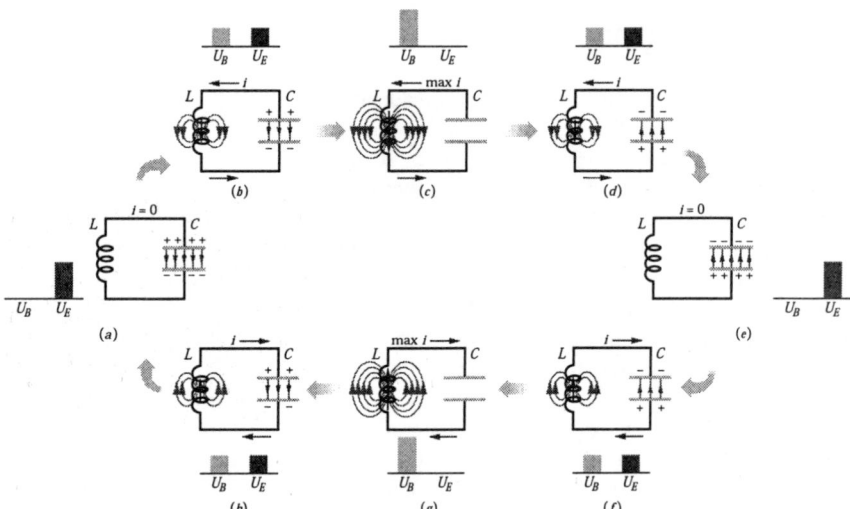

Fig. 33-1: Eight stages in a single cycle of oscillation of a resistanceless LC circuit. The bar graphs by each figure show the stored magnetic and electric energies. The magnetic field lines of the inductor and the electric field lines of the capacitor are shown. (*a*) Capacitor with maximum charge, no current. (*b*) Capacitor discharging, current increasing. (*c*) Capacitor fully discharged, current maximum. (*d*) Capacitor charging but with polarity opposite that in (*a*), current decreasing. (*e*) Capacitor with maximum charge having polarity opposite that in (*a*), no current. (*f*) Capacitor discharging, current increasing with direction opposite that in (*b*). (*g*) Capacitor fully discharged, current maximum. (*h*) Capacitor charging, current decreasing.

We now adopt the convention of representing *instantaneous values* of the electrical quantities of a sinusoidally oscillating circuit with small letters, such as q, and the *amplitudes* of those quantities with capital letters, such as Q. With this convention in mind, let us assume that initially the charge q on the capacitor in Fig. 33-1 is at its maximum value Q and that the current i through the inductor is zero. This initial state of the circuit is shown in Fig. 33-1*a*. The bar graphs for energy included there indicate that at this instant, with zero current through the inductor and maximum charge on the capacitor, the energy U_B of the magnetic field is zero and the energy U_E of the electric field is a maximum.

The capacitor now starts to discharge through the inductor, positive charge carriers moving counterclockwise, as shown in Fig. 33-1*b*. This means that a current i, given by dq/dt and pointing down in the inductor, is established. As the capacitor's charge decreases, the energy stored in the electric field within the capacitor also decreases. This energy is transferred to the magnetic field that appears around the inductor because of the current i that is building up there. Thus, the electric field decreases and the magnetic field builds up as energy is transferred from the electric field to the magnetic field.

The capacitor eventually loses all its charge (Fig. 33-1*c*) and thus also loses its electric field and the energy stored in that field. The energy has then been fully transferred to the magnetic field of the inductor. The magnetic field is then at its maximum magnitude, and the current through the inductor is then at its maximum value I.

Although the charge on the capacitor is now zero, the counterclockwise current must continue because the inductor does not allow it to change suddenly to zero. The current continues to transfer positive charge from the top plate to the bottom plate through the circuit (Fig. 33-1*d*). Energy now flows from the inductor back to the capacitor as the electric field within the capacitor builds up again. The current gradually decreases during this energy transfer. When, eventually, the energy has been transferred completely back to the capacitor (Fig. 33-1*e*), the current has decreased to zero (momentarily). The situation of Fig. 33-1*e* is like the initial situation, except that the capacitor is now charged oppositely.

The capacitor then starts to discharge again but now with a clockwise current (Fig. 33-1*f*). Reasoning as before, we see that the clockwise current builds to a maximum (Fig. 33-1*g*) and then decreases (Fig. 33-1*h*), until the circuit eventually returns to its initial situation (Fig. 33-1*a*). The process then repeats at some frequency f and thus at an angular frequency $\omega = 2\pi f$. In the ideal LC circuit with no resistance, there are no energy transfers other than that between the electric field of the capacitor and the magnetic field of the inductor. Owing to the conservation of energy, the oscillations continue indefinitely. The oscillations need not begin with the energy all in the electric field; the initial situation could be any other stage of the oscillation.

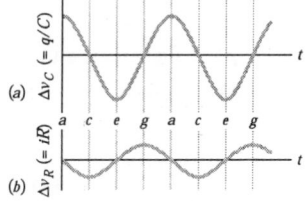

Fig. 33-2: (*a*) The potential difference across the capacitor of the circuit of Fig. 33-1 as a function of time. This quantity is proportional to the charge on the capacitor. (*b*) A potential proportional to the current in the circuit of Fig. 33-1. The letters refer to the correspondingly labeled oscillation stages in Fig. 33-1.

To find the charge q on the capacitor as a function of time, we can use a voltmeter to measure the time-varying potential difference (or *voltage*) $\Delta v_C(t)$ that exists across the capacitor C. From Eq. 28-1 we can write

$$\Delta v_C(t) = \left(\frac{1}{C}\right)q(t),$$

which allows us to find q. To measure the current, we can connect a small resistance R in series with the capacitor and inductor and measure the time-varying potential difference Δv_R across it; Δv_R is proportional to i through the relation

$$\Delta v_R(t) = i(t)R.$$

We assume here that R is so small that its effect on the behavior of the circuit is negligible. The variations in time of $\Delta v_C(t)$ and $\Delta v_R(t)$, and thus of $q(t)$ and $i(t)$, are shown in Fig. 33-2. All four quantities vary sinusoidally.

In an actual LC circuit, the oscillations will not continue indefinitely because there is always some resistance present that will drain energy from the electric and magnetic fields and dissipate it as thermal energy (the circuit may become warmer). The oscillations, once started, will die away as Fig. 33-3 suggests. Compare this figure with Fig. 16-16, which shows the decay of mechanical oscillations caused by frictional damping in a block-spring system.

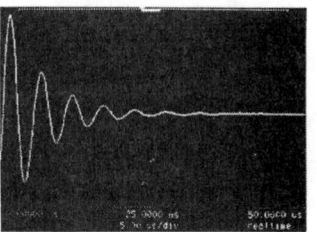

Fig. 33-3: An oscilloscope trace showing how the oscillations in an *RLC* circuit actually die away because energy is dissipated in the resistor as thermal energy.

READING EXERCISE 33-1: A charged capacitor and an inductor are connected in series at time $t = 0$. In terms of the period T of the resulting oscillations, determine how much later the following reach their maximums: (a) the charge on the capacitor; (b) the voltage across the capacitor, with its original polarity; (c) the energy stored in the electric field; and (d) the current.

Touchstone Example 33-2-1, at the end of this chapter, illustrates how to use what you learned in this section.

TE

33-3 The Electrical-Mechanical Analogy

Let us look a little closer at the analogy between the oscillating LC system of Fig. 33-1 and an oscillating block-spring system. Two kinds of energy are involved in the block-spring system. One is potential energy of the compressed or extended spring; the other is kinetic energy of the moving block. These two energies are given by the familiar formulas in the left energy column in Table 33-1.

TABLE 33-1: The Energy in Two Oscillating Systems Compared

Block-Spring System		LC Oscillator	
Element	Energy	Element	Energy
Spring	Potential, $\frac{1}{2}kx^2$	Capacitor	Electric, $\frac{1}{2}(1/C)q^2$
Block	Kinetic, $\frac{1}{2}mv^2$	Inductor	Magnetic, $\frac{1}{2}Li^2$
	$v = dx/dt$		$i = dq/dt$

The table also shows, in the right energy column, the two kinds of energy involved in LC oscillations. By looking across the table, we can see an analogy between the forms of the two pairs of energies—the mechanical energies of the block-spring system and the electromagnetic energies of the LC oscillator. The equations for v and i at the bottom of the table help us see the details of the analogy. They tell us that q corresponds to x, and i corresponds to v (in both equations, the former is differentiated to obtain the latter). These correspondences then suggest that, in the energy expressions, $1/C$ corresponds to k and L corresponds to m. Thus,

q corresponds to x, $1/C$ corresponds to k,

i corresponds to v,   and L corresponds to m.

These correspondences suggest that in an LC oscillator, the capacitor is mathematically like the spring in a block-spring system, and the inductor is like the block.

In Section 16-3 we saw that the angular frequency of oscillation of a (frictionless) block-spring system is

$$\omega = \sqrt{\frac{k}{m}} \qquad \text{(block-spring system).} \qquad (33\text{-}3)$$

The correspondences listed above suggest that to find the angular frequency of oscillation for a (resistanceless) LC circuit, k should be replaced by $1/C$ and m by L, yielding

$$\omega = \frac{1}{\sqrt{LC}} \qquad (LC \text{ circuit).} \qquad (33\text{-}4)$$

We derive this result in the next section.

33-4 *LC* Oscillations, Quantitatively

Here we want to show explicitly that Eq. 33-4 for the angular frequency of LC oscillations is correct. At the same time, we want to examine even more closely the analogy between LC oscillations and block-spring oscillations. We start by extending somewhat our earlier treatment of the mechanical block-spring oscillator.

The Block-Spring Oscillator

We analyzed block-spring oscillations in Chapter 16 in terms of energy transfers and did not—at that early stage—derive the fundamental differential equation that governs those oscillations. We do so now.

We can write, for the total energy U of a block-spring oscillator at any instant,

$$U = U_b + U_s = \tfrac{1}{2} mv^2 + \tfrac{1}{2} kx^2, \qquad (33\text{-}5)$$

where U_b and U_s are, respectively, the kinetic energy of the moving block and the potential energy of the stretched or compressed spring. If there is no friction—which we assume—the total energy U remains constant with time, even though v and x vary. In more formal language, $dU/dt = 0$. This leads to

$$\frac{dU}{dt} = \frac{d}{dt}\left(\tfrac{1}{2} mv^2 + \tfrac{1}{2} kx^2\right) = mv\frac{dv}{dt} + kx\frac{dx}{dt} = 0. \qquad (33\text{-}6)$$

However, $v = dx/dt$ and $dv/dt = d^2x/dt^2$. With these substitutions, Eq. 33-6 becomes

$$m\frac{d^2x}{dt^2} + kx = 0 \qquad \text{(block-spring oscillations).} \qquad (33\text{-}7)$$

Equation 33-7 is the fundamental differential equation that governs the frictionless block-spring oscillations. It involves the displacement x and its second derivative with respect to time.

The general solution to Eq. 33-7—that is, the function $x(t)$ that describes the block-spring oscillations—is (as we saw in Eq. 16-3)

$$x(t) = X \cos(\omega t + \phi) \qquad \text{(displacement),} \qquad (33\text{-}8)$$

in which X is the amplitude of the mechanical oscillations (represented by x_m in Chapter 16), ω is the angular frequency of the oscillations, and ϕ is a phase constant.

The *LC* Oscillator

Now let us analyze the oscillations of a resistanceless *LC* circuit, proceeding exactly as we just did for the block-spring oscillator. The total energy U present at any instant in an oscillating *LC* circuit is given by

$$U = U_B + U_E = \frac{Li^2}{2} + \frac{q^2}{2C}, \qquad (33\text{-}9)$$

in which U_B is the energy stored in the magnetic field of the inductor and U_E is the energy stored in the electric field of the capacitor. Since we have assumed the circuit resistance to be zero, no energy is transferred to thermal energy and U remains constant with time. In more formal language, dU/dt must be zero. This leads to

$$\frac{dU}{dt} = \frac{d}{dt}\left(\frac{Li^2}{2} + \frac{q^2}{2C} \right) = Li\frac{di}{dt} + \frac{q}{C}\frac{dq}{dt} = 0. \qquad (33\text{-}10)$$

However, $i = dq/dt$ and $di/dt = d^2q/dt^2$. With these substitutions, Eq. 33-10 becomes

$$L\frac{d^2q}{dt^2} + \frac{1}{C}q = 0 \qquad \text{(LC oscillations).} \qquad (33\text{-}11)$$

This is the differential equation that describes the oscillations of a resistanceless *LC* circuit. Careful comparison shows that Eqs. 33-11 and 33-7 are exactly of the same mathematical form, differing only in the symbols used.

Charge and Current Oscillations

Since the differential equations are mathematically identical, their solutions must also be mathematically identical. Because q corresponds to x, we can write the general solution of Eq. 33-11, giving $q(t)$ as a function of time, by analogy to Eq. 33-8 as

$$q(t) = Q\cos(\omega t + \phi) \qquad \text{(charge),} \qquad (33\text{-}12)$$

where Q is the amplitude of the charge variations, ω is the angular frequency of the electromagnetic oscillations, and ϕ is the phase constant.

Taking the first derivative of Eq. 33-12 with respect to time gives us the time varying current $i(t)$ of the *LC* oscillator:

$$i(t) = \frac{dq}{dt} = -\omega Q\sin(\omega t + \phi) \qquad \text{(current).} \qquad (33\text{-}13)$$

The amplitude I of this sinusoidally varying current is

$$I = \omega Q, \qquad (33\text{-}14)$$

so we can rewrite Eq. 33-13 as

$$i(t) = -I\sin(\omega t + \phi). \qquad (33\text{-}15)$$

Angular Frequencies

We can test whether Eq. 33-12 is a solution of Eq. 33-11 by substituting it and its second derivative with respect to time into Eq. 33-11. The first derivative of Eq. 33-12 is Eq. 33-13. The second derivative is then

$$\frac{d^2q}{dt^2} = -\omega^2 Q\cos(\omega t + \phi)$$

Substituting for q and d^2q/dt^2 into Eq. 33-11, we obtain

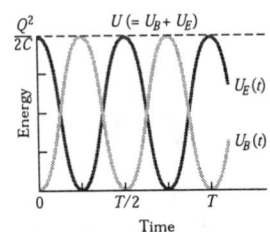

Fig. 33-4: The stored magnetic energy and electric energy in the circuit of Fig. 33-1 as a function of time. Note that their sum remains constant. T is the period of oscillation.

33-6

$$-L\omega^2 Q\cos(\omega t + \phi) + \frac{1}{C}Q\cos(\omega t + \phi) = 0.$$

Canceling $Q\cos(\omega t + \phi)$ and rearranging lead to

$$\omega = \frac{1}{\sqrt{LC}}.$$

Thus, Eq. 33-12 is indeed a solution of Eq. 33-11 if ω has the constant value $1/\sqrt{LC}$. Note that this expression for ω is exactly that given by Eq. 33-4, which we arrived at by examining correspondences.

The phase constant ϕ in Eq. 33-12 is determined by the conditions that prevail at any certain time, say, $t = 0$. If the conditions yield $\phi = 0$ at $t = 0$, Eq. 33-12 requires that $q = Q$ and Eq. 33-13 requires that $i = 0$; these are the initial conditions represented by Fig. 33-1a.

Electric and Magnetic Energy Oscillations

The electric energy stored in the LC circuit at any time t is, from Eqs. 33-1 and 33-12,

$$U_E = \frac{q^2}{2C} = \frac{Q^2}{2C}\cos^2(\omega t + \phi). \qquad (33\text{-}16)$$

The magnetic energy is, from Eqs. 33-2 and 33-13,

$$U_B = \tfrac{1}{2}Li^2 = \tfrac{1}{2}L\omega^2 Q^2 \sin^2(\omega t + \phi).$$

Substituting for ω from Eq. 33-4 then gives us

$$U_B = \frac{Q^2}{2C}\sin^2(\omega t + \phi). \qquad (33\text{-}17)$$

Figure 33-4 shows plots of $U_E(t)$ and $U_B(t)$ for the case of $\phi = 0$. Note that

1. The maximum values of U_E and U_B are both $Q^2/2C$.

2. At any instant the sum of U_E and U_B is equal to $Q^2/2C$, a constant.

3. When U_E is maximum, U_B is zero, and conversely.

READING EXERCISE 33-2: A capacitor in an LC oscillator has a maximum potential difference of 17 V and a maximum energy of $160\,\mu J$. When the capacitor has a potential difference of 5 V and an energy of $10\,\mu J$, what are (a) the emf across the inductor and (b) the energy stored in the magnetic field?

Touchstone Example 33-4-1, at the end of this chapter, illustrates how to use what you learned in this section.

TE

33-5 Damped Oscillations in an *RLC* Circuit

A circuit containing resistance, inductance, and capacitance is called an *RLC circuit*. We shall here discuss only *series RLC circuits* like that shown in Fig. 33-5. With a resistance R present, the total *electromagnetic energy* U of the circuit (the sum of the electric energy and magnetic energy) is no longer constant; instead, it decreases with time as energy is transferred to thermal energy in the resistance. Because of this loss of energy, the oscillations of charge, current, and potential difference continuously decrease in amplitude, and the oscillations are said to be damped. As you will see, they are *damped* in exactly the same way as those of the damped block-spring oscillator of Section 16-8.

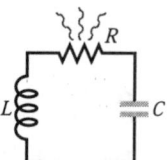

Fig. 33-5: A series *RLC* circuit. As the charge contained in the circuit oscillates back and forth through the resistance, electromagnetic energy is dissipated as thermal energy, damping (decreasing the amplitude of) the oscillations.

To analyze the oscillations of this circuit, we write an equation for the total electromagnetic energy U in the circuit at any instant. Because the resistance does not store electromagnetic energy, we can use Eq. 33-9:

$$U = U_B + U_E = \frac{Li^2}{2} + \frac{q^2}{2C}. \tag{33-18}$$

Now, however, this total energy decreases as energy is transferred to thermal energy. The rate of that transfer is, from Eq. 26-21,

$$\frac{dU}{dt} = -i^2 R, \tag{33-19}$$

where the minus sign indicates that U decreases. By differentiating Eq. 33-18 with respect to time and then substituting the result in Eq. 33-19, we obtain

$$\frac{dU}{dt} = Li\frac{di}{dt} + \frac{q}{C}\frac{dq}{dt} = -i^2 R.$$

Substituting dq/dt for i and d^2q/dt^2 for di/dt, we obtain

$$L\frac{d^2q}{dt^2} + R\frac{dq}{dt} + \frac{1}{C}q = 0 \qquad (RLC \text{ circuit}), \tag{33-20}$$

which is the differential equation that describes damped oscillations in an RLC circuit.

The solution to Eq. 33-20 is

$$q = Qe^{-Rt/2L} \cos(\omega' t + \phi), \tag{33-21}$$

in which

$$\omega' = \sqrt{\omega^2 - (R/2L)^2}, \tag{33-22}$$

where $\omega = 1/\sqrt{LC}$, as with an undamped oscillator. Equation 33-21 tells us how the charge on the capacitor oscillates in a damped RLC circuit; that equation is the electromagnetic counterpart of Eq. 16-40, which gives the displacement of a damped block-spring oscillator.

Equation 33-21 describes a sinusoidal oscillation (the cosine function) with an *exponentially decaying amplitude* $Qe^{-Rt/2L}$ (the factor that multiplies the cosine). The angular frequency ω' of the damped oscillations is always less than the angular frequency ω of the undamped oscillations; however, we shall here consider only situations in which R is small enough for us to replace ω' with ω.

Let us next find an expression for the total electromagnetic energy U of the circuit as a function of time. One way to do so is to monitor the energy of the electric field in the capacitor, which is given by Eq. 33-1 $(U_E = q^2/2C)$. By substituting Eq. 33-21 into Eq. 33-1, we obtain

$$U_E = \frac{q^2}{2C} = \frac{\left[Qe^{-Rt/2L}\cos(\omega' t + \phi)\right]^2}{2C} = \frac{Q^2}{2C}e^{-Rt/L}\cos^2(\omega' + \phi). \tag{33-23}$$

Thus, the energy of the electric field oscillates according to a cosine-squared term and the amplitude of that oscillation decreases exponentially with time.

Touchstone Example 33-5-1, at the end of this chapter, illustrates how to use what you learned in this section.

TE

33-6 Alternating Current

The oscillations in an *RLC* circuit will not damp out if an external emf device supplies enough energy to make up for the energy dissipated as thermal energy in the resistance *R*. Circuits in homes, offices, and factories, including countless *RLC* circuits, receive such energy from local power companies. In most countries the energy is supplied via oscillating emfs and currents—the current is said to be an **alternating current**, or **ac** for short. (The nonoscillating current from a battery is said to be a **direct current**, or **dc**.) These oscillating emfs and currents vary sinusoidally with time, reversing direction (in North America) 120 times per second and thus having frequency $f = 60$ Hz.

At first sight this may seem to be a strange arrangement. We have seen that the drift speed of the conduction electrons in household wiring may typically be 4×10^{-5} m/s. If we now reverse their direction every $\frac{1}{120}$ s, such electrons can move only about 3×10^{-7} m in a half-cycle. At this rate, a typical electron can drift past no more than about 10 atoms in the wiring before it is required to reverse its direction. How, you may wonder, can the electron ever get anywhere?

Although this question may be worrisome, it is a needless concern. The conduction electrons do not have to "get anywhere." When we say that the current in a wire is one ampere, we mean that charge passes through any plane cutting across that wire at the rate of one coulomb per second. The speed at which the charge carriers cross that plane does not matter directly; one ampere may correspond to many charge carriers moving very slowly or to a few moving very rapidly. Furthermore, the signal to the electrons to reverse directions—which originates in the alternating emf provided by the power company's generator—is propagated along the conductor at a speed close to that of light. All electrons, no matter where they are located, get their reversal instructions at about the same instant. Finally, we note that for many devices, such as lightbulbs and toasters, the direction of motion is unimportant as long as the electrons do move so as to transfer energy to the device via collisions with atoms in the device.

The basic advantage of alternating current is this: *As the current alternates, so does the magnetic field that surrounds the conductor.* This makes possible the use of Faraday's law of induction, which, among other things, means that we can step up (increase) or step down (decrease) the magnitude of an alternating potential difference at will, using a device called a transformer, as you will see later in this chapter. Moreover, alternating current is more readily adaptable to rotating machinery such as generators and motors than is (nonalternating) direct current.

Figure 33-6 shows a simple model of an ac generator. As the conducting loop is forced to rotate through the external magnetic field $\vec{B}$, a sinusoidally oscillating emf ε is induced in the loop:

$$\varepsilon = \varepsilon_m \sin \omega_d t. \qquad (33\text{-}24)$$

The *angular frequency* ω_d of the emf is equal to the angular speed with which the loop rotates in the magnetic field; the *phase* of the emf is $\omega_d t$; and the *amplitude* of the emf is ε_m (where the subscript stands for maximum). When the rotating loop is part of a closed conducting path, this emf produces (*drives*) a sinusoidal (alternating) current along the path with the same angular frequency ω_d, which then is called the **driving angular frequency**. We can write the current as

$$i = I \sin(\omega_d t - \phi), \qquad (33\text{-}25)$$

in which *I* is the amplitude of the driven current. (The phase $\omega_d t - \phi$ of the current is traditionally written with a minus sign instead of as $\omega_d t + \phi$.) We include a phase constant ϕ in Eq. 33-25 because the current *i* may not be in phase with the emf ε. (As you will see, the phase constant depends on the circuit to which the generator is connected.) We can also write the current *i* in terms of the **driving frequency** f_d of the emf, by substituting $2\pi f_d$ for ω_d in Eq. 33-25.

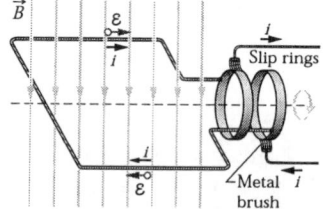

Fig. 33-6: The basic mechanism of an alternating-current generator is a conducting loop rotated in an external magnetic field. In practice, the alternating emf induced in a coil of many turns of wire is made accessible by means of slip rings attached to the rotating loop. Each ring is connected to one end of the loop wire and is electrically connected to the rest of the generator circuit by a conducting brush against which it slips as the loop (and it) rotates.

33-7 Forced Oscillations

We have seen that once started, the charge, potential difference, and current in both undamped LC circuits and damped RLC circuits (with small enough R) oscillate at angular frequency $\omega = 1/\sqrt{LC}$. Such oscillations are said to be *free oscillations* (free of any external emf), and the angular frequency ω is said to be the circuit's **natural angular frequency**.

When the external alternating emf of Eq. 33-24 is connected to an RLC circuit, the oscillations of charge, potential difference, and current are said to be *driven oscillations* or *forced oscillations*. These oscillations always occur at the driving angular frequency ω_d:

▶ Whatever the natural angular frequency ω of a circuit may be, forced oscillations of charge, current, and potential difference in the circuit always occur at the driving angular frequency ω_d.

However, as you will see in Section 33-9, the amplitudes of the oscillations very much depend on how close ω_d is to ω. When the two angular frequencies match—a condition known as resonance—the amplitude I of the current in the circuit is maximum.

33-8 Three Simple Circuits

Later in this chapter, we shall connect an external alternating emf device to a series RLC circuit as in Fig. 33-7. We shall then find expressions for the amplitude I and phase constant ϕ of the sinusoidally oscillating current in terms of the amplitude ε_m and angular frequency ω_d of the external emf. First, however, let us consider three simpler circuits, each having an external emf and only one other circuit element: R, C, or L. We start with a resistive element (a purely *resistive load*).

A Resistive Load

Figure 33-8a shows a circuit containing a resistance element of value R and an ac generator with the alternating emf of Eq. 33-24. By the loop rule, we have

$$\varepsilon - \Delta v_R = 0.$$

With Eq. 33-24, this gives us

$$\Delta v_R = \varepsilon_m \sin \omega_d t.$$

Because the amplitude ΔV_R of the alternating potential difference (or voltage) across the resistance is equal to the amplitude ε_m of the alternating emf, we can write this as

$$\Delta v_R = \Delta V_R \sin \omega_d t. \tag{33-26}$$

From the definition of resistance $(R = \Delta V / i)$, we can now write the current i_R in the resistance as

$$i_R = \frac{\Delta v_R}{R} = \frac{\Delta V_R}{R} \sin \omega_d t. \tag{33-27}$$

From Eq. 33-25, we can also write this current as

$$i_R = I_R \sin(\omega_d t - \phi), \tag{33-28}$$

where I_R is the amplitude of the current i_R in the resistance. Comparing Eqs. 33-27 and 33-28, we see that for a purely resistive load the phase constant $\phi = 0°$. We also see that the voltage amplitude and current amplitude are related by

$$\Delta V_R = I_R R \qquad \text{(resistor).} \tag{33-29}$$

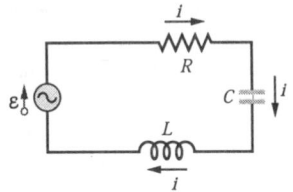

Fig. 33-7: A single-loop circuit containing a resistor, a capacitor, and an inductor. A generator, represented by a sine wave in a circle, produces an alternating emf that establishes an alternating current; the directions of the emf and current are indicated here at only one instant.

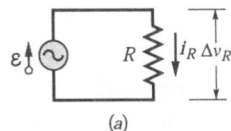

(a)

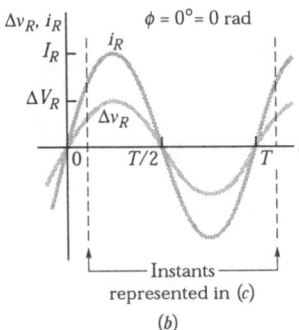

(b)

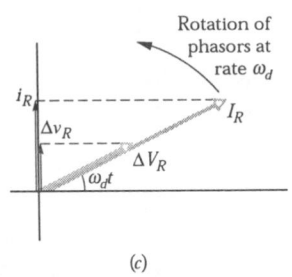

(c)

Fig. 33-8: (a) A resistor is connected across an alternating-current generator. (b) The current i_R and and the potential difference Δv_R across the resistor are plotted on the same graph, both versus time t. They are in phase and complete one cycle in one period T. (c) A phasor diagram shows the same thing as (b).

Although we found this relation for the circuit of Fig. 33-8a, it applies to any resistance in any ac circuit.

By comparing Eqs. 33-26 and 33-27, we see that the time-varying quantities Δv_R and i_R are both functions of $\sin \omega_d t$ with $\phi = 0°$. Thus, these two quantities are *in phase*, which means that their corresponding maxima (and minima) occur at the same times. Figure 33-8b, which is a plot of $\Delta v_R(t)$ and $i_R(t)$, illustrates this fact. Note that Δv_R and i_R do not decay here, because the generator supplies energy to the circuit to make up for the energy dissipated in R.

The time-varying quantities Δv_R and i_R can also be represented geometrically by *phasors*. Recall from Section 17-10 that phasors are vectors that rotate around an origin. Those that represent the voltage across and current in the resistor of Fig. 33-8a are shown in Fig. 33-8c at an arbitrary time t. Such phasors have the following properties:

Angular speed: Both phasors rotate counterclockwise about the origin with an angular speed equal to the angular frequency ω_d of Δv_R and i_R.

Length: The length of each phasor represents the amplitude of the alternating quantity: ΔV_R for the voltage and I_R for the current.

Projection: The projection of each phasor on the *vertical* axis represents the value of the alternating quantity at time t: Δv_R for the voltage and i_R for the current.

Rotation angle: The rotation angle of each phasor is equal to the phase of the alternating quantity at time t. In Fig. 33-8c, the voltage and current are in phase, so their phasors always have the same phase $\omega_d t$ and the same rotation angle, and thus they rotate together.

Mentally follow the rotation. Can you see that when the phasors have rotated so that $\omega_d t = 90°$ (they point vertically upward), they indicate that just then $\Delta v_R = \Delta V_R$? Equations 33-26 and 33-28 give the same results.

A Capacitive Load

Figure 33-9a shows a circuit containing a capacitance and a generator with the alternating emf of Eq. 33-24. Using the loop rule and proceeding as we did when we obtained Eq. 33-26, we find that the potential difference across the capacitor is

$$v_C = V_C \sin \omega_d t, \qquad (33-30)$$

where ΔV_C is the amplitude of the alternating voltage across the capacitor. From the definition of capacitance we can also write

$$q_C = C\Delta v_C = C\Delta V_C \sin \omega_d t. \qquad (33-31)$$

Our concern, however, is with the current rather than the charge. Thus, we differentiate Eq. 33-31 to find

$$i_C = \frac{dq_C}{dt} = \omega_d C\Delta V_C \cos \omega_d t. \qquad (33-32)$$

We now modify Eq. 33-32 in two ways. First, for reasons of symmetry of notation, we introduce the quantity X_C, called the capacitive reactance of a capacitor, defined as

$$X_C = \frac{1}{\omega_d C} \qquad \text{(capacitive reactance)}. \qquad (33-33)$$

Its value depends not only on the capacitance but also on the driving angular frequency ω_d. We know from the definition of the capacitive time constant ($\tau = RC$) that the SI unit for C can be expressed as seconds per ohm. Applying this to Eq. 33-33 shows that the SI unit of X_C is the *ohm*, just as for resistance R.

Second, we replace $\cos \omega_d t$ with a phase-shifted sine:

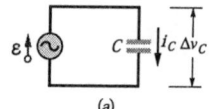

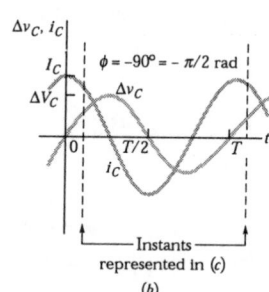

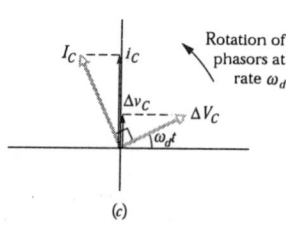

Fig. 33-9: (*a*) A capacitor is connected across an alternating-current generator. (*b*) The current in the capacitor leads the voltage by $90°(= \pi/2 \text{ rad})$. (*c*) A phasor diagram shows the same thing.

$$\cos\omega_d t = \sin(\omega_d t - 90°),$$

You can verify this identity by shifting a sine curve in the negative direction by 90°.

With these two modifications, Eq. 33-32 becomes

$$i_C = \left(\frac{\Delta V_C}{X_C}\right)\sin(\omega_d t + 90°). \qquad (33\text{-}34)$$

From Eq. 33-25, we can also write the current i_C in C as

$$i_C = I_C \sin(\omega_d t - \phi) \qquad (33\text{-}35)$$

where I_C is the amplitude of i_C. Comparing Eqs. 33-34 and 33-35, we see that for a purely capacitive load the phase constant ϕ for the current is –90°. We also see that the voltage amplitude and current amplitude are related by

$$\Delta V_C = I_C X_C \qquad \text{(capacitor).} \qquad (33\text{-}36)$$

Although we found this relation for the circuit of Fig. 33-9a, it applies to any capacitance in any ac circuit.

Comparison of Eqs. 33-30 and 33-34, or inspection of Fig. 33-9b, shows that the quantities Δv_C and i_C are 90°, or one-quarter cycle, out of phase. Furthermore, we see that i_C leads Δv_C, which means that, if you monitored the current i_C and the potential difference Δv_C in the circuit of Fig. 33-9a, you would find that i_C reaches its maximum before Δv_C does, by one-quarter cycle.

This relation between i_C and Δv_C is illustrated by the phasor diagram of Fig. 33-9c. As the phasors representing these two quantities rotate counterclockwise together, the phasor labeled I_C does indeed lead that labeled ΔV_C, and by an angle of 90°; that is, the phasor I_C coincides with the vertical axis one-quarter cycle before the phasor ΔV_C does. Be sure to convince yourself that the phasor diagram of Fig. 33-9c is consistent with Eqs. 33-30 and 33-34.

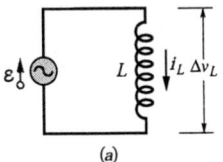

(a)

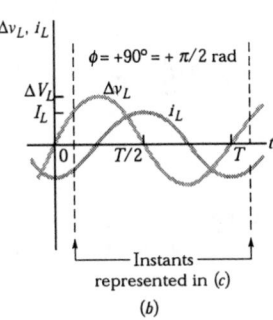

(b)

READING EXERCISE 33-3: The figure shows, in (a), a sine curve $S(t) = \sin(\omega_d t)$ and three other sinusoidal curves $A(t)$, $B(t)$, and $C(t)$, each of the form $\sin(\omega_d t - \phi)$. (a) Rank the three other curves according to the value of ϕ, most positive first and most negative last. (b) Which curve corresponds to which phasor in (b) of the figure? (c) Which curve leads the others?

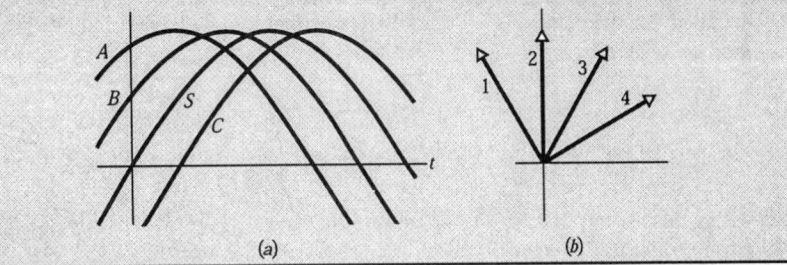

(a)   (b)

An Inductive Load

Figure 33-10a shows a circuit containing an inductance and a generator with the alternating emf of Eq. 33-24. Using the loop rule and proceeding as we did to obtain Eq. 33-26, we find that the potential difference across the inductance is

$$\Delta v_L = \Delta V_L \sin\omega_d t, \qquad (33\text{-}37)$$

where ΔV_L is the amplitude of Δv_L. From Eq. 31-32, we can write the potential difference across an inductance L, in which the current is changing at the rate di_L/dt, as

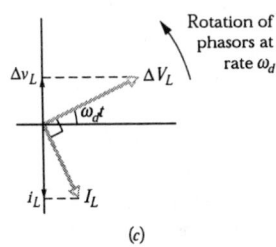

(c)

Fig. 33-10: (a) An inductor is connected across an alternating-current generator. (b) The current in the inductor lags the voltage by $90°(= \pi/2 \text{ rad})$. (c) A phasor diagram shows the same thing.

$$\Delta v_L = L\frac{di_L}{dt}. \qquad (33\text{-}38)$$

If we combine Eqs. 33-37 and 33-38 we have

$$\frac{di_L}{dt} = \frac{\Delta V_L}{L}\sin\omega_d t. \qquad (33\text{-}39)$$

Our concern, however, is with the current rather than with its time derivative. We find the former by integrating Eq. 33-39, obtaining

$$i_L = \int di_L = \frac{\Delta V_L}{L}\int \sin\omega_d t\, dt = -\left(\frac{\Delta V_L}{\omega_d L}\right)\cos\omega_d t. \qquad (33\text{-}40)$$

We now modify this equation in two ways. First, for reasons of symmetry of notation, we introduce the quantity X_L, called the **inductive reactance** of an inductor, which is defined as

$$X_L = \omega_d L \qquad \text{(inductive reactance).} \qquad (33\text{-}41)$$

The value of X_L depends on the driving angular frequency ω_d. The unit of the inductive time constant τ_L indicates that the SI unit of X_L is the *ohm*, just as it is for X_C and for R.

Second, we replace the function $-\cos\omega_d t$ in Eq. 33-40 with a phase-shifted sine—namely,

$$-\cos\omega_d t = \sin(\omega_d t - 90°).$$

You can verify this identity by shifting a sine curve in the positive direction by 90°.

With these two changes, Eq. 33-40 becomes

$$i_L = \left(\frac{\Delta V_L}{X_L}\right)\sin(\omega_d t - 90°). \qquad (33\text{-}42)$$

From Eq. 33-25, we can also write this current in the inductance as

$$i_L = I_L\sin(\omega_d t - \phi), \qquad (33\text{-}43)$$

where I_L is the amplitude of the current i_L. Comparing Eqs. 33-42 and 33-43, we see that for a purely inductive load the phase constant ϕ for the current is +90°. We also see that the voltage amplitude and current amplitude are related by

$$\Delta V_L = I_L X_L \qquad \text{(inductor).} \qquad (33\text{-}44)$$

Although we found this relation for the circuit of Fig. 33-10a, it applies to any inductance in any ac circuit.

Comparison of Eqs. 33-37 and 33-42, or inspection of Fig. 33-10b, shows that the quantities i_L and Δv_L are 90° out of phase. In this case, however, i_L *lags* Δv_L; that is, if you monitored the current i_L and the potential difference Δv_L in the circuit of Fig. 33-10a, you would find that i_L reaches its maximum value after Δv_L does, by one-quarter cycle.

The phasor diagram of Fig. 33-10c also contains this information. As the phasors rotate counterclockwise in the figure, the phasor labeled I_L does indeed lag that labeled ΔV_L, and by an angle of 90°. Be sure to convince yourself that Fig. 33-10c represents Eqs. 33-37 and 33-42.

Touchstone Examples 33-8-1, 33-8-2, and 33-8-3, at the end of this chapter, illustrate how to use what you learned in this section.

TE

33-9 The Series *RLC* Circuit

We are now ready to apply the alternating emf of Eq. 33-24,

$$\varepsilon = \varepsilon_m \sin \omega_d t \qquad \text{(applied emf)}, \qquad (33\text{-}45)$$

to the full *RLC* circuit of Fig. 33-7. Because R, L, and C are in series, the same current

$$i = I \sin(\omega_d t - \phi) \qquad (33\text{-}46)$$

is driven in all three of them. We wish to find the current amplitude I and the phase constant ϕ. The solution is simplified by the use of phasor diagrams.

Table 33-2 summarizes the relations between the current i and the voltage v for each of the three kinds of circuit elements we have considered. When an applied alternating voltage produces an alternating current in them, the current is in phase with the voltage across a resistor, leads the voltage across a capacitor, and lags the voltage across an inductor.

TABLE 33-2: Phase and Amplitude Relations for Alternating Currents and Voltages

Circuit Element	Symbol	Resistance or Reactance	Phase of the Current	Phase Constant (or Angle ϕ)	Amplitude Relation
Resistor	R	R	In phase with v_R	0° (= 0 rad)	$\Delta V_R = I_R R$
Capacitor	C	$X_C = 1/\omega_d C$	Leads Δv_C by 90° (= $\pi/2$ rad)	-90° (= $-\pi/2$ rad)	$\Delta V_C = I_C X_C$
Inductor	L	$X_L = \omega_d L$	Lags Δv_L by 90° (= $\pi/2$ rad)	+90° (= $+\pi/2$ rad)	$\Delta V_L = I_L X_L$

The Current Amplitude

We start with Fig. 33-11*a*, which shows the phasor representing the current of Eq. 33-46 at an arbitrary time t. The length of the phasor is the current amplitude I, the projection of the phasor on the vertical axis is the current i at time t, and the angle of rotation of the phasor is the phase $\omega_d t - \phi$ of the current at time t.

Figure 33-11*b* shows the phasors representing the voltages across R, L, and C at the same time t. Each phasor is oriented relative to the angle of rotation of current phasor I in Fig. 33-11*a*, based on the information in Table 33-2:

> *Resistor*: Here current and voltage are in phase, so the angle of rotation of voltage phasor ΔV_R is the same as that of phasor I.

> *Capacitor*: Here current leads voltage by 90°, so the angle of rotation of voltage phasor ΔV_C is 90° less than that of phasor I.

> *Inductor*: Here current lags voltage by 90°, so the angle of rotation of voltage phasor Δv_L is 90° greater than that of phasor I.

Figure 33-11*b* also shows the instantaneous voltages Δv_R, Δv_C, and Δv_L across R, C, and L at time t; those voltages are the projections of the corresponding phasors on the vertical axis of the figure.

Figure 33-11*c* shows the phasor representing the applied emf of Eq. 33-45. The length of the phasor is the emf amplitude ε_m, the projection of the phasor on the vertical axis is the emf ε at time t, and the angle of rotation of the phasor is the phase $\omega_d t$ of the emf at time t.

From the loop rule we know that at any instant the sum of the voltages Δv_R Δv_C, and Δv_L is equal to the applied emf ε :

$$\varepsilon = \Delta v_R + \Delta v_C + \Delta v_L. \qquad (33\text{-}47)$$

Thus, at time t the projection ε in Fig. 33-11*c* is equal to the algebraic sum of the projections Δv_R, Δv_C, and Δv_L in Fig. 33-11*b*. In fact, as the phasors rotate together, this equality always holds. This means that phasor ε_m in Fig. 33-11c must be equal to the vector sum of the three voltage phasors ΔV_R, ΔV_C, and ΔV_L in Fig. 33-11b.

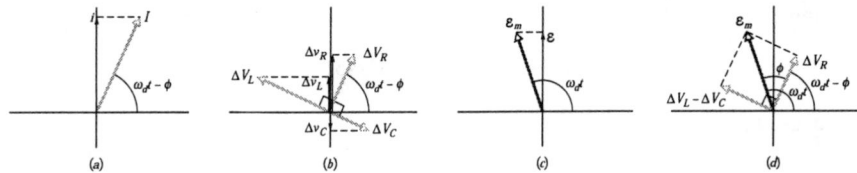

(a)   (b)   (c)   (d)

Fig. 33-11: (a) A phasor representing the alternating current in the driven RLC circuit of Fig. 33-7 at time t. The amplitude I, the instantaneous value i, and the phase $\omega_d t - \phi$ are shown. (b) Phasors representing the voltages across the inductor, resistor, and capacitor, oriented with respect to the current phasor in (a). (c) A phasor representing the alternating emf that drives the current of (a). (d) The emf phasor is equal to the vector sum of the three voltage phasors of (b). Here, voltage phasors ΔV_L and ΔV_C have been added to yield their net phasor ($\Delta V_L - \Delta V_C$).

That requirement is indicated in Fig. 33-11d, where phasor ε_m is drawn as the sum of phasors ΔV_R, ΔV_L, and ΔV_C. Because phasors ΔV_L and ΔV_C have opposite directions in the figure, we simplify the vector sum by first combining ΔV_L and ΔV_C to form the single phasor $\Delta V_L - \Delta V_C$. Then we combine that single phasor with ΔV_R to find the net phasor. Again, the net phasor must coincide with phasor ε_m, as shown.

Both triangles in Fig. 33-11d are right triangles. Applying the Pythagorean theorem to either one yields

$$\varepsilon_m^2 = \Delta V_R^2 + \left(\Delta V_L - \Delta V_C\right)^2. \tag{33-48}$$

From the amplitude information displayed in Table 33-2 we can rewrite this as

$$\varepsilon_m^2 = (IR)^2 + \left(IX_L - IX_C\right)^2, \tag{33-49}$$

and then rearrange it to the form

$$I = \frac{\varepsilon_m}{\sqrt{R^2 + \left(X_L - X_C\right)^2}}. \tag{33-50}$$

The denominator in Eq. 33-50 is called the **impedance** Z of the circuit for the driving angular frequency ω_d:

$$Z = \sqrt{R^2 + \left(X_L - X_C\right)^2} \qquad \text{(impedance defined).} \tag{33-51}$$

We can then write Eq. 33-50 as

$$I = \frac{\varepsilon_m}{Z}. \tag{33-52}$$

If we substitute for X_C and X_L from Eqs. 33-33 and 33-41, we can write Eq. 33-50 more explicitly as

$$I = \frac{\varepsilon_m}{\sqrt{R^2 + \left(\omega_d L - 1/\omega_d C\right)^2}} \qquad \text{(current amplitude).} \tag{33-53}$$

We have now accomplished half our goal: We have obtained an expression for the current amplitude I in terms of the sinusoidal driving emf and the circuit elements in a series RLC circuit.

The value of I depends on the difference between $\omega_d L$ and $1/\omega_d C$ in Eq. 33-53 or, equivalently, the difference between X_L and X_C in Eq. 33-50. In either equation, it does not matter which of the two quantities is greater because the difference is always squared.

The current that we have been describing in this section is the *steady-state current* that occurs after the alternating emf has been applied for some time. When the emf is first applied to a circuit, a brief *transient current* occurs. Its duration (before settling down into the steady-state current) is determined by the time constants $\tau_L = L/R$ and $\tau_C = RC$ as the inductive and capacitive elements "turn on." This transient current can be large and can, for example, destroy a motor on start-up if it is not properly taken into account in the motor's circuit design.

The Phase Constant

From the right-hand phasor triangle in Fig. 33-11*d* and from Table 33-2 we can write

$$\tan\phi = \frac{\Delta V_L - \Delta V_C}{\Delta V_R} = \frac{IX_L - IX_C}{IR}, \qquad (33\text{-}54)$$

which gives us

$$\tan\phi = \frac{X_L - X_C}{R} \qquad \text{(phase constant).} \qquad (33\text{-}55)$$

This is the other half of our goal: an equation for the phase constant ϕ in a sinusoidally driven series *RLC* circuit. In essence, it gives us three different results for the phase constant, depending on the relative values of X_L and X_C:

$X_L > X_C$: The circuit is said to be *more inductive than capacitive*. Equation 33-55 tells us that ϕ is positive for such a circuit, which means that phasor *I* rotates behind phasor ε_m (Fig. 33-12*a*). A plot of ε and *i* versus time is like that in Fig. 33-12*b*. (The phasors in Figs. 33-11*c* and *d* were drawn assuming $X_L > X_C$.)

$X_C > X_L$: The circuit is said to be *more capacitive than inductive*. Equation 33-55 tells us that ϕ is negative for such a circuit, which means that phasor *I* rotates ahead of phasor ε_m (Fig. 33-12*c*). A plot of ε and *i* versus time is like that in Fig. 33-12*d*.

$X_C = X_L$: The circuit is said to be in *resonance*, a state that is discussed next. Equation 33-55 tells us that $\phi = 0°$ for such a circuit, which means that phasors ε_m and *I* rotate together (Fig. 33-12*e*). A plot of ε and *i* versus time is like that in Fig. 33-12*f*.

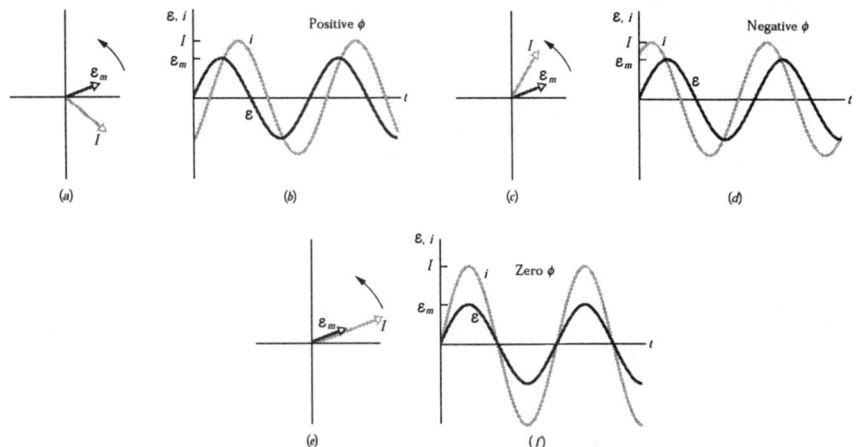

Fig. 33-12: Phasor diagrams and graphs of the alternating emf and current for the driven *RLC* circuit of Fig. 33-7. In the phasor diagram of (*a*) and the graph of (*b*), the current *i* lags the driving emf ε and the current's phase constant ϕ is positive. In (*c*) and (*d*), the current *i* leads the driving emf ε and its phase constant ϕ is negative. In (*e*) and (*f*), the current *i* is in phase with the driving emf ε and its phase constant ϕ is zero.

As illustration, let us reconsider two extreme circuits: In the *purely inductive circuit* of Fig. 33-10*a*, where X_L is nonzero and $X_C = R = 0$, Eq. 33-55 tells us that $\phi = +90°$ (the greatest value of ϕ), consistent with Fig. 33-10*c*. In the *purely capacitive circuit* of Fig. 33-9*a*, where X_C is nonzero and $X_L = R = 0$, Eq. 33-55 tells us that $\phi = -90°$ (the least value of ϕ), consistent with Fig. 33-9*c*.

Resonance

Equation 33-53 gives the current amplitude *I* in an *RLC* circuit as a function of the driving angular frequency ω_d of the external alternating emf. For a given resistance *R*, that amplitude is a maximum when the quantity $\omega_d L - 1/\omega_d C$ in the denominator is zero—that is, when

$$\omega_d L = \frac{1}{\omega_d C}$$

or $$\omega_d = \frac{1}{\sqrt{LC}} \qquad \text{(maximum } I\text{)}. \qquad (33\text{-}56)$$

Because the natural angular frequency ω of the *RLC* circuit is also equal to $1/\sqrt{LC}$, the maximum value of I occurs when the driving angular frequency matches the natural angular frequency—that is, at resonance. Thus, in an *RLC* circuit, resonance and maximum current amplitude I occur when

$$\omega_d = \omega = \frac{1}{\sqrt{LC}} \qquad \text{(resonance)}. \qquad (33\text{-}57)$$

Figure 33-13 shows three *resonance curves* for sinusoidally driven oscillations in three series *RLC* circuits differing only in R. Each curve peaks at its maximum current amplitude I when the ratio ω_d/ω is 1.00, but the maximum value of I decreases with increasing R. (The maximum I is always ε_m/R; to see why, combine Eqs. 33-51 and 33-52.) In addition, the curves increase in width (measured in Fig. 33-13 at half the maximum value of I) with increasing R.

To make physical sense of Fig. 33-13 consider how the reactances X_L and X_C change as we increase the driving angular frequency ω_d, starting with a value much less than the natural frequency ω. For small ω_d, reactance $X_L(=\omega_d L)$ is small and reactance $X_C(1/\omega_d C)$ is large. Thus, the circuit is mainly capacitive and the impedance is dominated by the large X_C, which keeps the current low.

As we increase ω_d, reactance X_C remains dominant but decreases while reactance X_L increases. The decrease in X_C decreases the impedance, allowing the current to increase, as we see on the left side of any resonance curve in Fig. 33-13. When the increasing X_L and the decreasing X_C reach equal values, the current is greatest and the circuit is in resonance, with $\omega_d = \omega$.

As we continue to increase ω_d, the increasing reactance X_L becomes progressively more dominant over the decreasing reactance X_C. The impedance increases because of X_L and the current decreases, as on the right side of any resonance curve in Fig. 33-13. In summary, then: The low-angular-frequency side of a resonance curve is dominated by the capacitor's reactance, the high-angular-frequency side is dominated by the inductor's reactance; and resonance occurs between the two regions.

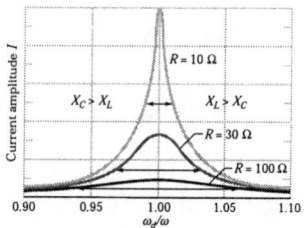

Fig. 33-13: *Resonance curves* for the driven *RLC* circuit of Fig. 33-7 with $L = 100\,\mu H$, $C = 100\,pF$, and three values of R. The current amplitude I of the alternating current depends on how close the driving angular frequency ω_d is to the natural angular frequency ω. The horizontal arrow on each curve measures the curve's width at the half-maximum level, a measure of the sharpness of the resonance. To the left of $\omega_d/\omega = 1.00$, the circuit is mainly capacitive, with $X_C > X_L$ to the right, it is mainly inductive, with $X_L > X_C$.

READING EXERCISE 33-4: Here are the capacitive reactance and inductive reactance, respectively, for three sinusoidally driven series *RLC* circuits: (1) $50\,\Omega$, $100\,\Omega$; (2) $100\,\Omega$, $50\,\Omega$; (3) $50\,\Omega$, $50\,\Omega$. (a) For each, does the current lead or lag the applied emf, or are the two in phase? (b) Which circuit is in resonance?

Touchstone Example 33-9-1, at the end of this chapter, illustrates how to use what you learned in this section.

33-10 Power in Alternating-Current Circuits

In the *RLC* circuit of Fig. 33-7, the source of energy is the alternating-current generator. Some of the energy that it provides is stored in the electric field in the capacitor, some is stored in the magnetic field in the inductor, and some is dissipated as thermal energy in the resistor. In steady-state operation—which we assume—the average energy stored in the capacitor and inductor together remains constant. The net transfer of energy is thus from the generator to the resistor, where electromagnetic energy is dissipated as thermal energy.

The instantaneous rate at which energy is dissipated in the resistor can be written, with the help of Eqs. 27-22 and 33-25, as

$$P = i^2 R = \left[I \sin(\omega_d t - \phi) \right]^2 R = I^2 R \sin^2(\omega_d t - \phi). \qquad (33\text{-}58)$$

The *average* rate at which energy is dissipated in the resistor, however, is the average of Eq. 33-58 over time. Although the average value of $\sin \theta$, where θ is any variable, is zero (Fig. 33-14a), the average value of $\sin^2 \theta$ over one complete cycle is $\frac{1}{2}$ (Fig. 33-14b). (Note in Fig. 33-14b how the shaded areas under the curve but above the horizontal line marked $+\frac{1}{2}$ exactly fill in the unshaded spaces below that line.) Thus, we can write, from Eq. 33-58,

$$P_{\text{avg}} = \frac{I^2 R}{2} = \left(\frac{I}{\sqrt{2}} \right)^2 R. \qquad (33\text{-}59)$$

The quantity $I/\sqrt{2}$ is called the **root-mean-square**, or **rms**, value of the current i:

$$I_{\text{rms}} = \frac{1}{\sqrt{2}} \qquad \text{(rms current)}. \qquad (33\text{-}60)$$

We can now rewrite Eq. 33-59 as

$$P_{\text{avg}} = I_{\text{rms}}^2 R \qquad \text{(average power)}. \qquad (33\text{-}61)$$

Equation 33-61 looks much like Eq. 26-21 $(P = i^2 R)$; the message is that if we switch to the rms current, we can compute the average rate of energy dissipation for alternating-current circuits just as for direct-current circuits.

We can also define rms values of voltages and emfs for alternating-current circuits:

$$\Delta V_{\text{rms}} = \frac{\Delta V}{\sqrt{2}} \quad \text{and} \quad \varepsilon_{\text{rms}} = \frac{\varepsilon_m}{\sqrt{2}} \qquad \text{(rms voltage; rms emf)}. \qquad (33\text{-}62)$$

Alternating-current instruments, such as ammeters and voltmeters, are usually calibrated to read I_{rms}, ΔV_{rms}, and ε_{rms}. Thus, if you plug an alternating-current voltmeter into a household electric outlet and it reads 120 V, that is an rms voltage. The maximum value of the potential difference at the outlet is $\sqrt{2} \times (120\,\text{V})$, or 170 V.

Because the proportionality factor $1/\sqrt{2}$ in Eqs. 33-60 and 33-62 is the same for all three variables, we can write Eqs. 33-52 and 33-50 as

$$I_{\text{rms}} = \frac{\varepsilon_{\text{rms}}}{Z} = \frac{\varepsilon_{\text{rms}}}{\sqrt{R^2 + (X_L - X_C)^2}}, \qquad (33\text{-}63)$$

and, indeed, this is the form that we almost always use.

We can use the relationship $I_{\text{rms}} = \varepsilon_{\text{rms}}/Z$ to recast Eq. 33-61 in a useful equivalent way. We write

$$P_{\text{avg}} = \frac{\varepsilon_{\text{rms}}}{Z} I_{\text{rms}} R = \varepsilon_{\text{rms}} I_{\text{rms}} \frac{R}{Z}. \qquad (33\text{-}64)$$

From Fig. 33-11d, Table 33-2, and Eq. 33-52, however, we see that R/Z is just the cosine of the phase constant ϕ:

$$\cos\phi = \frac{\Delta V_R}{\varepsilon_m} = \frac{IR}{IZ} = \frac{R}{Z}. \qquad (33\text{-}65)$$

Equation 33-64 then becomes

$$P_{\text{avg}} = \varepsilon_{\text{rms}} I_{\text{rms}} \cos\phi \qquad \text{(average power)}, \qquad (33\text{-}66)$$

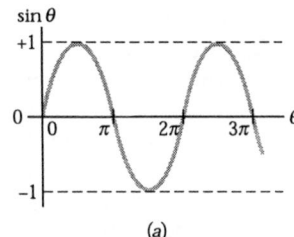

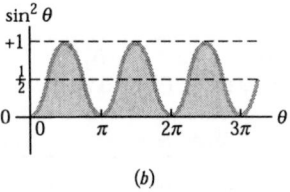

Fig. 33-14: (a) A plot of $\sin \theta$ versus θ. The average value over one cycle is zero. (b) A plot of $\sin^2 \theta$ versus θ. The average value over one cycle is $\frac{1}{2}$.

in which the term $\cos\phi$ is called the **power factor**. Because $\cos\phi = \cos(-\phi)$, Eq. 33-66 is independent of the sign of the phase constant ϕ.

To maximize the rate at which energy is supplied to a resistive load in an RLC circuit, we should keep the power factor $\cos \phi$ as close to unity as possible. This is equivalent to keeping the phase constant ϕ in Eq. 33-25 as close to zero as possible. If, for example, the circuit is highly inductive, it can be made less so by putting more capacitance in the circuit, connected in series. (Recall that putting an additional capacitance into a series of capacitances decreases the equivalent capacitance C_{eq} of the series.) Thus, the resulting decrease in C_{eq} in the circuit reduces the phase constant and increases the power factor in Eq. 33-66. Power companies place series-connected capacitors throughout their transmission systems to get these results.

READING EXERCISE 33-5: (a) If the current in a sinusoidally driven series RLC circuit leads the emf, would we increase or decrease the capacitance to increase the rate at which energy is supplied to the resistance? (b) Will this change bring the resonant angular frequency of the circuit closer to the angular frequency of the emf or put it further away?

Touchstone Example 33-10-1, at the end of this chapter, illustrates how to use what you learned in this section.

TE

33-11 Transformers

Energy Transmission Requirements

When an ac circuit has only a resistive load, the power factor in Eq. 33-66 is $\cos 0° = 1$ and the applied rms emf ε_{rms} is equal to the rms voltage ΔV_{rms} across the load. Thus, with an rms current I_{rms} in the load, energy is supplied and dissipated at the average rate of

$$P_{avg} = \varepsilon I = I\Delta V. \qquad (33\text{-}67)$$

(In Eq. 33-67 and the rest of this section, we follow conventional practice and drop the subscripts identifying rms quantities. Engineers and scientists assume that all time-varying currents and voltages are reported as rms values; that is what the meters read.) Equation 33-67 tells us that, to satisfy a given power requirement, we have a range of choices, from a relatively large current I and a relatively small voltage ΔV to just the reverse, provided only that the product $I\Delta V$ is as required.

In electric power distribution systems it is desirable for reasons of safety and for efficient equipment design to deal with relatively low voltages at both the generating end (the electric power plant) and the receiving end (the home or factory). Nobody wants an electric toaster or a child's electric train to operate at, say, 10 kV. On the other hand, in the transmission of electric energy from the generating plant to the consumer, we want the lowest practical current (hence the largest practical voltage) to minimize I^2R losses (often called *ohmic losses*) in the transmission line.

As an example, consider the 735 kV line used to transmit electric energy from the La Grande 2 hydroelectric plant in Quebec to Montreal, 1000 km away. Suppose that the current is 500 A and the power factor is close to unity. Then from Eq. 33-67, energy is supplied at the average rate

$$P_{avg} = \varepsilon I = (7.35 \times 10^5 \text{ V})(500 \text{ A}) = 368 \text{ MW}.$$

The resistance of the transmission line is about $0.220 \ \Omega/\text{km}$; thus, there is a total resistance of about $220 \ \Omega$ for the 1000 km stretch. Energy is dissipated owing to that resistance at a rate of about

$$P_{avg} = I^2 R = (500 \text{ A})^2 (220 \ \Omega) = 55.0 \text{ MW},$$

At 5:17 p.m. on November 9, 1965, a faulty relay in the power system near Niagara Falls opened a circuit breaker on a transmission line, automatically causing the current to switch to other lines, which overloaded those lines and made other circuit breakers open. Within minutes a runaway shutdown had blacked out much of New York, New England, and Ontario.

which is nearly 15% of the supply rate.

Imagine what would happen if we doubled the current and halved the voltage. Energy would be supplied by the plant at the same average rate of 368 MW as previously, but now energy would be dissipated at the rate of about

$$P_{avg} = I^2 R = (1000\,A)^2 (220\,\Omega) = 220\ \text{MW},$$

which is *almost 60% of the supply rate*. Hence the general energy transmission rule: Transmit at the highest possible voltage and the lowest possible current.

The Ideal Transformer

The transmission rule leads to a fundamental mismatch between the requirement for efficient high-voltage transmission and the need for safe low-voltage generation and consumption. We need a device with which we can raise (for transmission) and lower (for use) the ac voltage in a circuit, keeping the product current × voltage essentially constant. The **transformer** is such a device. It has no moving parts, operates by Faraday's law of induction, and has no simple direct-current counterpart.

The *ideal transformer* in Fig. 33-15 consists of two coils, with different numbers of turns, wound around an iron core. (The coils are insulated from the core.) In use, the primary winding, of N_p turns, is connected to an alternating-current generator whose emf ε at any time t is given by

$$\varepsilon = \varepsilon_m \sin \omega t. \tag{33-68}$$

The secondary winding, of N_s turns, is connected to load resistance R, but its circuit is an open circuit as long as switch S is open (which we assume for the present). Thus, there can be no current through the secondary coil. We assume further for this ideal transformer that the resistances of the primary and secondary windings are negligible, as are energy losses due to magnetic hysteresis in the iron core. Well-designed, high-capacity transformers can have energy losses as low as 1%, so our assumptions are reasonable.

For the assumed conditions, the primary winding (or *primary*) is a pure inductance, and the primary circuit is like that in Fig. 33-10a. Thus, the (very small) primary current, also called the *magnetizing current* I_{mag}, lags the primary voltage ΔV_p by 90°; the primary's power factor ($= \cos\phi$ in Eq. 33-66) is zero, so no power is delivered from the generator to the transformer.

However, the small alternating primary current I_{mag} induces an alternating magnetic flux Φ_B in the iron core. Because the core extends through the secondary winding (or *secondary*), this induced flux also extends through the turns of the secondary. From Faraday's law of induction (Eq. 31-6), the induced emf per turn ε_{turn} is the same for both the primary and the secondary. Also, the voltage ΔV_p across the primary is equal to the emf induced in the primary, and the voltage ΔV_s across the secondary is equal to the emf induced in the secondary. Thus, we can write

$$\varepsilon_{turn} = \frac{d\Phi_B}{dt} = \frac{\Delta V_p}{N_p} = \frac{\Delta V_s}{N_s}$$

and thus,

$$\Delta V_s = \Delta V_p \frac{N_s}{N_p} \qquad \text{(transformation of voltage).} \tag{33-69}$$

If $N_s > N_p$, the transformer is called a *step-up transformer* because it steps the primary's voltage ΔV_p up to a higher voltage ΔV_s. Similarly, if $N_s < N_p$, the device is a *step-down transformer*.

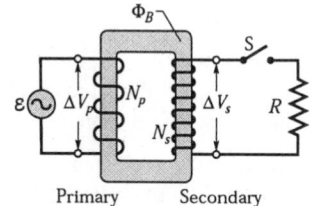

Fig. 33-15: An ideal transformer (two coils wound on an iron core) in a basic transformer circuit. An ac generator produces current in the coil at the left (the *primary*). The coil at the right (the *secondary*) is connected to the resistive load R when switch S is closed.

So far, with switch S open, no energy is transferred from the generator to the rest of the circuit. Now let us close S to connect the secondary to the resistive load R. (In general, the load would also contain inductive and capacitive elements, but here we consider just resistance R.) We find that now energy is transferred from the generator. Let us see why.

Several things happen when we close switch S.

1. An alternating current I_s appears in the secondary circuit, with corresponding energy dissipation rate $I_s^2 R (= (\Delta V_s^2)/R)$ in the resistive load.

2. This current produces its own alternating magnetic flux in the iron core, and this flux induces (from Faraday's law and Lenz's law) an opposing emf in the primary windings.

3. The voltage ΔV_p of the primary, however, cannot change in response to this opposing emf because it must always be equal to the emf ε that is provided by the generator; closing switch S cannot change this fact.

4. To maintain ΔV_p, the generator now produces (in addition to I_{mag}) an alternating current I_p in the primary circuit; the magnitude and phase constant of I_p are just those required for the emf induced by I_p in the primary to exactly cancel the emf induced there by I_s. Because the phase constant of I_p is not 90° like that of I_{mag}, this current I_p can transfer energy to the primary.

We want to relate I_s to I_p. However, rather than analyze the foregoing complex process in detail, let us just apply the principle of conservation of energy. The rate at which the generator transfers energy to the primary is equal to $I_p \Delta V_p$. The rate at which the primary then transfers energy to the secondary (via the alternating magnetic field linking the two coils) is $I_s \Delta V_s$. Because we assume that no energy is lost along the way, conservation of energy requires that

$$I_p \Delta V_p = I_s \Delta V_s.$$

Substituting for ΔV_s from Eq. 33-69, we find that

$$I_s = I_p \frac{N_p}{N_s} \qquad \text{(transformation of currents).} \qquad (33\text{-}70)$$

This equation tells us that the current I_s in the secondary can be greater than, less than, or the same as the current I_p in the primary, depending on the *turns ratio* N_p/N_s.

Current I_p appears in the primary circuit because of the resistive load R in the secondary circuit. To find I_p, we substitute $I_s = \Delta V_s / R$ into Eq. 33-70 and then we substitute for ΔV_s from Eq. 33-69. We find

$$I_p = \frac{1}{R} \left(\frac{N_s}{N_p} \right)^2 \Delta V_p. \qquad (33\text{-}71)$$

This equation has the form $I_p = \Delta V_p / R_{eq}$, where equivalent resistance R_{eq} is

$$R_{eq} = \left(\frac{N_p}{N_s} \right)^2 R. \qquad (33\text{-}72)$$

This R_{eq} is the value of the load resistance as "seen" by the generator; the generator produces the current I_p and voltage ΔV_p as if it were connected to a resistance R_{eq}.

Impedance Matching

Equation 33-72 suggests still another function for the transformer. For maximum transfer of energy from an emf device to a resistive load, the resistance of the emf device and the resistance of the load must be equal. The same relation holds for ac circuits except that the *impedance* (rather than just the resistance) of the generator must be matched to that of the load. Often this condition is not met. For example, in a music playing system, the amplifier has high impedance and the speaker set has low impedance. We can match the impedances of the two devices by coupling them through a transformer with a suitable turns ratio N_p/N_s.

READING EXERCISE 33-6: An alternating-current emf device has a smaller resistance than that of the resistive load; to increase the transfer of energy from the device to the load, a transformer will be connected between the two. (a) Should N_s be greater than or less than N_p? (b) Will that make it a step-up or step-down transformer?

Touchstone Example 33-2-1

A 1.5 μF capacitor is charged to 57 V. The charging battery is then disconnected, and a 12 mH coil is connected in series with the capacitor so that LC oscillations occur. What is the maximum current in the coil? Assume that the circuit contains no resistance.

SOLUTION: The **Key Ideas** here are these:

1. Because the circuit contains no resistance, the electromagnetic energy of the circuit is conserved as the energy is transferred back and forth between the electric field of the capacitor and the magnetic field of the coil (inductor).

2. At any time t, the energy $U_B(t)$ of the magnetic field is related to the current $i(t)$ through the coil by Eq. 32-2 ($U_B = Li^2/2$). When all the energy is stored as magnetic energy, the current is at its maximum value I and that energy is $U_{B,max} = LI^2/2$.

3. At any time t, the energy $U_E(t)$ of the electric field is related to the charge $q(t)$ on the capacitor by Eq. 33-1 ($U_E = q^2/2C$). When all the energy is stored as electric energy, the charge is at its maximum value Q and that energy is $U_{E,max} = Q^2/2C$.

With these ideas, we can now write the conservation of energy as

$$U_{B,max} = U_{E,max}$$

or

$$\frac{LI^2}{2} = \frac{Q^2}{2C}.$$

Solving for I gives us

$$I = \sqrt{\frac{Q^2}{LC}}.$$

We know L and C, but not Q. However, with Eq. 28-1 ($|q| = C|\Delta V|$) we can relate Q to the maximum potential difference ΔV across the capacitor, which is the initial potential difference of 57 V. Thus, substituting $Q = C\Delta V$ leads to a current magnitude of

$$|I| = |\Delta V|\sqrt{\frac{C}{L}} = (57 \text{ V}) \sqrt{\frac{1.5 \times 10^{-6} \text{ F}}{12 \times 10^{-3} \text{ H}}}$$

$$= 0.637 \text{ A} \approx 640 \text{ mA}. \qquad \text{(Answer)}$$

Touchstone Example 33-4-1

For the situation described in Touchstone Example 33-2-1, let the coil (inductor) be connected to the charged capacitor at time $t = 0$. The result is an LC circuit like that in Fig. 33-1.

(a) What is the potential difference $v_L(t)$ across the inductor as a function of time?

SOLUTION: One **Key Idea** here is that the current and potential differences of the circuit undergo sinusoidal oscillations. Another **Key Idea** is that we can still apply the loop rule to this oscillating circuit—just as we did for the nonoscillating circuits of Chapter 26. At any time t during the oscillations, the loop rule and Fig. 33-1 give us

$$\Delta v_L(t) = \Delta v_C(t); \qquad \text{(TE33-1)}$$

that is, the potential difference Δv_L across the inductor must always be equal to the potential difference v_C across the capacitor, so that the net potential difference around the circuit is zero. Thus, we will find $\Delta v_L(t)$ if we can find $\Delta v_C(t)$, and we can find $\Delta v_C(t)$ from $q(t)$ with Eq. 28-1 ($|q| = C\Delta V$).

Because the potential difference $\Delta v_C(t)$ is maximum when the oscillations begin at time $t = 0$, the charge q on the capacitor must also be maximum then. Thus, phase constant ϕ must be zero, so that Eq. 33-12 gives us

$$q = Q \cos \omega t. \qquad \text{(TE33-2)}$$

(Note that this cosine function does indeed yield maximum q ($= Q$) when $t = 0$.) To get the potential difference $v_C(t)$, we divide both sides of Eq. 33-2 by C to write

$$\frac{|q|}{C} = \frac{|Q|}{C} \cos \omega t,$$

and then use Eq. 28-1 to write

$$\Delta v_C = \Delta V_C \cos \omega t. \tag{TE33-3}$$

Here, ΔV_C is the amplitude of the oscillations in the potential difference Δv_C across the capacitor.
    Next, substituting $\Delta v_C = \Delta v_L$ from Eq. 33-1, we find

$$\Delta v_L = \Delta V_C \cos \omega t. \tag{TE33-4}$$

We can evaluate the right side of this equation by first noting that the amplitude ΔV_C is equal to the initial (maximum) potential difference of 57 V across the capacitor. Then, using the values of L and C from TE33-2-1, we find ω with Eq. 33-4:

$$\omega = \frac{1}{\sqrt{LC}} = \frac{1}{[(0.012 \text{ H})(1.5 \times 10^{-6} \text{ F})]^{0.5}}$$

$$= 7454 \text{ rad/s} \approx 7500 \text{ rad/s}.$$

Thus, Eq. TE33-4 becomes

$$\Delta v_L = (57 \text{ V}) \cos(7500 \text{ rad/s})t. \tag{Answer}$$

(b) What is the maximum rate $(di/dt)_{max}$ at which the current i changes in the circuit?

SOLUTION: The **Key Idea** here is that, with the charge on the capacitor oscillating as in Eq. 33-12, the current is in the form of Eq. 33-13. Because $\phi = 0$, that equation gives us

$$|i| = -\omega |Q| \sin \omega t.$$

Then

$$\left| \frac{di}{dt} \right| = \frac{d}{dt} (-\omega |Q| \sin \omega t) = -\omega^2 |Q| \cos \omega t.$$

We can simplify this equation by substituting $C|\Delta V_C|$ for $|Q|$ (because we know C and $|\Delta V_C|$ but not $|Q|$) and $1/\sqrt{LC}$ for ω according to Eq. 33-4. We get

$$\frac{di}{dt} = -\frac{1}{LC} C \Delta V_C \cos \omega t = -\frac{\Delta V_C}{L} \cos \omega t.$$

This tells us that the current changes at a varying (sinusoidal) rate, with its maximum rate of change being

$$\frac{|\Delta V_C|}{L} = \frac{57 \text{ V}}{0.012 \text{ H}} = 4750 \text{ A/s} \approx 4800 \text{ A/s}. \tag{Answer}$$

Touchstone Example 33-5-1

A series *RLC* circuit has inductance $L = 12$ mH, capacitance $C = 1.6$ μF, and resistance $R = 1.5$ Ω.

(a) At what time t will the amplitude of the charge oscillations in the circuit be 50% of its initial value?

SOLUTION: The **Key Idea** here is that the amplitude of the charge oscillations decreases exponentially with time t: According to Eq. 33-21, the charge amplitude at any time t is $|Q|e^{-Rt/2L}$, in which $|Q|$ is the amplitude at time $t = 0$. We want the time when the charge amplitude has decreased to $0.500|Q|$, that is, when

$$|Q|e^{-Rt/2L} = 0.50|Q|.$$

Canceling $|Q|$ and taking the natural logarithms of both sides, we have

$$-\frac{Rt}{2L} = \ln 0.50.$$

Solving for t and then substituting given data yield

$$t = -\frac{2L}{R} \ln 0.50 = -\frac{(2)(12 \times 10^{-3} \text{ H})(\ln 0.50)}{1.5 \ \Omega}$$

$$= 0.0111 \text{ s} \approx 11 \text{ ms.} \qquad \text{(Answer)}$$

(b) How many oscillations are completed within this time?

SOLUTION: The **Key Idea** here is that the time for one complete oscillation is the period $T = 2\pi/\omega$, where the angular frequency for LC oscillations is given by Eq. 33-4 ($\omega = 1/\sqrt{LC}$). Thus, in the time interval $\Delta t = 0.0111$ s, the number of complete oscillations is

$$\frac{\Delta t}{T} = \frac{\Delta t}{2\pi \sqrt{LC}}$$

$$= \frac{0.0111 \text{ s}}{2\pi[(12 \times 10^{-3} \text{ H})(1.6 \times 10^{-6} \text{ F})]^{1/2}} \approx 13. \qquad \text{(Answer)}$$

Thus, the amplitude decays by 50% in about 13 complete oscillations. This damping is less severe than that shown in Fig. 33-3, where the amplitude decays by a little more than 50% in one oscillation.

Touchstone Example 33-8-1

Purely resistive load. In Fig. 33-8a, resistance R is 200 Ω and the sinusoidal alternating emf device operates at amplitude $\varepsilon_m = 36.0$ V and frequency $f_d = 60.0$ Hz.

(a) What is the potential difference $\Delta v_R(t)$ across the resistance as a function of time t, and what is the amplitude ΔV_R of $\Delta v_R(t)$?

SOLUTION: When we applied the loop rule to the circuit of Fig. 33-8a, we found this **Key Idea**: In a circuit with a purely resistive load, the potential difference $\Delta v_R(t)$ across the resistance is always equal to the potential difference $\varepsilon(t)$ across the emf device. Thus, $\Delta v_R(t) = \varepsilon(t)$ and $\Delta V_R = \varepsilon_m$. Since ε_m is given, we can write

$$\Delta V_R = \varepsilon_m = 36.0 \text{ V.} \qquad \text{(Answer)}$$

To find $\Delta v_R(t)$, we use Eq. 33-24 to write

$$\Delta v_R(t) = \varepsilon(t) = \varepsilon_m \sin \omega_d t, \qquad \text{(TE33-5)}$$

and then substitute $\varepsilon_m = 36.0$ V and

$$\omega_d = 2\pi f_d = 2\pi(60 \text{ Hz}) = 120\pi$$

to obtain

$$\Delta v_R = (36.0 \text{ V}) \sin(120\pi t). \qquad \text{(Answer)}$$

We can leave the argument of the sine in this form for convenience, or we can write it as (377 rad/s)t or as (377 s^{-1})t.

(b) What are the current $i_R(t)$ in the resistance and the amplitude I_R of $i_R(t)$?

SOLUTION: The **Key Idea** here is that in an ac circuit with a purely resistive load, the alternating current $i_R(t)$ in the resistance is *in phase* with the alternating potential difference $\Delta v_R(t)$ across the resistance; that is, the phase constant ϕ for the current is zero. Thus, we can write

Eq. 33-25 as

$$i_R = |I_R| \sin(\omega_d t - \phi) = I_R \sin \omega_d t. \qquad \text{(TE33-6)}$$

From Eq. 33-29, the amplitude I_R is

$$|I_R| = \frac{|\Delta V_R|}{R} = \frac{36.0 \text{ V}}{200 \text{ }\Omega} = 0.180 \text{ A}. \qquad \text{(Answer)}$$

Substituting this and $\omega_d = 2\pi f_d = 120\pi$ into Eq. TE33-7, we have

$$|i_R| = (0.180 \text{ A}) \sin(120\pi t). \qquad \text{(Answer)}$$

Touchstone Example 33-8-2

Purely capacitive load. In Fig. 33-9a, capacitance C is 15.0 μF and the sinusoidal alternating emf device operates at amplitude $\varepsilon_m = 36.0$ V and frequency $f_d = 60.0$ Hz.

(a) What is the potential difference $\Delta v_C(t)$ across the capacitance and the amplitude ΔV_C of $\Delta v_C(t)$?

SOLUTION: If we apply the loop rule to the circuit of Fig. 33-9a, we find this **Key Idea**: In a circuit with a purely capacitive load, the potential difference $\Delta v_C(t)$ across the capacitance is always equal to the potential difference $\varepsilon(t)$ across the emf device. Thus, $\Delta v_C(t) = \varepsilon(t)$ and $\Delta V_C = \varepsilon_m$. Since ε_m is given, we have

$$|\Delta V_C| = |\varepsilon_m| = 36.0 \text{ V}. \qquad \text{(Answer)}$$

To find $\Delta v_C(t)$, we use Eq. 33-24 to write

$$\Delta v_C(t) = \varepsilon(t) = |\varepsilon_m| \sin \omega_d t. \qquad \text{(TE33-8)}$$

Then, substituting $\varepsilon_m = 36.0$ V and $\omega_d = 2\pi f_d = 120\pi$ into Eq. TE33-8, we have

$$\Delta v_C = (36.0 \text{ V}) \sin(120\pi t). \qquad \text{(Answer)}$$

(b) What is the current $i_C(t)$ in the circuit as a function of time and the amplitude I_C of $i_C(t)$?

SOLUTION: The **Key Idea** here is that in an ac circuit with a purely capacitive load, the alternating current $i_C(t)$ in the capacitance leads the alternating potential difference $\Delta v_C(t)$ by 90°; that is, the phase constant ϕ for the current is $-90°$ or $-\pi/2$ rad. Thus, we can write Eq. 33-25 as

$$i_C = |I_C| \sin(\omega_d t - \phi) = |I_C| \sin(\omega_d t + \pi/2). \qquad \text{(TE33-9)}$$

A second **Key Idea** is that we can find the amplitude I_C from Eq. 33-36 ($\Delta V_C = I_C X_C$) if we first find the capacitive reactance X_C. From Eq. 33-33 ($X_C = 1/\omega_d C$), with $\omega_d = 2\pi f_d$, we can write

$$X_C = \frac{1}{2\pi f_d C} = \frac{1}{(2\pi)(60.0 \text{ Hz})(15.0 \times 10^{-6} \text{ F})}$$
$$= 177 \text{ }\Omega.$$

Then Eq. 33-36 tells us that the current amplitude is

$$|I_C| = \frac{|\Delta V_C|}{X_C} = \frac{36.0 \text{ V}}{177 \text{ }\Omega} = 0.203 \text{ A}. \qquad \text{(Answer)}$$

Substituting this and $\omega_d = 2\pi f_d = 120\pi$ into Eq. TE33-9, we have

$$i_C = (0.203 \text{ A}) \sin(120\pi t + \pi/2). \qquad \text{(Answer)}$$

Touchstone Example 33-8-3

Purely inductive load. In Fig. 33-10*a*, inductance L is 230 mH and the sinusoidal alternating emf device operates at amplitude $\varepsilon_m = 36.0$ V and frequency $f_d = 60.0$ Hz.

(a) What is the potential difference $\Delta v_L(t)$ across the inductance and the amplitude ΔV_L of $\Delta v_L(t)$?

SOLUTION: If we apply the loop rule to the circuit in Fig. 33-10*a*, we find this **Key Idea**: In a circuit with a purely inductive load, the potential difference $\Delta v_L(t)$ across the inductance is always equal to the potential difference $\varepsilon(t)$ across the emf device. Thus, $\Delta v_L(t) = \varepsilon(t)$ and $\Delta V_L = \varepsilon_m$. Since ε_m is given, we have

$$|\Delta V_L| = |\varepsilon_m| = 36.0 \text{ V}. \qquad \text{(Answer)}$$

To find $\Delta v_L(t)$, we use Eq. 33-24 to write

$$\Delta v_L(t) = \varepsilon(t) = |\varepsilon_m| \sin \omega_d t. \qquad \text{(TE33-10)}$$

Then, substituting $\varepsilon_m = 36.0$ V and $\omega_d = 2\pi f_d = 120\pi$ into Eq. TE33-10, we have

$$\Delta v_L = (36.0 \text{ V}) \sin(120\pi t). \qquad \text{(Answer)}$$

(b) What is the current $i_L(t)$ in the circuit as a function of time and the amplitude I_L of $i_L(t)$?

SOLUTION: The **Key Idea** here is that in an ac circuit with a purely inductive load, the alternating current $i_L(t)$ in the inductance lags the alternating potential difference $\Delta v_L(t)$ by 90°. Thus, the phase constant ϕ for the current is +90° or $+\pi/2$ rad, and we can write Eq. 33-25 as

$$i_L = |I_L| \sin(\omega_d t - \phi) = |I_L| \sin(\omega_d t - \pi/2). \qquad \text{(TE33-11)}$$

A second **Key Idea** is that we can find the amplitude I_L from Eq. 33-44 ($|\Delta V_L| = |I_L||X_L|$) if we first find the inductive reactance X_L. From Eq. 33-41 ($X_L = \omega_d L$), with $\omega_d = 2\pi f_d$, we can write

$$X_L = 2\pi f_d L = (2\pi)(60.0 \text{ Hz})(230 \times 10^{-3} \text{ H})$$
$$= 86.7 \ \Omega.$$

Then Eq. 33-44 tells us that the current amplitude is

$$|I_L| = \frac{|\Delta V_L|}{X_L} = \frac{36.0 \text{ V}}{86.7 \ \Omega} = 0.415 \text{ A}. \qquad \text{(Answer)}$$

Substituting this and $\omega_d = 2\pi f_d = 120\pi$ into Eq. TE33-11, we have

$$i_L = (0.415 \text{ A}) \sin(120\pi t - \pi/2). \qquad \text{(Answer)}$$

Touchstone Example 33-9-1

In Fig. 33-7 let $R = 200 \ \Omega$, $C = 15.0 \ \mu$F, $L = 230$ mH, $f_d = 60.0$ Hz, and $\varepsilon_m = 36.0$ V. (These parameters are those used in Touchstone Examples 33-8-1, 33-8-2, and 33-8-3.)

(a) What is the current amplitude *I*?

SOLUTION: The **Key Idea** here is that current amplitude *I* depends on the amplitude ε_m of the driving emf and on the impedance *Z* of the circuit, according to Eq. 33-52 ($I = \varepsilon_m/Z$). Thus, we need to find *Z*, which depends on the circuit's resistance *R*, capacitive reactance X_C, and inductive reactance X_L.

The circuit's only resistance is the given resistance R. Its only capacitive reactance is due to the given capacitance and, from TE 33-8-2, $X_C = 177\ \Omega$. Its only inductive reactance is due to the given inductance and, from TE 33-8-3, $X_L = 86.7\ \Omega$. Thus, the circuit's impedance is

$$Z = \sqrt{R^2 + (X_L - X_C)^2}$$
$$= \sqrt{(200\ \Omega)^2 + (86.7\ \Omega - 177\ \Omega)^2}$$
$$= 219\ \Omega.$$

We then find

$$|I| = \frac{|\varepsilon_m|}{Z} = \frac{36.0\ \text{V}}{219\ \Omega} = 0.164\ \text{A}. \qquad \text{(Answer)}$$

(b) **What is the phase constant ϕ of the current in the circuit relative to the driving emf?**

SOLUTION: The **Key Idea** here is that the phase constant depends on the inductive reactance, the capacitive reactance, and the resistance of the circuit, according to Eq. 33-55. Solving that equation for ϕ leads to

$$\phi = \tan^{-1} \frac{X_L - X_C}{R} = \tan^{-1} \frac{86.7\ \Omega - 177\ \Omega}{200\ \Omega}$$
$$= -24.3° = -0.424\ \text{rad}. \qquad \text{(Answer)}$$

The negative phase constant is consistent with the fact that the load is mainly capacitive; that is, $X_C > X_L$.

Touchstone Example 33-10-1

A series *RLC* circuit, driven with $\varepsilon_{rms} = 120$ V at frequency $f_d = 60.0$ Hz, contains a resistance $R = 200\ \Omega$, an inductance with $X_L = 80.0\ \Omega$, and a capacitance with $X_C = 150\ \Omega$.

(a) **What are the power factor $\cos \phi$ and phase constant ϕ of the circuit?**

SOLUTION: The **Key Idea** here is that the power factor $\cos \phi$ can be found from the resistance R and impedance Z via Eq. 33-65 ($\cos \phi = R/Z$). To calculate Z, we use Eq. 33-51:

$$Z = \sqrt{R^2 + (X_L - X_C)^2}$$
$$= \sqrt{(200\ \Omega)^2 + (80.0\ \Omega - 150\ \Omega)^2} = 211.90\ \Omega.$$

Equation 33-65 then gives us

$$\cos \phi = \frac{R}{Z} = \frac{200\ \Omega}{211.90\ \Omega} = 0.9438 \approx 0.944. \qquad \text{(Answer)}$$

Taking the inverse cosine then yields

$$\phi = \cos^{-1} 0.944 = \pm 19.3°.$$

Both $+19.3°$ and $-19.3°$ have a cosine of 0.944. To determine which sign is correct, we must consider whether the current leads or lags the driving emf. Because $X_C > X_L$, this circuit is mainly capacitive, with the current leading the emf. Thus, ϕ must be negative:

$$\phi = -19.3°. \qquad \text{(Answer)}$$

We could, instead, have found ϕ with Eq. 33-55. A calculator would then have given us the complete answer, with the minus sign.

(b) **What is the average rate P_{avg} at which energy is dissipated in the resistance?**

SOLUTION: One way to answer this question is to use this **Key Idea**: Because the circuit is assumed to be in steady-state operation, the rate at which energy is dissipated in the resistance is equal to the rate at which energy is supplied to the circuit, as given by Eq. 33-66 ($P_{avg} = \varepsilon_{rms} I_{rms} \cos \phi$).

We are given the rms driving emf ε_{rms} and we know $\cos \phi$ from part (a). To find I_{rms} we use the **Key Idea** that the rms current is determined by the rms value of the driving emf and the circuit's impedance Z (which we know), according to Eq. 33-63:

$$|I_{rms}| = \frac{|\varepsilon_{rms}|}{Z}.$$

Substituting this into Eq. 33-66 then leads to

$$P_{avg} = |\varepsilon_{rms}||I_{rms}|\cos \phi = \frac{\varepsilon_{rms}^2}{Z} \cos \phi$$

$$= \frac{(120 \text{ V})^2}{211.90 \text{ } \Omega}(0.9438) = 64.1 \text{ W}. \qquad \text{(Answer)}$$

A second way to answer the question is to use the **Key Idea** that the rate at which energy is dissipated in a resistance R depends on the square of the rms current I_{rms} through it, according to Eq. 33-61. We then find

$$P_{avg} = I_{rms}^2 R = \frac{\varepsilon_{rms}^2}{Z^2} R$$

$$= \frac{(120 \text{ V})^2}{(211.90 \text{ } \Omega)^2}(200 \text{ } \Omega) = 64.1 \text{ W}. \qquad \text{(Answer)}$$

(c) What new capacitance C_{new} is needed to maximize P_{avg} if the other parameters of the circuit are not changed?

SOLUTION: One **Key Idea** here is that the average rate P_{avg} at which energy is supplied and dissipated is maximized if the circuit is brought into resonance with the driving emf. A second **Key Idea** is that resonance occurs when $X_C = X_L$. From the given data, we have $X_C > X_L$. Thus, we must decrease X_C to reach resonance. From Eq. 33-33 ($X_C = 1/\omega_d C$), we see that this means we must increase C to the new value C_{new}.

Using Eq. 33-33, we can write the condition $X_C = X_L$ as

$$\frac{1}{\omega_d C_{new}} = X_L.$$

Substituting $2\pi f_d$ for ω_d (because we are given f_d and not ω_d) and then solving for C_{new}, we find

$$C_{new} = \frac{1}{2\pi f_d X_L} = \frac{1}{(2\pi)(60 \text{ Hz})(80.0 \text{ } \Omega)}$$

$$= 3.32 \times 10^{-5} \text{ F} = 33.2 \text{ } \mu\text{F}. \qquad \text{(Answer)}$$

Following the procedure of part (b), you can show that with C_{new}, P_{avg} would then be at its maximum value of 72.0 W.

Exercises & Problems

Chapter 22

SEC. 22-8 Solving Problems Using Coulomb's Law

1E. What must be the distance between point charge $q_1 = 26.0 \ \mu C$ and point charge $q_2 = -47.0 \ \mu C$ for the electrostatic force between them to have a magnitude of 5.70 N?

2E. A point charge of $+3.00 \times 10^{-6}$ C is 12.0 cm distant from a second point charge of -1.50×10^{-6} C. Calculate the magnitude of the force on each charge.

3E. Two equally charged particles, held 3.2×10^{-3} m apart, are released from rest. The initial acceleration of the first particle is observed to be 7.0 m/s^2 and that of the second to be 9.0 m/s^2. If the mass of the first particle is 6.3×10^{-7} kg, what are (a) the mass of the second particle and (b) the magnitude of the charge of each particle?

4E. Identical isolated conducting spheres 1 and 2 have equal charges and are separated by a distance that is large compared with their diameters (Fig. 22-16a). The electrostatic force acting on sphere 2 due to sphere 1 is $\vec{F}$. Suppose now that a third identical sphere 3, having an insulating handle and initially neutral, is touched first to sphere 1 (Fig. 22-16b), then to sphere 2 (Fig. 22-16c), and finally removed (Fig. 22-16d). In terms of magnitude F, what is the magnitude of the electrostatic force $\vec{F}'$ that now acts on sphere 2?

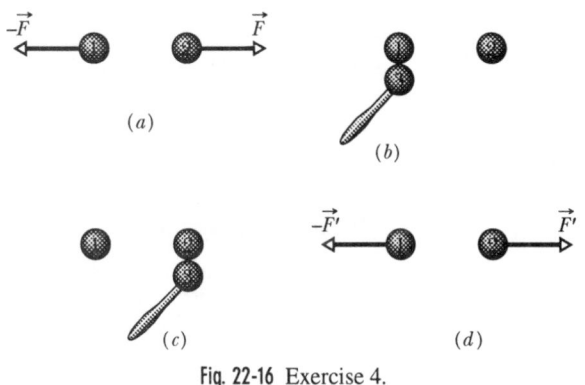

Fig. 22-16 Exercise 4.

5P. In Fig. 22-17, what are the (a) horizontal and (b) vertical components of the net electrostatic force on the charged particle in the lower left corner of the square if $q = 1.0 \times 10^{-7}$ C and $a = 5.0$ cm?

6P. Point charges q_1 and q_2 lie on the x axis at points $x = -a$ and $x = +a$, respectively. (a) How must q_1 and q_2 be related for the net electrostatic force on point charge $+Q$, placed at $x = +a/2$, to be zero? (b) Repeat (a) but with point charge $+Q$ now placed at $x = +3a/2$.

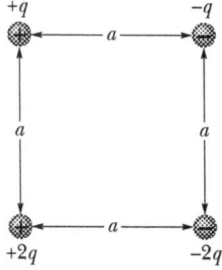

Fig. 22-17 Problem 5.

7P. Two identical conducting spheres, fixed in place, attract each other with an electrostatic force of 0.108 N when separated by 50.0 cm, center-to-center. The spheres are then connected by a thin conducting wire. When the wire is removed, the spheres repel each other with an electrostatic force of 0.0360 N. What were the initial charges on the spheres?

8P. In Fig. 22-18, three charged particles lie on a straight line and are separated by distances d. Charges q_1 and q_2 are held fixed. Charge q_3 is free to move but happens to be in equilibrium (no net electrostatic force acts on it). Find q_1 in terms of q_2.

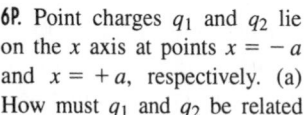

Fig. 22-18 Problem 8.

9P. Two free particles (that is, free to move) with charges $+q$ and $+4q$ are a distance L apart. A third charge is placed so that the entire system is in equilibrium. (a) Find the location, magnitude, and sign of the third charge. (b) Show that the equilibrium is unstable.

10P. Two fixed particles, of charges $q_1 = +1.0 \ \mu C$ and $q_2 = -3.0 \ \mu C$, are 10 cm apart. How far from each should a third charge be located so that no net electrostatic force acts on it?

11P. The charges and coordinates of two charged particles held fixed in the xy plane are $q_1 = +3.0 \ \mu C$, $x_1 = 3.5$ cm, $y_1 = 0.50$ cm, and $q_2 = -4.0 \ \mu C$, $x_2 = -2.0$ cm, $y_2 = 1.5$ cm. (a) Find the magnitude and direction of the electrostatic force on q_2. (b) Where could you locate a third charge $q_3 = +4.0 \ \mu C$ such that the net electrostatic force on q_2 is zero?

12P. A certain charge Q is divided into two parts q and $Q - q$, which are then separated by a certain distance. What must q be in terms of Q to maximize the electrostatic repulsion between the two charges?

13P. A particle with charge Q is fixed at each of two opposite corners of a square, and a particle with charge q is placed at each of the other two corners. (a) If the net electrostatic force on each particle with charge Q is zero, what is Q in terms of q? (b) Is there any value of q that makes the net electrostatic force on each of the four particles zero? Explain.

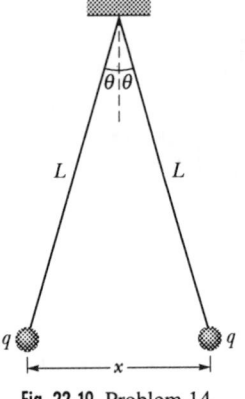

Fig. 22-19 Problem 14.

14P. In Fig. 22-19, two tiny conducting balls of identical mass m and identical charge q hang from nonconducting threads of length L. Assume that θ is so small that $\tan \theta$ can be replaced by its approximate equal, $\sin \theta$. (a) Show that, for equilibrium,

$$x = \left(\frac{q^2 L}{2\pi\varepsilon_0 mg} \right)^{1/3},$$

where x is the separation between the balls. (b) If $L = 120$ cm, $m = 10$ g, and $x = 5.0$ cm, what is q?

15P. Explain what happens to the balls of Problem 14b if one of them is discharged (loses its charge q to, say, the ground), and find the new equilibrium separation x, using the given values of L and m and the computed value of q.

16P. Figure 22-20 shows a long, nonconducting, massless rod of length L, pivoted at its center and balanced with a block of weight W at a distance x from the left end. At the left and right ends of the rod are attached small conducting spheres with positive charges q and $2q$, respectively. A distance h directly beneath each of these spheres is a fixed sphere with positive charge Q. (a) Find the distance x when the rod is horizontal and balanced. (b) What value should h have so that the rod exerts no vertical force on the bearing when the rod is horizontal and balanced?

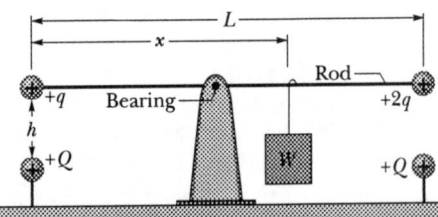

Fig. 22-20 Problem 16.

17E. What is the magnitude of the electrostatic force between a singly charged sodium ion (Na^+, of charge $+e$) and an adjacent singly charged chlorine ion (Cl^-, of charge $-e$) in a salt crystal if their separation is 2.82×10^{-10} m?

18E. What is the total charge in coulombs of 75.0 kg of electrons?

19E. How many megacoulombs of positive (or negative) charge are in 1.00 mol of neutral molecular-hydrogen gas (H_2)?

20E. The magnitude of the electrostatic force between two identical ions that are separated by a distance of 5.0×10^{-10} m is 3.7×10^{-9} N. (a) What is the charge of each ion? (b) How many electrons are "missing" from each ion (thus giving the ion its charge imbalance)?

21E. Two tiny, spherical water drops, with identical charges of -1.00×10^{-16} C, have a center-to-center separation of 1.00 cm. (a) What is the magnitude of the electrostatic force acting between them? (b) How many excess electrons are on each drop, giving it its charge imbalance?

22E. How many electrons would have to be removed from a coin to leave it with a charge of $+1.0 \times 10^{-7}$ C?

23P. Calculate the number of coulombs of positive charge in 250 cm^3 of (neutral) water (about a glassful).

24P. In the basic CsCl (cesium chloride) crystal structure, Cs^+ ions form the corners of a cube and a Cl^- ion is at the cube's center (Fig. 22-21). The edge length of the cube is 0.40 nm. The Cs^+ ions are each deficient by one electron (and thus each has a charge of $+e$), and the Cl^- ion has one excess electron (and thus has a charge of $-e$). (a) What is the magnitude of the net electrostatic force exerted on the Cl^- ion by the eight Cs^+ ions at the corners of the cube? (b) If one of the Cs^+ ions is missing, the crystal is said to have a *defect;* what is the magnitude of the net electrostatic force exerted on the Cl^- ion by the seven remaining Cs^+ ions?

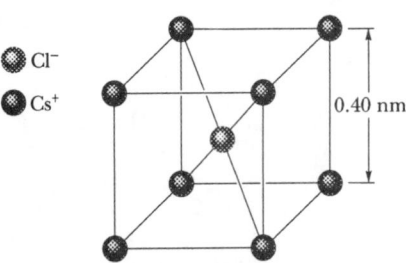

Fig. 22-21 Problem 24.

25P. We know that the negative charge on the electron and the positive charge on the proton are equal. Suppose, however, that these magnitudes differ from each other by 0.00010%. With what force would two copper coins, placed 1.0 m apart, repel each other? Assume that each coin contains 3×10^{22} copper atoms. (*Hint:* A neutral copper atom contains 29 protons and 29 electrons.) What do you conclude?

SEC. 22-9 Comparing Electrical and Gravitational Forces

26E. An electron is in a vacuum near the surface of Earth. Where should a second electron be placed so that the electrostatic force it exerts on the first electron balances the gravitational force on the first electron due to Earth?

27P. (a) What equal positive charges would have to be placed on Earth and on the Moon to neutralize their gravitational attraction? Do you need to know the lunar distance to solve this problem? Why or why not? (b) How many kilograms of hydrogen would be needed to provide the positive charge calculated in (a)?

Chapter 23

SEC. 23-5   Electric Field Plots and Lines

1E. In Fig. 23-19 the electric field lines on the left have twice the separation of those on the right.
(a) If the magnitude of the field at A is 40 N/C, what force acts on a proton at A? (b) What is the magnitude of the field at B?

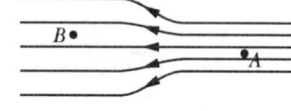

Fig. 23-19 Exercise 1.

2E. Sketch qualitatively the electric field lines both between and outside two concentric conducting spherical shells when a uniform positive charge q_1 is on the inner shell and a uniform negative charge $-q_2$ is on the outer. Consider the cases $q_1 > q_2$, $q_1 = q_2$, and $q_1 < q_2$.

3E. Sketch qualitatively the electric field lines for a thin, circular, uniformly charged disk of radius R. (*Hint:* Consider as limiting cases points very close to the disk, where the electric field is directed perpendicular to the surface, and points very far from it, where the electric field is like that of a point charge.)

SEC. 23-6   The Electric Field Due to Multiple Charges

4E. What is the magnitude of a point charge that would create an electric field of 1.00 N/C at points 1.00 m away?

5E. What is the magnitude of a point charge whose electric field 50 cm away has the magnitude of 2.0 N/C?

6E. Two particles with equal charge magnitudes 2.0×10^{-7} C but opposite signs are held 15 cm apart. What are the magnitude and direction of $\vec{E}$ at the point midway between the charges?

7E. An atom of plutonium-239 has a nuclear radius of 6.64 fm and the atomic number $Z = 94$. Assuming that the positive charge is distributed uniformly within the nucleus, what are the magnitude and direction of the electric field at the surface of the nucleus due to the positive charge?

8P. In Fig. 23-20, two fixed point charges $q_1 = +1.0 \times 10^{-6}$ C and $q_2 = +3.0 \times 10^{-6}$ C are separated by a distance $d = 10$ cm. Plot their net electric field $E(x)$ as a function of x for both positive and negative values of x, taking E to be positive when the vector $\vec{E}$ points to the right and negative when $\vec{E}$ points to the left.

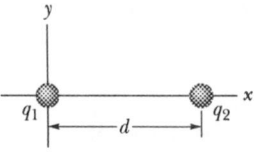

Fig. 23-20 Problems 8 and 10.

9P. Two point charges $q_1 = 2.1 \times 10^{-8}$ C and $q_2 = -4.0q_1$ are fixed in place 50 cm apart. Find the point along the straight line passing through the two charges at which the electric field is zero.

10P. (a) In Fig. 23-20, two fixed point charges $q_1 = -5q$ and $q_2 = +2q$ are separated by distance d. Locate the point (or points) at which the net electric field due to the two charges is zero. (b) Sketch the net electric field lines qualitatively.

11P. In Fig. 23-21, what is the magnitude of the electric field at point P due to the four point charges shown?

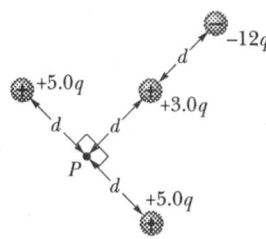

Fig. 23-21 Problem 11.

12P. Calculate the direction and magnitude of the electric field at point P in Fig. 23-22, due to the three point charges.

13P. What are the magnitude and direction of the electric field at the center of the square of Fig. 23-23 if $q = 1.0 \times 10^{-8}$ C and $a = 5.0$ cm?

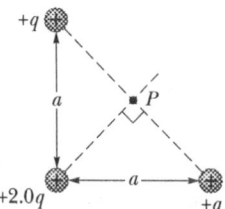

Fig. 23-22 Problem 12.

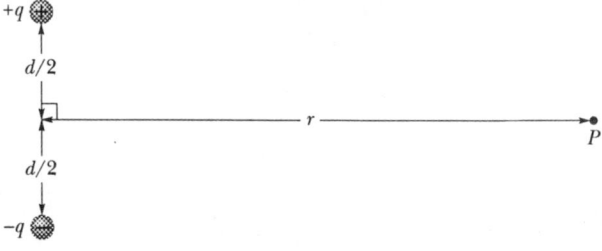

Fig. 23-23 Problem 13.

SEC. 23-7   The Electric Field Due to an Electric Dipole

14E. In Fig. 23-14, let both charges be positive. Assuming $z \gg d$, show that the magnitude of the vector $\vec{E}$ at point P in that figure is then given by

$$|\vec{E}| = \frac{1}{4\pi\varepsilon_0}\frac{2|q|}{z^2}.$$

15E. Calculate the electric dipole moment of an electron and a proton 4.30 nm apart.

16P. Find the magnitude and direction of the electric field at point P due to the electric dipole in Fig. 23-24. P is located at a distance $r \gg d$ along the perpendicular bisector of the line joining the charges. Express your answer in terms of the magnitude and direction of the electric dipole moment $\vec{p}$.

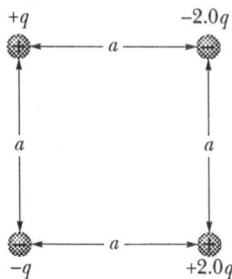

Fig. 23-24 Problem 16.

17P*. *Electric quadrupole.* Figure 23-25 shows an electric quadrupole. It consists of two dipoles with dipole moments that are equal in magnitude but opposite in direction. Show that the magnitude of the vector $\vec{E}$ on the axis of the quadrupole for a point P a distance z from its center (assume $z \gg d$) is given by

$$|\vec{E}| = \frac{3|Q|}{4\pi\varepsilon_0 z^4},$$

in which $Q(= 2qd^2)$ is known as the *quadrupole moment* of the charge distribution.

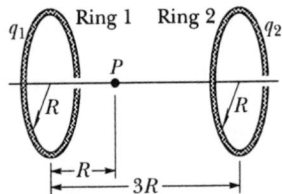

Fig. 23-25 Problem 17.

SEC. 23-8   The Electric Field Due to a Line of Charge

18E. Figure 23-26 shows two parallel nonconducting rings arranged with their central axes along a common line. Ring 1 has uniform charge q_1 and radius R; ring 2 has uniform charge q_2 and the same radius R. The rings are separated by a distance $3R$. The net electric field at point P on the common line, at distance R from ring 1, is zero. What is the ratio q_1/q_2?

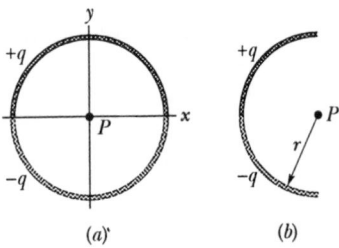

Fig. 23-26 Exercise 18.

19P. An electron is constrained to the central axis of the ring of charge of radius R discussed in Section 23-8. Show that the electrostatic force on the electron can cause it to oscillate through the center of the ring with an angular frequency

$$\omega = \sqrt{\frac{eq}{4\pi\varepsilon_0 mR^3}},$$

where q is the ring's charge and m is the electron's mass.

20P. In Fig. 23-27a, two curved plastic rods, one of charge $+q$ and the other of charge $-q$, form a circle of radius R in an xy plane. The x axis passes through their connecting points, and the charge is distributed uniformly on both rods. What are the magnitude and direction of the electric field $\vec{E}$ produced at P, the center of the circle?

Fig. 23-27 Problems 20 and 21.

21P. A thin glass rod is bent into a semicircle of radius r. A charge $+q$ is uniformly distributed along the upper half, and a charge $-q$ is uniformly distributed along the lower half, as shown in Fig.

23-27b. Find the magnitude and direction of the electric field $\vec{E}$ at P, the center of the semicircle.

22P. At what distance along the central axis of a ring of radius R and uniform charge is the magnitude of the electric field due to the ring's charge maximum?

23P. In Fig. 23-28, a nonconducting rod of length L has charge $-q$ uniformly distributed along its length. (a) What is the linear charge density of the rod? (b) What is the electric field at point P, a distance a from the end of the rod? (c) If P were very far from the rod compared to L, the rod would look like a point charge. Show that your answer to (b) reduces to the electric field of a point charge for $a \gg L$.

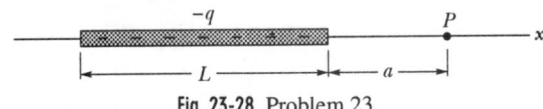

Fig. 23-28 Problem 23.

24P. A thin nonconducting rod of finite length L has a charge q spread uniformly along it. Show that the magnitude of the vector $\vec{E}$ is given by

$$|\vec{E}| = \frac{|q|}{2\pi\varepsilon_0 y} \frac{1}{(L^2 + 4y^2)^{1/2}}$$

gives the magnitude $|\vec{E}|$ of the electric field at point P on the perpendicular bisector of the rod (Fig. 23-29).

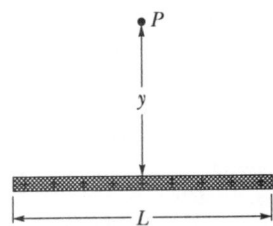

Fig. 23-30 Problem 24.

25P*. In Fig. 23-30, a "semi-infinite" nonconducting rod (that is, infinite in one direction only) has uniform linear charge density λ. Show that the electric field at point P makes an angle of 45° with the rod and that this result is independent of the distance R. (*Hint:* Separately find the parallel and perpendicular (to the rod) components of the electric field at P, and then compare those components.)

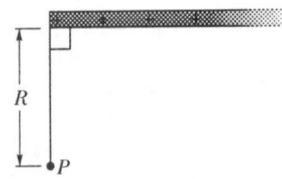

Fig. 23-30 Problem 25.

SEC. 23-9   A Point Charge in an Electric Field

26E. An electron is accelerated eastward at 1.80×10^9 m/s² by an electric field. Determine the magnitude and direction of the electric field.

27E. An electron is released from rest in a uniform electric field of magnitude 2.00×10^4 N/C. Calculate the acceleration of the electron. (Ignore gravitation.)

28E. An alpha particle (the nucleus of a helium atom) has a mass of 6.64×10^{-27} kg and a charge of $+2e$. What are the magnitude and direction of the electric field that will balance the gravitational force on it?

29E. Calculate the magnitude of the force, due to an electric dipole of dipole moment 3.6×10^{-29} C · m, on an electron 25 nm from the center of the dipole, along the dipole axis. Assume that this distance is large relative to the dipole's charge separation.

30E. Humid air breaks down (its molecules become ionized) in an electric field of 3.0×10^6 N/C. In that field, what is the magnitude of the electrostatic force on (a) an electron and (b) an ion with a single electron missing?

31E. A charged cloud system produces an electric field in the air near Earth's surface. A particle of charge -2.0×10^{-9} C is acted on by a downward electrostatic force of 3.0×10^{-6} N when placed in this field. (a) What is the magnitude of the electric field? (b) What are the magnitude and direction of the electrostatic force exerted on a proton placed in this field? (c) What is the gravitational force on the proton? (d) What is the ratio of the magnitude of the electrostatic force to the magnitude of the gravitational force in this case?

32E. An electric field $\vec{E}$ with an average magnitude of about 150 N/C points downward in the atmosphere near Earth's surface. We wish to "float" a sulfur sphere weighing 4.4 N in this field by charging the sphere. (a) What charge (both sign and magnitude) must be used? (b) Why is the experiment impractical?

33E. Beams of high-speed protons can be produced in "guns" using electric fields to accelerate the protons. (a) What acceleration would a proton experience if the gun's electric field were 2.00×10^4 N/C? (b) What speed would the proton attain if the field accelerated the proton through a distance of 1.00 cm?

34E. An electron with a speed of 5.00×10^8 cm/s enters an electric field of magnitude 1.00×10^3 N/C, traveling along the field lines in the direction that retards its motion. (a) How far will the electron travel in the field before stopping momentarily and (b) how much time will have elapsed? (c) If the region with the electric field is only 8.00 mm long (too short for the electron to stop within it), what fraction of the electron's initial kinetic energy will be lost in that region?

35P. A uniform electric field exists in a region between two oppositely charged plates. An electron is released from rest at the surface of the negatively charged plate and strikes the surface of the opposite plate, 2.0 cm away, in a time 1.5×10^{-8} s. (a) What is the speed of the electron as it strikes the second plate? (b) What is the magnitude of the electric field $\vec{E}$?

36P. At some instant the velocity components of an electron moving between two charged parallel plates are $v_x = 1.5 \times 10^5$ m/s and $v_y = 3.0 \times 10^3$ m/s. Suppose that the electric field between the plates is given by $\vec{E} = (120 \text{ N/C})\hat{j}$. (a) What is the acceleration of the electron? (b) What will be the velocity of the electron after its x coordinate has changed by 2.0 cm?

37P. Two large parallel copper plates are 5.0 cm apart and have a uniform electric field between them as depicted in Fig. 23-31. An electron is released from the negative plate at the same time that a proton is released from the positive plate. Neglect the force of the particles on each other and find their distance from the positive plate when they pass each other. (Does it surprise you that you need not know the electric field to solve this problem?)

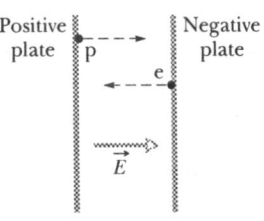
Positive plate p Negative plate e $\vec{E}$

Fig. 23-31 Problem 37.

38P. A 10.0 g block with a charge of $+8.00 \times 10^{-5}$ C is placed in electric field $\vec{E} = (3.00 \times 10^3)\hat{i} - 600\hat{j}$, where $\vec{E}$ is in newtons per coulomb. (a) What are the magnitude and direction of the force on the block? (b) If the block is released from rest at the origin at $t = 0$, what will be its coordinates at $t = 3.00$ s?

39P. In Fig. 22-32, a uniform, upward-directed electric field $\vec{E}$ of magnitude 2.00×10^3 N/C has been set up between two horizontal plates by charging the lower plate positively and the upper plate negatively. The plates have length $L = 10.0$ cm and separation $d = 2.00$ cm. An electron is then shot between the plates from the left edge of the lower plate. The initial velocity $\vec{v}_0$ of the electron makes an angle $\theta = 45.0°$ with the lower plate and has a magnitude of 6.00×10^6 m/s. (a) Will the electron strike one of the plates? (b) If so, which plate and how far horizontally from the left edge will the electron strike?

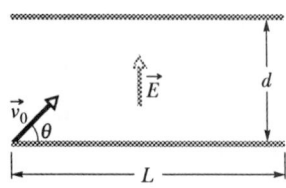
Fig. 23-32 Problem 39.

SEC. 23-10   A Dipole in an Electric Field

40E. An electric dipole, consisting of charges of magnitude 1.50 nC separated by 6.20 μm, is in an electric field of strength 1100 N/C. (a) What is the magnitude of the electric dipole moment? (b) What is the difference between the potential energies corresponding to dipole orientations parallel to and antiparallel to the field?

41E. An electric dipole consists of charges $+2e$ and $-2e$ separated by 0.78 nm. It is in an electric field of strength 3.4×10^6 N/C. Calculate the magnitude of the torque on the dipole when the dipole moment is (a) parallel to, (b) perpendicular to, and (c) antiparallel to the electric field.

42P. Find the work required to turn an electric dipole end for end in a uniform electric field $\vec{E}$, in terms of the magnitude $|\vec{p}|$ of the dipole moment, the magnitude $|\vec{E}|$ of the field, and the initial angle θ_0 between $\vec{p}$ and $\vec{E}$.

43P. Find the frequency of oscillation of an electric dipole, of dipole moment $\vec{p}$ and rotational inertia I, for small amplitudes of oscillation about its equilibrium position in a uniform electric field of magnitude $|\vec{E}|$.

Chapter 24

SEC. 24-2   Electric Flux

1E. The square surface shown in Fig. 24-18 measures 3.2 mm on each side. It is immersed in a uniform electric field with magnitude $|\vec{E}| = 1800$ N/C. The field lines make an angle of 35° with a normal to the surface, as shown. Take that normal to be directed "outward," as though

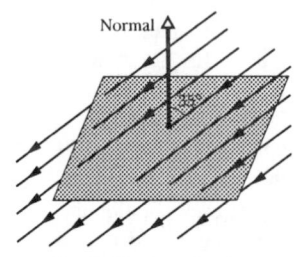
Normal

Fig. 24-18 Exercise 1.

the surface were one face of a box. Calculate the electric flux through the surface.

SEC. 24-3   Net Flux at a Closed Surface

2E. The cube in Fig. 24-19 has edge lengths of 1.40 m and is oriented as shown in a region of uniform electric field. Find the electric flux through the right face if the electric field, in newtons per coulomb, is given by (a) $6.00\hat{i}$, (b) $-2.00\hat{j}$, and (c) $-3.00\hat{i} + 4.00\hat{k}$. (d) What is the total flux through the cube for each of these fields?

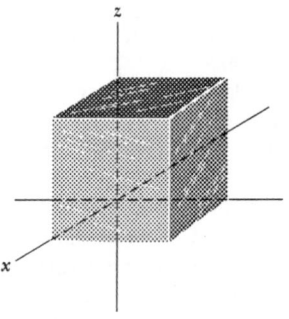

Fig. 24-19 Exercise 2 and Problems 6 and 9.

SEC. 24-4   Gauss's Law

3E. You have four point charges, $2q$, q, $-q$, and $-2q$. If possible, describe how you would place a closed surface that encloses at least the charge $2q$ (and perhaps other charges) and through which the net electric flux is (a) 0, (b) $+3q/\varepsilon_0$, and (c) $-2q/\varepsilon_0$.

4E. A point charge of 1.8 μC is at the center of a cubical Gaussian surface 55 cm on edge. What is the net electric flux through the surface?

5E. In Fig. 24-20, a butterfly net is in a uniform electric field of magnitude $\vec{E}$. The rim, a circle of radius a, is aligned perpendicular to the field. Find the electric flux through the netting.

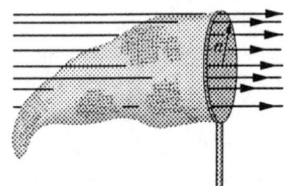

Fig. 24-20 Exercise 5.

6P. Find the net flux through the cube of Exercise 2 and Fig. 24-19 if the electric field is given by (a) $\vec{E} = (3.00y[\frac{N}{C \cdot m}])\hat{j}$ and (b) $\vec{E} = -(4.00[\frac{N}{C}])\hat{i} + (6.00[\frac{N}{C}] + 3.00y[\frac{N}{C \cdot m}])\hat{j}$. (c) In each case, how much charge is enclosed by the cube?

7P. When a shower is turned on in a closed bathroom, the splashing of the water on the bare tub can fill the room's air with negatively charged ions and produce an electric field in the air as great as 1000 N/C. Consider a bathroom with dimensions of 2.5 m × 3.0 m × 2.0 m. Along the ceiling, floor, and four walls, approximate the electric field in the air as being directed perpendicular to the surface and as having a uniform magnitude of 600 N/C. Also, treat those surfaces as forming a closed Gaussian surface around the room's air. What are (a) the volume charge density ρ and (b) the number of excess elementary charges e per cubic meter in the room's air?

8P. It is found experimentally that the electric field in a certain region of Earth's atmosphere is directed vertically down. At an altitude of 300 m the field has magnitude 60.0 N/C; at an altitude of 200 m, the magnitude is 100 N/C. Find the net amount of charge contained in a cube 100 m on edge, with horizontal faces at altitudes of 200 and 300 m. Neglect the curvature of Earth.

9P. At each point on the surface of the cube shown in Fig. 24-19, the electric field is in the positive direction of y. The length of each edge of the cube is 3.0 m. On the top surface of the

cube, $\vec{E} = -34\hat{j}$ N/C, and on the bottom face of the cube $\vec{E} = +20\hat{j}$ N/C. Determine the net charge contained within the cube.

10P. A point charge q is placed at one corner of a cube of edge a. What is the flux through each of the cube faces? (*Hint:* Use Gauss' law and symmetry arguments.)

SEC. 24-6   Application of Gauss' Law to Highly Symmetric Charge Distributions

11E. A point charge causes an electric flux of -750 N · m²/C to pass through a spherical Gaussian surface of 10.0 cm radius centered on the charge. (a) If the radius of the Gaussian surface were doubled, how much flux would pass through the surface? (b) What is the value of the point charge?

12E. A conducting sphere of radius 10 cm has an unknown charge. If the electric field 15 cm from the center of the sphere has the magnitude 3.0×10^3 N/C and is directed radially inward, what is the net charge on the sphere?

13E. Two charged concentric spheres have radii of 10.0 cm and 15.0 cm. The charge on the inner sphere is 4.00×10^{-8} C, and that on the outer sphere is 2.00×10^{-8} C. Find the electric field (a) at $r = 12.0$ cm and (b) at $r = 20.0$ cm.

14E. In a 1911 paper, Ernest Rutherford said: "In order to form some idea of the forces required to deflect an α particle through a large angle, consider an atom [as] containing a point positive charge Ze at its centre and surrounded by a distribution of negative electricity $-Ze$ uniformly distributed within a sphere of radius R. The electric field $E \ldots$ at a distance r from the center for a point inside the atom [is]

$$E = \frac{Ze}{4\pi\varepsilon_0}\left(\frac{1}{r^2} - \frac{r}{R^3}\right)."$$

Verify this equation.

15P. A proton with speed $v = 3.00 \times 10^5$ m/s orbits just outside a charged sphere of radius $r = 1.00$ cm. What is the charge on the sphere?

16P. A point charge $+q$ is placed at the center of an electrically neutral, spherical conducting shell with inner radius a and outer radius b. What charge appears on (a) the inner surface of the shell and (b) the outer surface? What is the net electric field at a distance r from the center of the shell if (c) $r < a$, (d) $b > r > a$, and (e) $r > b$? Sketch field lines for those three regions. For $r > b$, what is the net electric field due to (f) the central point charge plus the inner surface charge and (g) the outer surface charge? A point charge $-q$ is now placed outside the shell. Does this point charge change the charge distribution on (h) the outer surface and (i) the inner surface? Sketch the field lines now. (j) Is there an electrostatic force on the second point charge? (k) Is there a net electrostatic force on the first point charge? (l) Does this situation violate Newton's third law?

17P. A solid nonconducting sphere of radius R has a nonuniform charge distribution of volume charge density $\rho = \rho_s r/R$, where ρ_s is a constant and r is the distance from the center of the sphere. Show (a) that the total charge on the sphere is $Q = \pi\rho_s R^3$ and (b) that

$$|\vec{E}| = \frac{1}{4\pi\varepsilon_0}\frac{|Q|}{R^4}r^2$$

gives the magnitude of the electric field inside the sphere.

18P. A hydrogen atom can be considered as having a central point-like proton of positive charge $+e$ and an electron of negative charge $-e$ that is distributed about the proton according to the volume charge density $\rho = A\exp(-2r/a_0)$. Here A is a constant, $a_0 = 0.53 \times 10^{-10}$ m is the *Bohr radius,* and r is the distance from the center of the atom. (a) Using the fact that hydrogen is electrically neutral, find A. (b) Then find the electric field produced by the atom at the Bohr radius.

19P. In Fig. 24-21 a sphere, of radius a and charge $+q$ uniformly distributed throughout its volume, is concentric with a spherical conducting shell of inner radius b and outer radius c. This shell has a net charge of $-q$. Find expressions for the electric field, as a function of the radius r, (a) within the sphere $(r < a)$, (b) between the sphere and the shell $(a < r < b)$, (c) inside the shell $(b < r < c)$, and (d) outside the shell $(r > c)$. (e) What are the charges on the inner and outer surfaces of the shell?

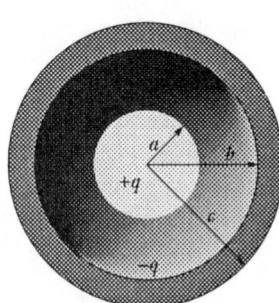

Fig. 24-21 Problem 19.

20P. Figure 24-22a shows a spherical shell of charge with uniform volume charge density ρ. Plot E due to the shell for distances r from the center of the shell ranging from zero to 30 cm. Assume that $\rho = 1.0 \times 10^{-6}$ C/m^3, $a = 10$ cm, and $b = 20$ cm.

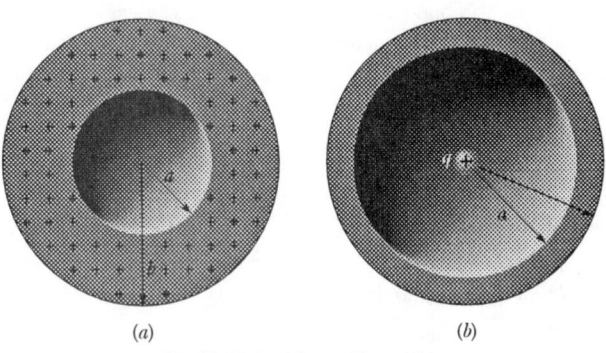

(a)   (b)

Fig. 24-22 Problems 20 and 21.

21P. In Fig. 24-22b, a nonconducting spherical shell, of inner radius a and outer radius b, has a positive volume charge density $\rho = A/r$ (within its thickness), where A is a constant and r is the distance from the center of the shell. In addition, a positive point charge q is located at that center. What value should A have if the electric field in the shell $(a \leq r \leq b)$ is to be uniform? (*Hint:* The constant A depends on a but not on b.)

22P*. A nonconducting sphere has a uniform volume charge density ρ. Let $\vec{r}$ be the vector from the center of the sphere to a general point P within the sphere. (a) Show that the electric field at

P is given by $\vec{E} = \rho\vec{r}/3\varepsilon_0$. (Note that the result is independent of the radius of the sphere.) (b) A spherical cavity is hollowed out of the sphere, as shown in Fig. 24-23. Using superposition concepts, show that the electric field at all points within the cavity is uniform and equal to $\vec{E} = \rho\vec{a}/3\varepsilon_0$, where $\vec{a}$ is the position vector from the center of the sphere to the center of the cavity. (Note that this result is independent of the radius of the sphere and the radius of the cavity.)

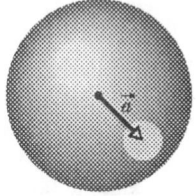

Fig. 24-23 Problem 22.

23P*. A spherically symmetrical but nonuniform volume distribution of charge produces an electric field of magnitude $|\vec{E}| = Kr^4$, directed radially outward from the center of the sphere. Here r is the radial distance from that center, and K is a positive constant. What is the volume density $|\rho|$ of the magnitude of the charge distribution?

24E. An infinite line of charge produces a field of 4.5×10^4 N/C at a distance of 2.0 m. Calculate the linear charge density.

25P. Figure 24-24 shows a section of a long, thin-walled metal tube of radius R, with a positive charge per unit length λ on its surface. Derive expressions for $|\vec{E}|$ in terms of the distance r from the tube axis, considering both (a) $r > R$ and (b) $r < R$. Plot your results for the range $r = 0$ to $r = 5.0$ cm, assuming that $\lambda = 2.0 \times 10^{-8}$ C/m and $R = 3.0$ cm. (*Hint:* Use cylindrical Gaussian surfaces, co-axial with the metal tube.)

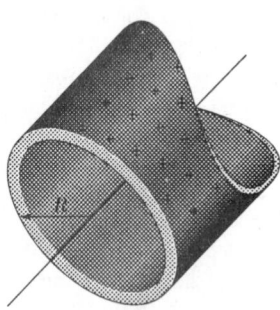

Fig. 24-24 Problem 25.

26P. A very long conducting cylindrical rod of length L with a total charge $+q$ is surrounded by a conducting cylindrical shell (also of length L) with total charge $-2q$, as shown in Fig. 24-25. Use Gauss' law to find (a) the electric field at points outside the conducting shell, (b) the distribution of charge on the shell, and (c) the electric field in the region between the shell and rod.

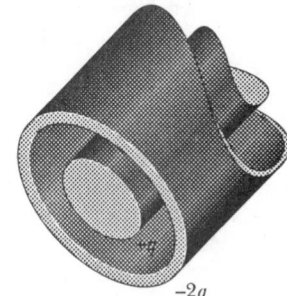

Fig. 24-25 Problem 26.

27P. A long, straight wire has fixed negative charge with a linear charge density of magnitude 3.6 nC/m. The wire is to be enclosed by a thin, nonconducting cylinder of outside radius 1.5 cm, coaxial with the wire. The cylinder is to have positive charge on its outside surface with a surface charge density σ such that the net external electric field is zero. Calculate the required σ.

28P. Two long, charged, concentric cylinders have radii of 3.0 and 6.0 cm. The charge per unit length is 5.0×10^{-6} C/m on the inner cylinder and -7.0×10^{-6} C/m on the outer cylinder. Find the electric field at (a) $r = 4.0$ cm and (b) $r = 8.0$ cm, where r is the radial distance from the common central axis.

29P. A long, nonconducting, solid cylinder of radius 4.0 cm has a nonuniform volume charge density ρ that is a function of the radial distance r from the axis of the cylinder, as given by $\rho = Ar^2$, with $A = 2.5\ \mu\text{C/m}^5$. What is the magnitude of the electric field at a radial distance of (a) 3.0 cm and (b) 5.0 cm from the axis of the cylinder?

30P. Figure 24-26 shows a Geiger counter, a device used to detect ionizing radiation (radiation that causes ionization of atoms). The counter consists of a thin, positively charged central wire surrounded by a concentric, circular, conducting cylinder with an equal negative charge. Thus, a strong radial electric field is set up inside the cylinder. The cylinder contains a low-pressure inert gas. When a particle of radiation enters the device through the cylinder wall, it ionizes a few of the gas atoms. The resulting free electrons (label e) are drawn to the positive wire. However, the electric field is so intense that, between collisions with other gas atoms, the free electrons gain energy sufficient to ionize these atoms also. More free electrons are thereby created, and the process is repeated until the electrons reach the wire. The resulting "avalanche" of electrons is collected by the wire, generating a signal that is used to record the passage of the original particle of radiation. Suppose that the radius of the central wire is 25 μm, the radius of the cylinder 1.4 cm, and the length of the tube 16 cm. If the electric field at the cylinder's inner wall is 2.9×10^4 N/C, what is the total positive charge on the central wire?

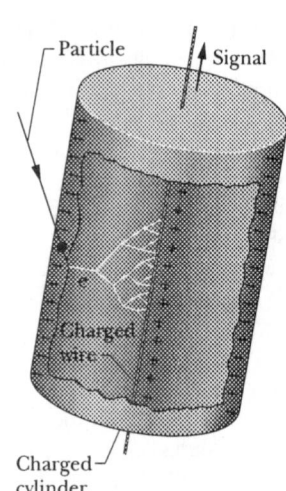

Fig. 24-26 Problem 30.

31P. Charge is distributed uniformly throughout the volume of an infinitely long cylinder of radius R. (a) Show that, at a distance r from the cylinder axis (for $r < R$),

$$|\vec{E}| = \frac{\rho r}{2\varepsilon_0},$$

where ρ is the volume charge density. (b) Write an expression for $|\vec{E}|$ when $r > R$.

32E. Figure 24-27 shows cross-sections through two large, parallel, nonconducting sheets with identical distributions of positive charge with surface charge density σ. What is $\vec{E}$ at points (a)

Fig. 24-27 Exercise 32.

above the sheets, (b) between them, and (c) below them?

33E. A square metal plate of edge length 8.0 cm and negligible thickness has a total charge of 6.0×10^{-6} C. (a) Estimate the magnitude E of the electric field just off the center of the plate (at,

say, a distance of 0.50 mm) by assuming that the charge is spread uniformly over the two faces of the plate. (b) Estimate E at a distance of 30 m (large relative to the plate size) by assuming that the plate is a point charge.

34P. In Fig. 24-28, a small, nonconducting ball of mass $m = 1.0$ mg and charge $q = 2.0 \times 10^{-8}$ C (distributed uniformly through its volume) hangs from an insulating thread that makes an angle $\theta = 30°$ with a vertical, uniformly charged nonconducting sheet (shown in cross section). Considering the gravitational force on the ball and assuming that the sheet extends far vertically and into and out of the page, calculate the surface charge density σ of the sheet.

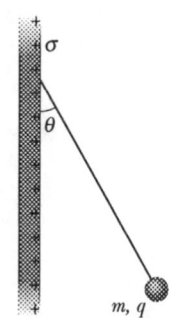

Fig. 24-28 Problem 34.

35P. Two large, thin metal plates are parallel and close to each other. On their inner faces, the plates have excess surface charge densities of opposite signs and with a magnitude 7.0×10^{-22} C/m^2, with the negatively charged plate on the left. What are the magnitude and direction of the electric field $\vec{E}$ (a) to the left of the plates, (b) to the right of the plates, and (c) between the plates?

36P. An electron is shot directly toward the center of a large metal plate that has excess negative charge with surface charge density 2.0×10^{-6} C/m^2. If the initial kinetic energy of the electron is 100 eV and if the electron is to stop (owing to electrostatic repulsion from the plate) just as it reaches the plate, how far from the plate must it be shot?

37P. Two large metal plates of area 1.0 m^2 face each other. They are 5.0 cm apart and have equal but opposite charges on their inner surfaces. If the magnitude $|\vec{E}|$ of the electric field between the plates is 55 N/C, what is the magnitude of the charge on each plate? Neglect edge effects.

38P*. A planar slab of thickness d has a uniform volume charge density ρ. Find the magnitude of the electric field at all points in space both (a) and (b) outside the slab, in terms of x, the distance measured from the central plane of the slab.

SEC. 24-8 A Charged Isolated Conductor

39E. The electric field just above the surface of the charged drum of a photocopying machine has a magnitude $|\vec{E}|$ of 2.3×10^5 N/C. What is the surface charge density on the drum, assuming that the drum is a conductor?

40E. A uniformly charged conducting sphere of 1.2 m diameter has a surface charge density of 8.1 μC/m^2. (a) Find the net charge on the sphere. (b) What is the total electric flux leaving the surface of the sphere?

41E. Space vehicles traveling through Earth's radiation belts can intercept a significant number of electrons. The resulting charge buildup can damage electronic components and disrupt operations. Suppose a spherical metallic satellite 1.3 m in diameter accumulates 2.4 μC of charge in one orbital revolution. (a) Find the resulting surface charge density. (b) Calculate the magnitude of the electric field just outside the surface of the satellite, due to the surface charge.

42P. An isolated conductor of arbitrary shape has a net charge of $+10 \times 10^{-6}$ C. Inside the conductor is a cavity within which is a point charge $q = +3.0 \times 10^{-6}$ C. What is the charge (a) on the cavity wall and (b) on the outer surface of the conductor?

Chapter 25

SEC. 25-3   Electric Potential

1E. The electric potential difference between the ground and a cloud in a particular thunderstorm is 1.2×10^9 V. What is the magnitude of the change in the electric potential energy (in multiples of the electron-volt) of an electron that moves between the ground and the cloud?

2P. In a given lightning flash, the potential difference between a cloud and the ground is 1.0×10^9 V and the quantity of charge transferred is 30 C. (a) What is the decrease in energy of that transferred charge? (b) If all that energy could be used to accelerate a 1000 kg automobile from rest, what would be the automobile's final speed? (c) If the energy could be used to melt ice, how much ice would it melt at 0°C? The heat of fusion of ice is 3.33×10^5 J/kg.

SEC. 25-4   Calculating the Potential from the Field

3E. When an electron moves from A to B along an electric field line in Fig. 25-22, the electric field does 3.94×10^{-19} J of work on it. What are the electric potential differences (a) $V_B - V_A$, (b) $V_C - V_A$, and (c) $V_C - V_B$?

4E. An infinite nonconducting sheet has a surface charge density $\sigma = 0.10$ $\mu C/m^2$ on one side. How far apart are equipotential surfaces whose potentials differ by 50 V?

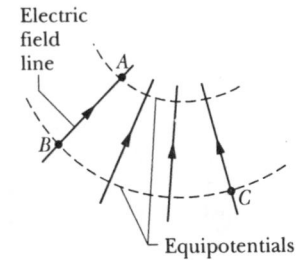

Fig. 25-22 Exercise 3.

5E. Two large, parallel, conducting plates are 12 cm apart and have charges of equal magnitude and opposite sign on their facing surfaces. An electrostatic force of 3.9×10^{-15} N acts on an electron placed anywhere between the two plates. (Neglect fringing.) (a) Find the electric field at the position of the electron. (b) What is the potential difference between the plates?

6P. A Geiger counter has a metal cylinder 2.00 cm in diameter along whose axis is stretched a wire 1.30×10^{-4} cm in diameter. If the potential difference between the wire and the cylinder is 850 V, what is the electric field at the surface of (a) the wire and (b) the cylinder? (*Hint:* Use the result of Problem 31 of Chapter 24.)

7P. The electric field inside a nonconducting sphere of radius R, with charge spread uniformly throughout its volume, is radially directed and has magnitude

$$|\vec{E}(r)| = \frac{|q|r}{4\pi\varepsilon_0 R^3}.$$

Here q (positive or negative) is the total charge within the sphere, and r is the distance from the sphere's center. (a) Taking $\Delta V = 0$ at the center of the sphere, find the electric potential $\Delta V(r)$ inside the sphere. (b) What is the difference in electric potential between a point on the surface and the sphere's center? (c) If q is positive, which of those two points is at the higher potential?

8P*. A charge q is distributed uniformly throughout a spherical volume of radius R. (a) Setting $\Delta V = 0$ at infinity, show that the potential at a distance r from the center, where $r < R$, is given by

$$V = \frac{q(3R^2 - r^2)}{8\pi\varepsilon_0 R^3}.$$

(*Hint:* See Section 24-6.) (b) Why does this result differ from that in (a) of Problem 7? (c) What is the potential difference between a point on the surface and the sphere's center? (d) Why doesn't this result differ from that of (b) of Problem 7?

9P. Figure 25-33 shows, edge-on, an infinite nonconducting sheet with positive surface charge density σ on one side. (a) Use Eq. 25-17 and Eq. 24-14 to show that the electric potential of an infinite sheet of charge can be written $\Delta V = \Delta V_0 - (\sigma/2\varepsilon_0)z$, where ΔV_0 is the electric potential at the surface of the sheet and z is the perpendicular distance from the sheet. (b) How much work is done by the electric field of the sheet as a small positive test charge q_0 is moved from an initial position on the sheet to a final position located a distance z from the sheet?

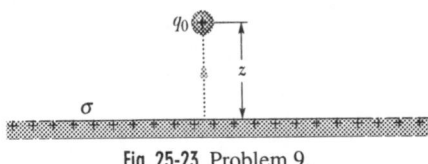

Fig. 25-23 Problem 9.

10P*. A thick spherical shell of charge Q and uniform volume charge density ρ is bounded by radii r_1 and r_2, where $r_2 > r_1$. With $\Delta V = 0$ at infinity, find the electric potential ΔV as a function of the distance r from the center of the distribution, considering the regions (a) $r > r_2$, (b) $r_2 > r > r_1$, and (c) $r < r_1$. (d) Do these solutions agree at $r = r_2$ and $r = r_1$? (*Hint:* See Section 24-6.)

SEC. 25-7   Potential Due to a Group of Point Charges

11E. As a space shuttle moves through the dilute ionized gas of Earth's ionosphere, its potential is typically changed by -1.0 V during one revolution. By assuming that the shuttle is a sphere of radius 10 m, estimate the amount of charge it collects.

12E. Consider a point charge $q = 1.0$ μC, point A at distance $d_1 = 2.0$ m from q, and point B at distance $d_2 = 1.0$ m. (a) If these points are diametrically opposite each other, as in Fig. 25-24a, what is the electric potential difference $\Delta V_A - \Delta V_B$? (b) What is that electric potential difference if points A and B are located as in Fig. 25-24b?

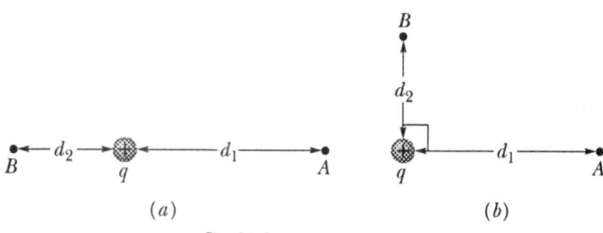

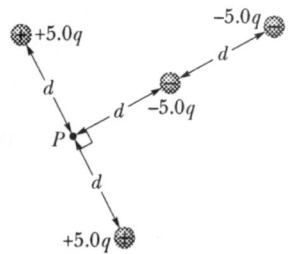

Fig. 25-24 Exercise 12.

Fig. 25-27 Problem 20.

13E. Figure 25-25 shows two charged particles on an axis. Sketch the electric field lines and the equipotential surfaces in the plane of the page for (a) $q_1 = +q$ and $q_2 = +2q$ and (b) $q_1 = +q$ and $q_2 = -3q$.

14E. In Fig. 25-25, set $\Delta V = 0$ at infinity and let the particles have charges $q_1 = +q$ and $q_2 = -3q$. Then locate (in terms of the separation distance d) any point on the

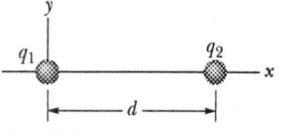

Fig. 25-25 Exercises 13, 14, 15.

x axis (other than at infinity) at which the net potential due to the two particles is zero.

15E. Two particles, of charges q_1 and q_2, are separated by distance d in Fig. 25-25. The net electric field of the particles is zero at $x = d/4$. With $\Delta V = 0$ at infinity, locate (in terms of d) any point on the x axis (other than at infinity) at which the electric potential due to the two particles is zero.

16P. A spherical drop of water carrying a charge of 30 pC has a potential of 500 V at its surface (with $\Delta V = 0$ at infinity). (a) What is the radius of the drop? (b) If two such drops of the same charge and radius combine to form a single spherical drop, what is the potential at the surface of the new drop?

17P. What are (a) the charge and (b) the charge density on the surface of a conducting sphere of radius 0.15 m whose potential is 200 V (with $\Delta V = 0$ at infinity)?

18P. An electric field of approximately 100 V/m is often observed near the surface of Earth. If this were the field over the entire surface, what would be the electric potential of a point on the surface? (Set $\Delta V = 0$ at infinity.)

19P. In Fig. 25-26, point P is at the center of the rectangle. With $\Delta V = 0$ at infinity, what is the net electric potential at P due to the six charged particles?

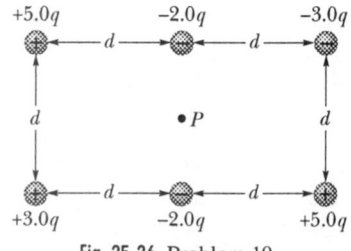

Fig. 25-26 Problem 19.

20P. In Fig. 25-27, what is the net potential at point P due to the four point charges, if $\Delta V = 0$ at infinity?

SEC. 25-8 Potential Due to an Electric Dipole

21E. The ammonia molecule NH_3 has a permanent electric dipole moment equal to 1.47 D, where 1 D = 1 debye unit = 3.34×10^{-30} C · m. Calculate the electric potential due to an ammonia molecule at a point 52.0 nm away along the axis of the dipole. (Set $V = 0$ at infinity.)

22P. Figure 25-28 shows three charged particles located on a horizontal axis. For points (such as P) on the axis with $r \gg d$, show that the electric potential $V(r)$ is given by

$$\Delta V = \frac{1}{4\pi\varepsilon_0}\frac{q}{r}\left(1 + \frac{2d}{r}\right).$$

(*Hint:* The charge configuration can be viewed as the sum of an isolated charge and a dipole.)

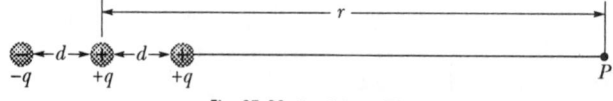

Fig. 25-28 Problem 22.

SEC. 25-9 Potential Due to a Continuous Charge Distribution

23E. (a) Figure 25-29a shows a positively charged plastic rod of length L and uniform linear charge density λ. Setting $\Delta V = 0$ at infinity and considering Fig. 25-14 and Eq. 25-31, find the electric potential at point P without written calculation. (b) Figure 25-29b shows an identical rod, except that it is split in half and the right half is negatively charged; the left and right halves have the same magnitude λ of uniform linear charge density. With ΔV still zero at infinity, what is the electric potential at point P in Fig. 25-29b?

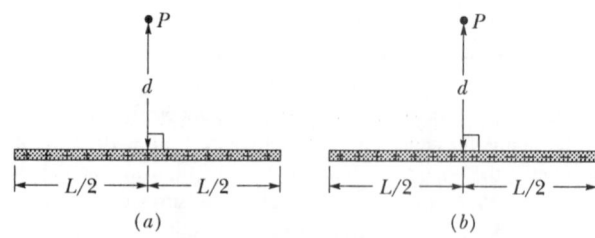

Fig. 25-29 Exercise 23.

24E. A plastic rod has been formed into a circle of radius R. It has a positive charge $+Q$ uniformly distributed along one-quarter of its circumference and a negative charge of $-6Q$ uniformly distributed along the rest of the circumference (Fig. 25-30). With $\Delta V = 0$ at infinity, what is the electric potential (a) at the center C of the circle and (b) at point P, which is on the central axis of the circle at distance z from the center?

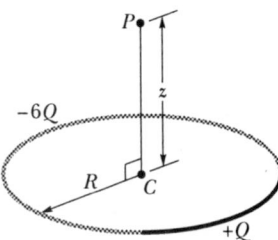

Fig. 25-30 Exercise 24.

25E. In Fig. 25-31, a plastic rod having a uniformly distributed charge $-Q$ has been bent into a circular arc of radius R and central angle $120°$. With $\Delta V = 0$ at infinity, what is the electric potential at P, the center of curvature of the rod?

26E. A plastic disk is charged on one side with a uniform surface charge density σ, and then three quadrants of the disk are removed. The remaining quadrant is shown in Fig. 25-32. With $\Delta V = 0$ at infinity, what is the potential due to the remaining quadrant at point P, which is on the central axis of the original disk at a distance z from the original center?

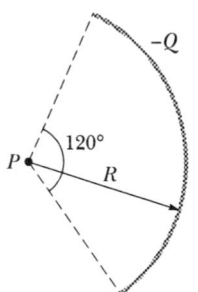

Fig. 25-31 Exercise 25.

27P. Figure 25-33 shows a plastic rod of length L and uniform positive charge Q lying on an x axis. With $\Delta V = 0$ at infinity, find the electric potential at point P_1 on the axis, at distance d from one end of the rod.

28P. The plastic rod shown in Fig. 25-33 has length L and a nonuniform linear charge density $\lambda = cx$, where c is a positive constant. With $\Delta V = 0$ at infinity, find the electric potential at point P_1 on the axis, at distance d from one end.

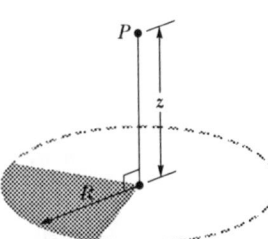

Fig. 25-32 Exercise 26.

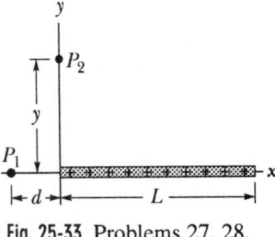

Fig. 25-33 Problems 27, 28, 32, and 33.

SEC. 25-10   Calculating the Field from the Potential

29E. Two large parallel metal plates are 1.5 cm apart and have equal but opposite charges on their facing surfaces. Take the potential of the negative plate to be zero. If the potential halfway between the plates if then $+5.0$ V, what is the electric field in the region between the plates?

30E. The electric potential at points in an xy plane is given by $\Delta V = (2.0\ \text{V/m}^2)x^2 - (3.0\ \text{V/m}^2)y^2$. What are the magnitude and direction of the electric field at the point $(3.0\ \text{m}, 2.0\ \text{m})$?

31P. (a) Using Eq. 25-28, show that the electric potential at a point on the central axis of a thin ring of charge of radius R and a distance z from the ring is

$$\Delta V = \frac{1}{4\pi\varepsilon_0} \frac{q}{\sqrt{z^2 + R^2}}.$$

(b) From this result, derive an expression for $|\vec{E}|$ at points on the ring's axis; compare your result with the calculation of $|\vec{E}|$ in Section 23-8.

32P. The plastic rod of length L in Fig. 25-33 has the nonuniform linear charge density $\lambda = cx$, where c is a positive constant. (a) With $\Delta V = 0$ at infinity, find the electric potential at point P_2 on the y axis, a distance y from one end of the rod. (b) From that result, find the electric field component E_y at P_2. (c) Why cannot the field component E_x at P_2 be found using the result of (a)?

33P. (a) Use the result of Problem 27 to find the electric field component E_x at point P_1 in Fig. 25-33. (*Hint:* First substitute the variable x for the distance d in the result.) (b) Use symmetry to determine the electric field component E_y at P_1.

SEC. 25-11   Potential of a Charged Isolated Conductor

34E. An empty hollow metal sphere has a potential of $+400$ V with respect to ground (defined to be at $\Delta V = 0$) and has a charge of 5.0×10^{-9} C. Find the electric potential at the center of the sphere.

35E. What is the excess charge on a conducting sphere of radius $r = 0.15$ m if the potential of the sphere is 1500 V and $\Delta V = 0$ at infinity?

36E. Consider two widely separated conducting spheres, 1 and 2, the second having twice the diameter of the first. The smaller sphere initially has a positive charge q, and the larger one is initially uncharged. You now connect the spheres with a long thin wire. (a) How are the final potentials ΔV_1 and ΔV_2 of the spheres related? (b) What are the final charges q_1 and q_2 on the spheres, in terms of q? (c) What is the ratio of the final surface charge density of sphere 1 to that of sphere 2?

37P. Two metal spheres, each of radius 3.0 cm, have a center-to-center separation of 2.0 m. One has a charge of $+1.0 \times 10^{-8}$ C; the other has a charge of -3.0×10^{-8} C. Assume that the separation is large enough relative to the size of the spheres to permit us to consider the charge on each to be uniformly distributed (the spheres do not affect each other). With $\Delta V = 0$ at infinity, calculate (a) the potential at the point halfway between their centers and (b) the potential of each sphere.

38P. A charged metal sphere of radius 15 cm has a net charge of 3.0×10^{-8} C. (a) What is the electric field at the sphere's surface? (b) If $\Delta V = 0$ at infinity, what is the electric potential at the sphere's surface? (c) At what distance from the sphere's surface has the electric potential decreased by 500 V?

39P. (a) If Earth had a net surface charge density of 1.0 electron per square meter (a very artificial assumption), what would its potential be? (Set $\Delta V = 0$ at infinity.) (b) What would be the electric field due to Earth just outside its surface?

40P. Two thin, isolated, concentric conducting spheres of radii R_1 and R_2 (with $R_1 < R_2$) have charges q_1 and q_2. With $\Delta V = 0$ at infinity, derive expressions for the electric field magnitude $|\vec{E}(r)|$ and the electric potential, and $\Delta V(r)$, where r is distance from the center of the spheres. Plot $|\vec{E}(r)|$ and $\Delta V(r)$ from $r = 0$ to $r = 4.0$ m for $R_1 = 0.50$ m, $R_2 = 1.0$ m, $q_1 = +2.0\ \mu C$, and $q_2 = +1.0\ \mu C$.

Chapter 26

SEC. 26-3   Electric Current

1P. A charged belt, 50 cm wide, travels at 30 m/s between a source of charge and a sphere. The belt carries charge into the sphere at a rate corresponding to 100 μA. Compute the surface charge density on the belt.

2P. An isolated conducting sphere has a 10 cm radius. One wire carries a current of 1.000 002 0 A into it. Another wire carries a current of 1.000 000 0 A out of it. How long would it take for the sphere to increase in potential by 1000 V?

3E. A current of 5.0 A exists in a 10 Ω resistor for 4.0 min. How many (a) coulombs and (b) electrons pass through any cross section of the resistor in this time?

SEC. 26-4   Potential Difference, Current and Resistance

4E. A human being can be electrocuted if a current as small as 50 mA passes near the heart. An electrician working with sweaty hands makes good contact with the two conductors he is holding, one in each hand. If his resistance is 2000 Ω, what might the fatal voltage be?

5P. An electrical cable consists of 125 strands of fine wire, each having 2.65 $\mu\Omega$ resistance. The same potential difference is applied between the ends of all the strands and results in a total current of 0.750 A. (a) What is the current in each strand? (b) What is the applied potential difference? (c) What is the resistance of the cable?

SEC. 26-5   Resistance and Resistivity

6E. A conducting wire has a 1.0 mm diameter, a 2.0 m length, and a 50 mΩ resistance. What is the resistivity of the material?

7E. A steel trolley-car rail has a cross-sectional area of 56.0 cm². What is the resistance of 10.0 km of rail? The resistivity of the steel is $3.00 \times 10^{-7}\ \Omega \cdot m$.

8E. A wire 4.00 m long and 6.00 mm in diameter has a resistance of 15.0 mΩ. A potential difference of 23.0 V is applied between the ends. (a) What is the current in the wire? (b) What is the current density? (c) Calculate the resistivity of the wire material. Identify the material. (Use Table 26-1.)

9E. A coil is formed by winding 250 turns of insulated 16-gauge copper wire (diameter = 1.3 mm) in a single layer on a cylindrical form of radius 12 cm. What is the resistance of the coil? Neglect the thickness of the insulation. (Use Table 26-1.)

10E. (a) At what temperature would the resistance of a copper conductor be double its resistance at 20.0°C? (Use 20.0°C as the reference point in Eq. 26-9; compare your answer with Fig. 26-22.) (b) Does this same "doubling temperature" hold for all copper conductors, regardless of shape or size?

11E. A wire with a resistance of 6.0 Ω is drawn out through a die so that its new length is three times its original length. Find the resistance of the longer wire, assuming that the resistivity and density of the material are unchanged.

12E. A certain wire has a resistance R. What is the resistance of a second wire, made of the same material, that is half as long and has half the diameter?

13P. Two conductors are made of the same material and have the same length. Conductor A is a solid wire of diameter 1.0 mm. Conductor B is a hollow tube of outside diameter 2.0 mm and inside diameter 1.0 mm. What is the resistance ratio R_A/R_B, measured between their ends?

14P. A common flashlight bulb is rated at 0.30 A and 2.9 V (the values of the current and voltage under operating conditions). If the resistance of the bulb filament at room temperature (20°C) is 1.1 Ω, what is the temperature of the filament when the bulb is on? The filament is made of tungsten.

15P. When a metal rod is heated, not only its resistance but also its length and its cross-sectional area change. The relation $R = \rho L/A$ suggests that all three factors should be taken into account in measuring ρ at various temperatures. (a) If the temperature changes by 1.0 C°, what percentage changes in R, L, and A occur for a copper conductor? (b) The coefficient of linear expansion for copper is $1.7 \times 10^{-5}/K$. What conclusion do you draw?

16P. If the gauge number of a wire is increased by 6, the diameter is halved; if a gauge number is increased by 1, the diameter decreases by the factor $2^{1/6}$ (see the table in Exercise 19). Knowing this, and knowing that 1000 ft of 10-gauge copper wire has a resistance of approximately 1.00 Ω, estimate the resistance of 25 ft of 22-gauge copper wire.

SEC. 26-8   Current Density

17E. A small but measurable current of 1.2×10^{-10} A exists in a copper wire whose diameter is 2.5 mm. Assuming the current is uniform, calculate (a) the current density and (b) the electron drift speed. (See TE26-8-1.)

18E. A beam contains 2.0×10^8 doubly charged positive ions per cubic centimeter, all of which are moving north with a speed of 1.0×10^5 m/s. (a) What are the magnitude and direction of the current density $\vec{J}$? (b) Can you calculate the total current i in this ion beam? If not, what additional information is needed?

19E. The (United States) National Electric Code, which sets maximum safe currents for insulated copper wires of various diameters, is given (in part) in the table. Plot the safe current density as a function of diameter. Which wire gauge has the maximum safe current density? ("Gauge" is a way of identifying wire diameters, and 1 mil = 10^{-3} in.)

Gauge	4	6	8	10	12	14	16	18
Diameter, mils	204	162	129	102	81	64	51	40
Safe current, A	70	50	35	25	20	15	6	3

20E. A fuse in an electric circuit is a wire that is designed to melt, and thereby open the circuit, if the current exceeds a predetermined value. Suppose that the material to be used in a fuse melts when the current density rises to 440 A/cm². What diameter of cylindrical wire should be used to make a fuse that will limit the current to 0.50 A?

21P. Near Earth, the density of protons in the solar wind (a stream of particles from the Sun) is 8.70 cm⁻³, and their speed is 470 km/s. (a) Find the current density of these protons. (b) If Earth's magnetic field did not deflect them, the protons would strike the planet. What total current would Earth then receive?

22P. A steady beam of alpha particles ($q = +2e$) traveling with constant kinetic energy 20 MeV carries a current of 0.25 μA. (a) If the beam is directed perpendicular to a plane surface, how many alpha particles strike the surface in 3.0 s? (b) At any instant, how many alpha particles are there in a given 20 cm length of the beam? (c) Through what potential difference is it necessary to accelerate each alpha particle from rest to bring it to an energy of 20 MeV?

23P. (a) The current density across a cylindrical conductor of radius R varies in magnitude according to the equation

$$|\vec{J}| = |\vec{J}_0|\left(1 - \frac{r}{R}\right),$$

where r is the distance from the central axis. Thus, the current density has a maximum magnitude of $|\vec{J}_0|$ at that axis ($r = 0$) and decreases linearly to zero at the surface ($r = R$). Calculate the current in terms of $|\vec{J}_0|$ and the conductor's cross-sectional area $A = \pi R^2$. (b) Suppose that, instead, the current density is a maximum J_0 at the cylinder's surface and decreases linearly to zero at the axis: $|\vec{J}| = |\vec{J}_0|r/R$. Calculate the magnitude of the current. Why is the result different from that in (a)?

24P. How long does it take electrons to get from a car battery to the starting motor? Assume the current is 300 A and the electrons travel through a copper wire with cross-sectional area 0.21 cm² and length 0.85 m. (See TE 26-6-1.)

25P. When 115 V is applied across a wire that is 10 m long and has a 0.30 mm radius, the current density is 1.4×10^4 A/m². Find the resistivity of the wire.

26P. A resistor has the shape of a truncated right-circular cone (Fig. 26-30). The end radii are a and b, and the altitude is L. If the taper is small, we may assume that the current density is uniform across any cross section. (a) Calculate the resistance of this object. (b) Show that your answer reduces to $\rho(L/A)$ for the special case of zero taper (that is, for $a = b$).

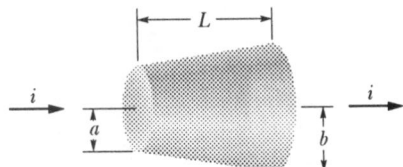

Fig. 26-30 Problem 26.

27E. A wire of Nichrome (a nickel–chromium–iron alloy commonly used in heating elements) is 1.0 m long and 1.0 mm² in cross-sectional area. It carries a current of 4.0 A when a 2.0 V potential difference is applied between its ends. Calculate the conductivity σ of Nichrome.

SEC. 26-9 A More General View of Resistivity, Ohm's Law and Drift Velocity

28P. A block in the shape of a rectangular solid has a cross-sectional area of 3.50 cm² across its width, a front-to-rear length of 15.8 cm, and a resistance of 935 Ω. The material of which the block is made has 5.33×10^{22} conduction electrons/m³. A potential difference of 35.8 V is maintained between its front and rear faces. (a) What is the current in the block? (b) If the current density is uniform, what is its value? (c) What is the drift velocity of the conduction electrons? (d) What is the magnitude of the electric field in the block?

29P. Earth's lower atmosphere contains negative and positive ions that are produced by radioactive elements in the soil and cosmic rays from space. In a certain region, the atmospheric electric field strength is 120 V/m, directed vertically down. This field causes singly charged positive ions, at a density of 620/cm³, to drift downward and singly charged negative ions, at a density of 550/cm³, to drift upward (Fig. 26-31). The measured conductivity of the air in that region is $2.70 \times 10^{-14}/\Omega \cdot$ m. Calculate (a) the ion drift speed, assumed to be the same for positive and negative ions, and (b) the current density.

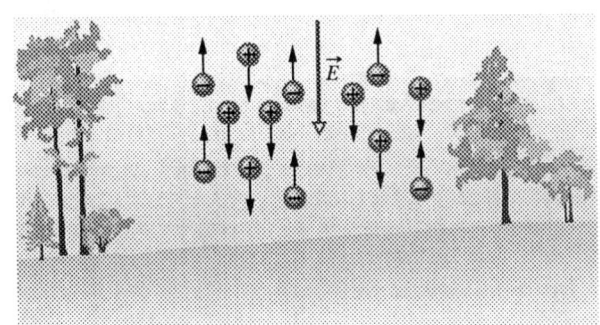

Fig. 26-31 Problem 29.

SEC. 26-10 Power in Electric Circuits

30E. A certain x-ray tube operates at a current of 7.0 mA and a potential difference of 80 kV. What is its power in watts?

31E. A student kept his 9.0 V, 7.0 W radio turned on at full volume from 9:00 p.m. until 2:00 a.m. How much charge went through it?

32E. A 120 V potential difference is applied to a space heater whose resistance is 14 Ω when hot. (a) At what rate is electric energy transferred to heat? (b) At 5.0¢/kW · h, what does it cost to operate the device for 5.0 h?

33E. Thermal energy is produced in a resistor at a rate of 100 W when the current is 3.00 A. What is the resistance?

34E. An unknown resistor is connected between the terminals of a 3.00 V battery. Energy is dissipated in the resistor at the rate of 0.540 W. The same resistor is then connected between the terminals of a 1.50 V battery. At what rate is energy now dissipated?

35E. A 120 V potential difference is applied to a space heater that dissipates 500 W during operation. (a) What is its resistance during operation? (b) At what rate do electrons flow through any cross section of the heater element?

36P. A 1250 W radiant heater is constructed to operate at 115 V. (a) What will be the current in the heater? (b) What is the resistance of the heating coil? (c) How much thermal energy is produced in 1.0 h by the heater?

37P. A heating element is made by maintaining a potential difference of 75.0 V across the length of a Nichrome wire that has a 2.60×10^{-6} m^2 cross section. Nichrome has a resistivity of 5.00×10^{-7} $\Omega \cdot$ m. (a) If the element dissipates 5000 W, what is its length? (b) If a potential difference of 100 V is used to obtain the same dissipation rate, what should the length be?

38P. A Nichrome heater dissipates 500 W when the applied potential difference is 110 V and the wire temperature is 800°C. What would be the dissipation rate if the wire temperature were held at 200°C by immersing the wire in a bath of cooling oil? The applied potential difference remains the same, and α for Nichrome at 800°C is 4.0×10^{-4}/K.

39P. A 100 W lightbulb is plugged into a standard 120 V outlet. (a) How much does it cost per month to leave the light turned on continuously? Assume electric energy costs 6¢/kW · h. (b) What is the resistance of the bulb? (c) What is the current in the bulb? (d) Is the resistance different when the bulb is turned off?

40P. A linear accelerator produces a pulsed beam of electrons. The pulse current is 0.50 A, and each pulse has a duration of 0.10 μs. (a) How many electrons are accelerated per pulse? (b) What is the average current for an accelerator operating at 500 pulses/s? (c) If the electrons are accelerated to an energy of 50 MeV, what are the average and peak powers of the accelerator?

41P. A cylindrical resistor of radius 5.0 mm and length 2.0 cm is made of material that has a resistivity of 3.5×10^{25} $\Omega \cdot$ m. What are (a) the current density and (b) the potential difference when the energy dissipation rate in the resistor is 1.0 W?

42P. A copper wire of cross-sectional area 2.0×10^{-6} m^2 and length 4.0 m has a current of 2.0 A uniformly distributed across that area. (a) What is the magnitude of the electric field along the wire? (b) How much electric energy is transferred to thermal energy in 30 min?

Chapter 27

SEC. 27-2 Current and Potential Difference in Single Loop Circuits

1E. Fig. 27-10 shows two ideal batteries with $\Delta V_{B_1} = 12$ V and $\Delta V_{B_2} = 8$ V. (a) What is the direction of the current in the resistor? (b) Which battery is doing positive work? (c) Which point, A or B, is at the higher potential?

2E. In Fig. 27-11, if the potential at point P is 100 V, what is the potential at point Q?

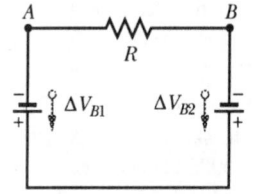

Fig. 27-10 Exercise 1.

3P. The current in a single-loop circuit with one resistance R is 5.0 A. When an additional resistance of 2.0 Ω is inserted in series with R, the current drops to 4.0 A. What is R?

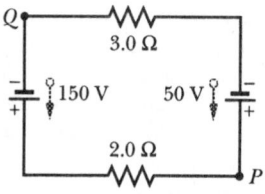

Fig. 27-11 Exercise 2.

SEC. 27-3 Series Resistance

4E. A simple ohmmeter is made by connecting an ideal 1.50 V flashlight battery in series with a resistance R and an ammeter that reads from 0 to 1.00 mA, as shown in Fig. 27-12. Resistance R is adjusted so that when the clip leads are shorted together, the meter deflects to its full-scale value of 1.00 mA. What external resistance across the leads results in a deflection of (a) 10%, (b) 50%, and (c) 90% of full scale? (d) If the ammeter has a resistance of 20.0 Ω and the internal resistance of the battery is negligible, what is the value of R?

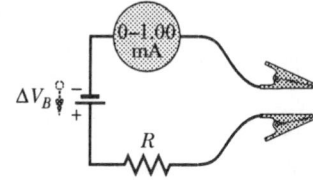

Fig. 27-12 Exercise 4.

SEC. 27-4 and 27-5 Multi Loop Circuits and Parallel Resistance

5P. (a) In Fig. 27-13, determine what the ammeter will read, assuming $\Delta V_B = 5.0$ V (for the ideal battery), $R_1 = 2.0$ Ω, $R_2 = 4.0$ Ω, and $R_3 = 6.0$ Ω. (b) The ammeter and the source of emf are now physically interchanged. Show that the ammeter reading remains unchanged.

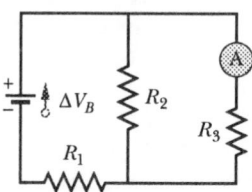

Fig. 27-13 Problem 5.

6E. In Fig. 27-14 find the current in each resistor and the potential difference between points a and b. Put $\Delta V_{B_1} = 6.0$ V, $\Delta V_{B_2} = 5.0$ V, $\Delta V_{B_3} = 4.0$ V, $R_1 = 100$ Ω, and $R_2 = 50$ Ω.

7E. Figure 27-15 shows a circuit containing three switches, labeled S_1, S_2, and S_3. Find the current at a for all possible combinations of switch settings. Put $\Delta V_B = 120$ V, $R_1 = 20.0$ V, and $R_2 = 10.0$ Ω. Assume that the battery has no resistance.

8E. In Fig. 27-5, calculate the potential difference between points c and d by as many paths as possible. Assume that $\Delta V_{B_1} = 4.0$ V, $\Delta V_{B_2} = 1.0$ V, $R_1 = R_2 = 10$ Ω, and $R_3 = 5.0$ Ω.

9P. In Fig. 27-16, find the equivalent resistance between points

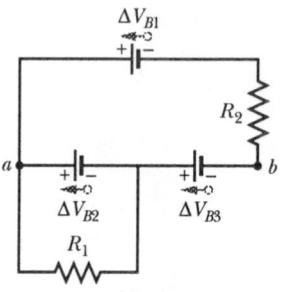

Fig. 27-14 Exercise 6.

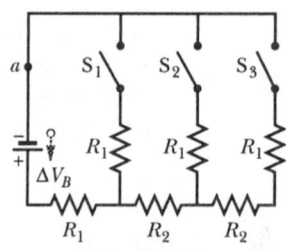

Fig. 27-15 Exercise 7.

(a) *F* and *H* and (b) *F* and *G*. (*Hint:* For each pair of points, imagine that a battery is connected across the pair.)

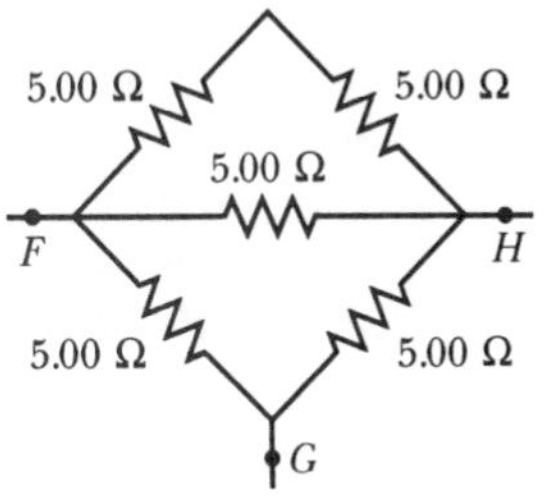

10P. In Fig. 27-17, R_s is to be adjusted in value by moving the sliding contact across it until points *a* and *b* are brought to the same potential. (One tests for this condition by momentarily connecting a sensitive ammeter between *a* and *b*; if these points are at the same potential, the ammeter will not deflect.) Show that when this adjustment is made, the following relation holds:

$$R_x = R_s\left(\frac{R_2}{R_1}\right).$$

An unknown resistance (R_x) can be measured in terms of a standard (R_s) using this device, which is called a Wheatstone bridge.

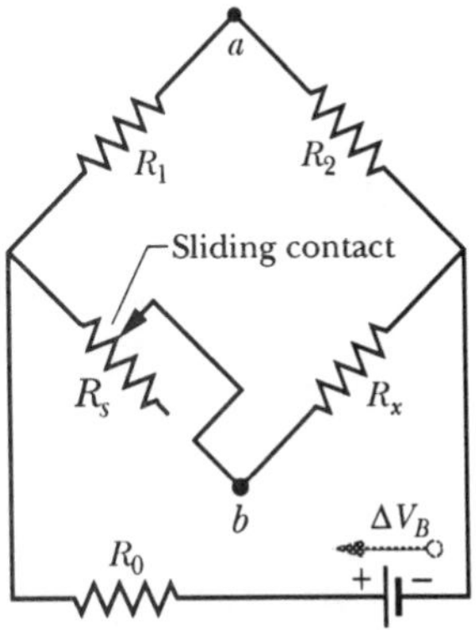

11E. By using only two resistors—singly, in series, or in parallel—you are able to obtain resistances of 3.0, 4.0, 12, and 16 Ω. What are the two resistances?

12E. Four 18.0 Ω resistors are connected in parallel across a 25.0 V battery. What is the current through the battery?

13E. In Fig. 28-18, find the equivalent resistance between points *D* and *E*. (*Hint:* Imagine that a battery is connected between points *D* and *E*.)

14E. Two lightbulbs, one of resistance R_1 and the other of resistance R_2, where $R_1 > R_2$, are connected to a battery (a) in parallel and (b) in series. Which bulb is brighter (dissipates more energy) in each case?

15E. Nine copper wires of length *l* and diameter *d* are connected in parallel to form a single composite conductor of resistance *R*. What must be the diameter *D* of a single copper wire of length *l* if it is to have the same resistance?

16P. You are given a number of 10 Ω resistors, each capable of dissipating only 1.0 W without being destroyed. What is the minimum number of such resistors that you need to combine in series or in parallel to make a 10 Ω resistance that is capable of dissipating at least 5.0 W?

17P. (a) In Fig. 27-19, what is the equivalent resistance of the network shown? (b) What is the current in each resistor? Put $R_1 = 100$ Ω, $R_2 = R_3 = 50$ Ω, $R_4 = 75$ Ω, and $\Delta V_B = 6.0$ V; assume the battery is ideal.

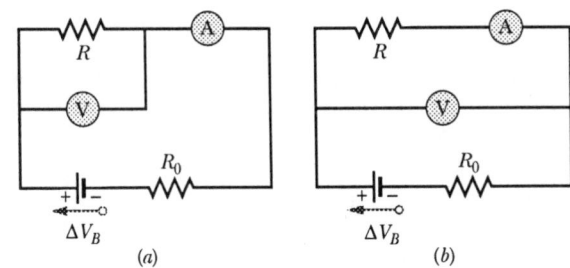
Fig. 27-16 Problem 9.

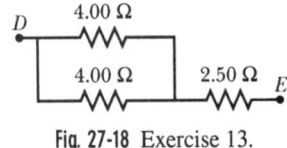
Fig. 27-17 Problem 10.

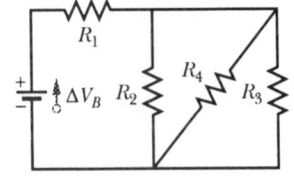
Fig. 27-18 Exercise 13.

Fig. 27-19 Problem 17.

18P. A copper wire of radius *a* = 0.250 mm has an aluminum jacket of outer radius *b* = 0.380 mm. (a) There is a current *i* = 2.00 A in the composite wire. Using Table 26-1, calculate the current in each material. (b) If a potential difference *V* = 12.0 V between the ends maintains the current, what is the length of the composite wire?

19P. A voltmeter (of resistance R_V) and an ammeter (of resistance R_A) are connected to measure a resistance *R*, as in Fig. 27-20*a*. The resistance is given by $R = \Delta V/i$, where ΔV is the voltmeter reading and *i* is the current in the resistance *R*. Some of the current *i'* registered by the ammeter goes through the voltmeter, so that the ratio of the meter readings (= $\Delta V/i'$) gives only an *apparent* resistance reading *R'*. Show that *R* and *R'* are related by

$$\frac{1}{R} = \frac{1}{R'} - \frac{1}{R_{\Delta V}}.$$

Note that as $R_{\Delta V} \to \infty$, $R' \to R$.

Fig. 27-20 Problems 19 to 21.

20P. (See Problem 19.) If an ammeter and a voltmeter are used to measure resistance, they may also be connected as in Fig. 27-20*b*. Again the ratio of the meter readings gives only an apparent resistance *R'*. Show that now *R'* is related to *R* by

$$R = R' - R_A,$$

in which R_A is the ammeter resistance. Note that as $R_A \to 0$, $R' \to R$.

21P. (See Problems 19 and 20.) In Fig. 27-20, the ammeter and voltmeter resistances are 3.00 and 300 Ω, respectively. Take $\Delta V_B = 12.0$ V for the ideal battery and $R_0 = 100$ Ω. If *R* = 85.0 V, (a) what will the meters read for the two different connections (Figs. 27-20*a* and *b*)? (b) What apparent resistance *R'* will be computed in each case?

SEC. 27-7 Internal Resistance and Power

22E. A standard flashlight battery can deliver about 2.0 W · h of energy before it runs down. (a) If a battery costs 80¢, what is the cost of operating a 100 W lamp for 8.0 h using batteries? (b) What is the cost if energy is provided at 6¢ per kilowatt-hour?

23E. A 5.0 A current is set up in a circuit for 6.0 min by a rechargeable battery with a 6.0 V emf. By how much is the chemical energy of the battery reduced?

24E. A certain car battery with a 12 V emf has an initial charge of 120 A · h. Assuming that the potential across the terminals stays constant until the battery is completely discharged, for how long can it deliver energy at the rate of 100 W?

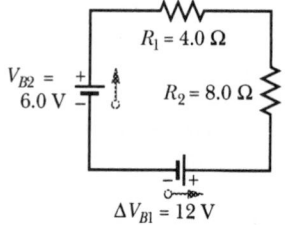

Fig. 27-21 Exercise 25.

25E. Assume that the batteries in Fig. 27-21 have negligible internal resistance. Find (a) the current in the circuit, (b) the power dissipated in each resistor, and (c) the power of each battery, stating whether energy is supplied by or absorbed by it.

26E. A wire of resistance 5.0 Ω is connected to a battery whose emf ε is 2.0 V and whose internal resistance is 1.0 Ω. In 2.0 min, (a) how much energy is transferred from chemical to electrical form? (b) How much energy appears in the wire as thermal energy? (c) Account for the difference between (a) and (b).

27E. A car battery with a 12 V emf and an internal resistance of 0.040 Ω is being charged with a current of 50 A. (a) What is the potential difference across its terminals? (b) At what rate is energy being dissipated as thermal energy in the battery? (c) At what rate is electric energy being converted to chemical energy? (d) What are the answers to (a) and (b) when the battery is used to supply 50 A to the starter motor?

28E. In Fig. 27-9a, put $\varepsilon = 2.0$ V and $r = 100$ Ω. Plot (a) the current and (b) the potential difference across R, as functions of R over the range 0 to 500 Ω. Make both plots on the same graph. (c) Make a third plot by multiplying together, for various values of R, the corresponding values on the two plotted curves. What is the physical significance of this third plot?

29E. In Fig. 27-22, circuit section AB absorbs energy at a rate of 50 W when a current $i = 1.0$ A passes through it in the indicated direction. (a) What is the potential difference between A and B? (b) Emf device X does not have internal resistance. What is its emf? (c) What is its *polarity* (the orientation of its positive and negative terminals)?

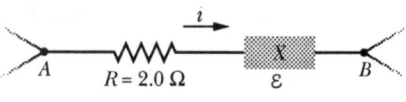

Fig. 27-22 Exercise 29.

30P. (a) In Fig. 27-23, what value must R have if the current in the circuit is to be 1.0 mA? Take $\varepsilon_1 = 2.0$ V, $\varepsilon_2 = 3.0$ V, and $r_1 = r_2 = 3.0$ Ω. (b) What is the rate at which thermal energy appears in R?

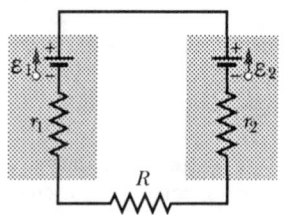

Fig. 27-23 Problem 30.

31P. The starting motor of an automobile is turning too slowly, and the mechanic has to decide whether to replace the motor, the cable, or the battery. The manufacturer's manual says that the 12 V battery should have no more than 0.020 Ω internal resistance, the motor no more than 0.200 Ω resistance, and the cable no more than 0.040 Ω resistance. The

mechanic turns on the motor and measures 11.4 V across the battery, 3.0 V across the cable, and a current of 50 A. Which part is defective?

32P. Two batteries having the same emf ε but different internal resistances r_1 and r_2 ($r_1 > r_2$) are connected in series to an external resistance R. (a) Find the value of R that makes the potential difference zero between the terminals of one battery. (b) Which battery is it?

33P. A solar cell generates a potential difference of 0.10 V when a 500 Ω resistor is connected across it, and a potential difference of 0.15 V when a 1000 Ω resistor is substituted. What are (a) the internal resistance and (b) the emf of the solar cell? (c) The area of the cell is 5.0 cm^2, and the rate per unit area at which it receives energy from light is 2.0 mW/cm^2. What is the efficiency of the cell for converting light energy to thermal energy in the 1000 Ω external resistor?

34P. (a) In Fig. 27-9a, show that the rate at which energy is dissipated in R as thermal energy is a maximum when $R = r$. (b) Show that this maximum power is $P = \varepsilon^2/4r$.

35P. Two batteries of emf ε and internal resistance r are connected in parallel across a resistor R, as in Fig. 27-24a. (a) For what value of R is the rate of electrical energy dissipation by the resistor a maximum? (b) What is the maximum energy dissipation rate?

36P. You are given two batteries of emf $\mathscr{E}$ and internal resistance r. They may be connected either in parallel (Fig. 27-24a) or in series (Fig. 27-24b) and are to be used to establish a current in a resistor R. (a) Derive expressions for the current in R for both arrangements. Which will yield the larger current (b) when $R > r$ and (c) when $R < r$?

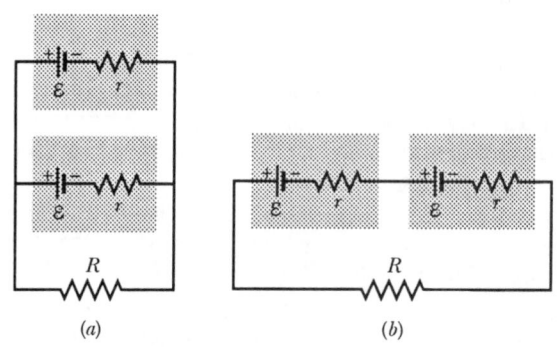

(a)   (b)

Fig. 27-24 Problems 35 and 36.

37P. In Fig. 27-25, $\varepsilon_1 = 3.00$ V, $\varepsilon_2 = 1.00$ V, $R_1 = 5.00$ Ω, $R_2 = 2.00$ Ω, $R_3 = 4.00$ Ω, and both batteries are ideal. What is the rate at which energy is dissipated in (a) R_1, (b) R_2, and (c) R_3? What is the power of (d) battery 1 and (e) battery 2?

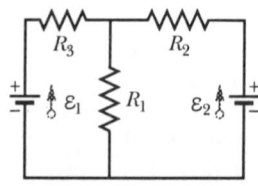

Fig. 27-25 Problem 31.

38P. In the circuit of Fig. 27-26, for what value of R will the ideal battery transfer energy to the resistors (a) at a rate of 60.0 W, (b) at the maximum possible rate, and (c) at the minimum possible rate? (d) What are those rates?

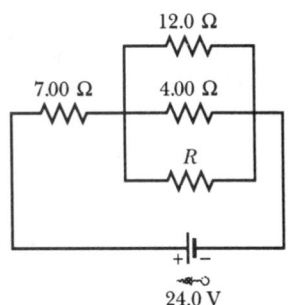

Fig. 27-26 Problem 38.

39P. (a) Calculate the current through each ideal battery in Fig. 27-27. Since the batteries are ideal, $\varepsilon = V_B$ in each case. Assume that $R_1 = 1.0\ \Omega$, $R_2 = 2.0\ \Omega$, $\varepsilon_1 = 2.0$ V, and $\varepsilon_2 = \varepsilon_3 = 4.0$ V. (b) Calculate $V_a - V_b$.

40P. In the circuit of Fig. 27-27, ε has a constant value but R can be varied. Find the value of R that results in the maximum heating in that resistor. The battery is ideal.

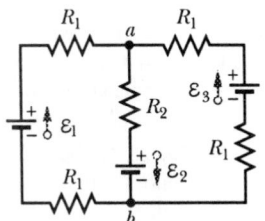

Fig. 27-27 Problem 39.

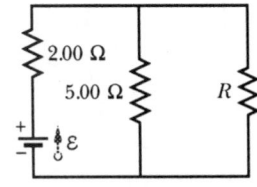

Fig. 27-28 Problem 40.

41P. When the lights of an automobile are switched on, an ammeter in series with them reads 10 A and a voltmeter connected across them reads 12 V. See Fig. 27-29. When the electric starting motor is turned on, the ammeter reading drops to 8.0 A and the lights dim somewhat. If the internal resistance of the battery is 0.050 Ω and that of the ammeter is negligible, what are (a) the emf of the battery and (b) the current through the starting motor when the lights are on?

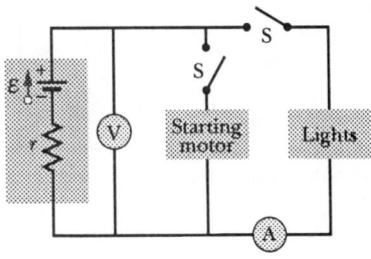

Fig. 27-29 Problem 41.

Chapter 28

SEC. 28-2   Capacitance

1E. An electrometer is a device used to measure static charge—an unknown charge is placed on the plates of the meter's capacitor, and the potential difference is measured. What minimum charge can be measured by an electrometer with a capacitance of 50 pF and a voltage sensitivity of 0.15 V?

2E. The two metal objects in Fig. 28-22 have net charges of +70 pC and −70 pC, which result in a 20 V potential difference between them. (a) What is the capacitance of the system? (b) If the charges are changed to +200 pC and −200 pC, what does the capacitance become? (c) What does the potential difference become?

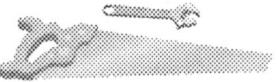

Fig. 28-22 Exercise 2.

3E. The capacitor in Fig. 28-23 has a capacitance of 25 μF and is initially uncharged. The battery provides a potential difference of 120 V. After switch S is closed, how much charge will pass through it?

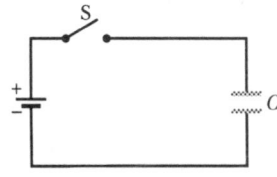

Fig. 28-23 Exercise 3.

SEC. 28-3   Calculating the Capacitance

4E. If we solve Eq. 28-9 for ε_0, we see that its SI unit is the farad per meter. Show that this unit is equivalent to that obtained earlier for ε_0—namely, the coulomb squared per newton-meter squared ($C^2/N \cdot m^2$).

5E. A parallel-plate capacitor has circular plates of 8.2 cm radius and 1.3 mm separation. (a) Calculate the capacitance. (b) What excess charge will appear on each of the plates if a potential difference of 120 V is applied?

6E. You have two flat metal plates, each of area 1.00 m^2, with which to construct a parallel-plate capacitor. If the capacitance of the device is to be 1.00 F, what must be the separation between the plates? Could this capacitor actually be constructed?

7E. A spherical drop of mercury of radius R has a capacitance given by $C = 4\pi\varepsilon_0 R$. If two such drops combine to form a single larger drop, what is its capacitance?

8E. The plates of a spherical capacitor have radii 38.0 mm and 40.0 mm. (a) Calculate the capacitance. (b) What must be the plate area of a parallel-plate capacitor with the same plate separation and capacitance?

9P. Suppose that the two spherical shells of a spherical capacitor have approximately equal radii. Under these conditions the device approximates a parallel-plate capacitor with $b - a = d$. Show that Eq. 28-17 does indeed reduce to Eq. 28-9 in this case.

SEC. 28-4   Capacitors in Parallel and in Series

10E. In Fig. 28-24, find the equivalent capacitance of the combination. Assume that $C_1 = 10.0\ \mu$F, $C_2 = 5.00\ \mu$F, and $C_3 = 4.00\ \mu$F.

11E. How many 1.00 μF capacitors must be connected in parallel to store a charge of 1.00 C with a potential of 110 V across the capacitors?

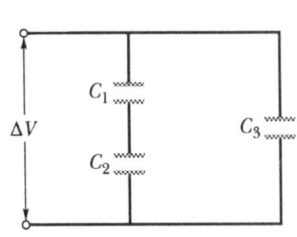

Fig. 28-24 Exercise 10 and Problem 30.

12E. Each of the uncharged capacitors in Fig. 28-25 has a capacitance of 25.0 μF. A potential difference of 4200 V is established when the switch is closed. How many coulombs of charge then pass through meter A?

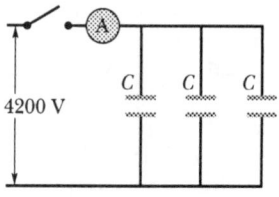

Fig. 28-25 Exercise 12.

13E. In Fig. 28-26 find the equivalent capacitance of the combination. Assume that $C_1 = 10.0$ μF, $C_2 = 5.00$ μF, and $C_3 = 4.00$ μF.

14P. In Fig. 28-26 suppose that capacitor 3 breaks down electrically, becoming equivalent to a conducting path. What *changes* in (a) the charge and (b) the potential difference occur for capacitor 1? Assume that $\Delta V = 100$ V.

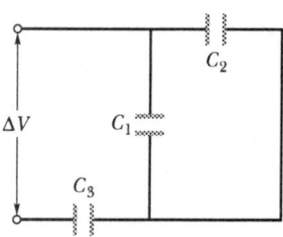

Fig. 28-26 Exercise 13, Problems 14 and 28.

15P. Figure 28-27 shows two capacitors in series; the center section of length b is movable vertically. Show that the equivalent capacitance of this series combination is independent of the position of the center section and is given by $C = \varepsilon_0 A/(a - b)$, where A is the plate area.

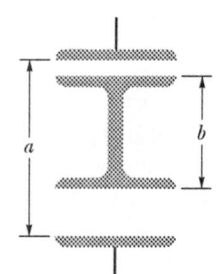

Fig. 28-27 Problem 15.

16P. In Fig. 28-28, the battery has a potential difference of 10 V and the five capacitors each have a capacitance of 10 μF. What is the charge on (a) capacitor 1 and (b) capacitor 2?

17P. A 100 pF capacitor is charged to a potential difference of 50 V, and the charging battery is disconnected. The capacitor is then connected in parallel with a second (initially uncharged) capacitor. If the potential difference across the first capacitor drops to 35 V, what is the capacitance of this second capacitor?

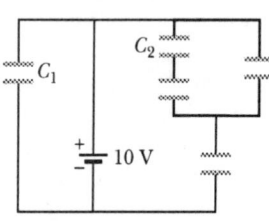

Fig. 28-28 Problem 16.

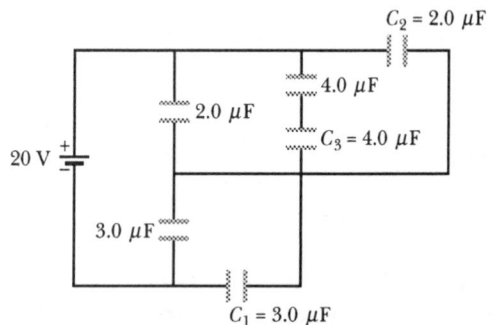

Fig. 28-29 Problem 18.

18P. In Fig. 28-29, the battery has a potential difference of 20 V. Find (a) the equivalent capacitance of all the capacitors and (b) the charge stored on that equivalent capacitance. Find the potential across and charge on (c) capacitor 1, (d) capacitor 2, and (e) capacitor 3.

19P. In Fig. 28-30, the capacitances are $C_1 = 1.0$ μF and $C_2 = 3.0$ μF and both capacitors are charged to a potential difference of $\Delta V = 100$ V but with opposite polarity as shown. Switches S_1 and S_2 are now closed. (a) What is now the potential difference between points a and b? What are now the charges on capacitors (b) 1 and (c) 2?

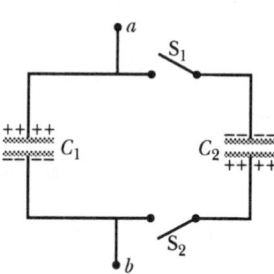

Fig. 28-30 Problem 19.

20P. In Fig. 28-31, battery B supplies 12 V. Find the charge on each capacitor (a) first when only switch S_1 is closed and (b) later when switch S_2 is also closed. Take $C_1 = 1.0$ μF, $C_2 = 2.0$ μF, $C_3 = 3.0$ μF, and $C_4 = 4.0$ μF.

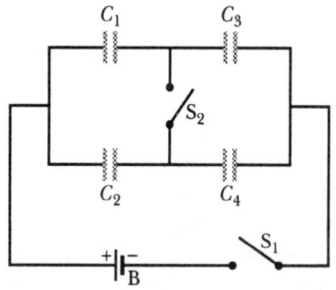

Fig. 28-31 Problem 20.

21P. When switch S is thrown to the left in Fig. 28-32, the plates of capacitor 1 acquire a potential difference ΔV_0. Capacitors 2 and 3 are initially uncharged. The switch is now thrown to the right. What are the final charges q_1, q_2, and q_3 on the capacitors?

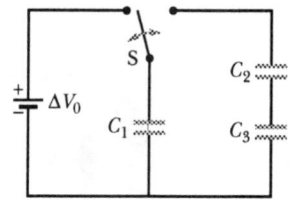

Fig. 28-32 Problem 21.

SEC. 28-5   Energy Stored in an Electric Field

22E. How much energy is stored in one cubic meter of air due to the "fair weather" electric field of magnitude 150 V/m?

23E. What capacitance is required to store an energy of 10 kW · h at a potential difference of 1000 V?

24E. A parallel-plate air-filled capacitor having area 40 cm^2 and plate spacing 1.0 mm is charged to a potential difference of 600 V. Find (a) the capacitance, (b) the magnitude of the charge on each plate, (c) the stored energy, (d) the electric field between the plates, and (e) the energy density between the plates.

25E. Two capacitors, of 2.0 and 4.0 μF capacitance, are connected in parallel across a 300 V potential difference. Calculate the total energy stored in the capacitors.

26P. A parallel-connected bank of 5.00 μF capacitors is used to store electric energy. What does it cost to charge the 2000 capacitors of the bank to 50 000 V, assuming 3.0¢/kW · h?

27P. One capacitor is charged until its stored energy is 4.0 J. A second uncharged capacitor is then connected to it in parallel. (a) If the charge distributes equally, what is now the total energy stored in the electric fields? (b) Where did the excess energy go?

28P. In Fig. 28-26 find (a) the charge, (b) the potential difference, and (c) the stored energy for each capacitor. Assume the numerical values of Exercise 13, with $\Delta V = 100$ V.

29P. A parallel-plate capacitor has plates of area A and separation d and is charged to a potential difference ΔV. The charging battery is then disconnected, and the plates are pulled apart until their separation is $2d$. Derive expressions in terms of A, d, and ΔV for (a) the new potential difference; (b) the initial and final stored energies, U_i and U_f; and (c) the work required to separate the plates.

30P. In Fig. 26-26, find (a) the charge, (b) the potential difference, and (c) the stored energy for each capacitor. Assume the numerical values of Exercise 10, with $\Delta V = 100$ V.

31P. A cylindrical capacitor has radii a and b as in Fig. 28-11. Show that half the stored electric potential energy lies within a cylinder whose radius is $r = \sqrt{ab}$.

32P. A charged isolated metal sphere of diameter 10 cm has a potential of 8000 V relative to $\Delta V = 0$ at infinity. Calculate the energy density in the electric field near the surface of the sphere.

33P. (a) Show that the plates of a parallel-plate capacitor attract each other with a force of magnitude given by $|\vec{F}| = q^2/2\varepsilon_0 A$. Do so by calculating the work needed to increase the plate separation from x to $x + dx$, with the charge q remaining constant. (b) Next show that the magnitude of the force per unit area (the *electrostatic stress*) acting on either capacitor plate is given by $\frac{1}{2}\varepsilon_0 E^2$. (Actually, this is the force per unit area on *any* conductor of *any* shape with an electric field $\vec{E}$ at its surface.)

SEC. 28-6   Capacitor with a Dielectric

34E. An air-filled parallel-plate capacitor has a capacitance of 1.3 pF. The separation of the plates is doubled and wax is inserted between them. The new capacitance is 2.6 pF. Find the dielectric constant of the wax.

35E. Give a 7.4 pF air-filled capacitor, you are asked to convert it to a capacitor that can store up to 7.4 μJ with a maximum potential difference of 652 V. What dielectric in Table 28-1 should you use to fill the gap in the air capacitor if you do not allow for a margin of error?

36E. A parallel-plate air-filled capacitor has a capacitance of 50 pF. (a) If each of its plates has an area of 0.35 m², what is the separation? (b) If the region between the plates is now filled with material having $\kappa = 5.6$, what is the capacitance?

37E. A coaxial cable used in a transmission line has an inner radius of 0.10 mm and an outer radius of 0.60 mm. Calculate the capacitance per meter for the cable. Assume that the space between the conductors is filled with polystyrene.

38P. You are asked to construct a capacitor having a capacitance near 1 nF and a breakdown potential in excess of 10 000 V. You think of using the sides of a tall Pyrex drinking glass as a dielectric, lining the inside and outside curved surfaces with aluminum foil to act as the plates. The glass is 15 cm tall with an inner radius of 3.6 cm and an outer radius of 3.8 cm. What are the (a) capacitance and (b) breakdown potential of this capacitor?

39P. A certain substance has a dielectric constant of 2.8 and a dielectric strength of 18 MV/m. If it is used as the dielectric material in a parallel-plate capacitor, what minimum area should the plates of the capacitor have to obtain a capacitance of 7.0×10^{-2} μF and to ensure that the capacitor will be able to withstand a potential difference of 4.0 kV?

40P. A parallel-plate capacitor of plate area A is filled with two dielectrics as in Fig. 28-33a. Show that the capacitance is

$$C = \frac{\varepsilon_0 A}{d}\frac{\kappa_1 + \kappa_2}{2}.$$

Check this formula for limiting cases. (*Hint:* Can you justify this arrangement as being two capacitors in parallel?)

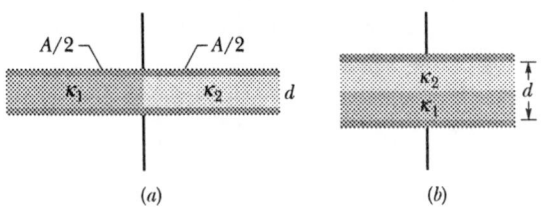

Fig. 28-33 Problems 40 and 41.

41P. A parallel-plate capacitor of plate area A is filled with two dielectrics as in Fig. 28-33b. Show that the capacitance is

$$C = \frac{2\varepsilon_0 A}{d}\frac{\kappa_1 \kappa_2}{\kappa_1 + \kappa_2}.$$

Check this formula for limiting cases. (*Hint:* Can you justify this arrangement as being two capacitors in series?)

42P. What is the capacitance of the capacitor, of plate area A, shown in Fig. 28-34? (*Hint:* See Problems 40 and 41.)

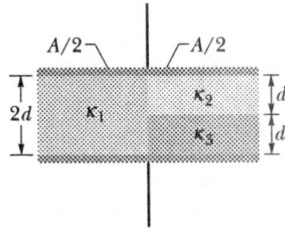

Fig. 28-34 Problem 42.

SEC. 28-8   Dielectrics and Gauss' Law

43E. A parallel-plate capacitor has a capacitance of 100 pF, a plate area of 100 cm², and a mica dielectric ($\kappa = 5.4$) completely filling the space between the plates. At 50 V potential difference, calculate (a) the electric field magnitude E in the mica, (b) the magnitude of the free charge on the plates, and (c) the magnitude of the induced surface charge on the mica.

44P. The space between two concentric conducting spherical shells of radii b and a (where $b > a$) is filled with a substance of dielectric constant κ. A potential difference ΔV exists between the inner and outer shells. Determine (a) the capacitance of the device, (b) the free charge q on the inner shell, and (c) the charge q' induced along the surface of the inner shell.

45P. Two parallel plates of area 100 cm^2 are given charges of equal magnitudes 8.9×10^{-7} C but opposite signs. The electric field within the dielectric material filling the space between the plates is 1.4×10^6 V/m. (a) Calculate the dielectric constant of the material. (b) Determine the magnitude of the charge induced on each dielectric surface.

SEC. 28-9   RC Circuits

46E. A capacitor with initial charge q_0 is discharged through a resistor. In terms of the time constant τ, how long is required for the capacitor to lose (a) the first one-third of its charge and (b) two-thirds of its charge?

47P. How many time constants must elapse for an initially uncharged capacitor in an RC series circuit to be charged to 99.0% of its equilibrium charge?

48E. A 15.0 kΩ resistor and a capacitor are connected in series and then a 12.0 V potential difference is suddenly applied across them. The potential difference across the capacitor rises to 5.00 V in 1.30 μs. (a) Calculate the time constant of the circuit. (b) Find the capacitance of the capacitor.

49P. The potential difference between the plates of a leaky (meaning that charges leaks from one plate to the other) 2.0 μF capacitor drops to one-fourth its initial value in 2.0 s. What is the equivalent resistance between the capacitor plates?

50P. A capacitor with an initial potential difference of 100 V is discharged through a resistor when a switch between them is closed at $t = 0$. At $t = 10.0$ s, the potential difference across the capacitor is 1.00 V. (a) What is the time constant of the circuit? (b) What is the potential difference across the capacitor at $t = 17.0$ s?

51P. Figure 28-35 shows the circuit of a flashing lamp, like those attached to barrels at highway construction sites. The fluorescent lamp L (of negligible capacitance) is connected in parallel across the capacitor C of an RC circuit. There is a current through the lamp only when the

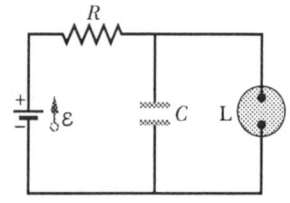

Fig. 28-35 Problem 51.

potential difference across it reaches the breakdown voltage V_L; in this event, the capacitor discharges completely through the lamp and the lamp flashes briefly. Suppose that two flashes per second are needed. For a lamp with breakdown voltage $\Delta V_L = 72.0$ V, wired to a 95.0 V ideal battery and a 0.150 μF capacitor, what should be the resistance R?

52P. A 1.0 μF capacitor with an initial stored energy of 0.50 J is discharged through a 1.0 MΩ resistor. (a) What is the initial charge on the capacitor? (b) What is the current through the resistor when the discharge starts? (c) Determine ΔV_C, the potential difference across the capacitor, and ΔV_R, the potential difference across the resistor, as functions of time. (d) Express the production rate of thermal energy in the resistor as a function of time.

53P. A controller on an electronics arcade game consists of a variable resistor connected across the plates of a 0.220 μF capacitor. The capacitor is charged to 5.00 V, then discharged through the resistor. The time for the potential difference across the plates to decrease to 0.800 V is measured by a clock inside the game. If the range of discharge times that can be handled effectively is from 10.0 μs to 6.00 ms, what should be the resistance range of the resistor?

Chapter 29

SEC. 29-2   The Definition of $\vec{B}$

1E. An alpha particle travels at a velocity $\vec{v}$ of magnitude 550 m/s through a uniform magnetic field $\vec{B}$ of magnitude 0.045 T. (An alpha particle has a charge of $+3.2 \times 10^{-19}$ C and a mass of 6.6×10^{-27} kg.) The angle between $\vec{v}$ and $\vec{B}$ is 52°. What are the magnitudes of (a) the force $\vec{F}_B$ acting on the particle due to the field and (b) the acceleration of the particle due to $\vec{F}_B$? (c) Does the speed of the particle increase, decrease, or remain equal to 550 m/s?

2E. An electron in a TV camera tube is moving at 7.20×10^6 m/s in a magnetic field of strength 83.0 mT. (a) Without knowing the direction of the field, what can you say about the greatest and least magnitudes of the force acting on the electron due to the field? (b) At one point the electron has an acceleration of magnitude 4.90×10^{14} m/s^2. What is the angle between the electron's velocity and the magnetic field?

3E. A proton traveling at 23.0° with respect to the direction of a magnetic field of strength 2.60 mT experiences a magnetic force of 6.50×10^{-17} N. Calculate (a) the proton's speed and (b) its kinetic energy in electron-volts.

4P. An electron that has velocity

$$\vec{v} = (2.0 \times 10^6 \text{ m/s})\hat{i} + (3.0 \times 10^6 \text{ m/s})\hat{j}$$

moves through the magnetic field $\vec{B} = (0.030 \text{ T})\hat{i} - (0.15 \text{ T})\hat{j}$. (a) Find the force on the electron. (b) Repeat your calculation for a proton having the same velocity.

5P. Each of the electrons in the beam of a television tube has a kinetic energy of 12.0 keV. The tube is oriented so that the electrons move horizontally from geomagnetic south to geomagnetic north. The vertical component of Earth's magnetic field points down and has a magnitude of 55.0 μT. (a) In what direction will the beam deflect? (b) What is the magnitude of the acceleration of a single electron due to the magnetic field? (c) How far will the beam deflect in moving 20.0 cm through the television tube?

SEC. 29-3   Crossed Fields: Discovery of the Electron

6E. A proton travels through uniform magnetic and electric fields. The magnetic field is $\vec{B} = (-2.5 \text{ mT})\hat{i}$. At one instant the velocity of the proton is $\vec{v} = (2000 \text{ m/s})\hat{j}$. At that instant, what is the magnitude of the net force acting on the proton if the electric field is (a) $(4.0 \text{ V/m})\hat{k}$, and (c) $(4.0 \text{ V/m})\hat{i}$?

7E. An electron with kinetic energy 2.5 keV moves horizontally into a region of space in which there is a downward-directed uniform electric field of magnitude 10 kV/m. (a) What are the magnitude and direction of the (smallest) uniform .nagnetic field that will cause the electron to continue to move horizontally? Ignore the gravitational force, which is rather small. (b) Is it possible for a proton to pass through the combination of fields undeflected? If so, under what circumstances?

8E. An electric field of magnitude 1.50 kV/m and a magnetic field of 0.400 T act on a moving electron to produce no net force. (a) Calculate the minimum speed $|\vec{v}|$ of the electron. (b) Draw a set of vectors $\vec{E}$, $\vec{B}$, and $\vec{v}$ that could yield the net force.

9P. An electron is accelerated through a potential difference of 1.0 kV and directed into a region between two parallel plates separated by 20 mm with a potential difference of 100 V between them. The electron is moving perpendicular to the electric field of the plates when it enters the region between the plates. What uniform magnetic field, applied perpendicular to both the electron path and the electric field, will allow the electron to travel in a straight line?

10P. An electron has an initial velocity of $(12.0 \text{ km/s})\hat{j} + (15.0 \text{ km/s})\hat{k}$ and a constant acceleration of $(2.00 \times 10^{12} \text{ m/s}^2)\hat{i}$ in a region in which uniform electric and magnetic fields are present. If $\vec{B} = (400 \ \mu\text{T})\hat{i}$, find the electric field $\vec{E}$.

11P. An ion source is producing ions of ^{6}Li (mass = 6.0 u), each with a charge of $+e$. The ions are accelerated by a potential difference of 10 kV and pass horizontally into a region in which there is a uniform vertical magnetic field of magnitude $|\vec{B}| = 1.2$ T. Calculate the strength of the smallest electric field, to be set up over the same region, that will allow the ^{6}Li ions to pass through undeflected.

SEC. 29-4   Crossed Fields: The Hall Effect

12E. A strip of copper 150 μm wide is placed in a uniform magnetic field $\vec{B}$ of magnitude 0.65 T, with $\vec{B}$ perpendicular to the strip. A current $i = 23$ A is then sent through the strip such that a Hall potential difference ΔV appears across the width of the strip. Calculate ΔV. (The number of charge carriers per unit volume for copper is 8.47×10^{28} electrons/m^3.)

13P. (a) In Fig. 29-7, show that the ratio of the Hall electric field E to the electric field E_C responsible for moving charge (the current) along the length of the strip is

$$\frac{|\vec{E}|}{|\vec{E_C}|} = \frac{|\vec{B}|}{ne\rho},$$

where ρ is the resistivity of the material and n is the number density of the charge carriers and e is the magnitude of the charge on the electron. (b) Compute this ratio numerically for Exercise 12. (See Table 26-1.)

14P. A metal strip 6.50 cm long, 0.850 cm wide, and 0.760 mm thick moves with constant velocity $\vec{v}$ through a uniform magnetic field of magnitude $|\vec{B}| = 1.20$ mT directed perpendicular to the strip, as shown in Fig. 29-19. A potential difference of 3.90 μV is measured between points x and y across the strip. Calculate the speed $|\vec{v}|$.

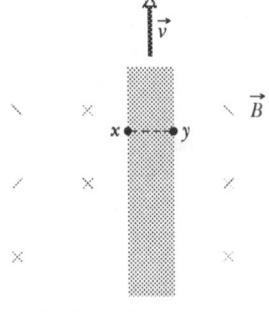

Fig. 29-19 Problem 14.

SEC. 29-5   A Circulating Charged Particle

15E. A uniform magnetic field is applied perpendicular to a beam of electrons moving at 1.3×10^6 m/s. What is the magnitude of the field if the electrons travel in a circular arc of radius 0.35 m?

16E. An electron is accelerated from rest by a potential difference of 350 V. It then enters a uniform magnetic field of magnitude 200 mT with its velocity perpendicular to the field. Calculate (a) the speed of the electron and (b) the radius of its path in the magnetic field.

17E. An electron with kinetic energy 1.20 keV circles in a plane perpendicular to a uniform magnetic field. The orbit radius is 25.0 cm. Find (a) the speed of the electron, (b) the magnetic field, (c) the frequency of circling, and (d) the period of the motion.

18E. Physicist S. A. Goudsmit devised a method for measuring the masses of heavy ions by timing their periods of revolution in a known magnetic field. A singly charged ion of iodine makes 7.00 rev in a field of 45.0 mT in 1.29 ms. Calculate its mass, in unified atomic mass units. (Actually, the method allows mass measurements to be carried out to much greater accuracy than these approximate data suggest.)

19E. (a) Find the frequency of revolution of an electron with an energy of 100 eV in a uniform magnetic field of magnitude 35.0 μT. (b) Calculate the radius of the path of this electron if its velocity is perpendicular to the magnetic field.

20E. An alpha particle ($q = +2e$, $m = 4.00$ u) travels in a circular path of radius 4.50 cm in a uniform magnetic field with magnitude $|\vec{B}| = 1.20$ T. Calculate (a) its speed, (b) its period of revolution, (c) its kinetic energy in electron-volts, and (d) the potential difference through which it would have to be accelerated to achieve this energy.

21E. A beam of electrons whose kinetic energy is K emerges from a thin-foil "window" at the end of an accelerator tube. There is a metal plate a distance d from this window and perpendicular to the direction of the emerging beam (Fig. 29-20). Show that we can prevent the beam from hitting the plate if we apply a uniform magnetic field $\vec{B}$ such that its magnitude is

$$|\vec{B}| \geq \sqrt{\frac{2mK}{e^2d^2}},$$

in which m and e are the electron mass and charge. How should $\vec{B}$ be oriented?

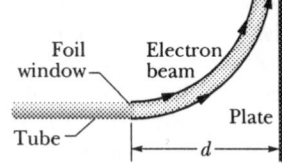

Fig. 29-20 Exercise 21.

22P. A source injects an electron of speed $|\vec{v}| = 1.5 \times 10^7$ m/s into a uniform magnetic field of magnitude $|\vec{B}| = 1.0 \times 10^{-3}$ T. The velocity of the electron makes an angle $\theta = 10°$ with the direction of the magnetic field. Find the distance d from the point of injection at which the electron next crosses the field line that passes through the injection point.

23P. In a nuclear experiment a proton with kinetic energy 1.0 MeV moves in a circular path in a uniform magnetic field. What energy must (a) an alpha particle ($q = +2e$, $m = 4.0$ u) and (b) a deuteron ($q = +e$, $m = 2.0$ u) have if they are to circulate in the same circular path?

24P. A proton, a deuteron ($q = +e$, $m = 2.0$ u), and an alpha particle ($q = +2e$, $m = 4.0$ u) with the same kinetic energies enter a region of uniform magnetic field $\vec{B}$, moving perpendicular to $\vec{B}$. Compare the radii of their circular paths.

25P. A certain commercial mass spectrometer (see Touchstone Example 29-5-1) is used to separate uranium ions of mass 3.92×10^{-25} kg and charge 3.20×10^{-19} C from related species. The ions are accelerated through a potential difference of 100 kV and then pass into a uniform magnetic field, where they are bent in a path of radius 1.00 m. After traveling through 180° and passing through a slit of width 1.00 mm and height 1.00 cm, they are collected in a cup. (a) What is the magnitude of the (perpendicular) magnetic field in the separator? If the machine is used to separate out 100 mg of material per hour, calculate (b) the current of the desired ions in the machine and (c) the thermal energy produced in the cup in 1.00 h.

26P. A proton of charge $+e$ and mass m enters a uniform magnetic field $\vec{B} = B\hat{i}$ with an initial velocity $\vec{v} = v_{0x}\hat{i} + v_{0y}\hat{j}$. Find an expression in unit-vector notation for its velocity $\vec{v}$ at any later time t.

27P. A positron with kinetic energy 2.0 keV is projected into a uniform magnetic field $\vec{B}$ of magnitude 0.10 T, with its velocity vector making an angle of 89° with $\vec{B}$. Find (a) the period, (b) the pitch p, and (c) the radius r of its helical path.

28P. In Fig. 29-21, a charged particle moves into a region of uniform magnetic field $|\vec{B}|$, goes through half a circle, and then exits that region. The particle is either a proton or an electron (you must decide which). It spends 130 ns within the region. (a) What is the magnitude of $|\vec{B}|$? (b) If the particle is sent back through the magnetic field (along the same initial path)

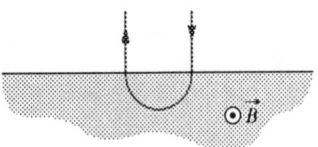

Fig. 29-21 Problem 28.

but with 2.00 times its previous kinetic energy, how much time does it spend within the field?

29P. A neutral particle is at rest in a uniform magnetic field $|\vec{B}|$. At time $t = 0$ it decays into two charged particles, each of mass m. (a) If the charge of one of the particles is $+q$, what is the charge of the other? (b) The two particles move off in separate paths, both of which lie in the plane perpendicular to $|\vec{B}|$. At a later time the particles collide. Express the time from decay until collision in terms of m, $|\vec{B}|$, and $|\vec{q}|$.

SEC. 29-6   Cyclotrons and Synchrotrons

30E. In a certain cyclotron a proton moves in a circle of radius 0.50 m. The magnitude of the magnetic field is 1.2 T. (a) What is the oscillator frequency? (b) What is the kinetic energy of the proton, in electron-volts?

SEC. 29-7   Magnetic Force on a Current-Carrying Wire

31E. A horizontal conductor that is part of a power line carries a current of 5000 A from south to north. The magnitude of the Earth's magnetic field is 60.0 μT. The field is directed toward the north and is inclined downward at 70° to the horizontal. Find the magnitude and direction of the magnetic force on 100 m of the conductor due to Earth's field.

32E. A wire 1.80 m long carries a current of 13.0 A and makes an angle of 35.0° with a uniform magnetic field of magnitude $|\vec{B}| = 1.50$ T. Calculate the magnitude of the magnetic force on the wire.

33E. A wire of 62.0 cm length and 13.0 g mass is suspended by a pair of flexible leads in a uniform magnetic field of magnitude 0.440 T (Fig. 29-22). What are the magnitude and direction of the current required to remove the tension in the supporting leads?

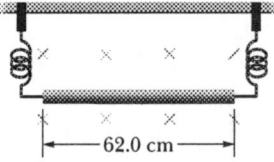

Fig. 29-22 Exercise 33.

34P. A wire 50 cm long lying along the x axis carries a current of 0.50 A in the positive x direction. It passes through a magnetic field $\vec{B} = (0.0030$ T$)\hat{j} + (0.010$ T$)\hat{k}$. Find the magnetic force on the wire.

35P. A 1.0 kg copper rod rests on two horizontal rails 1.0 m apart and carries a current of 50 A from one rail to the other. The coefficient of static friction between rod and rails is 0.60. What is the magnitude of the smallest magnetic field (not necessarily vertical) that would cause the rod to slide?

36P. Consider the possibility of a new design for an electric train. The engine is driven by the force on a conducting axle due to the vertical component of Earth's magnetic field. To produce the force, current is maintained down one rail, through a conducting wheel, through the axle, through another conducting wheel, and then back to the source via the other rail. (a) What magnitude of current is needed to provide a modest force of magnitude 10 kN? Take the vertical component of Earth's field to be 10 μT and the length of the axle to be 3.0 m. (b) At what rate would electric energy be lost for each ohm of resistance in the rails? (c) Is such a train totally or just marginally unrealistic?

SEC. 29-8   Torque on a Current Loop

37E. Figure 29-23 shows a rectangular 20-turn coil of wire, of dimensions 10 cm by 5.0 cm. It carries a current of 0.10 A and is hinged along one long side. It is mounted in the xy plane, at 30° to the direction of a uniform magnetic field of magnitude 0.50 T. Find the magnitude and direction of the torque acting on the coil about the hinge line.

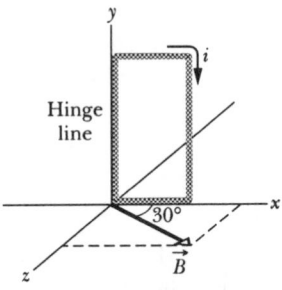

Fig. 29-23 Exercise 37.

38E. A single-turn current loop, carrying a current of 4.00 A, is in the shape of a right triangle with sides 50.0, 120, and 130 cm. The loop is in a uniform magnetic field of magnitude 75.0 mT whose direction is parallel to the current in the 130 cm side of the loop. (a) Find the magnitude of the magnetic force on each of the three sides of the loop. (b) Show that the total magnetic force on the loop is zero.

39E. A length L of wire carries a current i. Show that if the wire is formed into a circular coil, then the magnitude of the maximum torque in a given magnetic field is developed when the coil has one turn only. Also show that maximum torque has the magnitude $\tau = L^2iB/4\pi$.

40P. Prove that the relation $|\vec{\tau}| = |NiAB \sin \phi|$ holds for closed loops of arbitrary shape and not only for rectangular loops as in Fig. 29-17. (*Hint:* Replace the loop of arbitrary shape with an assembly of adjacent long, thin, approximately rectangular loops that are nearly equivalent to the loop of arbitrary shape as far as the distribution of current is concerned.)

41P. Figure 29-24 shows a wire ring of radius a that is perpendicular to the general direction of a radially symmetric, diverging magnetic field. The magnetic field at the ring is everywhere of the same magnitude $|\vec{B}|$, and its direction at the ring

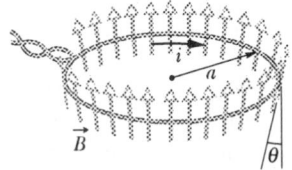

Fig. 29-24 Problem 41.

everywhere makes an angle θ with a normal to the plane of the ring. The twisted lead wires have no effect on the problem. Find the magnitude and direction of the force the field exerts on the ring if the ring carries a positive current i.

42P. A closed wire loop with current i is in a uniform magnetic field $\vec{B}$, with the plane of the loop at angle θ to the direction of $\vec{B}$. Show that the total magnetic force on the loop is zero. Does your proof also hold for a nonuniform magnetic field?

43P. A particle of charge q moves in a circle of radius a with speed $|\vec{v}|$. Find the maximum torque exerted on the loop by a uniform magnetic field of magnitude $|\vec{B}|$.

44P. Figure 29-25 shows a wooden cylinder with mass $m = 0.250$ kg and length $L = 0.100$ m, with $N = 10.0$ turns of wire wrapped around it longitudinally, so that the plane of the wire coil contains the axis of the cylinder. Also the plane of the coil is parallel to the inclined plane. There is a vertical, uniform magnetic field of magnitude 0.500 T. What is the least current magnitude $|i|$ through the coil that will prevent the cylinder from rolling down a plane inclined at an angle θ to the horizontal?

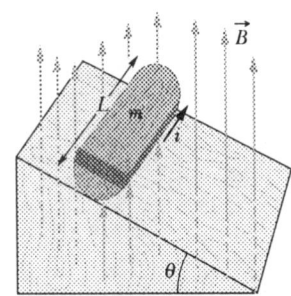

Fig. 29-25 Problem 44.

SEC. 29-9   The Magnetic Dipole Moment

45E. The magnitude of magnetic dipole moment of Earth is 8.00×10^{22} J/T. Assume that this is produced by charges flowing in Earth's molten outer core. If the radius of their circular path is 3500 km, calculate the magnitude of current each charge produces.

46E. A circular coil of 160 turns has a radius of 1.90 cm. (a) Calculate the current that results in a magnetic dipole moment of 2.30 A · m². (b) Find the maximum magnitude of torque that the coil, carrying this current, can experience in a uniform 35.0 mT magnetic field.

47E. A circular wire loop whose radius is 15.0 cm carries a current of magnitude 2.60 A. It is placed so that the normal to its plane makes an angle of 41.0° with a uniform magnetic field of magnitude 12.0 T. (a) Calculate the magnitude of the magnetic dipole

moment of the loop. (b) What is the magnitude of torque that acts on the loop?

48E. A current loop, carrying a current of magnitude 5.0 A, is in the shape of a right triangle with sides 30, 40, and 50 cm. The loop is in a uniform magnetic field of magnitude 80 mT whose direction is parallel to the current in the 50 cm side of the loop. Find the magnitude of (a) the magnetic dipole moment of the loop and (b) the torque on the loop.

49E. A stationary circular wall clock has a face with a radius of 15 cm. Six turns of wire are wound around its perimeter; the wire carries a current of 2.0 A in the clockwise direction. The clock is located where there is a constant, uniform external magnetic field of magnitude 70 mT (but the clock still keeps perfect time). At exactly 1:00 p.m., the hour hand of the clock points in the direction of the external magnetic field. (a) After how many minutes will the minute hand point in the direction of the torque on the winding due to the magnetic field? (b) Find the torque magnitude.

50E. Two concentric, circular wire loops, of radii 20.0 and 30.0 cm, are located in the xy plane; each carries a clockwise current of 7.00 A (Fig. 29-26). (a) Find the magnitude of the net magnetic dipole moment of this system. (b) Repeat for reversed current in the inner loop.

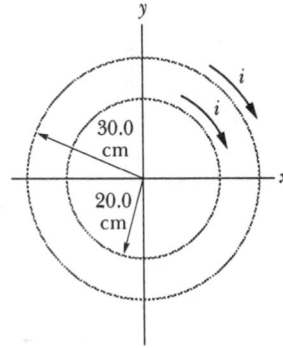

Fig. 29-26 Exercise 50.

51P. Figure 29-27 shows a current loop $ABCDEFA$ carrying a current $i = 5.00$ A. The sides of the loop are parallel to the coordinate axes, with $AB = 20.0$ cm, $BC = 30.0$ cm, and $FA = 10.0$ cm. Calculate the magnitude and direction of the magnetic dipole moment of this loop. (*Hint:* Imagine equal and opposite currents i in the line segment AD; then treat the two rectangular loops $ABCDA$ and $ADEFA$.)

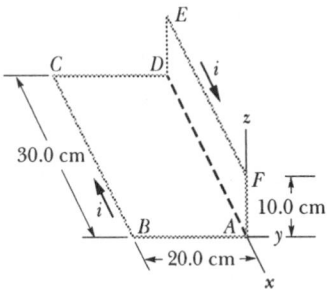

Fig. 29-27 Problem 51.

52P. A circular loop of wire having a radius of 8.0 cm carries a current of 0.20 A. A vector of unit length and parallel to the dipole moment $\vec{\mu}$ of the loop is given by $0.60\hat{i} - 0.80$ T$\hat{j}$. If the loop is located in a uniform magnetic field given by $|\vec{B}| = (0.25$ T$)\hat{i} + (0.30$ T$)\hat{k}$, find (a) the torque on the loop (in unit-vector notation) and (b) the magnetic potential energy of the loop.

Chapter 30

SEC. 30-1   Calculating the Magnetic Field Due to a Current

1E. A surveyor is using a magnetic compass 6.1 m below a power line in which there is a steady current of 100 A. (a) What is the magnitude of the magnetic field at the site of the compass due to the power line? (b) Will this interfere seriously with the compass reading? The horizontal component of Earth's magnetic field at the site is 20 μT.

2E. The electron gun in a traditional television tube fires electrons of kinetic energy 25 keV at the screen in a circular beam 0.22 mm in diameter; 5.6×10^{14} electrons arrive each second. Calculate the magnitude of the magnetic field produced by the beam at a point 1.5 mm from the beam axis.

3E. At a certain position in the Philippines, the magnitude of the Earth's magnetic field of 39 μT is horizontal and directed due north. Suppose the net field is zero exactly 8.0 cm above a long, straight, horizontal wire that carries a constant current. What are (a) the magnitude and (b) the direction of the current?

4E. A long wire carrying a current of 100 A is placed in a uniform external magnetic field of 5.0 mT. The wire is perpendicular to this magnetic field. Locate the points at which the net magnetic field is zero.

5E. A particle with positive charge q is a distance d from a long straight wire that carries a current i; the particle is traveling with speed $|\vec{v}|$ perpendicular to the wire. What are the direction and magnitude of the force on the particle if it is moving (a) toward and (b) away from the wire?

6E. A straight conductor carrying a current i splits into identical semicircular arcs as shown in Fig. 30-20. What is the magnitude of the magnetic field at the center C of the resulting circular loop?

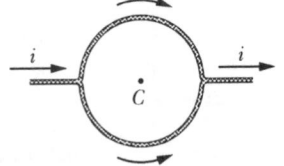

Fig. 30-20 Exercise 6.

7P. A wire carrying current i has the configuration shown in Fig. 30-21. Two semi-infinite straight sections, both tangent to the same circle, are connected by a circular arc, of central angle θ, along the circumference of the circle, with all sections lying in

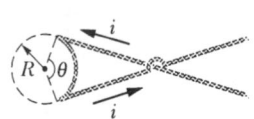

Fig. 30-21 Problem 7.

the same plane. What must θ be in order for $|\vec{B}|$ to be zero at the center of the circle?

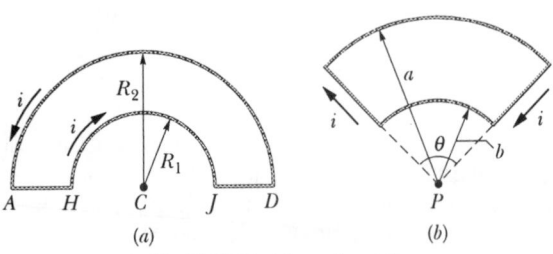

(a)

(b)

Fig. 30-22 Problems 8 and 9.

8P. Use the Biot–Savart law to calculate the magnitude and direction of the magnetic field $\vec{B}$ at C, the common center of the semicircular arcs AD and HJ in Fig. 30-22a. The two arcs, of radii R_2 and R_1, respectively, form part of the circuit $ADJHA$ carrying current i.

9P. In the circuit of Fig. 30-22b, the curved segments are arcs of circles of radii a and b with common center P. The straight segments are along radii. Find the magnitude and direction of the magnetic field $\vec{B}$ at point P, assuming a current i in the circuit.

10P. The wire shown in Fig. 30-23 carries current i. What are the magnitude and direction of the magnetic field $\vec{B}$ produced at the center C of the semicircle by (a) each straight segment of length L, (b) the semicircular segment of radius R, and (c) the entire wire?

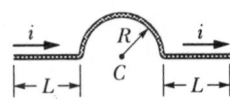

Fig. 30-23 Problem 10.

11P. In Fig. 30-24, a straight wire of length L carries current i. Show that the magnitude of the magnetic field $\vec{B}$ produced by this segment at P_1, a distance R from the segment along a perpendicular bisector, is

$$|\vec{B}| = \frac{\mu_0 |i|}{2\pi R} \frac{L}{(L^2 + 4R^2)^{1/2}}.$$

Show that this expression for $|\vec{B}|$ reduces to an expected result as $L \to \infty$.

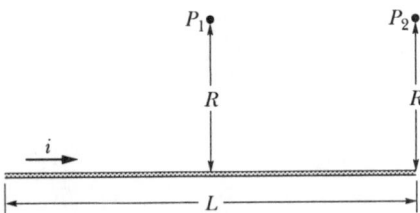

Fig. 30-24 Problems 11 and 13.

12P. A square loop of wire of edge length a carries current i. Using Problem 11, show that, at the center of the loop, the magnitude of the magnetic field produced by the current is

$$|\vec{B}| = \frac{2\sqrt{2}\mu_0 |i|}{\pi a}.$$

13P. In Fig. 30-24, a straight wire of length L carries current i. Show that

$$|\vec{B}| = \frac{\mu_0 |i|}{4\pi R} \frac{L}{(L^2 + 4R^2)^{1/2}}$$

gives the magnitude of the magnetic field $\vec{B}$ produced by the wire at P_2, a perpendicular distance R from one end of the wire.

14P. Using Problem 11, show that the magnitude of the magnetic field produced at the center of a rectangular loop of wire of length L and width W, carrying a current i, is

$$|\vec{B}| = \frac{2\mu_0 |i|}{\pi} \frac{(L^2 + W^2)^{1/2}}{LW}.$$

15P. A square loop of wire of edge length a carries current i. Using Problem 11, show that the magnitude of the magnetic field produced at a point on the axis of the loop and a distance x from its center is

$$|\vec{B}(x)| = \frac{4\mu_0|i|a^2}{\pi(4x^2 + a^2)(4x^2 + 2a^2)^{1/2}}.$$

Prove that this result is consistent with the result of Problem 12.

16P. In Fig. 30-25, a straight wire of length a carries a current i. Show that the magnitude of the magnetic field produced by the current at point P is $|\vec{B}| = \sqrt{2}\mu_0|i|/8\pi a$.

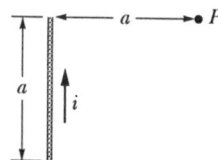

Fig. 30-25 Problem 16.

17P. Two wires, both of length L, are formed into a circle and a square, and each carries current i. Show that the square produces a greater magnetic field at its center than the circle produces at its center. (See Problem 12.)

18P. Find the magnitude and direction of the magnetic field $\vec{B}$ at point P in Fig. 30-26. (See Problem 16.)

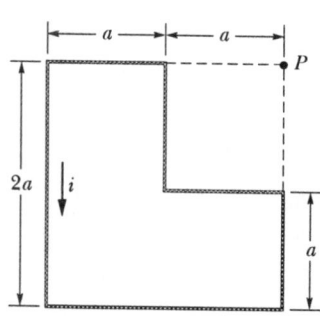

Fig. 30-26 Problem 18.

19P. Figure 30-27 shows a cross section of a long thin ribbon of width w that is carrying a uniformly distributed total current i into the page. Calculate the magnitude and direction of the magnetic field $\vec{B}$ at a point P in the plane of the ribbon at a distance d from its edge. (*Hint:* Imagine the ribbon to be constructed from many long, thin, parallel wires.)

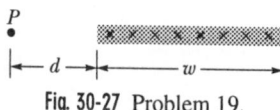

Fig. 30-27 Problem 19.

20P. Find the magnitude and direction of the magnetic field $\vec{B}$ at point P in Fig. 30-28, for $|i| = 10$ A and $a = 8.0$ cm. (See Problems 13 and 16.)

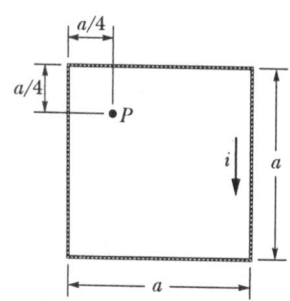

Fig. 30-28 Problem 20.

SEC. 30-2   Force Between Two Parallel Currents

21E. Two long parallel wires are 8.0 cm apart. What equal currents must be in the wires if the magnetic field halfway between them is to have a magnitude of 300 μT? Answer for both (a) parallel and (b) antiparallel currents.

22E. Two long parallel wires a distance d apart carry currents of i and $3i$ in the same direction. Locate the point or points at which their magnetic fields cancel.

23E. Two long, straight, parallel wires, separated by 0.75 cm, are perpendicular to the plane of the page as shown in Fig. 30-29. Wire 1 carries a current of 6.5 A into the page. What must be the current (magnitude and direction) in wire 2 for the resultant magnetic field at point P to be zero?

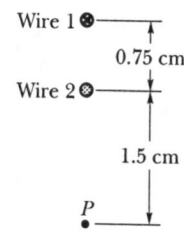

Fig. 30-29 Exercise 23.

24E. Figure 30-30 shows five long parallel wires in the xy plane. Each wire carries a current $i = 3.00$ A in the positive x direction. The separation between adjacent wires is $d = 8.00$ cm. In unit-vector notation, what are the magnitude and direction of the magnetic force per meter exerted on each of these five wires by the other wires?

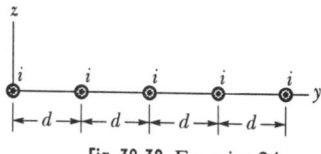

Fig. 30-30 Exercise 24.

25P. Four long copper wires are parallel to each other, their cross sections forming the corners of a square with sides $a = 20$ cm. A 20 A current exists in each wire in the direction shown in Fig. 30-31. What are the magnitude and direction of $\vec{B}$ at the center of the square?

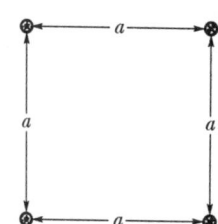

Fig. 30-31 Problems 25, 26, and 27.

26P. Four identical parallel currents i are arranged to form a square of edge length a as in Fig. 30-31, *except* that they are *all* out of the page. What is the force per unit length (magnitude and direction) on any one wire?

27P. In Fig. 30-31, what is the force per unit length acting on the lower left wire, in magnitude and direction, with the current directions as shown? The currents are i.

28P. Figure 30-32 is an idealized schematic drawing of a rail gun. Projectile P sits between two wide rails of circular cross section; a source of current sends current through the rails and through the (conducting) projectile itself (a fuse is not used). (a) Let w be the distance between the rails, R the radius of the rails, and i the current. Show that the magnitude of the force on the projectile is directed to the right along the rails and is given approximately by

$$|\vec{F}| = \frac{i^2 \mu_0}{2\pi} \ln \frac{w + R}{R}.$$

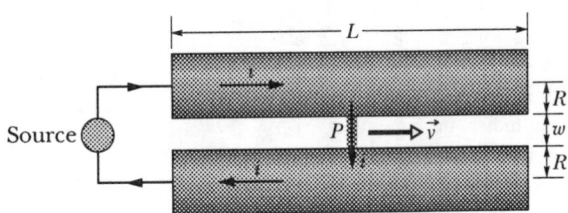

Fig. 30-32 Problem 28.

(b) If the projectile starts from the left end of the rails at rest, find the speed $|\vec{v}|$ at which it is expelled at the right. Assume that $|i| = 450$ kA, $w = 12$ mm, $R = 6.7$ cm, $L = 4.0$ m, and the mass of the projectile is $m = 10$ g.

29P. In Fig. 30-33, the long straight wire carries a current of 30 A and the rectangular loop carries a current of 20 A. Calculate the resultant force acting on the loop. Assume that $a = 1.0$ cm, $b = 8.0$ cm, and $L = 30$ cm.

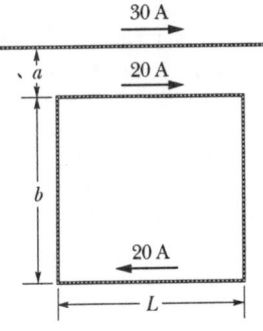

Fig. 30-33 Problem 29.

SEC. 30-3   Ampere's Law

30E. Eight wires cut the page perpendicularly at the points shown in Fig. 30-34. A wire labeled with the integer k ($k = 1$, $2, \ldots$, 8) carries the current ki. For those with odd k, the current is out of the page; for those with even k, it is into the page. Evaluate $\oint \vec{B} \cdot d\vec{s}$ along the closed path in the direction shown.

31E. Each of the eight conductors in Fig. 30-35 carries 2.0 A of current into or out of the page. Two paths are indicated for the line integral $\oint \vec{B} \cdot d\vec{s}$. What is the value of the integral for the path (a) at the left and (b) at the right?

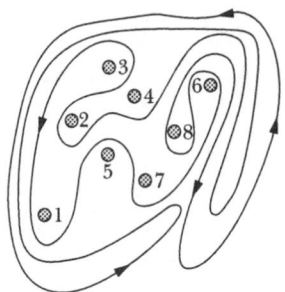

Fig. 30-34 Exercise 30.

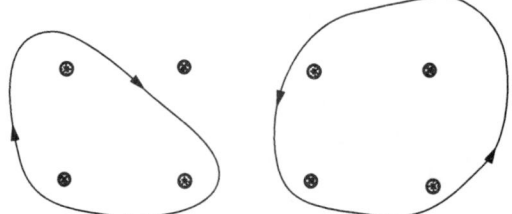

Fig. 30-35 Exercise 31.

32E. Figure 30-36 shows a cross section of a long cylindrical conductor of radius a, carrying a uniformly distributed current i. Assume that $a = 2.0$ cm and $i = 100$ A, and plot the magnitude of the magnetic field $|\vec{B}(r)|$ over the range $0 < r < 6.0$ cm.

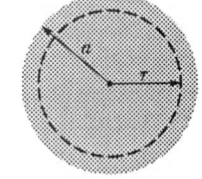

Fig. 30-36 Exercise 32.

33P. Show that a uniform magnetic field $\vec{B}$ cannot drop abruptly to zero (as is suggested by the lack of field lines to the right of point a in Fig. 30-37) as one moves perpendicular to $\vec{B}$, say along the horizontal arrow in the figure. (*Hint:* Apply Ampere's law to the rectangular path shown by the dashed lines.) In actual magnets "fringing" of the magnetic field lines always occurs, which means that $\vec{B}$ approaches zero in a gradual manner. Modify the field lines in the figure to indicate a more realistic situation.

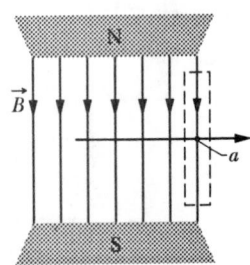

Fig. 30-37 Problem 33.

34P. Two square conducting loops carry currents of 5.0 and 3.0 A as shown in Fig. 30-38. What is the value of $\oint \vec{B} \cdot d\vec{s}$ for each of the two closed paths shown?

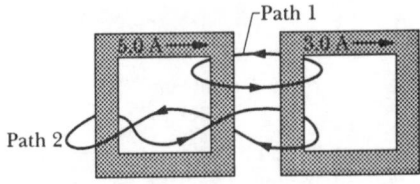

Fig. 30-38 Problem 34.

35P. The current density inside a long, solid, cylindrical wire of radius a is in the direction of the central axis and varies linearly with radial distance r from the axis according to $|\vec{J}| = |\vec{J}_0|r/a$. Find the magnitude and direction of the magnetic field inside the wire.

36P. A long straight wire (radius = 3.0 mm) carries a constant current distributed uniformly over a cross section perpendicular to the axis of the wire. If the magnitude of the current density is 100 A/m^2, what are the magnitudes of the magnetic fields (a) 2.0 mm from the axis of the wire and (b) 4.0 mm from the axis of the wire?

37P. Figure 30-39 shows a cross section of a long cylindrical conductor of radius a containing a long cylindrical hole of radius b. The axes of the cylinder and hole are parallel and are a distance d apart; a current i is uniformly distributed over the tinted area. (a) Use the superposition to show that the magnitude of the magnetic field at the center of the hole is

$$|\vec{B}| = \frac{\mu_0|i|d}{2\pi(a^2 - b^2)}.$$

(b) Discuss the two special cases $b = 0$ and $d = 0$. Use Ampere's law to show that the magnetic field in the hole is uniform. (*Hint:* Regard the cylindrical hole as resulting from the superposition of a complete cylinder (no hole) carrying a current in one direction and a cylinder of radius b carrying a current in the opposite direction, both cylinders having the same current density.)

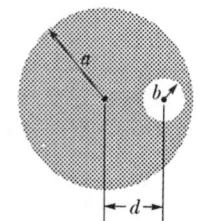

Fig. 30-39 Problem 37.

38P. A long circular pipe with outside radius R carries a (uniformly distributed) current i into the page as shown in Fig. 30-40. A wire runs parallel to the pipe at a distance of $3R$ from center to center. Find the magnitude and direction of the current in the wire such that the net magnetic field at point P has the same magnitude as the net magnetic field at the center of the pipe but is in the opposite direction.

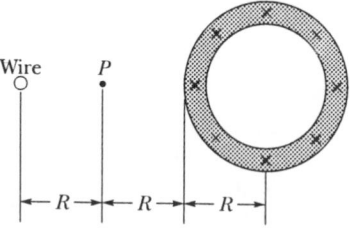

Fig. 30-40 Problem 38.

39P. Figure 30-41 shows a cross section of an infinite conducting sheet carrying a current per unit x-length of λ; the current emerges perpendicularly out of the page. (a) Use the Biot–Savart law and symmetry to show that for all points P above the sheet, and all points P' below it, the magnetic field $\vec{B}$ is parallel to the sheet and

directed as shown. (b) Use Ampere's law to prove that $|\vec{B}| = \frac{1}{2}\mu_0|\lambda|$ at all points P and P'.

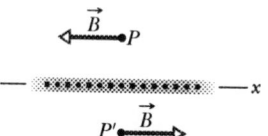

Fig. 30-41 Problems 39 and 44.

SEC. 30-4 Solenoids and Toroids

40E. A solenoid that is 95.0 cm long has a radius of 2.00 cm and a winding of 1200 turns; it carries a current of 3.60 A. Calculate the magnitude of the magnetic field inside the solenoid.

41E. A 200-turn solenoid having a length of 25 cm and a diameter of 10 cm carries a current of 0.30 A. Calculate the magnitude of the magnetic field $\vec{B}$ inside the solenoid.

42E. A solenoid 1.30 m long and 2.60 cm in diameter carries a current of 18.0 A. The magnitude of the magnetic field inside the solenoid is 23.0 mT. Find the length of the wire forming the solenoid.

43E. A toroid having a square cross section, 5.00 cm on a side, and an inner radius of 15.0 cm has 500 turns and carries a current of magnitude 0.800 A. (It is made up of a square solenoid—instead of a round one as in Fig. 30-16—bent into a doughnut shape.) What is the magnitude of the magnetic field inside the toroid at (a) the inner radius and (b) the outer radius of the toroid?

44P. Treat an ideal solenoid as a thin cylindrical conductor whose current per unit length, measured parallel to the cylinder axis, is λ. (a) By doing so, show that the magnitude of the magnetic field inside an ideal solenoid can be written as $|\vec{B}| = \mu_0|\lambda|$. This is the value of the *change* in $\vec{B}$ that you encounter as you move from inside the solenoid to outside, through the solenoid wall. (b) Show that the same change occurs as you move through an infinite flat current sheet such as that of Fig. 30-41 (see Problem 39). Does this equality surprise you?

45P. In Section 30-4, we showed that the magnitude of the magnetic field at any radius r *inside* a toroid is given by

$$|\vec{B}| = \frac{\mu_0|i|N}{2\pi r}.$$

Show that as you move from any point just inside a toroid to a point just outside, the magnitude of the *change* in $\vec{B}$ that you encounter is just $\mu_0\lambda$. Here λ is the current per unit length along a circumference of radius r within the toroid. Compare this with the similar result found in Problem 44. Isn't the equality surprising?

46P. A long solenoid has 100 turns/cm and carries current i. An electron moves within the solenoid in a circle of radius 2.30 cm perpendicular to the solenoid axis. The speed of the electron is $0.0460c$ (c = speed of light). Find the magnitude of the current i in the solenoid.

47P. A long solenoid with 10.0 turns/cm and a radius of 7.00 cm carries a current of 20.0 mA. A current of 6.00 A exists in a straight conductor located along the central axis of the solenoid. (a) At what radial distance from the axis will the direction of the resulting magnetic field be at 45.0° to the axial direction? (b) What is the magnitude of the magnetic field there?

SEC. 30-5   A Current-Carrying Coil as a Magnetic Dipole

48E. Figure 30-42a shows a length of wire carrying a current i and bent into a circular coil of one turn. In Fig. 30-42b the same length of wire has been bent more sharply, to give a coil of two turns, each of half the original radius. (a) If $|\vec{B}_a|$ and $|\vec{B}_b|$ are the magnitudes of the magnetic fields at the centers of the two coils, what is the ratio $|\vec{B}_b|/|\vec{B}_a|$? (b) What is the ratio of the magnitude of the dipole moments, $|\vec{\mu}_b|/|\vec{\mu}_a|$, of the coils?

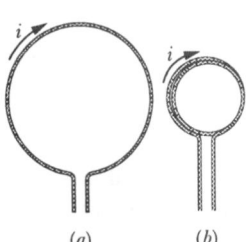

(a)   (b)

Fig. 30-42 Exercise 48.

49E. What is the magnetic dipole moment $\vec{\mu}$ of the solenoid described in Exercise 41?

50E. Figure 30-43 shows an arrangement known as a Helmholtz coil. It consists of two circular coaxial coils, each of N turns and radius R, separated by a distance R. The two coils carry equal currents i in the same direction. Find the magnitude of the net magnetic field at P, midway between the coils.

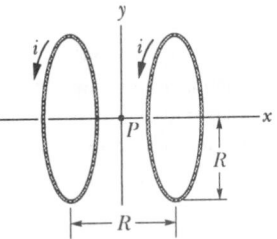

Fig. 30-43 Exercise 50; Problems 53 and 55.

51E. A student makes a short electromagnet by winding 300 turns of wire around a wooden cylinder of diameter $d = 5.0$ cm. The coil is connected to a battery producing a current of 4.0 A in the wire. (a) What is the magnetic moment of this device? (b) At what axial distance $z \gg d$ will the magnetic field of this dipole have the magnitude 5.0 μT (approximately one-tenth that of Earth's magnetic field)?

52E. The magnitude $|\vec{B}(x)|$ of the magnetic field at points on the axis of a square current loop of side a is given in Problem 15. (a) Show that the axial magnetic field of this loop, for $x \gg a$, is that of a magnetic dipole (see Eq. 30-29). (b) What is the magnitude of the magnetic dipole moment of this loop?

53P. Two 300-turn coils of radius R each carry a current i. They are arranged a distance R apart, as in Fig. 30-43. For $R = 5.0$ cm and $i = 50$ A, plot the magnitude $|\vec{B}(x)|$ of the net magnetic field as a function of distance x along the common x axis over the range $x = -5$ cm to $x = +5$ cm, taking $x = 0$ at the midpoint P. (Such coils provide an especially uniform field $\vec{B}$ near point P.) (Hint: See Eq. 30-28).

54P. A conductor carries a current of 6.0 A along the closed path $abcdefgha$ involving 8 of the 12 edges of a cube of side 10 cm as shown in Fig. 30-44. (a) Why can one regard this as the superposi-

tion of three square loops: $bcfgb$, $abgha$, and $cdefc$? (Hint: Draw currents around those square loops.) (b) Use this superposition to find the magnetic dipole moment $\vec{\mu}$ (magnitude and direction) of the closed path. (c) Calculate the magnitude and direction of the magnetic field $\vec{B}$ at the points $(x, y, z) = (0, 5.0$ m, $0)$ and $(5.0$ m, $0, 0)$.

55P. In Exercise 50 (Fig. 30-43), let the separation of the coils be a variable s (not necessarily equal to the coil radius R). (a) Show that the first derivative of the magnitude of the net magnetic field of the coils $(d|\vec{B}|dx)$ vanishes at the midpoint P regardless of the value of s. Why would you expect this to be true from symmetry? (b) Show that the second derivative (d^2B/dx^2) also vanishes at P, provided $s = R$. This accounts for the uniformity of B near P for this particular coil separation.

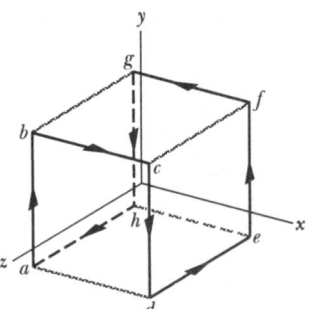

Fig. 30-44 Problem 54.

56P. A length of wire is formed into a closed circuit with radii a and b, as shown in Fig. 30-45, and carries a current i. (a) What are the magnitude and direction of $\vec{B}$ at point P? (b) Find the magnetic dipole moment of the circuit.

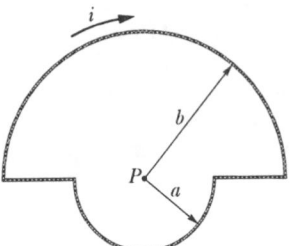

Fig. 30-45 Problem 56.

57P. A circular loop of radius 12 cm carries a current of 15 A. A flat coil of radius 0.82 cm, having 50 turns and a current of 1.3 A, is concentric with the loop. (a) What magnetic field $\vec{B}$ (magnitude and direction) does the loop produce at its center? (b) What torque acts on the coil? Assume that the planes of the loop and coil are perpendicular and that the magnetic field due to the loop is essentially uniform throughout the volume occupied by the coil.

58P. (a) A long wire is bent into the shape shown in Fig. 30-46, without the wire actually touching itself at P. The radius of the circular section is R. Determine the magnitude and direction of $\vec{B}$ at the center C of the circular section when the current i is as indicated. (b) Suppose the circular section of the wire is rotated without distortion about the indicated diameter, until the plane of the circle is perpendicular to the straight sections of the wire. The magnetic dipole moment associated with the circular section is

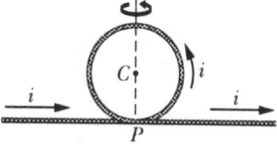

Fig. 30-46 Problem 58.

now in the direction of the current in the straight section of the wire. Determine $\vec{B}$ (magnitude and direction) at C in this case.

Chapter 31

SEC. 31-4 Lenz's Law

1E. A UHF television loop antenna has a diameter of 11 cm. The magnetic field of a TV signal is normal to the plane of the loop and, at one instant of time, its magnitude is changing at the rate 0.16 T/s. The magnetic field is uniform. What emf is induced in the antenna?

2E. A small loop of area A is inside of, and has its axis in the same direction as, a long solenoid of n turns per unit length and current i. If $i = i_0 \sin \omega t$, find the magnitude of the emf induced in the loop.

3E. The magnetic flux through the loop shown in Fig. 31-17 increases according to the relation $\Phi_B = (6.0 \text{ mW/s}^2)t^2 + (3.7 \text{ mW/s})t^2$. (a) What is the magnitude of the emf induced in the loop when $t = 2.0 \text{ } s$? (b) What is the direction of the current through R?

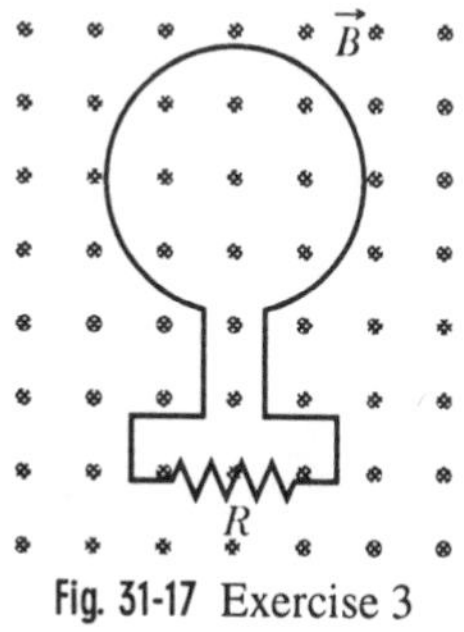

4E. The magnitude of the magnetic field through a single loop of wire, 12 cm in radius and of 8.5 Ω resistance, changes with

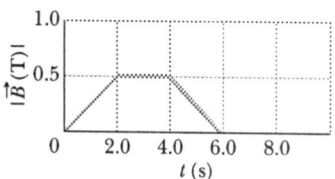

Fig. 31-17 Exercise 3 and Problem 11.

time as shown in Fig. 31-18. Calculate the magnitude of the emf in the loop as a function of time. Consider the time intervals (a) $t = 0$ to $t = 2.0$ s, (b) $t = 2.0$ s to $t = 4.0$ s, (c) $t = 4.0$ s to $t = 6.0$ s. The (uniform) magnetic field is perpendicular to the plane of the loop.

Fig. 31-18 Exercise 4.

5E. A uniform magnetic field is normal to the plane of a circular loop 10 cm in diameter and made of copper wire (of diameter 2.5 mm). (a) Calculate the resistance of the wire. (See Table 26-1.) (b) At what rate must the magnetic field change with time if an induced current of 10 A is to appear in the loop?

6P. The current in the solenoid of Touchstone Example 31-3-1 changes, not as stated there, but according to $i = (3.0 \text{ A/s})t + 1.0 \text{A/s}^2)t^2$. (a) Plot the induced emf in the coil from $t = 0.0$ to $t = 4.0$ s. (b) The resistance of the coil is 0.15 Ω. What is the current in the coil at $t = 2.0$ s?

7P. In Fig. 31-19 a 120-turn coil of radius 1.8 cm and resistance 5.3 Ω is placed *outside* a solenoid like that of Touchstone Example 31-3-1. If the current in the solenoid is changed as in that sample problem, what current appears in the coil while the solenoid current is being changed?

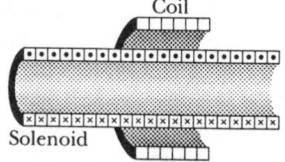

Fig. 31-19 Problem 7.

8P. An elastic conducting material is stretched into a circular loop of 12.0 cm radius. It is placed with its plane perpendicular to a uniform 0.800 T magnetic field. When released, the radius of the loop starts to shrink at an instantaneous rate of 75.0 cm/s. What magnitude of emf is induced in the loop at that instant?

9P. Figure 31-20 shows two parallel loops of wire having a common axis. The smaller loop (radius r) is above the larger loop (radius R) by a distance $x \gg R$. Consequently, the magnetic field due to the current i in the larger loop is nearly constant throughout the smaller loop. Suppose that x is increasing at the constant rate of $dx/dt = v$. (a) Determine the magnetic flux through the area bounded by the smaller loop as a function of x. (*Hint:* See Eq. 30-29.) In the smaller loop, find (b) the induced emf and (c) the direction of the induced current.

10P. In Fig. 31-21, a circular loop of wire 10 cm in diameter (seen edge-on) is placed with its normal $\vec{N}$ at an angle $\theta = 30°$ with the direction of a uniform magnetic field $\vec{B}$ of magnitude 0.50 T. The loop is then rotated such that $\vec{N}$ rotates in a cone about the field direction at the constant rate of 100 rev/min; the angle θ remains unchanged during the process. What is the emf induced in the loop?

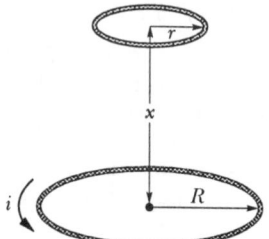

Fig. 31-20 Problem 9.

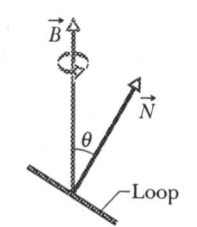

Fig. 31-21 Problem 10.

11P. In Fig. 31-17 let the flux through the loop be $\Phi_B(0)$ at time $t = 0$. Then let the magnetic field $\vec{B}$ vary in a continuous but unspecified way, in both magnitude and direction, so that at time t the flux is represented by $\Phi_B(t)$. (a) Show that the net charge $q(t)$ that has passed through resistor R in time t is

$$q(t) = \frac{1}{R}[\Phi_B(0) - \Phi_B(t)]$$

and is independent of the way $\vec{B}$ has changed. (b) If $\Phi_B(t) = \Phi_B(0)$ in a particular case, we have $q(t) = 0$. Is the induced current necessarily zero throughout the interval from 0 to t?

12P. A small circular loop of area 2.00 cm^2 is placed in the plane of, and concentric with, a large circular loop of radius 1.00 m. The current in the large loop is changed uniformly from 200 A to −200 A (a change in direction) in a time of 1.00 s, beginning at $t = 0$. (a) What is the magnitude of the magnetic field at the center of the small circular loop due to the current in the large loop at $t = 0$, $t = 0.500$ s, and $t = 1.00$ s? (b) What is the magnitude of the emf induced in the small loop at $t = 0.500$ s? (Since the inner loop is small, assume the field $\vec{B}$ due to the outer loop is uniform over the area of the smaller loop.)

13P. One hundred turns of insulated copper wire are wrapped around a wooden cylindrical core of cross-sectional area 1.20×10^{-3} m^2. The two terminals are connected to a resistor. The total resistance in the circuit is 13.0 Ω. If an externally applied uniform longitudinal magnetic field in the core changes from 1.60 T in one direction to 1.60 T in the opposite direction, how much charge flows through the circuit? (*Hint:* See Problem 11.)

14P. At a certain place, Earth's magnetic field has magnitude $|\vec{B}| = 0.590$ gauss and is inclined downward at an angle of 70.0° to the horizontal. A flat horizontal circular coil of wire with a radius of 10.0 cm has 1000 turns and a total resistance of 85.0 Ω. It is connected to a meter with 140 Ω resistance. The coil is flipped through a half-revolution about a diameter, so that it is again horizontal. How much charge flows through the meter during the flip? (*Hint:* See Problem 11.)

15P. A square wire loop with 2.00 m sides is perpendicular to a uniform magnetic field, with half the area of the loop in the field as shown in Fig. 31-22. The loop contains a 20.0 V battery with negligible internal resistance. If the magnitude of the field varies with time according to $\vec{B} = (0.0420$ T$) − (0.870$ T/s$)t$, what are (a) the magnitude of the emf in the circuit and (b) the direction of the current through the battery?

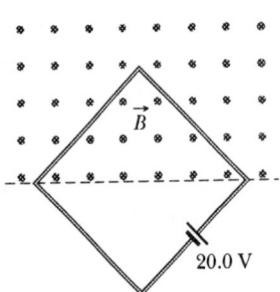

Fig. 31-22 Problem 15.

16P. A wire is bent into three circular segments, each of radius $r = 10$ cm, as shown in Fig. 31-23. Each segment is a quadrant of a circle, ab lying in the xy plane, bc lying in the yz plane, and ca lying in the zx plane. (a) If a uniform magnetic field $\vec{B}$ points in the positive x direction, what is the magnitude of the emf developed in the wire when $\vec{B}$ increases at the rate of 3.0 mT/s in the x direction? (b) What is the direction of the current in segment bc?

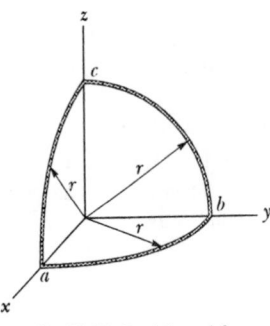

Fig. 31-23 Problem 16.

17P. A rectangular coil of N turns and of length a and width b is rotated at frequency f in a uniform magnetic field $\vec{B}$, as indicated in Fig. 31-24. The coil is connected to co-rotating cylinders, against which metal brushes slide to make contact. If we arbitrarily define ε as being positive during the first quarter-turn, (a) Show that the emf induced in the coil is given (as a function of time t) by

$$\varepsilon = 2\pi f Nab|\vec{B}|\sin(2\pi f t) = \varepsilon_0 \sin(2\pi f t).$$

This is the principle of the commercial alternating-current generator. (b) Design a loop that will produce an emf with $\varepsilon_0 = 150$ V when rotated at 60.0 rev/s in a uniform magnetic field of 0.500 T.

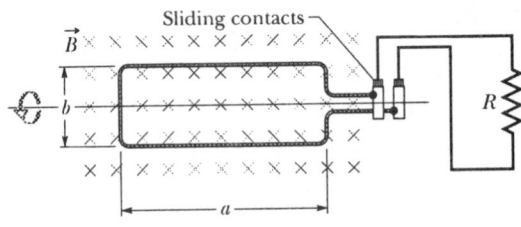

Fig. 31-24 Problem 17.

18P. A stiff wire bent into a semicircle of radius a is rotated with frequency f in a uniform magnetic field, as suggested in Fig. 31-25. What are (a) the frequency and (b) the amplitude of the varying emf induced in the loop?

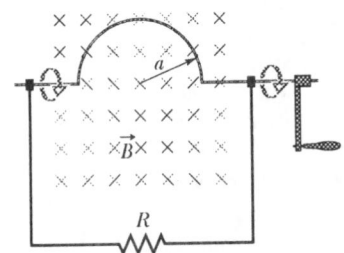

Fig. 31-25 Problem 18.

19P. An electric generator consists of 100 turns of wire formed into a rectangular loop 50.0 cm by 30.0 cm, placed entirely in a uniform magnetic field with magnitude $|\vec{B}| = 3.50$ T. What is the maximum value of the emf produced when the loop is spun at 1000 rev/min about an axis perpendicular to $\vec{B}$?

20P. In Fig. 31-26, a wire forms a closed circular loop, with radius $R = 2.0$ m and resistance 4.0 Ω. The circle is centered on a long straight wire; at time $t = 0$, the current in the long straight wire is 5.0 A rightward. Thereafter, the current changes according to $i = 5.0$ A $− (2.0$ A/s$^2)t^2$. (The straight wire is insulated, so there is no electrical contact between it and the wire of the loop.) What are the magnitude and direction of the current induced in the loop at times $t > 0$?

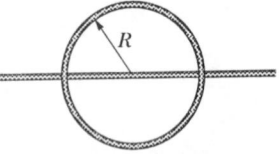

Fig. 31-26 Problem 20.

21P. In Fig. 31-27, the square loop of wire has sides of length 2.0 cm. A magnetic field is directed out of the page; its magnitude is given by $|\vec{B}| = 4.0t^2y$, where $|\vec{B}|$ is in teslas, t is in seconds, and y is in meters. Determine the emf around the square at $t =$ 2.5 s and indicate whether its direction is clockwise or counterclockwise.

22P. For the situation shown in Fig. 31-28, $a = 12.0$ cm and $b = 16.0$ cm. The current in the long straight wire is given by $i = 4.50t^2 - 10.0t$, where i is in amperes and t is in seconds. (a) Find the magnitude of the emf in the square loop at $t = 3.00$ s. (b) Indicate whether the direction of the induced current in the loop is clockwise or counterclockwise at $t = 3.00$ s.

23P*. Two long, parallel copper wires of diameter 2.5 mm carry currents of 10 A in opposite directions. (a) Assuming that their central axes are 20 mm apart, calculate the magnetic flux per meter of wire that exists in the space between those axes. (b) What fraction of this flux lies inside the wires? (c) Repeat part (a) for parallel currents.

24P. A rectangular loop of wire with length a, width b, and resistance R is placed near an infinitely long wire carrying current i, as shown in Fig. 31-29. The distance from the long wire to the center of the loop is r. Find (a) the magnitude of the magnetic flux through the loop and (b) the magnitude of the induced current in the loop as it moves away from the long wire with velocity $\vec{v}$. (c) Indicate whether the induced current is clockwise or counterclockwise.

SEC. 31-5   Induction and Energy Transfers

25E. If 50.0 cm of copper wire (diameter = 1.00 mm) is formed into a circular loop and placed perpendicular to a uniform magnetic field that is increasing at the constant rate of 10.0 mT/s, at what rate is thermal energy generated in the loop?

26E. A loop antenna of area A and resistance R is perpendicular to a uniform magnetic field $\vec{B}$. The field drops linearly to zero in a time interval Δt. Find an expression for the total thermal energy dissipated in the loop.

27E. A metal rod is forced to move with constant velocity $\vec{v}$ along two parallel metal rails, connected with a strip of metal at one end, as shown in Fig. 31-30. A magnetic field of magnitude $|\vec{B}| = 0.350$ T points out of the page. (a) If the rails are separated by 25.0 cm and the speed of the rod is 55.0 cm/s, what emf is generated? (b) If the rod has a

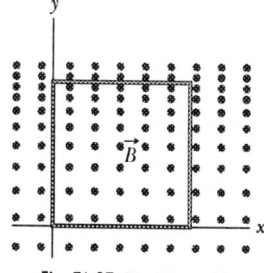
Fig. 31-27 Problem 21.

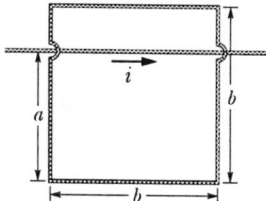
Fig. 31-28 Problem 22.

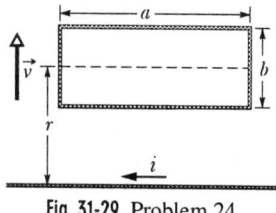
Fig. 31-29 Problem 24.

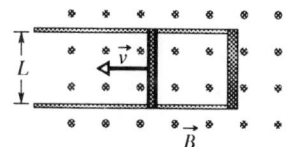
Fig. 31-30 Exercise 27 and Problem 29.

resistance of 18.0 Ω and the rails and connector have negligible resistance, what is the current in the rod? (c) At what rate is energy being transferred to thermal energy?

28P. In Fig. 31-31, a long rectangular conducting loop, of width L, resistance R, and mass m, is hung in a horizontal, uniform magnetic field $\vec{B}$ that is directed into the page and that exists only above line aa. The loop is then dropped; during its fall, it accelerates until it reaches a certain terminal speed $|\vec{v}_t|$. Ignoring air drag, find that terminal speed.

29P. The conducting rod shown in Fig. 31-30 has length L and is being pulled along horizontal, frictionless conducting rails at a constant velocity $\vec{v}$. The rails are connected at one end with a metal strip. A uniform magnetic field $\vec{B}$, directed out of the page, fills the region in which the rod moves. Assume that $L = 10$ cm, $|\vec{v}| = 5.0$ m/s, and $|\vec{B}| = 1.2$ T. (a) What are the magnitude and direction (clockwise or counterclockwise) of the emf induced in the rod? (b) What is the magnitude of the current in the conducting loop? Assume that the resistance of the rod is 0.40 Ω and that the resistance of the rails and metal strip is negligibly small. (c) At what rate is thermal energy being generated in the rod? (d) What magnitude of force must be applied to the rod by an external agent to maintain its motion? (e) At what rate does this external agent do work on the rod? Compare this answer with the answer to (c).

30P. Two straight conducting rails form a right angle where their ends are joined. A conducting bar in contact with the rails starts at the vertex at time $t = 0$ and moves with a constant velocity of magnitude 5.20 m/s along them, as shown in Fig. 31-32. A magnetic field of magnitude $|\vec{B}| = 0.350$ T is directed out of the page. Calculate (a) the flux through the triangle formed by the rails and bar at $t = 3.00$ s and (b) the magnitude of emf around the triangle at that time. (c) If we write the emf as $\varepsilon = at^n$, where a and n are constants, what is the value of n?

31P. Figure 31-33 shows a rod of length L caused to move at constant speed v along horizontal conducting rails. The magnetic field in which the rod moves is *not uniform* but is provided by a current i in a long wire parallel to the rails. Assume that $v = 5.00$ m/s, $a = 10.0$ mm, $L = 10.0$ cm, and $i = 100$ A. (a) Calculate the magnitude of the emf induced in the rod. (b) What is the magnitude of the current in

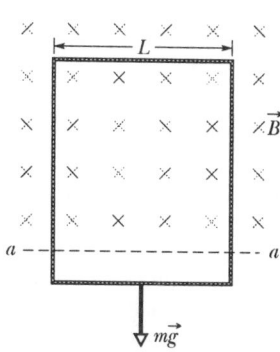
Fig. 31-31 Problem 28.

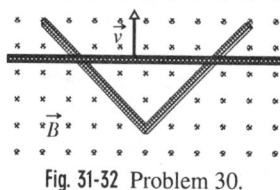
Fig. 31-32 Problem 30.

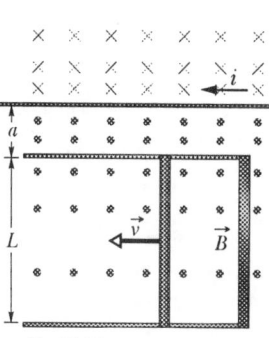
Fig. 31-33 Problem 31.

the conducting loop? Assume that the resistance of the rod is 0.400 Ω and that the resistance of the rails and the strip that connects them at the right is negligible. (c) At what rate is thermal energy being generated in the rod? (d) What magnitude of force must be applied to the rod by an external agent to maintain its motion? (e) At what rate does this external agent do work on the rod? Compare this answer to that for (c).

SEC. 31-6   Induced Electric Fields

32E. Figure 31-34 shows two circular regions R_1 and R_2 with radii $r_1 = 20.0$ cm and $r_2 = 30.0$ cm. In R_1 there is a uniform magnetic field of magnitude $|\vec{B}_1| = 50.0$ mT into the page, and in R_2 there is a uniform magnetic field of magnitude $|\vec{B}_2| = 75.0$ mT out of the page (ignore any fringing of these fields). Both fields are decreasing at the rate of 8.50 mT/s.

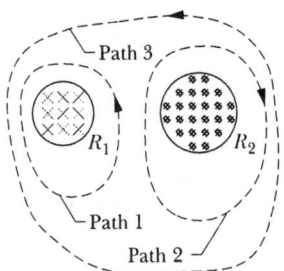

Fig. 31-34 Exercise 32.

Calculate the integral $\oint \vec{E} \cdot d\vec{s}$ for each of the three dashed paths.

33E. A long solenoid has a diameter of 12.0 cm. When a current i exists in its windings, a uniform magnetic field of magnitude $|\vec{B}| = 30.0$ mT is produced in its interior. By decreasing i, the field is caused to decrease at the rate of 6.50 mT/s. Calculate the magnitude of the induced electric field (a) 2.20 cm and (b) 8.20 cm from the axis of the solenoid.

34P. Early in 1981 the Francis Bitter National Magnet Laboratory at M.I.T. commenced operation of a 3.3-cm-diameter cylindrical magnet, which produces a 30 T field, then the world's largest steady-state field. The field magnitude can be varied sinusoidally between the limits of 29.6 and 30.0 T at a frequency of 15 Hz. When this is done, what is the maximum value of the magnitude of the induced electric field at a radial distance of 1.6 cm from the axis? (*Hint:* See Touchstone Example 31-6-1.)

35P. Prove that the electric field $\vec{E}$ in a charged parallel-plate capacitor cannot drop abruptly to zero (as is suggested at point *a* in Fig. 31-35), as one moves perpendicular to the field, say, along the horizontal arrow in the figure. Fringing of the field lines always occurs in actual capacitors, which means that $\vec{E}$ approaches zero in a continuous and gradual way (see Problem 33 in Chapter 30). (*Hint:* Apply Faraday's law to the rectangular path shown by the dashed lines.)

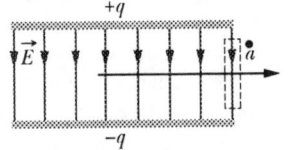

Fig. 31-35 Problem 35.

SEC. 31-7   Inductors and Inductance

36E. A circular coil has a 10.0 cm radius and consists of 30.0 closely wound turns of wire. An externally produced magnetic field of magnitude 2.60 mT is perpendicular to the coil. (a) If no current is in the coil, what is the magnitude of the magnetic flux that links its turns? (b) When the current in the coil is 3.80 A in a certain direction, the net flux through the coil is found to vanish. What is the inductance of the coil?

37E. The inductance of a close-packed coil of 400 turns is 8.0 mH. Calculate the magnetic flux through the coil when the current is 5.0 mA.

38P. A wide copper strip of width W is bent to form a tube of radius R with two parallel planar extensions, as shown in Fig. 31-36. There is a current i through the strip, distributed uniformly over its width. In this way a "one-turn solenoid" is formed. (a) Derive an expression for the magnitude of the magnetic field $\vec{B}$ in the tubular part (far away from the edges). (*Hint:* Assume that the magnetic field outside this one-turn solenoid is negligibly small.) (b) Find the inductance of this one-turn solenoid, neglecting the two planar extensions.

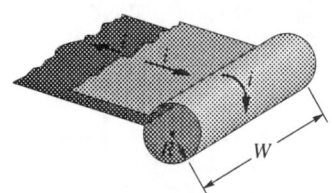

Fig. 31-36 Problem 38.

39P. Two long parallel wires, both of radius a and whose centers are a distance d apart, carry equal currents in opposite directions. Show that, neglecting the flux within the wires, the inductance of a length l of such a pair of wires is given by

$$L = \frac{\mu_0 l}{\pi} \ln \frac{d - a}{a}.$$

(*Hint:* Calculate the flux through a rectangle of which the wires form two opposite sides.)

SEC. 31-8   Self-Induction

40E. At a given instant the current and self-induced emf in an inductor are directed as indicated in Fig. 31-37. (a) Is the current increasing or decreasing? (b) The induced emf is 17 V and the

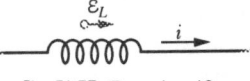

Fig. 31-37 Exercise 40.

rate of change of the current is 25 kA/s; find the inductance.

41E. A 12 H inductor carries a steady current of 2.0 A. How can a 60 V self-induced emf be made to appear in the inductor?

42P. The current i through a 4.6 H inductor varies with time t as shown by the graph of Fig. 31-38. The inductor has a resistance of 12 Ω. Find the magnitude of the induced emf ε during the time intervals (a) $t = 0$ to $t = 2$ ms, (b) $t = 2$ ms to $t = 5$ ms, (c) $t = 5$ ms to $t = 6$ ms. (Ignore the behavior at the ends of the intervals.)

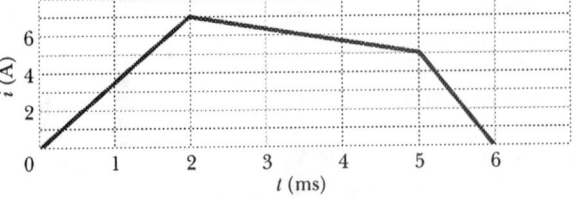

Fig. 31-38 Problem 42.

43P. *Inductors in series.* Two inductors L_1 and L_2 are connected in series and are separated by a large distance. (a) Show that the equivalent inductance is given by

$$L_{eq} = L_1 + L_2.$$

(*Hint:* Review the derivations for resistors in series and capacitors in series. Which is similar here?) (b) Why must their separation be large for this relationship to hold? (c) What is the generalization of (a) for N inductors in series?

44P. *Inductors in parallel.* Two inductors L_1 and L_2 are connected in parallel and separated by a large distance. (a) Show that the equivalent inductance is given by

$$\frac{1}{L_{eq}} = \frac{1}{L_1} + \frac{1}{L_2}.$$

(*Hint:* Review the derivations for resistors in parallel and capacitors in parallel. Which is similar here?) (b) Why must their separation be large for this relationship to hold? (c) What is the generalization of (a) for N inductors in parallel?

SEC. 31-9   *RL* Circuits

45E. In terms of τ_L, how long must we wait for the current in an RL circuit to build up to within 0.100% of its equilibrium value?

46E. The current in an RL circuit builds up to one-third of its steady-state value in 5.00 s. Find the inductive time constant.

47E. The current in an RL circuit drops from 1.0 A to 10 mA in the first second following removal of the battery from the circuit. If L is 10 H, find the resistance R in the circuit.

48E. Consider the RL circuit of Fig. 31-13. In terms of the battery emf ε, (a) what is the self-induced emf ε_L when the switch has just been closed on a, and (b) what is ε_L when $t = 2.0\tau_L$? (c) In terms of τ_L, when will ε_L be just one-half the battery emf ε?

49E. A solenoid having an inductance of 6.30 μH is connected in series with a 1.20 kΩ resistor. (a) If a 14.0 V battery is switched across the pair, how long will it take for the current through the resistor to reach 80.0% of its final value? (b) What is the current through the resistor at time $t = 1.0\tau_L$?

50P. Suppose the emf of the battery in the circuit of Fig. 31-14 varies with time t so that the current is given by $i(t) = 3.0 + 5.0t$, where i is in amperes and t is in seconds. Take $R = 4.0\ \Omega$ and $L = 6.0$ H, and find an expression for the battery emf as a function of time. (*Hint:* Apply the loop rule.)

51P. At time $t = 0$, a 45.0 V potential difference is suddenly applied to a coil with $L = 50.0$ mH and $R = 180\ \Omega$. At what rate is the current increasing at $t = 1.20$ ms?

52P. A wooden toroidal core with a square cross section has an inner radius of 10 cm and an outer radius of 12 cm. It is wound with one layer of wire (of diameter 1.0 mm and resistance per meter 0.020 Ω/m). What are (a) the inductance and (b) the inductive time constant of the resulting toroid? Ignore the thickness of the insulation on the wire.

53P. In Fig. 31-49, $\varepsilon = 100$ V, $R_1 = 10.0\ \Omega$, $R_2 = 20.0\ \Omega$, $R_3 = 30.0\ \Omega$, and $L = 2.00$ H. Find the values of i_1 and i_2 (a) immediately after the closing of switch S, (b) a long time later,

(c) immediately after the reopening of switch S, and (d) a long time after the reopening.

54P. In the circuit of Fig. 31-40, $\varepsilon = 10$ V, $R_1 = 5.0\ \Omega$, $R_2 = 10\ \Omega$, and $L = 5.0$ H. For the two separate conditions (I) switch S just closed and (II) switch S closed for a long time, calculate (a) the current i_1 through R_1, (b) the current i_2 through R_2, (c) the current i through the switch, (d) the potential difference across R_2, (e) the potential difference across L, and (f) the rate of change di_2/dt.

55P*. In the circuit shown in Fig. 31-41, switch S is closed at time $t = 0$. Thereafter, the constant current source, by varying its emf, maintains a constant current i out of its upper terminal. (a) Derive an expression for the current through the inductor as a function of time. (b) Show that the current through the resistor equals the current through the inductor at time $t = (L/R) \ln 2$.

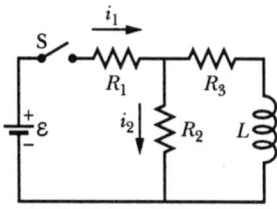

Fig. 31-39 Problem 53.

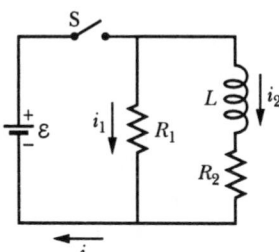

Fig. 31-40 Problem 54.

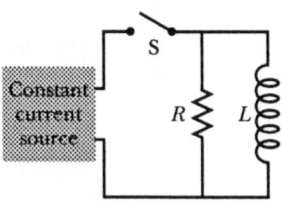

Fig. 31-41 Problem 55.

SEC. 31-10   Energy Stored in a Magnetic Field

56E. Consider the circuit of Fig. 31-18. In terms of the inductive time constant, at what instant after the battery is connected will the energy stored in the magnetic field of the inductor be half its steady-state value?

57E. Suppose that the inductive time constant for the circuit of Fig. 31-14 is 37.0 ms and the current in the circuit is zero at time $t = 0$. At what time does the rate at which energy is dissipated in the resistor equal the rate at which energy is being stored in the inductor?

58E. A coil with an inductance of 2.0 H and a resistance of 10 Ω is suddenly connected to a resistanceless battery with $\varepsilon = 100$ V. At 0.10 s after the connection is made, what are the rates at which (a) energy is being stored in the magnetic field, (b) thermal energy is appearing in the resistance, and (c) energy is being delivered by the battery?

59P. A coil is connected in series with a 10.0 kΩ resistor. A 50.0 V battery is applied across the two devices, and the current reaches a value of 2.00 mA after 5.00 ms. (a) Find the inductance of the coil. (b) How much energy is stored in the coil at this same moment?

60P. For the circuit of Fig. 31-14, assume that $\varepsilon = 10.0$ V, $R = 6.70\ \Omega$, and $L = 5.50$ H. The battery is connected at time $t = 0$. (a) How much energy is delivered by the battery during the first 2.00 s? (b) How much of this energy is stored in the magnetic field of the inductor? (c) How much of this energy is dissipated in the resistor?

61P. Prove that, after switch S in Fig. 31-13 has been thrown from *a* to *b*, all the energy stored in the inductor will ultimately appear as thermal energy in the resistor.

SEC. 31-11   Energy Density of a Magnetic Field

62E. A toroidal inductor with an inductance of 90.0 mH encloses a volume of 0.0200 m^3. If the average energy density in the toroid is 70.0 J/m^3, what is the current through the inductor?

63E. A solenoid that is 85.0 cm long has a cross-sectional area of 17.0 cm^2. There are 950 turns of wire carrying a current of 6.60 A. (a) Calculate the energy density of the magnetic field inside the solenoid. (b) Find the total energy stored in the magnetic field there (neglect end effects).

64E. The magnetic field in the interstellar space of our galaxy has a magnitude of about 10^{-10} T. How much energy is stored in this field in a cube 10 light-years on edge? (For scale, note that the nearest star is 4.3 light-years distant and the radius of our galaxy is about 8×10^4 light-years.)

65E. What must be the magnitude of a uniform electric field if it is to have the same energy density as that possessed by a 0.50 T magnetic field?

66E. A circular loop of wire 50 mm in radius carries a current of 100 A. (a) Find the magnetic field strength at the center of the loop. (b) Calculate the energy density at the center of the loop.

67P. A length of copper wire carries a current of 10 A, uniformly distributed through its cross section. Calculate the energy density of (a) the magnetic field and (b) the electric field at the surface of the wire. The wire diameter is 2.5 mm, and its resistance per unit length is 3.3 Ω/km.

SEC. 31-12   Mutual Induction

68E. Coil 1 has $L_1 = 25$ mH and $N_1 = 100$ turns. Coil 2 has $L_2 = 40$ mH and $N_2 = 200$ turns. The coils are rigidly positioned with respect to each other; their mutual inductance M is 3.0 mH. A 6.0 mA current in coil 1 is changing at the rate of 4.0 A/s. (a) What magnetic flux Φ_{12} links coil 1, and what self-induced emf appears there? (b) What magnetic flux Φ_{21} links coil 2, and what mutually induced emf appears there?

69E. Two coils are at fixed locations. When coil 1 has no current and the current in coil 2 increases at the rate 15.0 A/s, the emf in coil 1 is 25.0 mV. (a) What is their mutual inductance? (b) When coil 2 has no current and coil 1 has a current of 3.60 A, what is the flux linkage in coil 2?

70E. Two solenoids are part of the spark coil of an automobile. When the current in one solenoid falls from 6.0 A to zero in 2.5 ms, an emf of 30 kV is induced in the other solenoid. What is the mutual inductance M of the solenoids?

71P. Two coils, connected as shown in Fig. 31-42, separately have inductances L_1 and L_2. Their mutual inductance is M. (a) Show that this combination can be replaced by a single coil of equivalent inductance given by

$$L_{eq} = L_1 + L_2 + 2M.$$

(b) How could the coils in Fig. 31-52 be reconnected to yield an equivalent inductance of

$$L_{eq} = L_1 + L_2 - 2M?$$

(This problem is an extension of Problem 43, but the requirement that the coils be far apart has been removed.)

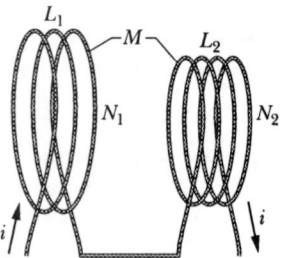

Fig. 31-42 Problem 71.

72P. A coil C of N turns is placed around a long solenoid S of radius R and n turns per unit length, as in Fig. 31-43. Show that the mutual inductance for the coil-solenoid combination is given by $M = \mu_0 \pi R^2 nN$. Explain why M does not depend on the shape, size, or possible lack of close-packing of the coil.

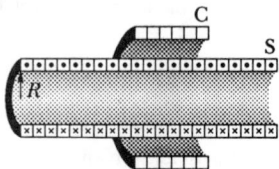

Fig. 31-43 Problem 72.

73P. Figure 31-44 shows, in cross section, two coaxial solenoids. Show that the mutual inductance M for a length l of this solenoid-solenoid combination is given by $M = \pi R_1^2 l \mu_0 n_1 n_2$, in which n_1 and n_2 are the respective numbers of turns per unit length and R_1 is the radius of the inner solenoid. Why does M depend on R_1 and not on R_2?

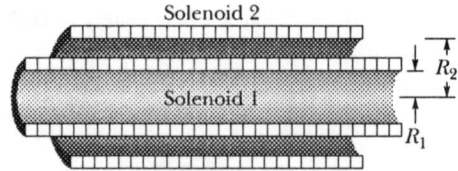

Fig. 31-44 Problem 73.

74P. Figure 31-45 shows a coil of N_2 turns wound as shown around part of a toroid of N_1 turns. The toroid's inner radius is a, its outer radius is b, and its height is h. Show that the mutual inductance M for the toroid-coil combination is

$$M = \frac{\mu_0 N_1 N_2 h}{2\pi} \ln \frac{b}{a}.$$

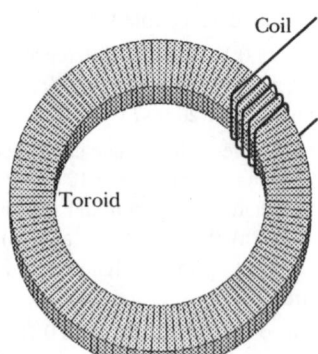

Fig. 31-45 Problem 74.

75P. A rectangular loop of N close-packed turns is positioned near a long straight wire as shown in Fig. 31-46. (a) What is the mutual inductance M for the loop–wire combination? (b) Evaluate M for $N = 100$, $a = 1.0$ cm, $b = 8.0$ cm, and $l = 30$ cm.

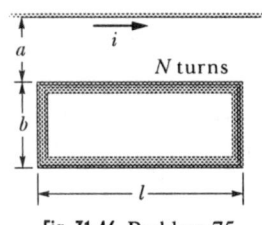

Fig. 31-46 Problem 75.

Chapter 32

SEC. 32-2   Gauss' Law for Magnetic Fields

1E. Imagine rolling a sheet of paper into a cylinder and placing a bar magnet near its end as shown in Fig. 32-17. (a) Sketch the magnetic field lines that pass through the surface of the cylinder. (b) What can you say about the sign of $\vec{B} \cdot d\vec{A}$ for every area $d\vec{A}$ on the surface? (c) Does this contradict Gauss' law for magnetism? Explain.

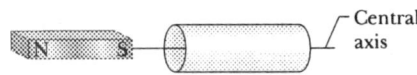

Fig. 32-17 Exercise 1.

2E. The magnetic flux through each of five faces of a die (singular of "dice") is given by $\Phi_B = \pm N$ Wb, where N (= 1 to 5) is the number of spots on the face. The flux is positive (outward) for N even and negative (inward) for N odd. What is the flux through the sixth face of the die? Is it directed in or out?

3P. A Gaussian surface in the shape of a right circular cylinder with end caps has a radius of 12.0 cm and a length of 80.0 cm. Through one end there is an inward magnetic flux of 25.0 μWb. At the other end there is a uniform magnetic field of 1.60 mT, normal to the surface and directed outward. What is the net magnetic flux through the curved surface?

SEC. 32-3   The Magnetism of Earth

4E. Assume the average value of the vertical component of Earth's magnetic field is 43 μT (downward) for all of Arizona, which has an area of 2.95×10^5 km^2, and calculate the net magnetic flux through the rest of Earth's surface (the entire surface excluding Arizona). Is that net magnetic flux outward or inward?

5E. In New Hampshire the average horizontal component of Earth's magnetic field in 1912 was 16 μT and the average inclination or "dip" was 73°. What was the corresponding magnitude of Earth's magnetic field?

6P. The magnetic field of Earth can be approximated as the magnetic field of a dipole, with horizontal and vertical components, at a point a distance r from Earth's center, given by

$$B_h = \frac{\mu_0 \mu}{4\pi r^3} \cos \lambda_m, \qquad B_v = \frac{\mu_0 \mu}{2\pi r^3} \sin \lambda_m,$$

where λ_m is the *magnetic latitude* (this type of latitude is measured from the geomagnetic equator toward the north or south geomagnetic pole). Assume that Earth's magnetic dipole moment is

$\mu = 8.00 \times 10^{22}$ A $\cdot$ m^2. (a) Show that the magnitude of Earth's field at latitude λ_m is given by

$$B = \frac{\mu_0 \mu}{4\pi r^3} \sqrt{1 + 3 \sin^2 \lambda_m}.$$

(b) Show that the inclination ϕ_i of the magnetic field is related to the magnetic latitude λ_m by

$$\tan \phi_i = 2 \tan \lambda_m .$$

7P. Use the results displayed in Problem 6 to predict Earth's magnetic field (both magnitude and inclination) at (a) the geomagnetic equator, (b) a point at geomagnetic latitude 60°, and (c) the north geomagnetic pole.

8P. Using the approximations given in Problem 6, find (a) the altitude above Earth's surface where the magnitude of its magnetic field is 50% of the surface value at the same latitude; (b) the maximum magnitude of the magnetic field at the core–mantle boundary, 2900 km below Earth's surface; and (c) the magnitude and inclination of Earth's magnetic field at the north geographic pole. Suggest why the values you calculated for (c) differ from measured values.

SEC. 32-4   Magnetism and Electrons

9E. What is the measured component of the orbital magnetic dipole moment of an electron with (a) $m_l = 1$ and (b) $m_l = -2$?

10E. What is the energy difference between parallel and antiparallel alignment of the z component of an electron's spin magnetic dipole moment with an external magnetic field of magnitude 0.25 T, directed parallel to the z axis?

11E. If an electron in an atom has an orbital angular momentum with $m_l = 0$, what are the components (a) $L_{orb,z}$ and (b) $\mu_{orb,z}$? If the atom is in an external magnetic field $\vec{B}$ of magnitude 35 mT and directed along the z axis, what are the potential energies associated with the orientations of (c) the electron's orbital magnetic dipole moment and (d) the electron's spin magnetic dipole moment? (e) Repeat (a) through (d) for $m_l = -3$.

SEC. 32-6   Diamagnetism

12E. Figure 32-18 shows a loop model (loop L) for a diagmagnetic material. (a) Sketch the magnetic field lines through and about the material due to the bar magnet. (b) What are the directions of the loop's net magnetic dipole moment $\vec{\mu}$ and the conventional current i in the loop? (c) What is the direction of the magnetic force on the loop?

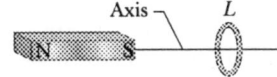

Fig. 32-18 Exercises 12 and 16.

13P*. Assume that an electron of mass m and charge magnitude e moves in a circular orbit of radius r about a nucleus. A uniform magnetic field $\vec{B}$ is then established perpendicular to the plane of the orbit. Assuming also that the radius of the orbit does not change and that the change in the speed of the electron due to field $\vec{B}$ is small, find an expression for the change in the orbital magnetic dipole moment of the electron due to the field.

SEC. 32-7   Paramagnetism

14E. A magnetic field of magnitude 0.50 T is applied to a paramagnetic gas whose atoms have an intrinsic magnetic dipole moment of magnitude 1.0×10^{-23} J/T. At what temperature will the mean kinetic energy of translation of the gas atoms be equal to the energy required to reverse such a dipole end for end in this magnetic field?

15E. A magnet in the form of a cylindrical rod has a length of 5.00 cm and a diameter of 1.00 cm. It has a uniform magnetization of 5.30×10^3 A/m. What is the magnitude of its magnetic dipole moment?

16E. Repeat Exercise 12 for the case in which loop L is the model for a paramagnetic material.

17E. A sample of the paramagnetic salt to which the magnetization curve of Fig. 32-9 applies is to be tested to see whether it obeys Curie's law. The sample is placed in a uniform 0.50 T magnetic field that remains constant throughout the experiment. The magnetization M is then measured at temperatures ranging from 10 to 300 K. Will it be found that Curie's law is valid under these conditions?

18E. A sample of the paramagnetic salt to which the magnetization curve of Fig. 32-9 applies is held at room temperature (300 K). At what applied magnetic field will the degree of magnetic saturation of the sample be (a) 50% and (b) 90%? (c) Are these fields attainable in the laboratory?

19P. An electron with kinetic energy K_e travels in a circular path that is perpendicular to a uniform magnetic field, the electron's motion subject only to the force due to the field. (a) Show that the magnetic dipole moment of the electron due to its orbital motion has magnitude $\mu = K_e/|\vec{B}|$ and that it is in the direction opposite that of $\vec{B}$. (b) What are the magnitude and direction of the magnetic dipole moment of a positive ion with kinetic energy K_i under the same circumstances? (c) An ionized gas consists of 5.3×10^{21} electrons/m³ and the same number density of ions. Take the average electron kinetic energy to be 6.2×10^{-20} J and the average ion kinetic energy to be 7.6×10^{-21} J. Calculate the magnetization of the gas when it is in a magnetic field of 1.2 T.

SEC. 32-8   Ferromagnetism

20E. Measurements in mines and boreholes indicate that Earth's interior temperature increases with depth at the average rate of 30 C°/km. Assuming a surface temperature of 10°C, at what depth does iron cease to be ferromagnetic? (The Curie temperature of iron varies very little with pressure.)

21E. The dipole moment associated with an atom of iron in an iron bar has magnitude 2.1×10^{-23} J/T. Assume that all the atoms in the bar, which is 5.0 cm long and has a cross-sectional area of 1.0 cm², have their dipole moments aligned. (a) What is the magnitude of the dipole moment of the bar? (b) What is the magnitude of the torque that must be exerted to hold this magnet perpendicular to an external field of 1.5 T? (The density of iron is 7.9 g/cm³.)

22E. The saturation magnetization M_{max} of the ferromagnetic metal nickel is 4.70×10^5 A/m. Calculate the magnetic moment of a single nickel atom. (The density of nickel is 8.90 g/cm³ and its molar mass is 58.71 g/mol.)

23P. Figure 32-19 shows the apparatus used in a lecture demonstration of para- and diamagnetism. A sample of the magnetic material is suspended by a long string in the nonuniform field ($d = 2$ cm) between the poles of a powerful electromagnet. Pole P_1 is sharply pointed and pole P_2 is rounded as indicated. Any deflection of the string from the vertical is visible to the audience by means of an optical projection system (not shown). (a) First a bismuth (highly diamagnetic) sample is used. When the electromagnet is turned on, the sample is observed to deflect slightly (about 1 mm) toward one of the poles. What is the direction of this deflection? (b) Next an aluminum (paramagnetic, conducting) sample is used. When the electromagnet is turned on, the sample is observed to deflect strongly (about 1 cm) toward one pole for about a second and then deflect moderately (a few millimeters) toward the other pole. Explain and indicate the direction of these deflections. (*Hint:* The aluminum sample is a conductor, for which Lenz's law applies.) (c) What would happen if a ferromagnetic sample were used?

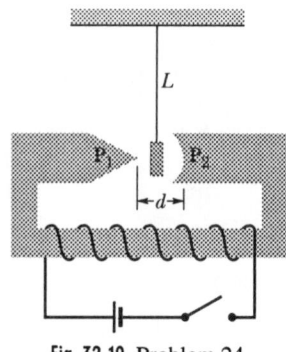

Fig. 32-19 Problem 24.

24P. The magnetic dipole moment of Earth has magnitude 8.0×10^{22} J/T. (a) If the origin of this magnetism were a magnetized iron sphere at the center of Earth, what would be its radius? (b) What fraction of the volume of Earth would such a sphere occupy? Assume complete alignment of the dipoles. The density of Earth's inner core is 14 g/cm³. The magnetic dipole moment of an iron atom is 2.1×10^{-23} J/T. (*Note:* Earth's inner core is in fact thought to be in both liquid and solid forms and partly iron, but a permanent magnet as the source of Earth's magnetism has been ruled out by several considerations. For one, the temperature is certainly above the Curie point.)

SEC. 32-9   Induced Magnetic Fields

25E. Touchstone Example 32-9-1 describes the charging of a parallel-plate capacitor with circular plates of radius 55.0 mm. At what two radii r from the central axis of the capacitor is the magnitude of the induced magnetic field equal to 50% of its maximum value?

26E. The induced magnetic field 6.0 mm from the central axis of a circular parallel-plate capacitor and between the plates has magnitude of 2.0×10^{-7} T. The plates have radius 3.0 mm. At what rate $|d\vec{E}/dt|$ is the electric field magnitude between the plates changing?

27P. Suppose that a parallel-plate capacitor has circular plates with radius $R = 30$ mm and a plate separation of 5.0 mm. Suppose also that a sinusoidal potential difference with a maximum value of 150 V and a frequency of 60 Hz is applied across the plates; that is,

$$\Delta V = (150 \text{ V}) \sin[2\pi(60 \text{ Hz})t].$$

(a) Find $|\vec{B}_{max}(R)|$, the maximum value of the magnitude of the induced magnetic field that occurs at $r = R$. (b) Plot $B_{max}(r)$ for $0 < r < 10$ cm.

SEC. 32-10   Displacement Current

28E. Prove that the displacement current in a parallel-plate capacitor of capacitance C can be written as $i_d = C(d\Delta V/dt)$, where ΔV is the potential difference between the plates.

29E. At what rate must the potential difference between the plates of a parallel-plate capacitor with a 2.0 μF capacitance be changed to produce a displacement current of 1.5 A?

30E. For the situation of Touchstone Example 32-9-1, show that the magnitude of the current density of the displacement current is $|\vec{J}_d| = \varepsilon_0(|d\vec{E}/dt|)$ for $r \leq$ R.

31E. A parallel-plate capacitor with circular plates of radius 0.10 m is being discharged. A circular loop of radius 0.20 m is concentric with the capacitor and halfway between the plates. The displacement current through the loop is 2.0 A. At what rate is the magnitude of the electric field between the plates changing?

32P. As a parallel-plate capacitor with circular plates 20 cm in diameter is being charged, the current density of the displacement current in the region between the plates is uniform and has a magnitude of 20 A/m². (a) Calculate the magnitude $|\vec{B}|$ of the magnetic field at a distance $r = 50$ mm from the axis of symmetry of this region. (b) Calculate $|d\vec{E}/dt|$ in this region.

33P. The magnitude of the electric field between the two circular parallel plates in Fig. 32-20 is $|\vec{E}| = (4.0 \times 10^5) - (6.0 \times 10^4 t)$, with $|\vec{E}|$ in volts per meter and t in seconds. At $t = 0$, the field is upward as shown. The plate area is 4.0×10^{-2} m². For $t \geq 0$, (a) what are the magnitude and direction of the displacement current between the plates and (b) is the direction of the induced magnetic field clockwise or counterclockwise around the plates?

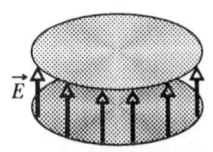

Fig. 32-20 Problem 34.

34P. The magnitude of a uniform electric field collapses to zero from an initial strength of 6.0×10^5 N/C in a time of 15 μs in the manner shown in Fig. 32-21. Calculate the magnitude of the displacement current, through a 1.6 m² area perpendicular to the field, during each of the time intervals a, b, and c shown on the graph. (Ignore the behavior at the ends of the intervals.)

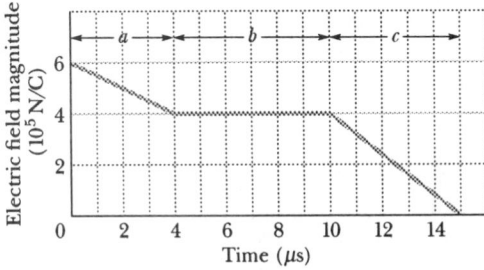

Fig. 32-21 Problem 35.

35P. A parallel-plate capacitor with circular plates is being charged. Consider a circular loop centered on the central axis between the plates. The loop radius is 0.20 m; the plate radius is 0.10 m; and the displacement current through the loop is 2.0 A. What is the rate at which the magnitude of the electric field between the plates is changing?

36P. A parallel-plate capacitor has square plates 1.0 m on a side as shown in Fig. 32-22. A current of 2.0 A charges the capacitor, producing a uniform electric field $\vec{E}$ between the plates, with $\vec{E}$ perpendicular to the plates. (a) What is the displacement current i_d through the region between the plates? (b) What is $|d\vec{E}|/dt$ in this region? (c) What is the displacement current through the square dashed path between the plates? (d) What is $\oint \vec{B} \cdot d\vec{s}$ around this square dashed path?

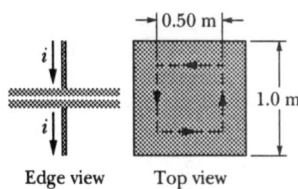

Fig. 32-22 Problem 37.

37P. A capacitor with parallel circular plates of radius R is discharging via a current of 12.0 A. Consider a loop of radius $R/3$ that is centered on the central axis between the plates. (a) How much displacement current is encircled by the loop? The maximum induced magnetic field has a magnitude of 12.0 mT. (b) At what radial distance from the central axis of the plate is the magnitude of the induced magnetic field 3.00 mT?

Chapter 33

SEC. 33-2   *LC* Oscillations, Qualitatively

1E. What is the capacitance of an oscillating *LC* circuit if the maximum charge on the capacitor is 1.60 μC and the total energy is 140 μJ?

2E. In an oscillating *LC* circuit, $L = 1.10$ mH and $C = 4.00$ μF. The maximum charge on the capacitor is 3.00 μC. Find the maximum current.

3E. An oscillating *LC* circuit consists of a 75.0 mH inductor and a 3.60 μF capacitor. If the maximum charge on the capacitor is 2.90 μC, (a) what is the total energy in the circuit and (b) what is the maximum current?

4E. In a certain oscillating *LC* circuit the total energy is converted from electric energy in the capacitor to magnetic energy in the inductor in 1.50 μs. (a) What is the period of oscillation? (b) What is the frequency of oscillation? (c) How long after the magnetic energy is a maximum will it be a maximum again?

5P. The frequency of oscillation of a certain *LC* circuit is 200 kHz. At time $t = 0$, plate A of the capacitor has maximum positive charge. At what times $t > 0$ will (a) plate A again have maximum positive charge, (b) the other plate of the capacitor have maximum positive charge, and (c) the inductor have maximum magnetic field?

SEC. 33-3 The Electrical—Mechanical Analogy

6E. A 0.50 kg body oscillates in SHM on a spring that, when extended 2.0 mm from its equilibrium, has an 8.0 N restoring force. (a) What is the angular frequency of oscillation? (b) What is the period of oscillation? (c) What is the capacitance of an LC circuit with the same period if L is chosen to be 5.0 H?

7P. The energy in an oscillating LC circuit containing a 1.25 H inductor is 5.70 μJ. The maximum charge on the capacitor is 175 μC. Find (a) the mass, (b) the spring constant, (c) the maximum displacement, and (d) the maximum speed for a mechanical system with the same period.

SEC. 33-4 *LC* Oscillations, Quantitatively

8E. LC oscillators have been used in circuits connected to loudspeakers to create some of the sounds of electronic music. What inductance must be used with a 6.7 μF capacitor to produce a frequency of 10 kHz, which is near the middle of the audible range of frequencies?

9E. In an oscillating LC circuit with $L = 50$ mH and $C = 4.0$ μF, the current is initially a maximum. How long will it take before the capacitor is fully charged for the first time?

10E. A single loop consists of inductors ($L_1, L_2, \ldots$), capacitors ($C_1, C_2, \ldots$), and resistors ($R_1, R_2, \ldots$) connected in series as shown, for example, in Fig. 33-16a. Show that regardless of the sequence of these circuit elements in the loop, the behavior of this circuit is identical to that of the simple LC circuit shown in Fig. 33-16b. (*Hint:* Consider the loop rule and see Problem 43 in Chapter 31.

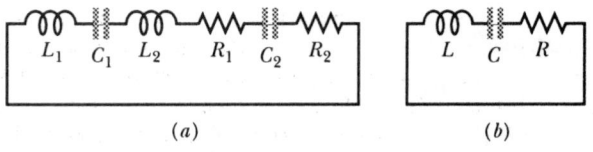

(a)   (b)

Fig. 33-16 Exercise 10.

11P. An oscillating LC circuit consisting of a 1.0 nF capacitor and a 3.0 mH coil has a maximum voltage of 3.0 V. (a) What is the maximum charge on the capacitor? (b) What is the maximum current through the circuit? (c) What is the maximum energy stored in the magnetic field of the coil?

12P. In an oscillating LC circuit in which $C = 400$ μF, the maximum potential difference across the capacitor during the oscillations is 1.50 V and the maximum current through the inductor is 50.0 mA. (a) What is the inductance L? (b) What is the frequency of the oscillations? (c) How much time is required for the charge on the capacitor to rise from zero to its maximum value?

13P. In the circuit shown in Fig. 33-17 the switch is kept in position a for a long time. It is

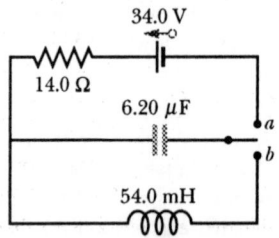

34.0 V

14.0 Ω

6.20 μF

a

b

54.0 mH

Fig. 33-17 Problem 13.

then thrown to position b. (a) Calculate the frequency of the resulting oscillating current. (b) What is the amplitude of the current oscillations?

14P. You are given a 10 mH inductor and two capacitors, of 5.0 μF and 2.0 μF capacitance. List the oscillation frequencies that can be generated by connecting these elements in various combinations.

15P. A variable capacitor with a range from 10 to 365 pF is used with a coil to form a variable-frequency LC circuit to tune the input to a radio. (a) What ratio of maximum to minimum frequencies may be obtained with such a capacitor? (b) If this circuit is to obtain frequencies from 0.54 MHz to 1.60 MHz, the ratio computed in (a) is too large. By adding a capacitor in parallel to the variable capacitor, this range may be adjusted. What should be the capacitance of this added capacitor, and what inductance should be used to obtain the desired range of frequencies?

16P. In an oscillating LC circuit, 75.0% of the total energy is stored in the magnetic field of the inductor at a certain instant. (a) In terms of the maximum charge on the capacitor, what is the charge there at that instant? (b) In terms of the maximum current in the inductor, what is the current there at that instant?

17P. In an oscillating LC circuit, $L = 25.0$ mH and $C = 7.80$ μF. At time $t = 0$ the current is 9.20 mA, the charge on the capacitor is 3.80 μC, and the capacitor is charging. (a) What is the total energy in the circuit? (b) What is the maximum charge on the capacitor? (c) What is the maximum current? (d) If the charge on the capacitor is given by $q = |Q| \cos(\omega t + \phi)$, what is the phase angle ϕ? (e) Suppose the data are the same, except that the capacitor is discharging at $t = 0$. What then is ϕ?

18P. An inductor is connected across a capacitor whose capacitance can be varied by turning a knob. We wish to make the frequency of oscillation of this LC circuit vary linearly with the angle of rotation of the knob, going from 2×10^5 to 4×10^5 Hz as the knob turns through 180°. If $L = 1.0$ mH, plot the required capacitance C as a function of the angle of rotation of the knob.

19P. In an oscillating LC circuit, $L = 3.00$ mH and $C = 2.70$ μF. At $t = 0$ the charge on the capacitor is zero and the current is 2.00 A. (a) What is the maximum charge that will appear on the capacitor? (b) In terms of the period T of oscillation, how much time will elapse after $t = 0$ until the energy stored in the capacitor will be increasing at its greatest rate? (c) What is this greatest rate at which energy is transferred to the capacitor?

20P. A series circuit containing inductance L_1 and capacitance C_1 oscillates at angular frequency ω. A second series circuit, containing inductance L_2 and capacitance C_2, oscillates at the same angular frequency. In terms of ω, what is the angular frequency of oscillation of a series circuit containing all four of these elements? Neglect resistance. (*Hint:* Use the formulas for equivalent capacitance and equivalent inductance; see Section 28-4 and Problem 43 in Chapter 31.)

21P. In an oscillating LC circuit with $C = 64.0$ μF, the current as a function of time is given by $i = (1.60) \sin(2500t + 0.680)$, where t is in seconds, i in amperes, and the phase constant in radians. (a) How soon after $t = 0$ will the current reach its maximum value? What are (b) the inductance L and (c) the total energy?

22P. Three identical inductors L and two identical capacitors C are connected in a two-loop circuit as shown in Fig. 33-18. (a) Suppose the currents are as shown in Fig. 33-18a. What is the current in the middle inductor? Write the loop equations and show that they are satisfied if the current oscillates with angular frequency $\omega = 1/\sqrt{LC}$. (b) Now suppose the currents are as shown in Fig. 33-18b. What is the current in the middle inductor? Write the loop equations and show that they are satisfied if the current oscillates with angular frequency $\omega = 1/\sqrt{3LC}$. Because the circuit can oscillate at two different frequencies, we cannot find an equivalent single-loop LC circuit to replace it.

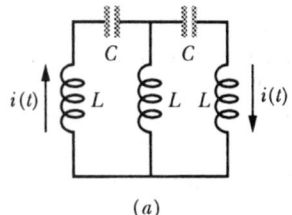

(a)

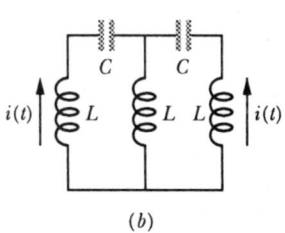

(b)

Fig. 33-18 Problem 22.

23P*. In Fig. 33-19, capacitor 1 with $C_1 = 900 \ \mu\text{F}$ is initially charged to 100 V and capacitor 2 with $C_2 = 100 \ \mu\text{F}$ is uncharged. The inductor has an inductance of 10.0 H. Describe in detail how one might charge capacitor 2 to 300 V by manipulating switches S_1 and S_2.

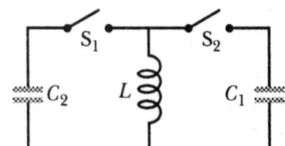

Fig. 33-19 Problem 23

SEC. 33-5 Damped Oscillations in an *RLC* Circuit

24E. Consider a damped LC circuit. (a) Show that the damping term $e^{-Rt/2L}$ (which involves L but not C) can be rewritten in a more symmetric manner (involving L and C) as $e^{-\pi R(\sqrt{C/L})t/T}$. Here T is the period of oscillation (neglecting resistance). (b) Using (a), show that the SI unit of $\sqrt{L/C}$ is the ohm. (c) Using (a), show that the condition that the fractional energy loss per cycle be small is $R \ll \sqrt{L/C}$.

25E. What resistance R should be connected in series with an inductance $L = 220$ mH and capacitance $C = 12.0 \ \mu\text{F}$ for the maximum charge on the capacitor to decay to 99.0% of its initial value in 50.0 cycles? (Assume $\omega' \approx \omega$.)

26P. A single-loop circuit consists of a $7.20 \ \Omega$ resistor, a 12.0 H inductor, and a $3.20 \ \mu\text{F}$ capacitor. Initially the capacitor has a charge of $6.20 \ \mu\text{C}$ and the current is zero. Calculate the charge on the capacitor N complete cycles later for $N = 5$, 10, and 100.

27P. In an oscillating series RLC circuit, find the time required for the maximum energy present in the capacitor during an oscillation to fall to half its initial value. Assume $q = Q$ at $t = 0$.

28P. At time $t = 0$ there is no charge on the capacitor of a series RLC circuit but there is current I through the inductor. (a) Find the phase constant ϕ in Eq. 33-21 for the circuit. (b) Write an expression for the charge q on the capacitor as a function of time t and in terms of the current amplitude and angular frequency ω' of the oscillations.

29P*. In an oscillating series RLC circuit, show that the fraction of the energy lost per cycle of oscillation, $\Delta U/U$, is given to a close approximation by $2\pi R/\omega L$. The quantity $\omega L/R$ is often called the Q of the circuit (for *quality*). A high-Q circuit has low resistance and a low fractional energy loss ($= 2\pi/Q$) per cycle.

SEC. 33-8 Three Simple Circuits

30E. A $1.50 \ \mu\text{F}$ capacitor is connected as in Fig. 33-9a to an ac generator with $|\varepsilon_m| = 30.0$ V. What is the amplitude of the resulting alternating current if the frequency of the emf is (a) 1.00 kHz and (b) 8.00 kHz?

31E. A 50.0 mH inductor is connected as in Fig. 33-10a to an ac generator with $|\varepsilon_m| = 30.0$ V. What is the amplitude of the resulting alternating current if the frequency of the emf is (a) 1.00 kHz and (b) 8.00 kHz?

32E. A $50 \ \Omega$ resistor is connected as in Fig. 33-8a to an ac generator with $|\varepsilon_m| = 30.0$ V. What is the amplitude of the resulting alternating current if the frequency of the emf is (a) 1.00 kHz and (b) 8.00 kHz?

33E. (a) At what frequency would a 6.0 mH inductor and a $10 \ \mu\text{F}$ capacitor have the same reactance? (b) What would the reactance be? (c) Show that this frequency would be the natural frequency of an oscillating circuit with the same L and C.

34P. An ac generator has emf $\varepsilon = |\varepsilon_m| \sin \omega_d t$, with $\mathcal{E}_m = 25.0$ V and $\omega_d = 377$ rad/s. It is connected to a 12.7 H inductor. (a) What is the maximum value of the current? (b) When the current is a maximum, what is the emf of the generator? (c) When the emf of the generator is -12.5 V and increasing in magnitude, what is the current?

35P. An ac generator has emf $\varepsilon = |\varepsilon_m| \sin(\omega_d t - \pi/4)$, where $|\varepsilon_m| = 30.0$ V and $\omega_d = 350$ rad/s. The current produced in a connected circuit is $i(t) = I \sin(\omega_d t - 3\pi/4)$, where $I = 620$ mA. (a) At what time after $t = 0$ does the generator emf first reach a maximum? (b) At what time after $t = 0$ does the current first reach a maximum? (c) The circuit contains a single element other than the generator. Is it a capacitor, an inductor, or a resistor? Justify your answer. (d) What is the value of the capacitance, inductance, or resistance, as the case may be?

36P. The ac generator of Problem 34 is connected to a $4.15 \ \mu\text{F}$ capacitor. (a) What is the maximum value of the current? (b) When the current is a maximum, what is the emf of the generator? (c) When the emf of the generator is -12.5 V and increasing in magnitude, what is the current?

SEC. 33-9 The Series *RLC* Circuit

37E. (a) Find Z, ϕ, and I for the situation of Touchstone Example 33-9-1 with the capacitor removed from the circuit, all other parameters remaining unchanged. (b) Draw to scale a phasor diagram like that of Fig. 33-11d for this new situation.

38E. (a) Find Z, ϕ, and I for the situation of Touchstone Example 33-9-1 with the inductor removed from the circuit, all other parameters remaining unchanged. (b) Draw to scale a phasor diagram like that of Fig. 33-11d for this new situation.

39E. (a) Find Z, ϕ, and I for the situation of Touchstone Example 33-9-1 with $C = 70.0$ μF, the other parameters remaining unchanged. (b) Draw a phasor diagram like that of Fig. 33-11d for this new situation and compare the two diagrams closely.

40P. In Fig. 33-20, a generator with an adjustable frequency of oscillation is connected to a variable resistance R, a capacitor of $C = 5.50$ μF, and an inductor of inductance L. The amplitude of the current produced in the circuit by the generator is at half-maximum level when the generator's frequency is 1.30 or 1.50 kHz. (a) What is L? (b) If R is increased, what happens to the frequencies at which the current amplitude is at half-maximum level?

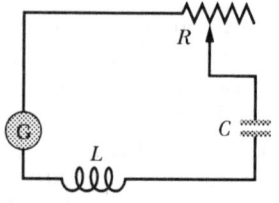

Fig. 33-20 Problem 40.

41P. In an RLC circuit, can the amplitude of the voltage across an inductor be greater than the amplitude of the generator emf? Consider an RLC circuit with $\varepsilon_m = 10$ V, $R = 10$ Ω, $L = 1.0$ H, and $C = 1.0$ μF. Find the amplitude of the voltage across the inductor at resonance.

42P. When the generator emf in Touchstone Example 33-9-1 is a maximum, what is the voltage across (a) the generator, (b) the resistance, (c) the capacitance, and (d) the inductance? (e) By summing these with appropriate signs, verify that the loop rule is satisfied.

43P. A coil of inductance 88 mH and unknown resistance and a 0.94 μF capacitor are connected in series with an alternating emf of frequency 930 Hz. If the phase constant between the applied voltage and the current is 75°, what is the resistance of the coil?

44P. An ac generator with $\varepsilon_m = 220$ V and operating at 400 Hz causes oscillations in a series RLC circuit having $R = 220$ Ω, $L = 150$ mH, and $C = 24.0$ μF. Find (a) the capacitive reactance X_C, (b) the impedance Z, and (c) the current amplitude I. A second capacitor of the same capacitance is then connected in series with the other components. Determine whether the values of (d) X_C, (e) Z, and (f) I increase, decrease, or remain the same.

45P. An RLC circuit such as that of Fig. 33-7 has $R = 5.00$ Ω, $C = 20.0$ μF, $L = 1.00$ H, and $\varepsilon_m = 30.0$ V. (a) At what angular frequency ω_d will the current amplitude have its maximum value, as in the resonance curves of Fig. 33-13? (b) What is this maximum value? (c) At what two angular frequencies ω_{d1} and ω_{d2} will the current amplitude be half this maximum value? (d) What is the fractional half-width $[= (\omega_{d1} - \omega_{d2})/\omega]$ of the resonance curve for this circuit?

46P. An ac generator is to be connected in series with an inductor of $L = 2.00$ mH and a capacitance C. You are to produce C by using capacitors of capacitances $C_1 = 4.00$ μF and $C_2 = 6.00$ μF, either singly or together. What resonant frequencies can the circuit have, depending on how you use C_1 and C_2?

47P. Show that the fractional half-width (see Problem 45) of a resonance curve is given by

$$\frac{\Delta\omega_d}{\omega} = \sqrt{\frac{3C}{L}} R,$$

in which ω is the angular frequency at resonance and $\Delta\omega_d$ is the width of the resonance curve at half-amplitude. Note that $\Delta\omega_d/\omega$ increases with R, as Fig. 33-13 shows. Use this formula to check the answer to Problem 45d.

48P. In Fig. 33-21, a generator with an adjustable frequency of oscillation is connected to resistance $R = 100$ Ω, inductances $L_1 = 1.70$ mH and $L_2 = 2.30$ mH, and capacitances $C_1 = 4.00$ μF, $C_2 = 2.50$ μF, and $C_3 = 3.50$ μF. (a) What is the resonant frequency of the circuit? (*Hint:* See Problem 43 in Chapter 31.) What happens to the resonant frequency if (b) the value of R is increased, (c) the value of L_1 is increased, and (d) capacitance C_3 is removed from the circuit?

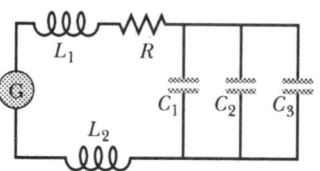

Fig. 33-21 Problem 48.

SEC. 33-10   Power in Alternating-Current Circuits

49E. What direct current will produce the same amount of thermal energy, in a particular resistor, as an alternating current that has a maximum value of 2.60 A?

50E. An ac voltmeter with large impedance is connected in turn across the inductor, the capacitor, and the resistor in a series circuit having an alternating emf of 100 V (rms); it gives the same reading in volts in each case. What is this reading?

51E. What is the maximum value of an ac voltage whose rms value is 100 V?

52E. (a) For the conditions in Problem 34c, is the generator supplying energy to or taking energy from the rest of the circuit? (b) Repeat for the conditions of Problem 36c.

53E. Calculate the average rate of energy dissipation in the circuits of Exercises 31, 32, 37, and 38.

54E. Show that the average rate at which energy is supplied to the circuit of Fig. 33-7 can also be written as $P_{\text{avg}} = \varepsilon_{\text{rms}}^2 R/Z^2$. Show that this expression for average power gives reasonable results for a purely resistive circuit, for an RLC circuit at resonance, for a purely capacitive circuit, and for a purely inductive circuit.

55E. An air conditioner connected to a 120 V rms ac line is equivalent to a 12.0 Ω resistance and a 1.30 Ω inductive reactance in series. (a) Calculate the impedance of the air conditioner. (b) Find the average rate at which energy is supplied to the appliance.

56P. In a series oscillating RLC circuit, $R = 16.0$ Ω, $C = 31.2$ μF, $L = 9.20$ mH, and $\varepsilon = |\varepsilon_m| \sin \omega_d t$ with $|\varepsilon_m| = 45.0$ V and $\omega_d = 3000$ rad/s. For time $t = 0.442$ ms find (a) the rate at which energy is being supplied by the generator, (b) the rate at which the energy in the capacitor is changing, (c) the rate at which the energy in the inductor is changing, and (d) the rate at which energy is being dissipated in the resistor. (e) What is the meaning of a negative result for any of (a), (b), and (c)? (f) Show that the results of (b), (c), and (d) sum to the result of (a).

57P. Figure 33-22 shows an ac generator connected to a "black box" through a pair of terminals. The box contains an *RLC* circuit, possibly even a multiloop circuit, whose elements and connections we do not know. Measurements outside the box reveal that

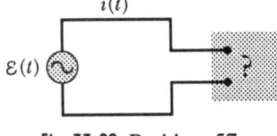

Fig. 33-22 Problem 57.

$$\varepsilon(t) = (75.0 \text{ V}) \sin \omega_d t$$

and $$i(t) = (1.20 \text{ A}) \sin(\omega_d t + 42.0°).$$

(a) What is the power factor? (b) Does the current lead or lag the emf? (c) Is the circuit in the box largely inductive or largely capacitive? (d) Is the circuit in the box in resonance? (e) Must there be a capacitor in the box? An inductor? A resistor? (f) At what average rate is energy delivered to the box by the generator? (g) Why don't you need to know the angular frequency ω_d to answer all these questions?

58P. In Fig. 33-23 show that the average rate at which energy is dissipated in resistance R is a maximum when R is equal to the internal resistance r of the ac generator. (In the text discussion we have tacitly assumed that $r = 0$.)

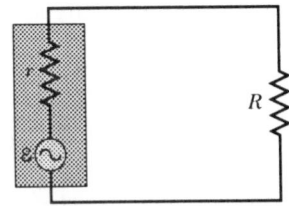

Fig. 33-23 Problems 58 and 65.

59P. In an *RLC* circuit such as that of Fig. 33-7 assume that $R = 5.00 \ \Omega$, $L = 60.0$ mH, $f_d = 60.0$ Hz, and $|\varepsilon_m| = 30.0$ V. For what values of the capacitance would the average rate at which energy is dissipated in the resistance be (a) a maximum and (b) a minimum? (c) What are these maximum and minimum energy dissipation rates? What are (d) the corresponding phase angles and (e) the corresponding power factors?

60P. A typical "light dimmer" used to dim the stage lights in a theater consists of a variable inductor L (whose inductance is adjustable between zero and L_{max}) connected in series with the lightbulb B as shown in Fig.

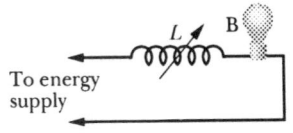

Fig. 33-24 Problem 60.

33-24. The electrical supply is 120 V (rms) at 60.0 Hz; the lightbulb is rated as "120 V, 1000 W." (a) What L_{max} is required if the rate of energy dissipation in the lightbulb is to be varied by a factor of 5 from its upper limit of 1000 W? Assume that the resistance of the lightbulb is independent of its temperature. (b) Could one use a variable resistor (adjustable between zero and R_{max}) instead of an inductor? If so, what R_{max} is required? Why isn't this done?

61P. In Fig. 33-25, $R = 15.0 \ \Omega$, $C = 4.70 \ \mu$F, and $L = 25.0$ mH. The generator provides a sinusoidal voltage of 75.0 V (rms) and frequency $f = 550$ Hz. (a) Calculate the rms current. (b) Find the rms voltages ΔV_{ab},

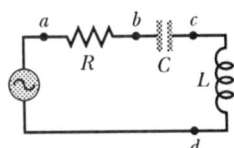

Fig. 33-25 Problem 61.

ΔV_{bc}, ΔV_{cd}, ΔV_{bd}, ΔV_{ad}. (c) At what average rate is energy dissipated by each of the three circuit elements?

SEC. 33-11   Transformers

62E. A generator supplies 100 V to the primary coil of a transformer of 50 turns. If the secondary coil has 500 turns, what is the secondary voltage?

63E. A transformer has 500 primary turns and 10 secondary turns. (a) If ΔV_p is 120 V (rms), what is ΔV_s with an open circuit? (b) If the secondary now has a resistive load of 15 Ω, what are the currents in the primary and secondary?

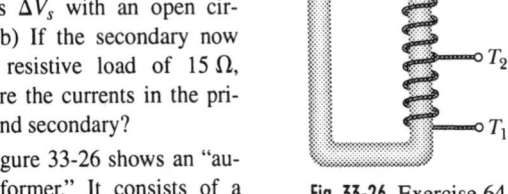

Fig. 33-26 Exercise 64.

64E. Figure 33-26 shows an "autotransformer." It consists of a single coil (with an iron core). Three taps T_i are provided. Between taps T_1 and T_2 there are 200 turns, and between taps T_2 and T_3 there are 800 turns. Any two taps can be considered the "primary terminals" and any two taps can be considered the "secondary terminals." List all the ratios by which the primary voltage may be changed to a secondary voltage.

65P. In Fig. 33-23 let the rectangular box on the left represent the (high-impedance) output of an audio amplifier, with $r = 1000 \ \Omega$. Let $R = 10 \ \Omega$ represent the (low-impedance) coil of a loudspeaker. For maximum transfer of energy to the load R we must have $R = r$, and that is not true in this case. However, a transformer can be used to "transform" resistances, making them behave electrically as if they were larger or smaller than they actually are. Sketch the primary and secondary coils of a transformer that can be introduced between the amplifier and the speaker in Fig. 33-29 to match the impedances. What must be the turns ratio?

APPENDIX A
The International System of Units (SI)*

1. SI Base Units

1. The SI Base Units

Quantity	Name	Symbol	Definition
length	meter	m	". . . the length of the path traveled by light in vacuum in 1/299,792,458 of a second." (1983)
mass	kilogram	kg	". . . this prototype [a certain platinum–iridium cylinder] shall henceforth be considered to be the unit of mass." (1889)
time	second	s	". . . the duration of 9,192,631,770 periods of the radiation corresponding to the transition between the two hyperfine levels of the ground state of the cesium-133 atom." (1967)
electric current	ampere	A	". . . that constant current which, if maintained in two straight parallel conductors of infinite length, of negligible circular cross section, and placed 1 meter apart in vacuum, would produce between these conductors a force equal to 2×10^{-7} newton per meter of length." (1946)
thermodynamic temperature	kelvin	K	". . . the fraction 1/273.16 of the thermodynamic temperature of the triple point of water." (1967)
amount of substance	mole	mol	". . . the amount of substance of a system which contains as many elementary entities as there are atoms in 0.012 kilogram of carbon-12." (1971)
luminous intensity	candela	cd	". . . the luminous intensity, in the perpendicular direction, of a surface of 1/600,000 square meter of a blackbody at the temperature of freezing platinum under a pressure of 101.325 newtons per square meter." (1967)

2. The SI Supplementary Units

2. The SI Supplementary Units

Quantity	Name of Unit	Symbol
plane angle	radian	rad
solid angle	steradian	sr

*Adapted from "The International System of Units (SI)," National Bureau of Standards Special Publication 330, 1972 edition. The definitions above were adopted by the General Conference of Weights and Measures, an international body, on the dates shown. In this book we do not use the candela.

3.   Some SI Derivations

3.   Some SI Derived Units

Quantity	Name of Unit	Symbol	
area	square meter	m^2	
volume	cubic meter	m^3	
frequency	hertz	Hz	s^{-1}
mass density (density)	kilogram per cubic meter	kg/m^3	
speed, velocity	meter per second	m/s	
angular velocity	radian per second	rad/s	
acceleration	meter per second per second	m/s^2	
angular acceleration	radian per second per second	rad/s^2	
force	newton	N	$kg \cdot m/s^2$
pressure	pascal	Pa	N/m^2
work, energy, quantity of heat	joule	J	$N \cdot m$
power	watt	W	J/s
quantity of electric charge	coulomb	C	$A \cdot s$
potential difference, electromotive force	volt	V	W/A
electric field strength	volt per meter (or newton per coulomb)	V/m	N/C
electric resistance	ohm	Ω	V/A
capacitance	farad	F	$A \cdot s/V$
magnetic flux	weber	Wb	$V \cdot s$
inductance	henry	H	$V \cdot s/A$
magnetic flux density	tesla	T	Wb/m^2
magnetic field strength	ampere per meter	A/m	
entropy	joule per kelvin	J/K	
specific heat	joule per kilogram kelvin	$J/(kg \cdot K)$	
thermal conductivity	watt per meter kelvin	$W/(m \cdot K)$	
radiant intensity	watt per steradian	W/sr	

4.   Significant Figures and the Precision of Numerical Results

Quoting the result of a calculation or a measurement to the correct number of significant figures is merely a way of telling your reader how precise you believe the result to be. Quoting too many significant figures overstates the precision of your result while quoting too few implies less precision than the result may actually possess. So how many significant figures should you quote when reporting the result of a measurement or calculation?

Determining Significant Figures

Before answering the question of how many significant figures to quote, we need to have a clear method for determining how many significant figures a reported number has. The standard method is quite simple:

> **Method for Counting Significant Figures:** Read the number from left to right, and count the first non-zero digit and all the digits (zero or not) to the right of it as significant.

Using this rule, 350 mm, 0.000350 km, and 0.350 m each has *three* significant figures. In fact, each of these numbers merely represents the same distance, expressed in different units. As you can see from this example, the number of *decimal places* that a number has is *not* the same as its number of *significant figures*. The first of these distances has zero decimal places, the second has six decimal places, while the third has three, yet all three of these numbers have three significant figures.

One consequence of this method is especially worth noting. Trailing zeros count as significant figures. For example, 2700 m/s has four significant figures. If you really meant it to have only three significant figures, you would have to write it either as 2.70 km/s (changing the unit) or 2.70×10^3 m/s (using scientific notation.)

A Simple Rule for Reporting Significant Figures in a Calculated Result

Now that you know how to count significant figures, how many should the result of a calculation have? A simple rule that will work for each of the steps in most calculations is:

> **Significant Figures in a Calculated Result:** Unless addition is involved, no final result should be quoted to *more* significant figures than the original data from which it was derived.

In introductory physics you will only rarely encounter data that are known to better than two, three, or four significant figures. This simple rule then tells you that you can't go very far wrong if you round off all your final results to three significant figures. There are two situations in which the simple rule should not be applied to a calculation. One is when an exact number is involved in the calculation and another is when a calculation is done in parts so that intermediate results are used.

1. *Using Exact Data*   There are some obvious situations in which a number used in a calculation is exact. Numbers based on counting items are exact. For example, if you are told that there are 5 people on an elevator, there are exactly 5 people, not 4.7 or 5.1. Another situation arises when a number is exact to a certain number of significant figures by definition. For example, the conversion factor 2.54 cm/ inch does *not* have three significant figures because the inch is *defined* to be exactly 2.5400000 . . . cm. *Data that are known exactly should not be included when deciding which of the original data has the fewest significant figures.*

2. *Significant Figures in Intermediate Results*   Only the final result that you quote at the end of your calculation should be rounded using the simple rule. Intermediate results should never be rounded at all. Modern spreadsheet software takes care of this for you, as does your calculator if you store your intermediate results in its memory rather than writing them down and then rekeying them. If you must write down intermediate results, always keep a few more significant figures than your final result will have.

Understanding and Refining the Simple Significant Figure Rule

Since quoting the result of a calculation or a measurement to the correct number of significant figures is merely a way of indicating its precision, you need to understand what limits the precision of data before you can acquire a better understanding of the simple rule and its exceptions.

Absolute Precision   There are two ways of talking about precision. The first of these is *absolute precision*, which tells you explicitly the smallest scale division of the measurement. It's always quoted in the same units as the quantity being measured. For example, saying "I measured the length of the table to the nearest centimeter" states the absolute precision of the measurement. Knowing the absolute precision tells you how many *decimal places* the measurement has; it alone does not determine the number of significant figures. For example, if the table is 235 cm long, then 1 cm of absolute precision translates into three significant figures. On the other hand, if the table is for a doll's house and is only 8 cm long, then the same 1 cm of absolute precision yields only one significant figure.

Relative Precision   Because of this disadvantage of absolute precision, scientists often prefer to describe the precision of data *relative* to the size of the quantity being measured. To use the previous examples, the *relative precision* of the length of the real table is 1 cm out of 235 cm. This is usually stated as a ratio (1 part in 235) or, more conveniently, as a percentage ($1/235 = 0.004255 \approx 0.4\%$). In the case of the toy table, the same 1 cm of absolute precision yields a relative precision of only 1 part in 8 or $1/8 = 0.125 = 12.5\%$.

Inconsistencies between Significant Figures and Relative Precision   You may have noticed an inconsistency that goes with using a certain number of significant figures to express relative precision. Quoted to the same number of significant figures, the relative precision of results can be quite different. For example, 13 cm and 94 cm both have two significant figures. Yet the first is specified to only 1 part in 13 or $1/13 \approx 10\%$, while the second is known to 1 part in 94 or $1/94 \approx 1\%$. This bias toward greater relative precision for results with larger first significant figures is one weakness of using significant figures to track the precision of calculated results. To partially address this problem, you can include one additional significant figure than the simple rule suggests, when the final result of a calculation has a 1 as its first significant figure.

Multiplying and Dividing   When multiplying or dividing numbers, the *relative* precision of the result can not exceed that of the least precise number used. Since the number of significant figures in the result tells us its relative precision, the simple rule is all that you need when you multiply or divide. For example, the area of a strip of paper whose measured size is 280 cm by 2.5 cm would be correctly reported, according to the simple rule, as 7.0×10^2 cm^2. This result has only two significant figures since the less precise mesurement, 2.5 cm, that went into the calculation had only two significant figures. Reporting this result as 700 cm^2 would not be correct since this reported result has three significant figures, exceeding the relative precision of the 2.5 cm measurement.

Addition and Subtraction   When adding or subtracting, you line up the decimal points before you add or subtract. This means that it's the *absolute* precision of the least precise number that limits the precision of the sum or the difference. This can lead to some exceptions to the simple rule. For example, adding 957 cm and 878 cm yields 1835 cm. Here the result is reliable to an absolute precision of about 1 cm since both of the original distances had this reliability. But the result then has four significant figures while each of the original numbers had only three. If, on the other hand, you take the difference betwen these two distances you get 79 cm. The difference is still reliable to about 1 cm, but that absolute precision now translates into only two significant figures worth of relative precision. So, you should be careful when adding or subtracting, since addition can actually increase the relative precision of your result and, more importantly, subtraction can reduce it.

Evaluating Functions   What about the evaluation of functions? For example, how many significant figures does the sin(88.2°) have? You can also take an empirical approach to answering this question. First use your calculator to note that sin(88.2°) = 0.999506. Now add 1 to the least significant decimal place of the argument of the function and evaluate it again. Here this gives sin(88.3°) = 0.999559. Take the last significant figure in the result to be *the first one from the left that changed* when you repeated the calculation. In this example the first digit that changed was the 0; it became a 5 (the second 5) in the recalculation. So, using the empirical approach gives you 5 significant figures.

APPENDIX B
Some Fundamental Constants of Physics*

Constant	Symbol	Computational Value	Best (1998) Value	
			Value[a]	Uncertainty[b]
Speed of light in a vacuum	c	3.00×10^8 m/s	2.997 924 58	exact
Elementary charge	e	1.60×10^{-19} C	1.602 176 462	0.039
Gravitational constant	G	6.67×10^{-11} m^3/s$^2 \cdot$ kg	6.673	1500
Universal gas constant	R	8.31 J/mol $\cdot$ K	8.314 472	1.7
Avogadro constant	N_A	6.02×10^{23} mol^{-1}	6.022 141 99	0.079
Boltzmann constant	k	1.38×10^{-23} J/K	1.380 650 3	1.7
Stefan–Boltzmann constant	σ	5.67×10^{-8} W/m$^2 \cdot$ K^4	5.670 400	7.0
Molar volume of ideal gas at STP[d]	V_m	2.27×10^{-2} m^3/mol	2.271 098 1	1.7
Permittivity constant	ϵ_0	8.85×10^{-12} F/m	8.854 187 817 62	exact
Permeability constant	μ_0	1.26×10^{-6} H/m	1.256 637 061 43	exact
Planck constant	h	6.63×10^{-34} J $\cdot$ s	6.626 068 76	0.078
Electron mass[c]	m_e	9.11×10^{-31} kg	9.109 381 88	0.079
		5.49×10^{-4} u	5.485 799 110	0.0021
Proton mass[c]	m_p	1.67×10^{-27} kg	1.672 621 58	0.079
		1.0073 u	1.007 276 466 88	1.3×10^{-4}
Ratio of proton mass to electron mass	m_p/m_e	1840	1836.152 667 5	0.0021
Electron charge-to-mass ratio	e/m_e	1.76×10^{11} C/kg	1.758 820 174	0.040
Neutron mass[c]	m_n	1.68×10^{-27} kg	1.674 927 16	0.079
		1.0087 u	1.008 664 915 78	5.4×10^{-4}
Hydrogen atom mass[c]	m_{1H}	1.0078 u	1.007 825 031 6	0.0005
Deuterium atom mass[c]	m_{2H}	2.0141 u	2.014 101 777 9	0.0005
Helium atom mass[c]	m_{4He}	4.0026 u	4.002 603 2	0.067
Muon mass	m_μ	1.88×10^{-28} kg	1.883 531 09	0.084
Electron magnetic moment	μ_e	9.28×10^{-24} J/T	9.284 763 62	0.040
Proton magnetic moment	μ_p	1.41×10^{-26} J/T	1.410 606 663	0.041
Bohr magneton	μ_B	9.27×10^{-24} J/T	9.274 008 99	0.040
Nuclear magneton	μ_N	5.05×10^{-27} J/T	5.050 783 17	0.040
Bohr radius	r_B	5.29×10^{-11} m	5.291 772 083	0.0037
Rydberg constant	R	1.10×10^7 m^{-1}	1.097 373 156 854 8	7.6×10^{-6}
Electron Compton wavelength	λ_C	2.43×10^{-12} m	2.426 310 215	0.0073

[a]Values given in this column should be given the same unit and power of 10 as the computational value.

[b]Parts per million.

[c]Masses given in u are in unified atomic mass units, where 1 u = 1.660 538 73 $\times 10^{-27}$ kg.

[d]STP means standard temperature and pressure: 0°C and 1.0 atm (0.1 MPa).

*The values in this table were selected from the 1998 CODATA recommended values (www.physics.nist.gov).

Some Astronomical Data

Some Distances from Earth

To the Moon*	3.82×10^8 m	To the center of our galaxy	2.2×10^{20} m
To the Sun*	1.50×10^{11} m	To the Andromeda Galaxy	2.1×10^{22} m
To the nearest star (Proxima Centauri)	4.04×10^{16} m	To the edge of the observable universe	$\sim 10^{26}$ m

*Mean distance.

The Sun, Earth, and the Moon

Property	Unit	Sun	Earth	Moon
Mass	kg	1.99×10^{30}	5.98×10^{24}	7.36×10^{22}
Mean radius	m	6.96×10^8	6.37×10^6	1.74×10^6
Mean density	kg/m³	1410	5520	3340
Free-fall acceleration at the surface	m/s²	274	9.81	1.67
Escape velocity	km/s	618	11.2	2.38
Period of rotation[a]	—	37 d at poles[b] 26 d at equator[b]	23 h 56 min	27.3 d
Radiation power[c]	W	3.90×10^{26}		

[a]Measured with respect to the distant stars.
[b]The Sun, a ball of gas, does not rotate as a rigid body.
[c]Just outside Earth's atmosphere solar energy is received, assuming normal incidence, at the rate of 1340 W/m².

Some Properties of the Planets

	Mercury	Venus	Earth	Mars	Jupiter	Saturn	Uranus	Neptune	Pluto
Mean distance from Sun, 10^6 km	57.9	108	150	228	778	1430	2870	4500	5900
Period of revolution, y	0.241	0.615	1.00	1.88	11.9	29.5	84.0	165	248
Period of rotation,[a] d	58.7	−243[b]	0.997	1.03	0.409	0.426	−0.451[b]	0.658	6.39
Orbital speed, km/s	47.9	35.0	29.8	24.1	13.1	9.64	6.81	5.43	4.74
Inclination of axis to orbit	<28°	≈3°	23.4°	25.0°	3.08°	26.7°	97.9°	29.6°	57.5°
Inclination of orbit to Earth's orbit	7.00°	3.39°		1.85°	1.30°	2.49°	0.77°	1.77°	17.2°
Eccentricity of orbit	0.206	0.0068	0.0167	0.0934	0.0485	0.0556	0.0472	0.0086	0.250
Equatorial diameter, km	4880	12 100	12 800	6790	143 000	120 000	51 800	49 500	2300
Mass (Earth = 1)	0.0558	0.815	1.000	0.107	318	95.1	14.5	17.2	0.002
Density (water = 1)	5.60	5.20	5.52	3.95	1.31	0.704	1.21	1.67	2.03
Surface value of g,[c] m/s²	3.78	8.60	9.78	3.72	22.9	9.05	7.77	11.0	0.5
Escape velocity,[c] km/s	4.3	10.3	11.2	5.0	59.5	35.6	21.2	23.6	1.1
Known satellites	0	0	1	2	16 + ring	18 + rings	17 + rings	8 + rings	1

[a]Measured with respect to the distant stars.
[b]Venus and Uranus rotate opposite their orbital motion.
[c]Gravitational acceleration measured at the planet's equator.

APPENDIX D
Conversion Factors

Conversion factors may be read directly from these tables. For example, 1 degree = 2.778×10^{-3} revolutions, so $16.7° = 16.7 \times 2.778 \times 10^{-3}$ rev. The SI units are fully capitalized. Adapted in part from G. Shortley and D. Williams, *Elements of Physics,* 1971, Prentice-Hall, Englewood Cliffs, NJ.

Plane Angle

	°	′	″	RADIAN	rev
1 degree =	1	60	3600	1.745×10^{-2}	2.778×10^{-3}
1 minute =	1.667×10^{-2}	1	60	2.909×10^{-4}	4.630×10^{-5}
1 second =	2.778×10^{-4}	1.667×10^{-2}	1	4.848×10^{-6}	7.716×10^{-7}
1 RADIAN =	57.30	3438	2.063×10^{5}	1	0.1592
1 revolution =	360	2.16×10^{4}	1.296×10^{6}	6.283	1

Solid Angle

1 sphere = 4π steradians = 12.57 steradians

Length

	cm	METER	km	in.	ft	mi
1 centimeter =	1	10^{-2}	10^{-5}	0.3937	3.281×10^{-2}	6.214×10^{-6}
1 METER =	100	1	10^{-3}	39.37	3.281	6.214×10^{-4}
1 kilometer =	10^{5}	1000	1	3.937×10^{4}	3281	0.6214
1 inch =	2.540	2.540×10^{-2}	2.540×10^{-5}	1	8.333×10^{-2}	1.578×10^{-5}
1 foot =	30.48	0.3048	3.048×10^{-4}	12	1	1.894×10^{-4}
1 mile =	1.609×10^{5}	1609	1.609	6.336×10^{4}	5280	1

1 angström = 10^{-10} m   1 fermi = 10^{-15} m   1 fathom = 6 ft   1 rod = 16.5 ft

1 nautical mile = 1852 m   1 light-year = 9.460×10^{12} km   1 Bohr radius = 5.292×10^{-11} m   1 mil = 10^{-3} in.

= 1.151 miles = 6076 ft   1 parsec = 3.084×10^{13} km   1 yard = 3 ft   1 nm = 10^{-9} m

Area

	METER^2	cm^2	ft^2	in.^{2}
1 SQUARE METER =	1	10^{4}	10.76	1550
1 square centimeter =	10^{-4}	1	1.076×10^{-3}	0.1550
1 square foot =	9.290×10^{-2}	929.0	1	144
1 square inch =	6.452×10^{-4}	6.452	6.944×10^{-3}	1

1 square mile = 2.788×10^{7} ft^2 = 640 acres   1 acre = 43 560 ft^2

1 barn = 10^{-28} m^2 1 hectare = 10^{4} m^2 = 2.471 acres

Volume

	METER^3	cm^3	L	ft^3	in.^{3}
1 CUBIC METER = 1		10^6	1000	35.31	6.102×10^4
1 cubic centimeter = 10^{-6}		1	1.000×10^{-3}	3.531×10^{-5}	6.102×10^{-2}
1 liter = 1.000×10^{-3}		1000	1	3.531×10^{-2}	61.02
1 cubic foot = 2.832×10^{-2}		2.832×10^4	28.32	1	1728
1 cubic inch = 1.639×10^{-5}		16.39	1.639×10^{-2}	5.787×10^{-4}	1

1 U.S. fluid gallon = 4 U.S. fluid quarts = 8 U.S. pints = 128 U.S. fluid ounces = 231 in.^{3}
1 British imperial gallon = 277.4 in.^{3} = 1.201 U.S. fluid gallons

Mass

Quantities in the colored areas are not mass units but are often used as such. When we write, for example, 1 kg "=" 2.205 lb, this means that a kilogram is a *mass* that *weighs* 2.205 pounds at a location where g has the standard value of 9.80665 m/s^2.

	g	KILOGRAM	slug	u	oz	lb	ton
1 gram = 1	0.001		6.852×10^{-5}	6.022×10^{23}	3.527×10^{-2}	2.205×10^{-3}	1.102×10^{-6}
1 KILOGRAM = 1000	1		6.852×10^{-2}	6.022×10^{26}	35.27	2.205	1.102×10^{-3}
1 slug = 1.459×10^4	14.59		1	8.786×10^{27}	514.8	32.17	1.609×10^{-2}
1 atomic mass unit = 1.661×10^{-24}	1.661×10^{-27}		1.138×10^{-28}	1	5.857×10^{-26}	3.662×10^{-27}	1.830×10^{-30}
1 ounce = 28.35	2.835×10^{-2}		1.943×10^{-3}	1.718×10^{25}	1	6.250×10^{-2}	3.125×10^{-5}
1 pound = 453.6	0.4536		3.108×10^{-2}	2.732×10^{26}	16	1	0.0005
1 ton = 9.072×10^5	907.2		62.16	5.463×10^{29}	3.2×10^4	2000	1

1 metric ton = 1000 kg

Density

Quantities in the colored areas are weight densities and, as such, are dimensionally different from mass densities. See note for mass table.

	slug/ft^3	KILOGRAM/ METER^3	g/cm^3	lb/ft^3	lb/in.^{3}
1 slug per foot^3 = 1		515.4	0.5154	32.17	1.862×10^{-2}
1 KILOGRAM per METER^3 = 1.940×10^{-3}		1	0.001	6.243×10^{-2}	3.613×10^{-5}
1 gram per centimeter^3 = 1.940		1000	1	62.43	3.613×10^{-2}
1 pound per foot^3 = 3.108×10^{-2}		16.02	16.02×10^{-2}	1	5.787×10^{-4}
1 pound per inch^3 = 53.71		2.768×10^4	27.68	1728	1

Time

	y	d	h	min	SECOND
1 year = 1		365.25	8.766×10^3	5.259×10^5	3.156×10^7
1 day = 2.738×10^{-3}		1	24	1440	8.640×10^4
1 hour = 1.141×10^{-4}		4.167×10^{-2}	1	60	3600
1 minute = 1.901×10^{-6}		6.944×10^{-4}	1.667×10^{-2}	1	60
1 SECOND = 3.169×10^{-8}		1.157×10^{-5}	2.778×10^{-4}	1.667×10^{-2}	1

Speed

	ft/s	km/h	METER/SECOND	mi/h	cm/s
1 foot per second = 1	1.097	0.3048	0.6818	30.48
1 kilometer per hour = 0.9113	1	0.2778	0.6214	27.78
1 METER per SECOND = 3.281	3.6	1	2.237	100
1 mile per hour = 1.467	1.609	0.4470	1	44.70
1 centimeter per second = 3.281×10^{-2}	3.6×10^{-2}	0.01	2.237×10^{-2}	1

1 knot = 1 nautical mi/h = 1.688 ft/s   1 mi/min = 88.00 ft/s = 60.00 mi/h

Force

Force units in the colored areas are now little used. To clarify: 1 gram-force (= 1 gf) is the force of gravity that would act on an object whose mass is 1 gram at a location where g has the standard value of 9.80665 m/s^2.

	dyne	NEWTON	lb	pdl	gf	kgf
1 dyne = 1	10^{-5}	2.248×10^{-6}	7.233×10^{-5}	1.020×10^{-3}	1.020×10^{-6}
1 NEWTON = 10^5	1	0.2248	7.233	102.0	0.1020
1 pound = 4.448×10^5	4.448	1	32.17	453.6	0.4536
1 poundal = 1.383×10^4	0.1383	3.108×10^{-2}	1	14.10	1.410×10^2
1 gram-force = 980.7	9.807×10^{-3}	2.205×10^{-3}	7.093×10^{-2}	1	0.001
1 kilogram-force = 9.807×10^5	9.807	2.205	70.93	1000	1

1 ton = 2000 lb

Pressure

	atm	dyne/cm^2	inch of water	cm Hg	PASCAL	lb/in.^{2}	lb/ft^2
1 atmosphere = 1	1.013×10^6	406.8	76	1.013×10^5	14.70	2116
1 dyne per centimeter^2 = 9.869×10^{-7}	1	4.015×10^{-4}	7.501×10^{-5}	0.1	1.405×10^{-5}	2.089×10^{-3}
1 inch of water^a at 4°C = 2.458×10^{-3}	2491	1	0.1868	249.1	3.613×10^{-2}	5.202
1 centimeter of mercury^a at 0°C = 1.316×10^{-2}	1.333×10^4	5.353	1	1333	0.1934	27.85
1 PASCAL = 9.869×10^{-6}	10	4.015×10^{-3}	7.501×10^{-4}	1	1.450×10^{-4}	2.089×10^{-2}
1 pound per inch^2 = 6.805×10^{-2}	6.895×10^4	27.68	5.171	6.895×10^3	1	144
1 pound per foot^2 = 4.725×10^{-4}	478.8	0.1922	3.591×10^{-2}	47.88	6.944×10^{-3}	1

^{a}Where the acceleration of gravity has the standard value of 9.80665 m/s^2.

1 bar = 10^6 dyne/cm^2 = 0.1 MPa   1 millibar = 10^3 dyne/cm^2 = 10^2 Pa   1 torr = 1 mm Hg

Energy, Work, Heat

Quantities in the colored areas are not energy units but are included for convenience. They arise from the relativistic mass–energy equivalence formula $E = mc^2$ and represent the energy released if a kilogram or unified atomic mass unit (u) is completely converted to energy (bottom two rows) or the mass that would be completely converted to one unit of energy (rightmost two columns).

	Btu	erg	ft·lb	hp·h	JOULE	cal	kW·h	eV	MeV	kg	u
1 British thermal unit =	1	1.055×10^{10}	777.9	3.929×10^{-4}	1055	252.0	2.930×10^{-4}	6.585×10^{21}	6.585×10^{15}	1.174×10^{-14}	7.070×10^{12}
1 erg =	9.481×10^{-11}	1	7.376×10^{-8}	3.725×10^{-14}	10^{-7}	2.389×10^{-8}	2.778×10^{-14}	6.242×10^{11}	6.242×10^{5}	1.113×10^{-24}	670.2
1 foot-pound =	1.285×10^{-3}	1.356×10^{7}	1	5.051×10^{-7}	1.356	0.3238	3.766×10^{-7}	8.464×10^{18}	8.464×10^{12}	1.509×10^{-17}	9.037×10^{9}
1 horsepower-hour =	2545	2.685×10^{13}	1.980×10^{6}	1	2.685×10^{6}	6.413×10^{5}	0.7457	1.676×10^{25}	1.676×10^{19}	2.988×10^{-11}	1.799×10^{16}
1 JOULE =	9.481×10^{-4}	10^{7}	0.7376	3.725×10^{-7}	1	0.2389	2.778×10^{-7}	6.242×10^{18}	6.242×10^{12}	1.113×10^{-17}	6.702×10^{9}
1 calorie =	3.969×10^{-3}	4.186×10^{7}	3.088	1.560×10^{-6}	4.186	1	1.163×10^{-6}	2.613×10^{19}	2.613×10^{13}	4.660×10^{-17}	2.806×10^{10}
1 kilowatt-hour =	3413	3.600×10^{13}	2.655×10^{6}	1.341	3.600×10^{6}	8.600×10^{5}	1	2.247×10^{25}	2.247×10^{19}	4.007×10^{-11}	2.413×10^{16}
1 electron-volt =	1.519×10^{-22}	1.602×10^{-12}	1.182×10^{-19}	5.967×10^{-26}	1.602×10^{-19}	3.827×10^{-20}	4.450×10^{-26}	1	10^{-6}	1.783×10^{-36}	1.074×10^{-9}
1 million electron-volts =	1.519×10^{-16}	1.602×10^{-6}	1.182×10^{-13}	5.967×10^{-20}	1.602×10^{-13}	3.827×10^{-14}	4.450×10^{-20}	10^{-6}	1	1.783×10^{-30}	1.074×10^{-3}
1 kilogram =	8.521×10^{13}	8.987×10^{23}	6.629×10^{16}	3.348×10^{10}	8.987×10^{16}	2.146×10^{16}	2.497×10^{10}	5.610×10^{35}	5.610×10^{29}	1	6.022×10^{26}
1 unified atomic mass unit =	1.415×10^{-13}	1.492×10^{-3}	1.101×10^{-10}	5.559×10^{-17}	1.492×10^{-10}	3.564×10^{-11}	4.146×10^{-17}	9.320×10^{8}	932.0	1.661×10^{-27}	1

Power

	Btu/h	ft·lb/s	hp	cal/s	kW	WATT
1 British thermal unit per hour =	1	0.2161	3.929×10^{-4}	6.998×10^{-2}	2.930×10^{-4}	0.2930
1 foot-pound per second =	4.628	1	1.818×10^{-3}	0.3239	1.356×10^{-3}	1.356
1 horsepower =	2545	550	1	178.1	0.7457	745.7
1 calorie per second =	14.29	3.088	5.615×10^{-3}	1	4.186×10^{-3}	4.186
1 kilowatt =	3413	737.6	1.341	238.9	1	1000
1 WATT =	3.413	0.7376	1.341×10^{-3}	0.2389	0.001	1

Magnetic Field

	gauss	TESLA	milligauss
1 gauss =	1	10^{-4}	1000
1 TESLA =	10^{4}	1	10^{7}
1 milligauss =	0.001	10^{-7}	1

1 tesla = 1 weber/meter^2

Magnetic Flux

	maxwell	WEBER
1 maxwell =	1	10^{-8}
1 WEBER =	10^{8}	1

APPENDIX E
Mathematical Formulas

Geometry

Circle of radius r: circumference $= 2\pi r$; area $= \pi r^2$.

Sphere of radius r: area $= 4\pi r^2$; volume $= \frac{4}{3}\pi r^3$.

Right circular cylinder of radius r and height h:
   area $= 2\pi r^2 + 2\pi rh$; volume $= \pi r^2 h$.

Triangle of base a and altitude h: area $= \frac{1}{2}ah$.

Quadratic Formula

If $ax^2 + bx + c = 0$, then $x = \dfrac{-b \pm \sqrt{b^2 - 4ac}}{2a}$.

Trigonometric Functions of Angle θ

$\sin \theta = \dfrac{y}{r}$ $\cos \theta = \dfrac{x}{r}$

$\tan \theta = \dfrac{y}{x}$ $\cot \theta = \dfrac{x}{y}$

$\sec \theta = \dfrac{r}{x}$ $\csc \theta = \dfrac{r}{y}$

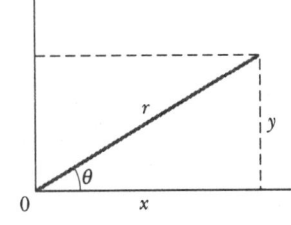

Pythagorean Theorem

In this right triangle,
   $a^2 + b^2 = c^2$

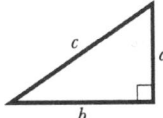

Triangles

Angles are A, B, C

Opposite sides are a, b, c

Angles $A + B + C = 180°$

$\dfrac{\sin A}{a} = \dfrac{\sin B}{b} = \dfrac{\sin C}{c}$

$c^2 = a^2 + b^2 - 2ab \cos C$

Exterior angle $D = A + C$

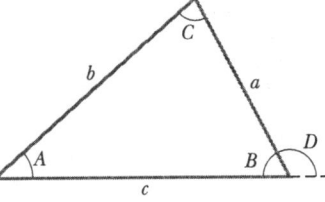

Mathematical Signs and Symbols

$=$ equals

$\approx$ equals approximately

$\sim$ is the order of magnitude of

$\neq$ is not equal to

$\equiv$ is identical to, is defined as

$>$ is greater than ($\gg$ is much greater than)

$<$ is less than ($\ll$ is much less than)

$\geq$ is greater than or equal to (or, is no less than)

$\leq$ is less than or equal to (or, is no more than)

$\pm$ plus or minus

$\propto$ is proportional to

Σ the sum of

x_{avg} the average value of x

Trigonometric Identities

$\sin(90° - \theta) = \cos \theta$

$\cos(90° - \theta) = \sin \theta$

$\sin \theta / \cos \theta = \tan \theta$

$\sin^2 \theta + \cos^2 \theta = 1$

$\sec^2 \theta - \tan^2 \theta = 1$

$\csc^2 \theta - \cot^2 \theta = 1$

$\sin 2\theta = 2 \sin \theta \cos \theta$

$\cos 2\theta = \cos^2 \theta - \sin^2 \theta = 2\cos^2 \theta - 1 = 1 - 2\sin^2 \theta$

$\sin(\alpha \pm \beta) = \sin \alpha \cos \beta \pm \cos \alpha \sin \beta$

$\cos(\alpha \pm \beta) = \cos \alpha \cos \beta \mp \sin \alpha \sin \beta$

$\tan(\alpha \pm \beta) = \dfrac{\tan \alpha \pm \tan \beta}{1 \mp \tan \alpha \tan \beta}$

$\sin \alpha \pm \sin \beta = 2 \sin \frac{1}{2}(\alpha \pm \beta) \cos \frac{1}{2}(\alpha \mp \beta)$

$\cos \alpha + \cos \beta = 2 \cos \frac{1}{2}(\alpha + \beta) \cos \frac{1}{2}(\alpha - \beta)$

$\cos \alpha - \cos \beta = -2 \sin \frac{1}{2}(\alpha + \beta) \sin \frac{1}{2}(\alpha - \beta)$

Binomial Theorem

$$(1 + x)^n = 1 + \frac{nx}{1!} + \frac{n(n-1)x^2}{2!} + \cdots \qquad (x^2 < 1)$$

Exponential Expansion

$$e^x = 1 + x + \frac{x^2}{2!} + \frac{x^3}{3!} + \cdots$$

Logarithmic Expansion

$$\ln(1 + x) = x - \tfrac{1}{2}x^2 + \tfrac{1}{3}x^3 - \cdots \qquad (|x| < 1)$$

Trigonometric Expansions
(θ in radians)

$$\sin \theta = \theta - \frac{\theta^3}{3!} + \frac{\theta^5}{5!} - \cdots$$

$$\cos \theta = 1 - \frac{\theta^2}{2!} + \frac{\theta^4}{4!} - \cdots$$

$$\tan \theta = \theta + \frac{\theta^3}{3} + \frac{2\theta^5}{15} + \cdots$$

Cramer's Rule

Two simultaneous equations in unknowns x and y,

$$a_1x + b_1y = c_1 \quad \text{and} \quad a_2x + b_2y = c_2,$$

have the solutions

$$x = \frac{\begin{vmatrix} c_1 & b_1 \\ c_2 & b_2 \end{vmatrix}}{\begin{vmatrix} a_1 & b_1 \\ a_2 & b_2 \end{vmatrix}} = \frac{c_1b_2 - c_2b_1}{a_1b_2 - a_2b_1}$$

and

$$y = \frac{\begin{vmatrix} a_1 & c_1 \\ a_2 & c_2 \end{vmatrix}}{\begin{vmatrix} a_1 & b_1 \\ a_2 & b_2 \end{vmatrix}} = \frac{a_1c_2 - a_2c_1}{a_1b_2 - a_2b_1}.$$

Products of Vectors

Let $\hat{\imath}$, $\hat{\jmath}$, and $\hat{k}$ be unit vectors in the x, y, and z directions. Then

$$\hat{\imath} \cdot \hat{\imath} = \hat{\jmath} \cdot \hat{\jmath} = \hat{k} \cdot \hat{k} = 1, \qquad \hat{\imath} \cdot \hat{\jmath} = \hat{\jmath} \cdot \hat{k} = \hat{k} \cdot \hat{\imath} = 0,$$

$$\hat{\imath} \times \hat{\imath} = \hat{\jmath} \times \hat{\jmath} = \hat{k} \times \hat{k} = 0,$$

$$\hat{\imath} \times \hat{\jmath} = \hat{k}, \qquad \hat{\jmath} \times \hat{k} = \hat{\imath}, \qquad \hat{k} \times \hat{\imath} = \hat{\jmath}.$$

Any vector $\vec{a}$ with components a_x, a_y, and a_z along the x, y, and z axes can be written as

$$\vec{a} = a_x\hat{\imath} + a_y\hat{\jmath} + a_z\hat{k}.$$

Let $\vec{a}$, $\vec{b}$, and $\vec{c}$ be arbitrary vectors with magnitudes a, b, and c. Then

$$\vec{a} \times (\vec{b} + \vec{c}) = (\vec{a} \times \vec{b}) + (\vec{a} \times \vec{c})$$

$$(s\vec{a}) \times \vec{b} = \vec{a} \times (s\vec{b}) = s(\vec{a} \times \vec{b}) \quad (s = \text{a scalar}).$$

Let θ be the smaller of the two angles between $\vec{a}$ and $\vec{b}$. Then

$$\vec{a} \cdot \vec{b} = \vec{b} \cdot \vec{a} = a_xb_x + a_yb_y + a_zb_z = ab \cos \theta$$

$$\vec{a} \times \vec{b} = -\vec{b} \times \vec{a} = \begin{vmatrix} \hat{\imath} & \hat{\jmath} & \hat{k} \\ a_x & a_y & a_z \\ b_x & b_y & b_z \end{vmatrix}$$

$$= \hat{\imath} \begin{vmatrix} a_y & a_z \\ b_y & b_z \end{vmatrix} - \hat{\jmath} \begin{vmatrix} a_x & a_z \\ b_x & b_z \end{vmatrix} + \hat{k} \begin{vmatrix} a_x & a_y \\ b_x & b_y \end{vmatrix}$$

$$= (a_yb_z - b_ya_z)\hat{\imath} + (a_zb_x - b_za_x)\hat{\jmath} + (a_xb_y - b_xa_y)\hat{k}$$

$$|\vec{a} \times \vec{b}| = ab \sin \theta$$

$$\vec{a} \cdot (\vec{b} \times \vec{c}) = \vec{b} \cdot (\vec{c} \times \vec{a}) = \vec{c} \cdot (\vec{a} \times \vec{b})$$

$$\vec{a} \times (\vec{b} \times \vec{c}) = (\vec{a} \cdot \vec{c})\vec{b} - (\vec{a} \cdot \vec{b})\vec{c}$$

Derivatives and Integrals

In what follows, the letters u and v stand for any functions of x, and a and m are constants. To each of the indefinite integrals should be added an arbitrary constant of integration. The *Handbook of Chemistry and Physics* (CRC Press Inc.) gives a more extensive tabulation.

1. $\dfrac{dx}{dx} = 1$

2. $\dfrac{d}{dx}(au) = a\dfrac{du}{dx}$

3. $\dfrac{d}{dx}(u + v) = \dfrac{du}{dx} + \dfrac{dv}{dx}$

4. $\dfrac{d}{dx}x^m = mx^{m-1}$

5. $\dfrac{d}{dx}\ln x = \dfrac{1}{x}$

6. $\dfrac{d}{dx}(uv) = u\dfrac{dv}{dx} + v\dfrac{du}{dx}$

7. $\dfrac{d}{dx}e^x = e^x$

8. $\dfrac{d}{dx}\sin x = \cos x$

9. $\dfrac{d}{dx}\cos x = -\sin x$

10. $\dfrac{d}{dx}\tan x = \sec^2 x$

11. $\dfrac{d}{dx}\cot x = -\csc^2 x$

12. $\dfrac{d}{dx}\sec x = \tan x \sec x$

13. $\dfrac{d}{dx}\csc x = -\cot x \csc x$

14. $\dfrac{d}{dx}e^u = e^u\dfrac{du}{dx}$

15. $\dfrac{d}{dx}\sin u = \cos u\dfrac{du}{dx}$

16. $\dfrac{d}{dx}\cos u = -\sin u\dfrac{du}{dx}$

1. $\displaystyle\int dx = x$

2. $\displaystyle\int au\,dx = a\int u\,dx$

3. $\displaystyle\int (u + v)\,dx = \int u\,dx + \int v\,dx$

4. $\displaystyle\int x^m\,dx = \dfrac{x^{m+1}}{m + 1}\quad (m \neq -1)$

5. $\displaystyle\int \dfrac{dx}{x} = \ln |x|$

6. $\displaystyle\int u\dfrac{dv}{dx}\,dx = uv - \int v\dfrac{du}{dx}\,dx$

7. $\displaystyle\int e^x\,dx = e^x$

8. $\displaystyle\int \sin x\,dx = -\cos x$

9. $\displaystyle\int \cos x\,dx = \sin x$

10. $\displaystyle\int \tan x\,dx = \ln |\sec x|$

11. $\displaystyle\int \sin^2 x\,dx = \tfrac{1}{2}x - \tfrac{1}{4}\sin 2x$

12. $\displaystyle\int e^{-ax}\,dx = -\dfrac{1}{a}e^{-ax}$

13. $\displaystyle\int xe^{-ax}\,dx = -\dfrac{1}{a^2}(ax + 1)e^{-ax}$

14. $\displaystyle\int x^2 e^{-ax}\,dx = -\dfrac{1}{a^3}(a^2x^2 + 2ax + 2)e^{-ax}$

15. $\displaystyle\int_0^\infty x^n e^{-ax}\,dx = \dfrac{n!}{a^{n+1}}$

16. $\displaystyle\int_0^\infty x^{2n} e^{-ax^2}\,dx = \dfrac{1\cdot 3\cdot 5\,\cdots\,(2n - 1)}{2^{n+1}a^n}\sqrt{\dfrac{\pi}{a}}$

17. $\displaystyle\int \dfrac{dx}{\sqrt{x^2 + a^2}} = \ln(x + \sqrt{x^2 + a^2})$

18. $\displaystyle\int \dfrac{x\,dx}{(x^2 + a^2)^{3/2}} = -\dfrac{1}{(x^2 + a^2)^{1/2}}$

19. $\displaystyle\int \dfrac{dx}{(x^2 + a^2)^{3/2}} = \dfrac{x}{a^2(x^2 + a^2)^{1/2}}$

20. $\displaystyle\int_0^\infty x^{2n+1} e^{-ax^2}\,dx = \dfrac{n!}{2a^{n+1}}\quad (a > 0)$

21. $\displaystyle\int \dfrac{x\,dx}{x + d} = x - d\ln(x + d)$

APPENDIX F
Properties of the Elements

All physical properties are for a pressure of 1 atm unless otherwise specified.

Element	Symbol	Atomic Number Z	Molar Mass, g/mol	Density, g/cm³ at 20°C	Melting Point, °C	Boiling Point, °C	Specific Heat, J/(g·°C) at 25°C
Actinium	Ac	89	(227)	10.06	1323	(3473)	0.092
Aluminum	Al	13	26.9815	2.699	660	2450	0.900
Americium	Am	95	(243)	13.67	1541	—	—
Antimony	Sb	51	121.75	6.691	630.5	1380	0.205
Argon	Ar	18	39.948	1.6626×10^{-3}	−189.4	−185.8	0.523
Arsenic	As	33	74.9216	5.78	817 (28 atm)	613	0.331
Astatine	At	85	(210)	—	(302)	—	—
Barium	Ba	56	137.34	3.594	729	1640	0.205
Berkelium	Bk	97	(247)	14.79	—	—	—
Beryllium	Be	4	9.0122	1.848	1287	2770	1.83
Bismuth	Bi	83	208.980	9.747	271.37	1560	0.122
Bohrium	Bh	107	262.12	—	—	—	—
Boron	B	5	10.811	2.34	2030	—	1.11
Bromine	Br	35	79.909	3.12 (liquid)	−7.2	58	0.293
Cadmium	Cd	48	112.40	8.65	321.03	765	0.226
Calcium	Ca	20	40.08	1.55	838	1440	0.624
Californium	Cf	98	(251)	—	—	—	—
Carbon	C	6	12.01115	2.26	3727	4830	0.691
Cerium	Ce	58	140.12	6.768	804	3470	0.188
Cesium	Cs	55	132.905	1.873	28.40	690	0.243
Chlorine	Cl	17	35.453	3.214×10^{-3} (0°C)	−101	−34.7	0.486
Chromium	Cr	24	51.996	7.19	1857	2665	0.448
Cobalt	Co	27	58.9332	8.85	1495	2900	0.423
Copper	Cu	29	63.54	8.96	1083.40	2595	0.385
Curium	Cm	96	(247)	13.3	—	—	—
Dubnium	Db	105	262.114	—	—	—	—
Dysprosium	Dy	66	162.50	8.55	1409	2330	0.172
Einsteinium	Es	99	(254)	—	—	—	—
Erbium	Er	68	167.26	9.15	1522	2630	0.167
Europium	Eu	63	151.96	5.243	817	1490	0.163
Fermium	Fm	100	(237)	—	—	—	—
Fluorine	F	9	18.9984	1.696×10^{-3} (0°C)	−219.6	−188.2	0.753
Francium	Fr	87	(223)	—	(27)	—	—
Gadolinium	Gd	64	157.25	7.90	1312	2730	0.234
Gallium	Ga	31	69.72	5.907	29.75	2237	0.377

Element	Symbol	Atomic Number Z	Molar Mass, g/mol	Density, g/cm³ at 20°C	Melting Point, °C	Boiling Point, °C	Specific Heat, J/(g · °C) at 25°C
Germanium	Ge	32	72.59	5.323	937.25	2830	0.322
Gold	Au	79	196.967	19.32	1064.43	2970	0.131
Hafnium	Hf	72	178.49	13.31	2227	5400	0.144
Hassium	Hs	108	(265)	—	—	—	—
Helium	He	2	4.0026	0.1664×10^{-3}	−269.7	−268.9	5.23
Holmium	Ho	67	164.930	8.79	1470	2330	0.165
Hydrogen	H	1	1.00797	0.08375×10^{-3}	−259.19	−252.7	14.4
Indium	In	49	114.82	7.31	156.634	2000	0.233
Iodine	I	53	126.9044	4.93	113.7	183	0.218
Iridium	Ir	77	192.2	22.5	2447	(5300)	0.130
Iron	Fe	26	55.847	7.874	1536.5	3000	0.447
Krypton	Kr	36	83.80	3.488×10^{-3}	−157.37	−152	0.247
Lanthanum	La	57	138.91	6.189	920	3470	0.195
Lawrencium	Lr	103	(257)	—	—	—	—
Lead	Pb	82	207.19	11.35	327.45	1725	0.129
Lithium	Li	3	6.939	0.534	180.55	1300	3.58
Lutetium	Lu	71	174.97	9.849	1663	1930	0.155
Magnesium	Mg	12	24.312	1.738	650	1107	1.03
Manganese	Mn	25	54.9380	7.44	1244	2150	0.481
Meitnerium	Mt	109	(266)	—	—	—	—
Mendelevium	Md	101	(256)	—	—	—	—
Mercury	Hg	80	200.59	13.55	−38.87	357	0.138
Molybdenum	Mo	42	95.94	10.22	2617	5560	0.251
Neodymium	Nd	60	144.24	7.007	1016	3180	0.188
Neon	Ne	10	20.183	0.8387×10^{-3}	−248.597	−246.0	1.03
Neptunium	Np	93	(237)	20.25	637	—	1.26
Nickel	Ni	28	58.71	8.902	1453	2730	0.444
Niobium	Nb	41	92.906	8.57	2468	4927	0.264
Nitrogen	N	7	14.0067	1.1649×10^{-3}	−210	−195.8	1.03
Nobelium	No	102	(255)	—	—	—	—
Osmium	Os	76	190.2	22.59	3027	5500	0.130
Oxygen	O	8	15.9994	1.3318×10^{-3}	−218.80	−183.0	0.913
Palladium	Pd	46	106.4	12.02	1552	3980	0.243
Phosphorus	P	15	30.9738	1.83	44.25	280	0.741
Platinum	Pt	78	195.09	21.45	1769	4530	0.134
Plutonium	Pu	94	(244)	19.8	640	3235	0.130
Polonium	Po	84	(210)	9.32	254	—	—
Potassium	K	19	39.102	0.862	63.20	760	0.758
Praseodymium	Pr	59	140.907	6.773	931	3020	0.197
Promethium	Pm	61	(145)	7.22	(1027)	—	—
Protactinium	Pa	91	(231)	15.37 (estimated)	(1230)	—	—
Radium	Ra	88	(226)	5.0	700	—	—
Radon	Rn	86	(222)	9.96×10^{-3} (0°C)	(−71)	−61.8	0.092
Rhenium	Re	75	186.2	21.02	3180	5900	0.134

Element	Symbol	Atomic Number Z	Molar Mass, g/mol	Density, g/cm³ at 20°C	Melting Point, °C	Boiling Point, °C	Specific Heat, J/(g · °C) at 25°C
Rhodium	Rh	45	102.905	12.41	1963	4500	0.243
Rubidium	Rb	37	85.47	1.532	39.49	688	0.364
Ruthenium	Ru	44	101.107	12.37	2250	4900	0.239
Rutherfordium	Rf	104	261.11	—	—	—	—
Samarium	Sm	62	150.35	7.52	1072	1630	0.197
Scandium	Sc	21	44.956	2.99	1539	2730	0.569
Seaborgium	Sg	106	263.118	—	—	—	—
Selenium	Se	34	78.96	4.79	221	685	0.318
Silicon	Si	14	28.086	2.33	1412	2680	0.712
Silver	Ag	47	107.870	10.49	960.8	2210	0.234
Sodium	Na	11	22.9898	0.9712	97.85	892	1.23
Strontium	Sr	38	87.62	2.54	768	1380	0.737
Sulfur	S	16	32.064	2.07	119.0	444.6	0.707
Tantalum	Ta	73	180.948	16.6	3014	5425	0.138
Technetium	Tc	43	(99)	11.46	2200	—	0.209
Tellurium	Te	52	127.60	6.24	449.5	990	0.201
Terbium	Tb	65	158.924	8.229	1357	2530	0.180
Thallium	Tl	81	204.37	11.85	304	1457	0.130
Thorium	Th	90	(232)	11.72	1755	(3850)	0.117
Thulium	Tm	69	168.934	9.32	1545	1720	0.159
Tin	Sn	50	118.69	7.2984	231.868	2270	0.226
Titanium	Ti	22	47.90	4.54	1670	3260	0.523
Tungsten	W	74	183.85	19.3	3380	5930	0.134
Un-named	Uun	110	(269)	—	—	—	—
Un-named	Uuu	111	(272)	—	—	—	—
Un-named	Uub	112	(264)	—	—	—	—
Un-named	Uut	113	—	—	—	—	—
Un-named	Unq	114	(285)	—	—	—	—
Un-named	Uup	115	—	—	—	—	—
Un-named	Uuh	116	(289)	—	—	—	—
Un-named	Uus	117	—	—	—	—	—
Un-named	Uuo	118	(293)	—	—	—	—
Uranium	U	92	(238)	18.95	1132	3818	0.117
Vanadium	V	23	50.942	6.11	1902	3400	0.490
Xenon	Xe	54	131.30	5.495×10^{-3}	−111.79	−108	0.159
Ytterbium	Yb	70	173.04	6.965	824	1530	0.155
Yttrium	Y	39	88.905	4.469	1526	3030	0.297
Zinc	Zn	30	65.37	7.133	419.58	906	0.389
Zirconium	Zr	40	91.22	6.506	1852	3580	0.276

The values in parentheses in the column of molar masses are the mass numbers of the longest-lived isotopes of those elements that are radioactive. Melting points and boiling points in parentheses are uncertain.

The data for gases are valid only when these are in their usual molecular state, such as H_2, He, O_2, Ne, etc. The specific heats of the gases are the values at constant pressure.

Source: Adapted from J. Emsley, *The Elements,* 3rd ed., 1998, Clarendon Press, Oxford. See also www.webelements.com for the latest values and newest elements.

Periodic Table
of the Elements

	Metals
	Metalloids
	Nonmetals

Alkali metals
IA

Noble gases 0

THE HORIZONTAL PERIODS

1																	2
1 H	IIA										IIIA	IVA	VA	VIA	VIIA		2 He
3 Li	4 Be			Transition metals								5 B	6 C	7 N	8 O	9 F	10 Ne
11 Na	12 Mg	IIIB	IVB	VB	VIB	VIIB		VIIIB		IB	IIB	13 Al	14 Si	15 P	16 S	17 Cl	18 Ar
19 K	20 Ca	21 Sc	22 Ti	23 V	24 Cr	25 Mn	26 Fe	27 Co	28 Ni	29 Cu	30 Zn	31 Ga	32 Ge	33 As	34 Se	35 Br	36 Kr
37 Rb	38 Sr	39 Y	40 Zr	41 Nb	42 Mo	43 Tc	44 Ru	45 Rh	46 Pd	47 Ag	48 Cd	49 In	50 Sn	51 Sb	52 Te	53 I	54 Xe
55 Cs	56 Ba	57-71 *	72 Hf	73 Ta	74 W	75 Re	76 Os	77 Ir	78 Pt	79 Au	80 Hg	81 Tl	82 Pb	83 Bi	84 Po	85 At	86 Rn
87 Fr	88 Ra	89-103 †	104 Rf	105 Db	106 Sg	107 Bh	108 Hs	109 Mt	110	111	112	113	114	115	116	117	118

Inner transition metals

Lanthanide series *

57 La	58 Ce	59 Pr	60 Nd	61 Pm	62 Sm	63 Eu	64 Gd	65 Tb	66 Dy	67 Ho	68 Er	69 Tm	70 Yb	71 Lu

Actinide series †

89 Ac	90 Th	91 Pa	92 U	93 Np	94 Pu	95 Am	96 Cm	97 Bk	98 Cf	99 Es	100 Fm	101 Md	102 No	103 Lr

The names for elements 104 through 109 (Rutherfordium, Dubnium, Seaborgium, Bohrium, Hassium, and Meitnerium, respectively) were adopted by the International Union of Pure and Applied Chemistry (IUPAC) in 1997. Elements 110, 111, 112, 114, 116, and 118 have been discovered but, as of 2000, have not yet been named. See www.webelements.com for the latest information and newest elements.

PHOTO CREDITS

Chapter 22
Opener: ©Fundamental Photographs. Fig. 22-1: Courtesy Swedish Amber Museum. Fig. 22-9: Johann Gabriel Doppelmayr, *Neuentdeckte Phaenomena von Bewünderswurdigen Würckungen der Natur*, Nuremberg, 1744. Fig. 22-12: Courtesy PASCO Scientific.

Chapter 23
Opener: Tsuyoshi Nishiinoue/Orion Press. Fig. 23-15: Russ Kinne/Comstock, Inc.

Chapter 24
Opener: Ralph H. Wetmore II/Stone. Fig. TE24-4: ©C. Johnny Autery. Fig. TE24-5: Courtesy E. Philip Krider, Institute for Atmospheric Physics, University of Arizona, Tucson.

Chapter 25
Opener and Fig. 25-20: Courtesy NOAA. Fig. 25-3: Courtesy PASCO scientific. Fig. 25-21: Courtesy Westinghouse Corporation.

Chapter 26
Opener: ©UPI/Corbis Images. Fig. 26-2b: Courtesy PASCO scientific. Pg. 26-4: Courtesy Southern California Edison Company. Fig. 26-18: The Image Works. Pg. 26-21: ©Laurie Rubin. Pg. 26-23: Courtesy Shoji Tonaka, International Superconductivity Technology Center, Tokyo, Japan.

Chapter 27
Opener: Hans Reinhard/Bruce Coleman Inc.

Chapter 28
Opener: Bruce Ayres/Tony Stone Images/New York, Inc. Fig. 28-2: Paul Silvermann/Fundamental Photographs. Fig. 28-3: Courtesy Priscilla Laws. Fig. 28-4: Courtesy Priscilla Laws. Fig. 28-10: Courtesy Priscilla Laws. Fig. 28-14: ©Estate of Harold Edgerton, courtesy of Palm Press, Inc. Fig. 28-15: Courtesy The Royal Institute, England.

Chapter 29
Opener: Johnny Johnson/Tony Stone Images/New York, Inc. Fig. 29-1: Ray Pfortner/Peter Arnold, Inc. Fig. 29-3: Lawrence Berkeley Laboratory/Photo Researchers. Fig. 29-4: Courtesy Dr. Richard Cannon, Southeast Missouri State University, Cape Girardeau. Fig. 29-8: Courtesy John Le P. Webb, Sussex University, England. Fig. 29-11: Courtesy Dr. L. A. Frank, University of Iowa.

Chapter 30
Opener: Michael Brown/ Gamma Liaison. Fig. 30-3: Courtesy Education Development Center.

Chapter 31
Opener: Dan McCoy/Black Star. Fig. 31-5: Courtesy Fender Musical Instruments Corporation. Page 31-12: Courtesy The Royal Institute, England.

Chapter 32
Opener: Courtesy A. K. Geim, High Field Magnet Laboratory, University of Nijmegen, The Netherlands. Fig. 32-1: Runk/Schoenberger/Grant Heilman Photography. Pg. 32-10: Peter Lerman. Fig. 32-12: Courtesy Ralph W. DeBlois.

Chapter 33:
Opener: Photo by Rick Diaz, provided courtesy Haverfield Helicopter Co. Fig. 33-3: Courtesy Agilent Technologies. Pg. 33-19: Ted Cowell/Black Star.